Oscillatory Event-Related Brain Dynamics

NATO ASI Series

Advanced Science Institutes Series

A series presenting the results of activities sponsored by the NATO Science Committee, which aims at the dissemination of advanced scientific and technological knowledge, with a view to strengthening links between scientific communities.

The series is published by an international board of publishers in conjunction with the NATO Scientific Affairs Division

A	**Life Sciences**	Plenum Publishing Corporation
B	**Physics**	New York and London
C	**Mathematical and Physical Sciences**	Kluwer Academic Publishers
D	**Behavioral and Social Sciences**	Dordrecht, Boston, and London
E	**Applied Sciences**	
F	**Computer and Systems Sciences**	Springer-Verlag
G	**Ecological Sciences**	Berlin, Heidelberg, New York, London,
H	**Cell Biology**	Paris, Tokyo, Hong Kong, and Barcelona
I	**Global Environmental Change**	

Recent Volumes in this Series

Series A: Life Sciences

Oscillatory Event-Related Brain Dynamics

Edited by

Christo Pantev
Thomas Elbert
Bernd Lütkenhöner

Institute for Experimental Audiology
University of Münster
Münster, Germany

Plenum Press
New York and London
Published in cooperation with NATO Scientific Affairs Division

Proceedings of a NATO Advanced Research Workshop on
Oscillatory Event-Related Brain Dynamics,
held September 1–5, 1993,
in Tecklenburg, Germany

NATO-PCO-DATA BASE

The electronic index to the NATO ASI Series provides full bibliographical references (with keywords and/or abstracts) to more than 30,000 contributions from international scientists published in all sections of the NATO ASI Series. Access to the NATO-PCO-DATA BASE is possible in two ways:

—via online FILE 128 (NATO-PCO-DATA BASE) hosted by ESRIN, Via Galileo Galilei, I-00044 Frascati, Italy

—via CD-ROM "NATO Science and Technology Disk" with user-friendly retrieval software in English, French, and German (©WTV GmbH and DATAWARE Technologies, Inc. 1989). The CD-ROM also contains the AGARD Aerospace Database.

The CD-ROM can be ordered through any member of the Board of Publishers or through NATO-PCO, Overijse, Belgium.

Library of Congress Cataloging-in-Publication Data

On file

ISBN 0-306-44894-7

©1994 Plenum Press, New York
A Division of Plenum Publishing Corporation
233 Spring Street, New York, N.Y. 10013

Printed in the United States of America

PREFACE

How does the brain code and process incoming information, how does it recognize a certain object, how does a certain Gestalt come into our awareness? One of the key issues to conscious realization of an object, of a Gestalt is the attention devoted to the corresponding sensory input which evokes the neural pattern underlying the Gestalt. This requires that the attention be devoted to one set of objects at a time. However, the attention may be switched quickly between different objects or ongoing input processes. It is to be expected that such mechanisms are reflected in the neural dynamics: Neurons or neuronal assemblies which pertain to one object may fire, possibly in rapid bursts at a time. Such firing bursts may enhance the synaptic strength in the corresponding cell assembly and thereby form the substrate of short-term memory.

However, we may well become aware of two different objects at a time. How can we avoid that the firing patterns which may relate to say a certain type of movement (columns in V5) or to a color (V4) of one object do not become mixed with those of another object? Such a blend may only happen if the presentation times become very short (below 20-30 ms).

One possibility is that neurons pertaining to one cell assembly fire synchronously. Then different cell assemblies firing at different rates may code different information. Work summarized in the present book shows that such correlated coherent firing indeed may exist and that these oscillations can be observed not only in various animals, but that similar phenomena may be recorded noninvasively from humans.

Neurons can fire with any frequency between one and one hundred times per second. Information may be coded by any deviance from an idling rhythm. While such idling rhythms dominate the low frequencies, perhaps to save metabolic costs, firing bursts with frequencies above 20 to 30 Hz (i.e., oscillations in the gamma-band) may specifically reflect ongoing information processing. On the single cell level, such oscillations have been related to such information processing, e.g. to the binding of different features into one and the same object. The cell assembly which represents this object is formed by the simultaneous firing of all of its members. Consequently, one should expect that the more neurons there are to participate in such a mass oscillation, the more complex the activity under consideration will be. Is it possible to evoke macroscopically large cell assemblies such that we will be able to detect oscillatory firing patterns in EEG and MEG? What are the magnitudes to be expected? Can we increase the signal to noise ratio by averaging in the frequency domain, may chaos provide a useful tool for these analyses? Can we enhance the signal by involving a larger fraction of the receptive fields in one channel? Will multisensory stimulation evoke oscillations with a macroscopic dimension? What is the relation between the attention-related slow brain potentials and oscillatory phenomena?

The present book was stimulated by such questions and by the increasing exploration of oscillations in the EEG and MEG gamma-band in particular. Work in our own laboratory has led us to explore the extent to which oscillatory phenomena in neural mass activity can be related to oscillatory activity in local field potentials or even single cells.

A number of researchers active in this field were brought together with the specific aim of discussing such questions on the background of their own research. The chapters in this book result from the presentations and discussions during this workshop, which was held in Tecklenburg, Westfalia in September 1993.

The workshop was held in honor of the founder and current director of our Institute and research group, Prof. Dr. Manfried Hoke. His scientific enthusiasm and his untiring efforts in providing premium research tools laid the foundations onto which our own scientific achievements were built. We are grateful to have such a fine scientist as our friend and teacher.

The workshop was made possible through funds from the Deutsche Forschungsgemeinschaft and the NATO Scientific Affairs division. We acknowledge the assistance of Drs. Olivier Bertrand, Scott Makeig and Larry Roberts in writing the respective grant applications.

The workshop relied on the organizational talent of Christina Robert and Annette Sütfeld. In addition, Helga Janutta assisted in preparing the present book. We would like to thank them for their energetic efforts which made the present project possible.

The Westfälische-Wilhems-Universität, represented by M. Gotthardt, chief of the administration of the medical faculty, and the Ministerium für Wissenschaft und Forschung des Landes NRW represented by H. Thomas have continuously provided the conditions which allow for outstanding research to blossom, despite the many difficulties involved. We are grateful to have such an exceptional administrative support.

The Editors
Münster, Germany

CONTENTS

INTRODUCTORY TALK

Manfried Hoke

Institute of Experimental Audiology
University of Münster
Münster, Germany

I'd like to cordially welcome you to Germany and to the small village of Teck-lenburg which has been chosen for this symposium.

I assume that nobody expects me to give a scientific talk. My personal contri-bution to oscillatory event-related brain dynamics is negligible, at least as far as my personal scientific activity is concerned. My contribution is rather **in**direct, which does not necessarily mean **in**significant. It is related to the story of my life, or, to be more precise, to the scientific part of it. So I'd kindly ask you not to consider me to be immodest if I address some aspects of my life rather than dealing with oscillatory phenomena.

My way to auditory physiology and audiology was not a straightforward one. It was not planned from the very beginning. On the contrary, it was rather tortuous, and now that I've learned something about chaos, it reminds me of the famous "but-terfly effect", that is, that minimal changes in the initial conditions – and of course not only in the initial conditions – do have a tremendous effect on what is going to happen later on, and such minimal changes have also caused dramatic changes in my curriculum, at least and especially in its earlier stages.

During the last school years I had begun to develop a distinct preference for physics. It was basically my physics teacher, whom I very much revered, who elicited this inclination. What appears very strange retrospectively is that, in spite of my deep reverence, I had then been unable to understand why his favourite field in physics was acoustics, which appeared to me one of the most boring fields. At that time, I also felt another inclination, an inclination to medicine, but when I had to decide which subject I should study I chose physics. It was just a few months after I had begun to study physics that again the butterfly's flapping his wings became ob-vious. The result was that I changed the subject of my studies: I changed from phys-ics to medicine. The butterfly's flutter was the fixed idea that medicine is a sort of general education rather than a science, and my intention was that I first should complete my "general education" before studying a real science: namely physics. Political and personal circumstances prohibited me for years from studying physics as well, and when I was in a position to do so, I felt that it was too late. This is not at

Oscillatory Event-Related Brain Dynamics, Edited
by C. Pantev, Plenum Press, New York, 1994

all a melancholic reminiscence; on the contrary, I feel that, in the end, it turned out to be not that bad, and who knows which career I would have made as a physicist?

Similar to my winding way to medicine, the way which led me to audiology was tortuous too. During my studies, I had no clear-cut preference for any medical discipline, especially not for a clinical one, and certainly not for otorhinolaryngology. Looking for a suitable place for doing my doctoral thesis I finally ended up in pharmacology, and my work fascinated me so much that I decided to remain in this discipline. However, I had not reckoned with my host, the director of the Institute of Pharmacology, whose intentions were quite different from my expectations. The reason was my fateful inclination to physics. I preferred to develop electronic devices for new techniques of investigation rather than to investigate novel drugs with conventional techniques, for which my chief received an honorarium from a pharmaceutical company. Exactly at that time, it was a German congress of gynecology where again the butterfly's wings flapped. Still working in the Institute of Pharmacology – and instead of testing drugs – I had developed a device for monitoring the fetal pulse. Today a routine instrument, at that time there didn't exist any such device. There were two instruments presented at this congress, and the inventor of the other instrument – I think he really was an inventor because he got a patent for his principle – urged me to come to his place, namely to Münster, to the Institute of Physiology, where he was an assistant professor. The reasons for his advice were evident: In the Institute of Pharmacology, I had been testing – if I had not been developing and constructing a new device – cardiac glycosides. The director of the Institute of Physiology in Münster was a famous cardiac physiologist, so it was almost the same field in which I would continue to work, and, furthermore, he told me that I could expect full support from my chief-to-be. This was convincing, and I followed his advice immediately. However, as it turned out, it was, on the short term, a wrong decision. It was very soon that I happened to realize that the director of the institute **had been** a famous cardiac physiologist. Since the instruments necessary for my scientific work– oscilloscope, negative-capacitance amplifiers etc. – had to be purchased and the delivery time was relatively long, he urged me to begin the investigation with a string galvanometer, the very instrument with which he had collected his scientific merits. So, since he was unable to convince me that studies with such an outdated instrument could yield novel results, he recommended me to do – until my instruments had been delivered – some research with a professor whose small lab was called "Physiological Acoustics", which meant auditory physiology. This appeared to me as jumping out of the frying pan into the fire. On the one hand investigations with an outdated instrument, on the other hand working in the boring field of acoustics. I finally decided to jump into the fire, and I've never regretted this decision.

The professor who headed this lab of Physiological Acoustics, Prof. Eberhard Lerche, to whom I owe a debt of gratitude, succeeded to arouse my interest for acoustics, for auditory physiology and audiology, and not only interest, but also fascination and, probably, obsession. Again I indulged in my passion, developing instruments, in this case a device for investigating the middle ear transfer function, which consisted of a narrow-band control amplifier to keep the sound pressure level of a gliding tone constant. The quantity which I measured were cochlear microphonics, which means that auditory evoked potentials originating at the very periphery of the auditory system were the first ones that I investigated.

And again the butterfly's wings flapped. My working in the Institute of Physiology appeared to me to be a dead end, with no attractive perspective, and it was my mentor, Professor Lerche, who convinced me that any other occupation than in the field of research would never satisfy me. He gave me the advice to begin a training in otorhinolaryngology, where I would be able to continue my scientfic work if I ever wanted to. I followed his advice, and he was right. The director of the ENT department of the University of Münster, Professor Karl Mündnich, was interested in having me as an assistant professor and offered me an open position.

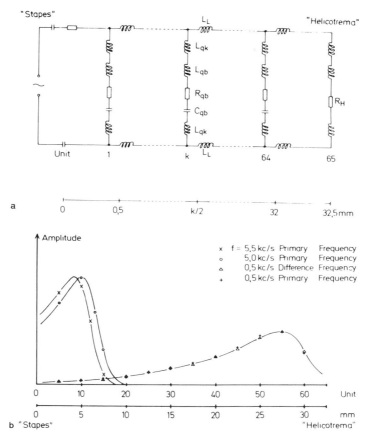

Figure 1. Top: The linear electric chain network model of the cochlear partition after Oetinger and Hauser (1965) which was used, after adding a quadratic nonlinearity, for the simulation of the generation of auditory combination tones. Bottom: Distribution along the model cochlear partition of the amplitude of the travelling waves of two primary frequencies f_1 and f_2 (5.0 and 5.5 kHz, resp.), of the travelling wave of the resulting difference frequency f_2-f_1 (0.5 kHz), and of a primary of the same frequency (0.5 kHz). The amplitude distributions of the latter two travelling waves are identical. (From Landwehr et al., 1970).

Again – as already in both the Institute of Pharmacology and the Institute of Physiology – I had to begin with setting up the prerequisites for my experimental work, because not even the simplest instruments – except a screwdriver – existed. But Professor Mündnich gave me all the support he could give, and I also owe a great debt of gratitude especially to him. When I took up my work in the ENT department, I still took a particular interest in cochlear microphonics (CM), and my primary goal was to investigate whether or not CM do have a significance in the objective assessment of hearing in humans. When it turned out that they do not, my interest began to ascend from the periphery of the auditory system to its centre, the auditory cortex. But before I abandoned my work on CM, I still did some experimental work in animals. At that time – around 1970 – it was generally accepted that the basilar membrane (BM) of the cochlea vibrates linearly, and that it possesses only a very limited frequency selectivity, basically owing to the convincing experimental results and the scientific authority of Georg von Békésy (1943). To account for the high frequency selectivity and also for certain properties of the cochlear combination tone, $2f_1-f_2$, and the two-tone suppression observed in the cochlear nerve, a two-filter system was proposed, and an essential, i.e. level-independent nonlinear process was envisaged to be "sandwiched" between a broadly tuned input (basilar membrane) and a sharply tuned "second" filter which was projected into the auditory nerve fibres (Evans, 1973).

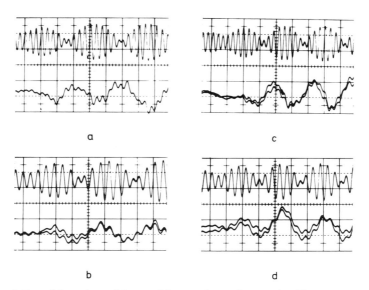

Figure 2. Stimulation of the guinea pig's ear with two primary frequencies. Top traces: sound pressure wave forms measured in front of the entrance of the outer ear canal; bottom traces: cochlear microphonics recorded from the third cochlear turn. Difference frequency 300 Hz elicited with two primary tones of 2.7 and 3.0 kHz (a), and 1.8 and 2.1 kHz (b), resp.; difference frequency 400 Hz elicited with two primary tones of 2.8 and 3.2 kHz (c), and 2.0 and 2.4 kHz (d). (From Hoke and Landwehr, 1970).

The first paper reporting nonlinearities at the basilar membrane level, from William S. Rhode, appeared in 1971. Using the Mössbauer technique he was able to demonstrate that the basilar membrane indeed exhibits a distinct nonlinearity of the saturating type in the region of the characteristic frequency, and that the highly tuned portion of the frequency tuning curves (FTC) of auditory nerve fibres is already present in the tuning characteristics of the basilar membrane. He could demonstrate that both the nonlinearity and the highly tuned portion of the basilar membrane FTC vanish rapidly after death, and several hours after death linearity and tuning characteristics of the BM resemble those measured by G. v. Békésy.

It was in 1969 that we began our research on the nonlinearity of the basilar membrane. We modified an electric chain network model of the basilar membrane in such a manner that a quadratic nonlinearity was added in parallel to the circuit elements representing the stiffness of the basilar membrane (Landwehr et al., 1970). With this model we were able to demonstrate that, if two primary frequencies are fed into the network, a travelling wave of the difference frequency f_2-f_1 develops in the region of the amplitude maxima of the two primary frequencies and assumes its amplitude maximum at exactly the same position where a single primary of identical frequency would do (Fig. 1). The only difference was – as could be expected – the longer travel time of the difference frequency, owing to the additional travel time of the two primaries until they assumed their amplitude maximum. That this concept does not hold true for our nonlinear model only, could then be proved in animal experiments. We were the first to demonstrate (Hoke and Landwehr, 1970) that indeed a travelling wave of the difference frequency originates from that region where the two primaries assume their amplitude maximum, and then assumes its own maximum at that place where a primary of the same frequency would do (Fig. 2).

FNA-Spektrogramm

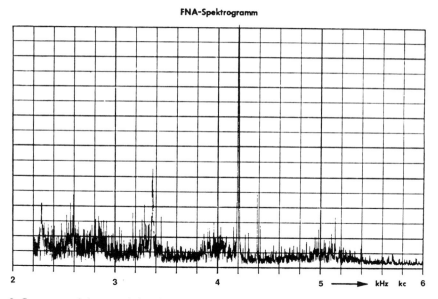

Figure 3. Spectrum of the sound signal recorded from the outer ear canal of a patient suffering from sensorineural deafness. The narrow peak at a frequency of 4.2 kHz indicates that a practically pure tone is radiated. (From Kumpf and Hoke, 1970).

Also the travel time of the difference frequency travelling wave was prolonged by the travel time which the primaries need to arrive at their place of maximal amplitude. Since our results were published in German only, they fell into oblivion. We also did not follow this line of research, except in a study in which we demonstrated that the psychophysical tuning curves of the difference tones $2f_1-f_2$ and $2f_2-f_1$ are identical with those of single primary tones of the same frequency

(Hoke et al., 1980). The concept of the linear, broadly tuned cochlea finally began to sway with the first measurement of tuning curves of single hair cells (Russell and Sellick, 1977), the first reports about measurements in auditory nerve fibres that the spatiotemporal distribution patterns of the distortion products $2f_1$-f_2 and $2f_2$-f_1 resemble those generated by a single tone of corresponding frequency (Kim et al., 1979) – a finding first reported on a symposium organized by our group here in Münster –, and, finally, the first detection and identification of otoacoustic emissions (Kemp, 1978).

Nowadays it is generally accepted that the basilar membrane-haircell system is highly nonlinear, that this system possesses the same tuning characteristics which were formerly attributed to auditory nerve fibres, which means that the so-called second filter does not exist, and that this system even produces an acoustic signal which can be recorded in the outer ear canal: the so-called otoacoustic emissions.

Otoacoustic emissions are another episode in my scientific life. As early as 1970, that is, in the same year in which we detected the travelling wave of distortion products, we reported a pure tone of about 4200 Hz (Fig. 3) which could repeatedly be measured with a sensitive microphone in the outer ear canal of a patient suffering from a sensorineural deafness (Kumpf and Hoke, 1970). At that time, we were unable to explain this finding, and it took almost one decade until spontaneous and evoked otoacoustic emissions were detected and their cochlear origin verified.

I mentioned earlier that my interest began to ascend from the periphery of the auditory system to its centre, the auditory cortex. After abandoning the research on cochlear microphonics, brainstem auditory evoked potentials became the focus of interest for many years, and we especially developed powerful techniques for the frequency-specific assessment of the auditory threshold. Again I indulged in my old passion, and I developed hard- and software of a computer-controlled system for the measurement of auditory evoked potentials. I even found an easy to grasp acronym for this system: AERIS, which stands for Auditory Evoked Response Investigation System. At present, my collaborator Bernhard Ross is developing the fourth generation system on a SUN workstation.

We then extended our interest to the slow cortical auditory evoked potentials, and it was especially my collaborator Bernd Lütkenhöner who devoted several years of research to this topic. Clinical routine in my institute would be unthinkable without the AEP detection system he developed together with Bernhard Ross.

One of the first proposals which I wrote for the head of the ENT department as early as 1968 was a proposal for the erection of a new building for the sections of Clinical and Experimental Audiology. It took exactly ten years until the foundation stone was layed. From now on, the development of my career as well as that of the Experimental Audiology Group became less unpredictable; the flutter of the butterfly's wings less effective. Two years later, in 1980, we moved into the new building, which offered our group a usable space of more than 1,000 m^2. The new premises comprised fully equipped electronic and mechanical workshops, each headed by a master craftsman; two anechoic, soundproof, electrically shielded chambers; a fully integrated computer center etc. The budget for the initial equipment amounted to more than 1 million DM. Since auditory evoked potentials had been the main field of research for more than 10 years, it was only consequent to extend our field of interest to auditory evoked magnetic fields, and part of that budget was used for the purchase of a single-channel SQUID system. An Alexander von Humboldt research fellow, Christo Pantev, came and began to establish, together with two physics diploma students and the other scientific coworkers of the group, our first setup for the investigation of auditory evoked magnetic fields. In 1986, the Institute of Experimental Audiology was founded, advancing the large de facto department to which the one-man business had grown to an independent institute of the University of Münster, and I was appointed the director of this new institute. This Institute of Experimental Audiology is unique in Germany, and there are only few institutions of this kind in Europe.

The success continued. Owing to the great scientific achievements of our group we were able to establish, in 1989, the Clinical Research Unit "Biomagnetismus und Biosignalanalyse", the most expensive unit which has ever been supported by the DFG, the German Research Association. The total of research grants received amounts to approximately 20 million DM, more than half of it being spent for the Clinical Research Unit. In 1990/91, another 2 million DM were spent to erect a special building for the Biomagnetism Centre, which now belongs to the best equipped centres of this kind in the world. There are now more than fifteen scientists currently working at the Biomagnetism Centre who hold university degrees in medicine, physics or engineering. The measurements with High Tc SQUIDs that were performed in Münster stand as current world records in this field and are one example that demonstrates the strong technological background of this Centre. The Münster Biomagnetism Centre, located in the heart of Central Europe, has become a famous address for scientists from all around the world: It's become not a strange, but a strong attractor.

With the establishment of the Research Unit, a position for a full professor was also provided, and Professor Thomas Elbert who received the call for this position acted as a stimulant to our group, since he enlarged the areas of our research by novel fields of interest. One of these fields is the dynamics of nonlinear systems, and here the wheel comes full circle: an arc extending from the research on the nonlinearity of the peripheral auditory system to the dynamics of nonlinear systems.

Speaking of things coming full circle: It was only few years after my birth when Berger (1937) first detected those EEG waves which are the topic of this symposium. Later, I took up studying medicine at the same University, the University of Jena, where he detected the EEG.

I hope that I have been able to illustrate that, though my personal contribution to oscillatory event-related brain dynamics is negligible, my indirect contribution, that is, the creation of the Institute of Experimental Audiology with its Biomagnetism Centre, is certainly not insignificant. I wish you a pleasant stay in Germany and the Symposium the highest possible success.

REFERENCES

Békésy, G. von, 1943, Über die Resonanzkurve und die Abklingzeit der verschiedenen Stellen der Schneckentrennwand, Akust. Z. 8:66-76 (On the resonance curve and the decay period at various points of the cochlear partition, *J. Acoust. Soc. Amer.* 21:245-254, 1949).

Berger, H., 1937, *Arch. Psychiatr.* 106:165-187.

Evans, E.F., Wilson, J.P., 1973, The frequency selectivity of the cochlea, *in*: "Basic Mechanisms in Hearing", A. Møller, ed, Academic Press, New York, San Francisco, London, pp. 519-554.

Goldstein, J.L., 1972, Evidence from aural combination tones and musical notes against classical temporal periodicity theory, Proc Symp on Hearing Theory, IPO Eindhoven, pp. 186-208.

Hoke, M., Landwehr, F.J., 1970, Tierexperimenteller Nachweis der Intermodulationswanderwelle, *Arch. klin. exp. Ohren-, Nasen- u. Kehlkopfheilk.* 196/2:148-153.

Hoke, M., Lütkenhöner, B., Bappert, E., 1980, Psychophysical tuning curves for the difference tones $2f_1-f_2$ and $2f_2-f_1$, *in*: "Psychophysical, Physiological and Behavioural Studies in Hearing", G. van den Brink, F.A. Bilsen, eds, Delft University Press, Delft, pp. 278-282.

Kemp, D.T., 1978, Stimulated acoustic emissions from the human auditory system, *J. Acoust. Soc. Amer.* 64:1386-1391.

Kim, D.O., Siegel, J.H., Molnar, C.E., 1979, Cochlear nonlinear phenomena in two-tone responses, *in*: "Models of the Auditory System and Related Signal Processing Techniques", M. Hoke, E. de Boer, eds, *Scand. Audiol. Suppl.* 9:63-82.

Landwehr, F.J., Bally, G. von, Hoke, M., 1970, Untersuchungen über die Innenohrmechanik an einem nichtlinearen elektronischen Modell, *Arch. klin. exp. Ohren-, Nasen- u. Kehlkopfheilk.* 196/2:142-148.

Rhode, W.S., 1971, Observations of the vibration of the basilar membrane in squirrel monkeys using the Mössbauer technique, *J. Acoust. Soc. Amer.* 49:1218-1231.

Russell, I.J., Sellick, P.M., 1977, Tuning properties of cochlear hair cells, *Nature* 267:858-860.

INTRODUCTORY REMARKS

Robert Galambos and Theodore H. Bullock

University of California, San Diego
La Jolla, California, U.S.A. 92093-0201

Our task is twofold: to restate the themes and aims of this conference and to explain why it is so appropriate that it is dedicated to Manfried Hoke on the occasion of his 60th birthday.

The organizers first stated the main goal of the meeting and published collection of papers as the discussion of the "significance and theoretical basis of coherent oscillatory activity, which has been found to occur in local field potentials, as well as in scalp EEG and MEG, when organisms are presented with perceptual, cognitive and behavioral tasks. One example of this activity is "gamma band" oscillations (20-50 Hz) which may bind features to form objects in vision and in somesthesis. An emerging consensus maintains that oscillatory rhythms play an important role in perception and attention, and may support the encoding of spatial and temporal order during behavioral adaptation. We also plan on discussing how recent developments in the recording and analysis of brain potentials and neuromagnetic fields (MEG) may assist the study of oscillatory phenomena in conjunction with other methods including neural and dynamic modelling of psychological functions."

We would only add to this formulation of aims what may be assumed to be understood, namely that some workers wish to reexamine the proposition that event related brain oscillations are likely to be important in cognitive achievements such as "binding" and attention, or at least whether the evidence so far justifies this conclusion.

The phrase "oscillatory event related brain dynamics" is obviously intended to broaden the topic beyond ERPs - event related potentials, to include magnetic fields generated by the brain. At the same time it invites the consideration of other signs of activity, such as optical changes and impedance changes. The extremely useful methods of localizing brain activity such as local blood flow measurement, 2-deoxyglocose uptake, positron emission tomography and magnetic resonance imaging are only excluded because their time resolution does not permit the detection of oscillations of tens of Herz.

Oscillatory Event-Related Brain Dynamics, Edited
by C. Pantev, Plenum Press, New York, 1994

Event related brain oscillations have come into prominence in recent years, due especially to the findings of the Singer group in Frankfurt and the Eckhorn group in Marburg that neuronal units and clusters in the visual cortex of anesthetized cats show bursts of spikes and correlated slow field potential oscillations of ca. 35-70 Hz for a second or so after adequate stimulation such as a stripe in the appropriate orientation moving in the right direction. Particularly intriguing has been the further finding that two such loci several millimeters apart oscillate in a congruent fashion if the stripe segments that stimulate each of them are moving together but not if the movements are uncorrelated. A major role in calling attention to this class of phenomena was then played by several findings, in different laboratories, of gamma-band responses to omitted stimuli and to acoustic stimuli. Less known but perhaps relevant are reports of event related oscillations in fish and turtles after omitted stimuli and simple stimuli, as well as older examples from the literature. The conclusion is abundantly justified that a class of oscillatory responses bespeaking synchrony among a set of neurons is widespread, both to simple stimuli and to events with a more cognitive component, and furthermore, that this class deserves new experiments to clarify its significance and to uncover its mechanisms.

The phenomena under scrutiny here overlap extensively with those treated in a recent volume on "Induced rhythms in the brain" (Basar and Bullock 1992, Birkhäuser, Boston), which can be recommended as useful background, with extensive literature review.

It is fitting that this symposium and collection of papers should be in honor of Manfried Hoke. First, not only his personal friends but a world community of scientists honor his creation - the laboratory, the group and the inspired work they have contributed to the fields of audiology and of brain physiology. The techniques of magnetic field recording from the brain, in which he has been such an active leader, play a great role in the most recent findings reported in this meeting and hold promise of being key players in the future study of brain dynamics. For those who don't recall, we should point out that this is but the latest in a long succession of campaigns for improvements in the methodology of electrophysiological and electrodiagnostic procedures that Professor Hoke has waged. It is not the case, however, that Hoke's reputation rests primarily on methodological advances. He has contributed notably to both observational and theoretical-interpretive audiology. He early (1968) recognized and emphasized the nonlinearities in the inner ear mechanics. Already in 1970 he and his coworker, W. Kumpf reported a case of objective tinnitus, when only a few others were known, having properties that conform to what we call today otoacoustic emission with very narrow band, continuous spontaneity even during anesthesia. His publications range from the acoustic periphery to the cortex, from "decoding" human compound action potentials, through summing potentials, distortion products, auditory brainstem responses, and Wiener filtering to organization of the auditory cortex and gamma band activity there. We, speaking for the members of this symposium, who have enjoyed an unusually stimulating and successful meeting, wish Professor Hoke many more years of satisfying insights into the wonders of hearing!

A COMPARATIVE SURVEY OF EVENT RELATED BRAIN OSCILLATIONS

Theodore H. Bullock and Jerzy Z. Achimowicz

University of California, San Diego
La Jolla, California, U.S.A. 92093-0201

INTRODUCTION

Nature loves to oscillate. Ongoing oscillations of a wide range of periods are familiar in animals, from the circannual, circalunar and circadian, to the so-called minute rhythms of Galambos and Makeig (1988), from the respiratory, cardiac, and EEG rhythms of delta, theta and alpha frequencies, to the 40 Hz cerebral, 200 Hz cerebellar, and the medullary pacemakers of electric organ discharges in certain electric fish which run at 1000-2000 Hz, night and day. In addition, many living systems - or parts of them - love to show event-related oscillation. The gamma band of frequencies is popular from invertebrates to mammals, especially for transient oscillations, such as event-related rhythms, which have recently come to prominence and have been called induced rhythms in a recent book of that title (Basar and Bullock 1992).

The purpose of this paper is to shed some perspective on the topic of the symposium by surveying the literature for examples of event-related oscillations, particularly those in the gamma band. We ask, for each example, four kinds of questions. (i) Are the events with which they are related similar? (ii) Are they suggestive of a common meaning, or (iii) of a common mechanism? (iv) Are the dynamics of the oscillation basically common or diverse?

Actually, there are so many examples in the literature that we have had to make a selection. The bottom line is that the examples known *do not* fall into a single or a few clear categories, and do not appear to have a common mechanism or meaning. We conclude that anytime we think of a candidate hypothesis for the function or mechanism of induced rhythms, it would be wise to look around for cases that might test the hypothesis or at least its generality. If a rhythm correlates with something, for example a behavior or some aspect of the stimulus, in some species or preparation, there are at least two kinds of pitfalls to drawing conclusions. First,

Oscillatory Event-Related Brain Dynamics, Edited
by C. Pantev, Plenum Press, New York, 1994

there may be counter examples, easily overlooked. Second, other aspects of the stimulus or response may be the significant correlates. It is not very safe to propose a certain aspect of the behavior, e.g. its cognitive aspect, is the relevant correlate, or to propose a causal relation, especially the polarity - which event is necessary for the other. However, we are personally thankful that some of our colleagues are courageous enough to make the proposals, since they, more than the physiological facts, have attracted the great interest to these oscillations.

First, then, let us define some terms. An oscillation is a rhythm or fairly regular fluctuation in some measure of activity. That means it is not simply a portion of broadband activity isolated by a bandpass filter or caused by the ringing of a filter by a transient. Filtered waves may, of course, be a true oscillation but if, for example, a power spectrum reveals a broad elevation, such as 25-50 Hz, it should be called merely a "gamma band component of the power spectrum," not a rhythm. Imagine hitting all notes on the piano over a whole octave! It is, in fact, quite difficult to establish that a true rhythm exists, in the presence of considerable wideband activity, unless there is a clear and narrow power peak in the wideband spectrum.

Cole (1957) and Enright (1965, 1989) have reminded us, in historically important papers, of several pitfalls in "finding" spurious rhythms. Caution is important in making statements and models about attributing functions to presumed gamma oscillations.

Besides the definition and validation of a rhythm, we should, for the purposes of this book, distinguish (i) *ongoing*, background gamma-band activity from (ii) *induced* and from (iii) *driven* activity. Because we are concerned, by our title, mainly with event-related oscillations, we are mainly talking about induced rhythms. These are defined as oscillations caused or modulated by stimuli, by events or by changes in brain state that do not directly drive successive cycles (Bullock 1992).

Some authors have made a distinction between *evoked* and *induced* rhythms or activity. Consistent with the definitions and historical introduction in the recent volume on "Induced Rhythms in the Brain" (Basar and Bullock 1992), the distinction recommended is the following. Induced rhythms form the broader category, within which evoked *rhythms* are a subset; evoked *potentials* are often nonrhythmic. The term induced rhythms was introduced to call attention to a large variety of oscillatory responses that follow either clearly timed stimuli or less sharply timed state changes such as attention, sleep, expectations and seizures. Induced rhythms may be tightly or loosely time- and/or phase-locked to the triggering event and hence may appear in averaged responses or only in single trials. Evoked rhythms are the subset following clearly timed stimuli or events; they may also be more or less tightly time-locked and survive averaging well or poorly. The term evoked is generally not used for responses to less sharply timed state changes such as those just named. It is used for responses that are either rhythmic or, more commonly, episodic such as one or more peaks of different duration. Nonrhythmic, episodic peaks or humps are not generally called induced except as a common word like elicited. The terms evoked and induced thus overlap. It is wise to add adjectives when some feature is important, such as "phase-locked" or "nonphase-locked," "averaged" or "not averaged," rather than to set up still more jargon by arbitrary distinctions not widely understood.

Finally, the definition of gamma band we use is 25-50 Hz, more or less, for mammals, or 15-35 Hz for exothermic species, such as insects, fish and turtles, at 15-

22°C but the numerical value of the limits of the gamma band are not rigorously circumscribed.

SELECTED EXAMPLES FROM NONMAMMALIAN SPECIES AND MAMMALIAN SUBCORTICAL LEVELS

Isolated axons vary widely; some respond to a pulse or step stimulus with no oscillation, others with a few damped cycles, others with rhythmic discharge on the crests of subthreshold, local oscillations lasting up to dozens of cycles. Each is a characteristic of that type of fiber (Arvanitaki et al. 1936, Arvanitaki 1938, 1939a,b, Hodgkin 1948, Wright and Adelman 1954, Tasaki and Terakawa 1982). The frequency can be up to >100 Hz. Isolated muscle fibers and other types of cells (for example, egg cells, endocrine glands) also show rhythmic fluctuations related to impinging events.

Sense organs of various modalities show oscillations induced by adequate stimuli. Especially interesting are the electroreceptors that "ring" at the characteristic or best frequency for sensitivity, whether as a result of their sharp tuning or part of the mechanism of being tuned, we don't know. The oscillations of the stretch receptors of crayfish have been particularly well analyzed (Erxleben 1989, Morris 1990) to the level of patch clamp data on single channel openings and closings.

Retinas have labile oscillations with changes in illumination. Fröhlich (1913) described 30-90 Hz rhythms in the isolated octopus eye after ON and 20-45 Hz waves after OFF. Adrian and Matthews (1928) described a rhythmic succession of large waves, after a long latency, in whole nerve recordings from the eel *(Conger)*, under certain conditions, especially illumination of a large part of the retina. The frequency was usually between 5-15 Hz at the low temperature of this preparation. A series of authors have added observations on other species, including mammals (Bullock 1992). From the properties and the combinations of conditions conducive to these rhythms it seems unlikely that they represent the same physiological process in these different eyes. Figure 1, although chosen to illustrate oscillations in the optic tectum, actually shows some forms of retinal rhythms, since the tectal waves follow quite closely waves recorded with semimicroelectrodes in the retina, which persist even after the optic nerve is cut.

Ganglion cells like the giant neurons of *Aplysia* give diverse responses not only to synaptic input but to a standard event such as a depolarizing pulse; some respond rhythmically (Arvanitaki and Cardot 1941, Arvanitaki and Chalazonitis 1955, 1961). Lobster stomatogastric ganglion cells which have a spontaneous rhythm modulate that rhythm to suitable input (Ayers and Selverston 1979, Selverston 1993). This preparation has been intensively studied and illustrates how difficult it can be to decide whether a rhythm should be attributed to pacemaker cells that others follow, with modifications, or to the whole circuit or some essential subsets with particular time constants and interaction strengths.

Insect optic lobes were found quite early (Jahn & Wulff 1942, Crescitelli and Jahn 1942, Bernhard 1942) to be capable, if critical conditions are just right, including partial light adaptation, of compound field potential oscillations lasting for a second or two at 20-30 Hz, triggered by the onset of a steady light. Unit spikes tend to cluster at one phase of this 30-50 ms sine wave, riding on it rather than summating to cause it. This particular case has the special interest that it is one of a very few which show invertebrate central ganglia are capable, under special conditions, not yet understood, of slow waves like the vertebrate EEG.

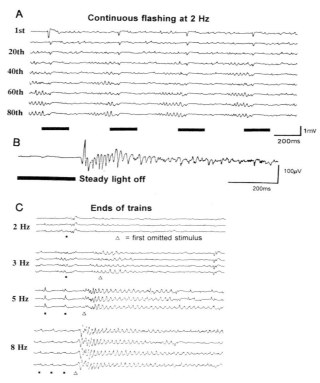

Figure 1. Induced rhythms with simple visual stimuli. Single sweeps recorded from the optic tectum of a ray (*Platyrhinoidis*) after whole-field flash stimuli (black bars). **A.** Continuous flashing at 2 Hz gradually induces a late oscillation following a few hundred milliseconds of suppression of EEG; the next onset of light causes a small VEP and suppression of the oscillation. **B.** The offset of a steady light, providing it has lasted at least 500 ms, causes a long oscillation in good preparations. The latter part of this record is normal EEG. **C.** The ends of 30 second trains of flashes (black squares) at different frequencies, with the due-time of the first omitted stimuli indicated by the triangles. The Omitted Stimulus Potentials include an oscillation of fixed frequency but variable duration, depending on the conditioning frequency. Most of the features seen in the midbrain can already be seen in recordings from the retina with the optic nerve cut. (Kindness of M.H. Hofmann)

Vertebrate brainstem structures, including primary sensory nuclei of the cranial nerves, reticular nuclei, and tectum, as well as the cerebellum, each have been found to employ induced rhythms upon occasion. It seems likely that this array of exemplars is not homogeneous in mechanism or meaning. The medullary nuclei for primary afferents from electroreceptors in elasmobranch fish are in the same column as lateral line mechanoreceptive and octaval nuclei. They show a labile or facultative synchronized oscillation at 5-20 Hz at 15-20°C for several hundred milliseconds after a single physiological stimulus, e.g. a 10 ms current pulse of a few microvolts in the water. In the tectum of fish oscillations of 15-25 Hz (15-20°C) can be seen after a single light flash or part of the VEP, or after the OFF, or after the end of a train of light flashes in the frequency range of 5 Hz or higher - as part of the omitted stimulus potential (OSP) (Fig. 1C). This echoes and modulates the OSP that we noted above already in the retina.

In neurons of the mammalian inferior olive, substantia nigra and other nuclei induced rhythms in the 10 Hz range are modulated by various impinging influences, especially calcium channel blockers (Llinás and Yarom 1986, Llinás 1988, Fujimura and Matsuda 1989).

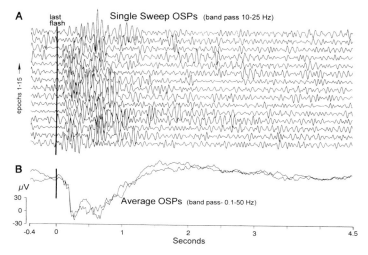

Figure 2. Flash induced rhythms in a reptile. **A**. Single sweeps recorded in the dorsal cortex of the cerebrum of a turtle (*Pseudemys scripta*) in the visual area at the end of a 10 s conditioning train of 20 Hz flashes. Slow components of the OSP have been filtered out to show the not strictly phase-locked 15-20 Hz oscillations. **B**. Average OSPs (two subsets) with wideband filter, showing that the oscillations are much attenuated in the average. (Kindness of J.C. Prechtl)

Thalamic nuclei have repeatedly been seen to display oscillations in response to transient stimuli. One of the first reports was by Chang in 1950, who invoked the corticothalamic "reverberating circuit," popular at that time. Galambos et al. in 1952 saw repetitive or oscillatory firing in the medial geniculate; Bishop et al. in 1953 in the lateral geniculate and, skipping intervening years, Leresche et al. in 1990 in thalamocortical cells.

Hippocampal induced rhythms in the gamma band have been described after acoustic and other stimuli (Basar and coworkers 1980, Basar et al. 1992, Basar-Eroglu et al. 1992). The much studied theta rhythm or RSA of the hippocampus arises in synchronized firing of septal cells and spreads widely through the limbic and related structures (Lopes da Silva 1992) in response to certain kinds of stimuli, certain drugs or certain forms of behavior, such as walking in rats.

The olfactory bulb and associated structures are well known substrates for oscillations brought on by physiological odor stimuli (references in Bullock 1992). Even in this much studied phenomenon it is not clear whether we have to do with a single, common physiological process or whether more than one alternative or sequential mechanism may be involved in inducing the rhythms.

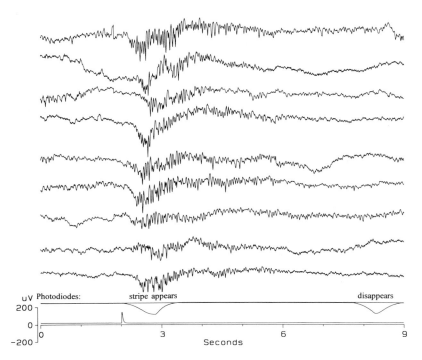

Figure 3. Moving bar induced rhythm in a reptile. Single sweeps recorded from the dorsal cortex of the cerebrum of a turtle (*Pseudemys scripta*) in the visual area, at the depth of the pyramidal cell somata during 7 s of horizontal movement (11°/s) of a single 9° vertical stripe across 77° of visual field; 5 s of rest separated the sweeps. Photodiodes at each side of the screen show the appearance and disappearance of the stripe. A slow evoked negative-positive wave begins before the oscillation grows out of the background EEG. (Kindness of J.C. Prechtl)

SELECTED EXAMPLES FROM THE CORTEX

Visual cortex has been found to oscillate under steady or transient photic stimulation by many workers, going back more than 30 years (Chatrian et al. 1960, Grüsser and Grüsser-Cornehls 1962, Doty and Kimura 1963, Hughes 1964, Regan 1968, Sturr and Shansky 1971, Abdullaev et al. 1977, Whittaker and Siegfried 1983, Friedlander 1983, Pöppel and Logothetis 1986, Freeman and vanDijk 1987). We mention these studies, without reviewing them, just to remind ourselves that the elegant recent studies reported elsewhere in this volume had a premonitory background of diverse observations - and quite likely diverse phenomena. Prechtl and Bullock (1994) have found induced rhythms to visual stimuli in the turtle cortex and dorsal ventricular ridge - to ON, OFF and omitted flashes (Fig. 2) and (Prechtl and Bullock 1993) to moving bar stimuli (Fig. 3). Once again, we are led to suppose that the event related oscillations are a widespread and heterogeneous class of responses.

Cortex other than visual has also given evidence of induced rhythms under such a variety of conditions that we will mention only two. Already in 1938 Loomis, Harvey and Hobart described the "K complex" in sleeping humans following acoustic stimuli; after long latency negative and then positive swings a series of 8-14 Hz oscillations lasts for a second or more. A decade later, Bremer in 1949 showed rhythmic afterpotentials in the cat auditory cortex after a click; sometimes it could also be seen in the medial geniculate, especially in the strychninized state. Other examples are mentioned in the introduction to the book on induced rhythms (Basar and Bullock 1992).

SOME LESS FAMILIAR DESCRIPTORS THAT MIGHT DISCRIMINATE OR DISTINGUISH BETWEEN CASES

Thus far we have offered simply a selection of literature. The following is a modest selection of suggestions for possibly heuristic, new or less familiar ways to look at data from preparations or cases, interesting because they exhibit apparent oscillations that are event-related. The assumption we are making is that there is more *evidence of deterministic dynamical structure*, and of cooperativity available in the usual data we record than is visible by inspection of the voltage against time record, and of the ordinary power spectrum that discards temporal structure and assumes linearity as well as independence of frequencies.

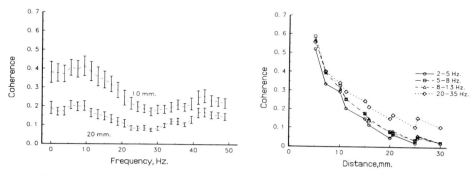

Figure 4. Coherence distribution in space; average decline with distance. Pairs of subdural electrodes recording human EEG; average behavior of pooled pairs of the same separation. *Left graph*, coherence at different frequencies, means of many pairs 10 mm and 20 mm apart, with standard error bars, showing modest drop above 15 Hz. *Right graph*, coherence as a function of separation of electrodes, for several bands; means of many pairs of the given separation (fewer pairs available at the greater distances, hence the standard errors are larger). At 5 mm coherence has already dropped to <0.6 from the near 1.0 it has to show at very small separations where electrodes see almost completely overlapping volumes of tissue; in another 5 mm it drops another half. In this case some higher bands drop less than lower frequency bands but this is not always true.

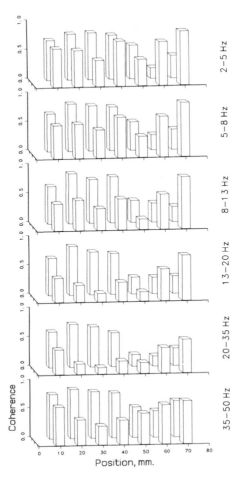

Figure 5. Coherence distribution in space; local differentiation. Single pairs of adjacent electrodes in 2 rows of an 8X8 array resting on the pia mater of the lateral aspect of the right inferior occipital and posterior temporal lobe of an epileptic patient, during slow wave sleep. Left to right the front row pairs are #1-2, 2-3, ... 10 mm apart; the back row pairs are #9-10,#10-11... Electrode #1 is close to the tentorium, #8 is lateral, #9-16 are 20 mm superior. Bars show the mean coherence for the given band over 18 samples of 5 s each. Note that coherence can be quite local; adjoining pairs, although sharing one site, can be quite different. Note also that the profile of higher and lower values tends to repeat in different bands, that is, when coherence is high or low, this is correlated over a wide range of Fourier component frequencies. The same is often true of fluctuations over time (see Figs. 6 and 7).

We shall confine our remarks to the *compound field potentials* seen by micro- or semimicroelectrodes in or on the surface of the brain, passing over for this occasion the single unit records that see only the spike form of neuronal activity from some selection of units. At the same time we emphasize two methodological options. One is the opportunity to study many loci at the same time, by multichannel recording with closely spaced electrodes that permit some insight into the spatial microstructure and cooperativity of the assembly activity - particularly by special computations that quantify degrees of congruence versus independence. The other option we underline is the analysis of each sweep or trial of event related activity, before averaging, with attention to the state of the ongoing EEG at the same locus before the defined event.

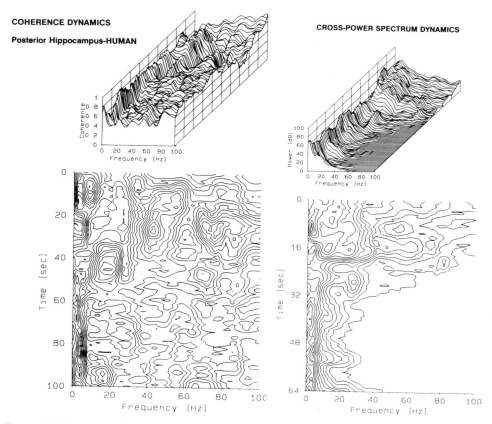

Figure 6. Coherence and power spectra changing in time. Examples chosen from depth recordings in the posterior hippocampus of a human epileptic patient during ca. 50s of preictal EEG and 50s of seizure. (Data kindness of S.S. Spencer and R.B. Duckrow) **Left**, coherence between two electrodes 6.5 mm apart, computed every second and displayed as a 5-point weighted moving average. Time proceeds from the nearer to the farther spectra for 100 s. 3-D display above and topographic display below. Note the tendency to sustained but fluctuating ridges at ca. 5, 20, 40 and 70 Hz and the elevation of coherence at all frequencies above ca.15 Hz in the later stages. Note that, with few exceptions (5-8 Hz), coherence is similar in neighboring frequencies, i.e. the Fourier components are not independent. It also changes with time similarly for a band of several Hz to an octave or more. Sections for any frequency or band show the largest fluctuations have periods of ca. 5-10s. **Right**, the same for cross-power between the same electrodes, but only for 64s. Note the poor correlation between power and coherence.

Microstructure of coherence

The first form of analysis to mention is the classical coherence - not often used, especially for intracranial recordings, and not widely familiar as to what it shows.

Coherence is a linear, pairwise comparison, in that respect like cross-correlation, but it adds the feature that a number is computed for the degree of similarity for each frequency, at the resolution selected. It provides a display of the fraction of energy at each Fourier frequency component that is in phase throughout the defined epoch, whatever the phase. Obviously, for short epochs there is a strong likelihood that the two time series, even if entirely independent, will have by chance some of the energy at each frequency in the same phase and this is called the bias; it only goes down to values <0.1 for independent time series of the usual EEG type when the epoch is longer than some seconds, but the bias is known, so that confidence levels can be calculated for any sample.

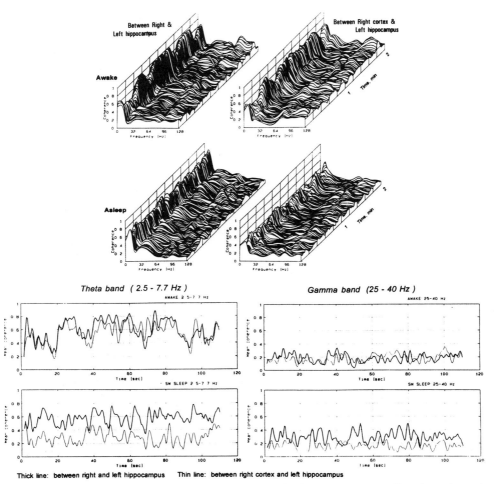

Figure 7. Coherence over time. Examples chosen from freely moving rat recordings from electrodes in right and left hippocampus and dorsal cortex, 2 min samples. (Data kindness of J.M. Gaztelu) Spectra over time are shown in 3D plots, awake and asleep, between points on the right and left hippocampus and between a point in the right cortex and a point in the left hippocampus. Sections of such plots are shown on the right for the theta and gamma bands. Note that the coherence is generally higher for the left to right hippocampus, especially in the sleeping state and that the gamma band coherence is much lower at these separations than the delta-plus-theta band. Note also that coherence is typically waxing and waning with periods in the range of 2-10 s.

What we have found that seems not to have been appreciated before is that coherence tends strongly to be *not very different for different frequencies*, even for wide bands of one or two octaves or more; even from 1 to 50 Hz there is usually only a slight decline in coherence. When alpha or thete is strong, there may be a modest high of coherence at that frequency but in general the *power spectrum does not predict* or correlate well with the coherence spectrum, when we study subdural EEG from the cortical surface or depths and pairs of loci a few millimeters apart, rather than scalp recordings, whose peculiarities we will not deal with in this account.

Under such conditions, coherence tends on average, quite reliably to *fall with distance* such that, from a value of 1.0 when the loci are so close together that the

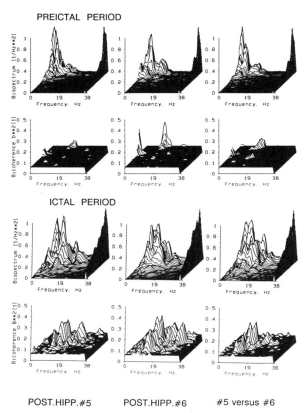

POST.HIPP.#5   POST.HIPP.#6   #5 versus #6

Figure 8. Nonlinear, nonGaussian higher moments of ongoing EEG. Bispectra (upper panels of each set) and bicoherence (lower panels of each set) of all pairs of frequencies, X and Y, from 1-38 Hz within each of two channels (left and middle, #5 and #6), and between the same channels (cross-spectra, right column, #5 vs. #6). *Upper set*, just before onset of an electrical seizure. *Lower set*, after onset of seizure. The left and middle upper panels, with auto-bispectra, also show the power spectra along the right margin; the right hand, upper panel, with cross-bispectra, also shows cross-power for the two channels. The prevailing plane of each graph is the 5% significance level; all values not exceeding that significance have been converted to this plane so that all peaks and bumps shown are at least that significant. Note that only one pair of frequencies shows preictal bicoherence, ca. 3 and 14 Hz, whereas many pairs in the range of 8-30 Hz show a substantial ictal mountain range. Same data set as Figure 6.

electrodes see virtually the same volume of tissue, it drops by a significant amount, e.g. to 0.9 at ca. 0.5-1 mm, then monotonically to ca. 0.5 at 4-8 mm and to values insignificantly different from the chance level, e.g. 0.1 at ca. 20-30 mm (Fig. 4B).

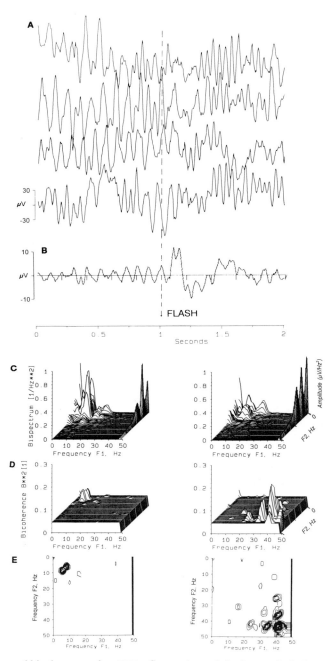

Figure 9. Bispectra and bicoherence of an ERP. Comparison of the 1 s epochs before and after a light flash stimulus, recorded from a parietal electrode (P4) on the human scalp, repeated every 3 to 5 s

with randomly varying intervals. **A.** Four single sweeps as samples of the data, showing the large alpha activity in this subject. **B.** Average of 20 sweeps showing the VEP with P50-70, N110 and P250 waves. **C.** Auto-bispectrum of the frequencies in the same channel, before (*left*) and after (*right*) the stimulus. The conventional amplitude spectrum is along the right margin. **D.** Auto-bicoherence before the stimulus has only a small peak at 7 and 10 Hz; after the stimulus this disappears and three new peaks appear representing quadratic phase coupling at ca. 23 and 41, 34 and 42, and 41 and 43 Hz - each shown symmetrically duplicated on two sides of the diagonal. **E.** The same in topographic form.

This is almost always true in large scale averages of many pairs over many tens of seconds from all parts of the neocortex we have sampled in rats, rabbits and humans, whether awake or in slow wave sleep. The variance, however, is large and can be found to result largely from maintained local highs and lows - in short, from microstructure on the fractional centimeter scale (Fig. 5).

Since J.Z.A. introduced to our laboratory the **autoregressive methods** of analysis that permit useful short epoch calculations we have learned something new about the *time trends* of coherence. We had known that in short epochs variability is very high, both in successive samples and in simultaneous samples from different but adjacent pairs, even those that share one electrode, such as AB and BC. Now we can characterize them reliably, in spite of the variation. The *two kinds of variability* are quite distinct - spatial, simultaneous and temporal, successive sampling. The former tends to be stable over minutes at least; the latter shows a significant tendency to drift slowly up and down with preferred periods of several seconds to several tens of seconds (Figs. 6, 7).

The various kinds of rises and falls of coherence tend to *affect all frequencies* together in the 1-50 Hz range. This brief summary of the characteristics of this descriptor is enough already to suggest that the classical model of EEG as a series of independent rhythms, such as is assumed by the linear Fourier analysis can be excluded. It also suggests that there is a *good deal of microstructure* in the EEG, in space in the millimeter domain and in time in the domain of seconds or fractions of seconds - and that there is a lot more information in these time series, from a number of loci, than we have known about with our normal power spectral analyses. We have begun to plot coherence in the last second or two before and in the first second or two of an event-related response, for chosen bands of frequency, for single sweeps as well as averages, for many electrode pairs, close together. With sliding window displays of the time trend of coherence we can now follow the changes in this measure with a resolution of less than one second. Our experience is still too limited to report that consistent peaks of coherence appear during the ERP or which band of frequencies will be affected but we expect this approach to produce new insights.

The conclusion we draw is that we have yet to discover the main features of the EEG and ERP in respect to this measure of cooperativity, in the dimensions that matter functionally -in millimeters and seconds, in the various parts of the brain, its behavioral and pathological states and its stages of ontogeny, maturity and senility and of evolution.

Bispectrum and bicoherence

The last point we will make here is that we should not stop with linear, second order measures. One example of the available, well developed methods of nonlinear analysis of third order moments is the bispectrum and its derivatives, especially bicoherence. These are much used in engineering for system identification. They are independent of the Gaussian fluctuations but quantify separately the departures from Gaussianity due to skewness around the time axis, asymmetry around the voltage axis and quadratic phase coupling between every pair of frequencies. The last mentioned, the phase coupling, is measured, normalized to the total energy at each frequency, by the bicoherence, a number between zero and one for each pair of frequencies. Much of the time the normal EEG shows no bicoherence above the chance level but episodically during normal alert periods and less often during slow wave sleep, without any obvious sign in the raw data, peaks or rugged mountains appear at some part of the plane defined by F_1 and F_2 - frequency pairs in the same electrode, and they may appear also in the cross-bicoherence plot between two electrodes, if they are not far apart. Our experience is still limited and applies mainly to human subdural recordings from epileptic patients, with some depth recordings from patients (Fig. 8), as well as rats and turtles. During electrical seizures the bicoherence is usually very mountainous, with many peaks and high valleys, sometimes confined to the high-high quadrant of the plot, sometimes spreading to all parts of the F_1-F_2 plane. Interictal periods can be quite flat or may have a small peak, usually in the low-low quadrant. We have begun to apply these methods to event related responses and find that new peaks can appear in the bicoherence plot during the ERP (Fig. 9). It is too early to make statements about consistent tendencies but the methods are sensitive to the changes in dynamical properties of the responses buried in the ongoing, event-unrelated background activity.

We conclude that we have yet to discover the main features of the EEG and ERP in respect to this measure of cooperativity, as in the coherence estimate, especially in the dimensions of millimeters and seconds, in the parts of the brain, behavioral and pathological states, stages of ontogeny, maturity and senility - and of evolution.

We do not doubt that still other descriptors will turn out to be useful, discriminating between places, states and stages - if we will but widen our habits of analysis to try some of the powerful tools that can disclose mutual dependencies and meaningful fine structure in time and space.

Acknowledgements

This work has been supported by grants from the National Institute for Neurological Diseases and Stroke.

REFERENCES

Abdullaev, G.B., Gadzhieva, N.A., Rzaeva, N.M., Alekperova, S.A., Kambarli, E.I., Dimitrenko, A.I., and Gasanova, S.A., 1977, Oscillatory potentials in the structures of visual system, *Fiziol. Zh. SSSR* 12:1653-1661.

Adrian, E.D. and Matthews, R., 1928, The action of light on the eye. Part III. The interaction of retinal neurones, *J. Physiol.* 65:273-298.

Arvanitaki, A. 1938, "Les variations graduées de la polarisation des systèmes excitables," Thesis, Univ. Lyons, Hermann et cie, Paris.

Arvanitaki, A., 1939, Recherche sur la réponse oscillatoire locale de l'axone géant isolé de *Sepia*, *Arch. Int. Physiol.* 49:209-256.

Arvanitaki, A., 1939, Contributions à l'étude analytique de la réponse électrique oscillatoire locale de l'axone isolé de *Sepia*, *C. R. Soc. Biol. (Paris)* 131:1117-1120.

Arvanitaki, A. and Cardot, H., 1941, Réponses rhytmiques ganglionnaires, graduées en fonction de la polarisation appliquée. Lois des latences et des fréquences, *C. R. Soc. Biol. (Paris)* 135:1211-1216.

Arvanitaki, A. and Chalazonitis, N., 1955, Les potentiels bioélectriques endocytaires du neurone géant d'*Aplysia* en activité autorhytmique, *C. R. Acad. Sci. (Paris)* 240:349-351.

Arvanitaki, A. and Chalazonitis, N., 1961, Excitatory and inhibitory processes initiated by light and infra red radiations in single identifiable nerve cells (giant ganglion cells of *Aplysia*), *in:* "Nervous Inhibition," E. Florey, ed., Pergamon Press, Oxford.

Arvanitaki, A., Fessard, A., and Kruta, V., 1936, Mode répétitif de la réponse électrique des nerfs visceraux et étoilés chez *Sepia officinalis*, *C. R. Soc. Biol. (Paris)* 122:1203-1204.

Ayers, J.L.Jr. and Selverston, A.I., 1979, Monosynaptic entrainment of an endogenous pacemaker network: a cellular mechanism for von Holst's magnet effect, *J. Comp. Physiol.* 129:5-17.

Basar, E. 1980, "EEG-Brain Dynamics," Elsevier,Amsterdam.

Basar, E. and Bullock, T.H. 1992, "Induced Rhythms in the Brain," Birkhäuser, Boston.

Basar, E., Basar-Eroglu, C., Parnefjord,R., Rahn, E., and Schürmann, M., 1992, Evoked potentials: ensembles of brain induced rhythmicities in the alpha, theta and gamma ranges, *in:* "Induced Rhythms in the Brain," E. Basar and T.H. Bullock, eds., Birkhäuser, Boston, pp. 155-181.

Basar-Eroglu, C., Basar, E., Demiralp, T., and Schurmann, M., 1992, P300-response: possible psychophysiological correlates in delta and theta frequency channels. A review, *Int. J. Psychophysiol.* 13:161-179.

Bernhard, C.G., 1942, Isolation of retinal from optic ganglion response in the eye of *Dytiscus*, *J. Neurophysiol.* 5:32-48.

Bishop, P.O., Jeremy, D., and McLeod, J.G., 1953, Phenomenon of repetitive firing in lateral geniculate of cat, *J. Neurophysiol.* 16:437-447.

Bremer, F., 1949, Considérations sur l'origine et la nature des "ondes" cérébrales, *Electroencephalogr. Clin. Neurophysiol.* 1:177-193.

Bullock, T.H. 1992, Introduction to induced rhythms: a widespread, heterogeneous class of oscillations, *in:* "Induced Rhythms in the Brain," E. Basar and T.H. Bullock, eds., Birkhäuser, Boston, pp. 1-26.

Chang, H.-t., 1950, The repetitive discharges of corticothalamic reverberating circuit, *J. Neurophysiol.* 13:235-257.

Chatrian, G.E., Bickford, R.G., and Uihlein, A., 1960, Depth electrographic study of a fast rhythm evoked from the human calcarine region by steady illumination, *Electroencephalogr. Clin. Neurophysiol.* 12:167-176.

Cole, L.C., 1957, Biological clock in the unicorn, *Science* 125:874-876.

Crescitelli, F. and Jahn, T.L., 1942, Oscillatory electrical activity from the insect compound eye, *J. Cell. Comp. Physiol.* 19:47-66.

Doty, R.W. and Kimura, D.S., 1963, Oscillatory potentials in the visual system of cats and monkeys, *J. Physiol.* 168:205-218.

Enright, J.T., 1965, The search for rhythmicity in biological time series, *J. Theor. Biol.* 8:426-468.

Enright, J.T., 1989, The parallactic view, statistical testing, and circular reasoning, *J. Biol. Rhythms* 4:295-304.

Erxleben, C., 1989, Stretch-activated current through single ion channels in the abdominal stretch receptor organ of the crayfish, *J. Gen. Physiol.* 94:1071-1083.

Freeman, W.J. and van Dijk, B.W., 1987, Spatial patterns of visual cortical fast EEG during conditioned reflex in a rhesus monkey, *Brain Res.* 422:267-276.

Friedlander, M.J. 1983, The visual prosencephalon of teleosts, *in:* "Fish Neurobiology, Vol. 2: Higher Brain Areas and Functions," R.E. Davis and R.G. Northcutt, eds., Univ. of Michigan Press, Ann Arbor, pp. 91-115.

Fröhlich, F.W., 1913, Beiträge zur allgemeinen Physiologie der Sinnesorgane, *Z. Sinnesphysiol.* 48:28-164.

Fujimura, K. and Matsuda, Y., 1989, Autogenous oscillatory potentials in neurons of the guinea pig substantia nigra pars compacta in vitro, *Neurosci. Lett.* 104:53-57.

Galambos, R. and Makeig, S., 1988, Dynamic changes in steady-state responses, *in:* "Dynamics of Sensory and Cognitive Processing by the Brain," E. Basar, ed, Springer-Verlag, Berlin, pp. 103-122.

Galambos, R., Rose, J.E., Bromiley, R.B., and Hughes, J.R., 1952, Microelectrode studies on medial geniculate body of cat. II. Response to clicks, *J. Neurophysiol.* 15:359-380.

Grüsser, O.-J. and Grüsser-Cornehls, U., 1962, Periodische Aktivierungsphasen visueller Neurone nach kurzen Lichtreizen verschiedener Dauer, *Pflügers Arch.* 275:291-311.

Hodgkin, A.L., 1948, The local electric changes associated with repetitive action in a non-medullated axon, *J. Physiol.* 107:165-181.

Hughes, J.R., 1964, Responses from the visual cortex of unanesthetized monkeys, *Int. Rev. Neurobiol.* 7:99-152.

Jahn, T.L. and Wulff, V.J., 1942, Allocation of electrical responses from the compound eye of grasshoppers, *J. Gen. Physiol.* 26:75-88.

Leresche, N., Jassik-Gerschenfeld, D., Haby, M., Soltesz, I., and Crunelli, V., 1990, Pacemaker-like and other types of spontaneous membrane potential oscillations of thalamocortical cells, *Neurosci. Lett.* 113:72-77.

Llinás, R.R., 1988, The intrinsic electrophysiological properties of mammalian neurons: insights into central nervous system function, *Science* 242:1654-1664.

Llinás, R.R. and Yarom, Y., 1986, Oscillatory properties of guinea-pig inferior olivary neurones and their pharmacological modulation: an *in vitro* study, *J. Physiol. (Lond.)* 376:163-182.

Loomis, A.L., Harvey, E.N., and Hobart III, G.A., 1938, Distribution of disturbance-patterns in the human electroencephalogram, with special reference to sleep, *J. Neurophysiol.* 1:413-430.

Lopes da Silva, F., 1992, The rhythmic slow activity (theta) of the limbic cortex: an oscillation in search of a function, *in:* "Induced Rhythms in the Brain," E. Basar and T.H. Bullock, eds., Birkhäuser, Boston, pp. 83-102.

Morris, C., 1990, Mechanosensitive ion channels, *J. Membr. Biol.* 113:93-107.

Pöppel, E. and Logothetis, N., 1986, Neuronal oscillations in the human brain, *Naturwissenschaften* 73:267-268.

Prechtl, J.C. and Bullock, T.H., 1994, Event-related potentials to omitted visual stimuli in a reptile, *Electroencephalogr. Clin. Neurophysiol. (in press)*

Regan, D., 1968, A high frequency mechanism which underlies visual evoked potentials, *Electroencephalogr. Clin. Neurophysiol.* 25:231-237.

Selverston, A.I., 1993, Modeling of neural circuits: what have we learned? *Annu. Rev. Neurosci.* 16:531-546.

Sturr, J.F. and Shansky, M.S., 1971, Cortical and subcortical responses to flicker in cats, *Exp. Neurol.* 33:279-290.

Tasaki, I. and Terakawa, S. 1982, Oscillatory miniature responses in the squid giant axon: origin of rhythmical activities in the nerve membrane, *in:* "Cellular Pacemakers," Vol. 1, D. Carpenter, ed., John Wiley & Sons, Inc., pp. 163-186.

Whittaker, S.G. and Siegfried, J.B., 1983, Origin of wavelets in the visual evoked potential, *Electroencephalogr. Clin. Neurophysiol.* 55:91-101.

Wright, E.B. and Adelman, W.J., 1954, Accommodation in three single motor axons of the crayfish claw, *J. Cell. Comp. Physiol.* 43:119-132.

PHYSIOLOGIC AND EPILEPTIC OSCILLATIONS IN A SMALL INVERTEBRATE NETWORK

Ulrich Altrup[1], Michael Madeja[2], Martin Wiemann[1] and
Erwin-Josef Speckmann[1,2]

[1]Institut für Experimentelle Epilepsieforschung
[2]Institut für Physiologie
Universität Münster 48149 Münster, Germany

INTRODUCTION

When compared to vertebrate nervous systems, invertebrate ganglia are studied easily because of several technical advantages. The relative simplicity of these ganglia enabled e.g. the study of the ionic bases of bioelectrical activity of neurons and the study of mechanisms underlying learning and memory. Furthermore, there are several examples of oscillatory events in invertebrate nervous systems which are well understood on the single neuron level (cf. Kandel, 1976; Fentress, 1976). Thus, the buccal ganglia of snails which are part of the snail central nervous system contain a neuronal generator of the motor program for food intake i.e. for biting and swallowing. The generator is spontaneously active even when the ganglia are isolated from the animal. The physiologic oscillatory activity of this generator has been studied in several snail species (cf. Siegler, 1977; Bulloch and Dorsett, 1979; Benjamin and Elliott, 1989).

To get further information on the intimate mechanisms underlying the physiologic oscillatory events, another type of oscillatory processes has been superimposed. To this purpose, rhythmic epileptic activity was elicited by application of epileptogenic drugs. Then, the physiologic oscillations in the buccal ganglia of *Helix pomatia* resulting from the activities of the intraganglionic generator for food intake are compared with epileptic oscillations in these ganglia.

METHODS

The investigations were performed on the isolated buccal ganglia or on semi-in-

Oscillatory Event-Related Brain Dynamics, Edited
by C. Pantev, Plenum Press, New York, 1994

tact preparations of the snail *Helix pomatia*. Isolated ganglia were prepared from the animal after cutting the nerves peripherally. Semi-intact preparations consisted of the buccal ganglia connected to pharynx, oesophagus and salivary glands. Methods are described in more detail elsewhere (Altrup, 1987; Altrup et al., 1990; Peters and Altrup, 1984).

Stable intracellular recordings for several days were done on several neurons simultaneously. For current injections a second microelectrode was impaled into the respective neuronal soma. Intracellular stainings were performed with intracellular injection of the dye cobalt-lysine (Altrup and Peters, 1982). Microelectrodes contained 150 mmol/l KCl. Concentrations of free K^+ and of Ca^{2+} in the extracellular space were measured by ion-selective microelectrodes. Electrical stimuli were applied to nerves of the buccal ganglia and consisted of bipolar pulses of double threshold intensity (duration: 1.5 ms). Field potentials were recorded via fire polished glass electrodes. Movements of the pharynx were detected by means of a light beam and photo diode.

The preparations were kept in an experimental chamber containing a snail "Ringer" solution (NaCl:130, KCl:4.5, CaCl$_2$:9, Tris-Cl:5 mmol/l). The chamber (volume: 1 ml) was continuously perfused (rate: 1 ml/min). Temperature was adjusted to 20°C and pH to 7.4. In some experiments CaCl$_2$ was reduced to 2 mmol/l and MgCl$_2$ was added to give 9 mmol/l ("high Mg-low Ca solution").

Epileptic activity appeared in neurons B2 and B3 when pentylenetetrazol (40 to 50 mmol/l) or etomidate (0.5 to 0.7 mmol/l) was added to the bath solution (Altrup et al., 1991).

RESULTS

Buccal ganglia and their target structures

The buccal ganglia of the snail are situated on the pharynx of the animal. They are connected to the cerebral ganglia via cerebrobuccal-connectives. The pharynx is positioned between mouth and oesophagus. It contains the muscular machinery for food intake movements. Fig. 1 demonstrates that the buccal ganglia modulate oesophageal and salivary gland activities and that the ganglia generate the pharyngeal motor activities. It can be seen that a cut of the nerves between the buccal ganglia and their peripheral targets altered the rhythms of peristaltic activities of oesophagus and of salivary glands and blocked the pharyngeal movements.

Mechanisms underlying pharyngeal motor activities have been studied (Peters and Altrup, 1984). Simultaneous recordings from several neurons showed that an intraganglionic network of neurons is pre-switched to the efferent motorneurons. The activation of the motorneurons was spontaneous and nearly simultaneous. It appears in a rhythmic manner with a repetition rate of about 1/min. These oscillations of activities could be shown in all neurons studied. In the semi-intact preparation in Fig. 2 it is demonstrated that the movements of the pharynx were coupled to depolarizations in motorneurons. To show the depolarizations without action potentials superimposed, a respective example was chosen. The coordinated peripheral muscular activities resulted from the structure of the synaptic depolarizations, i.e. synergistic motorneurons, partly synergistic/antagonistic and antagonistic motorneurons were

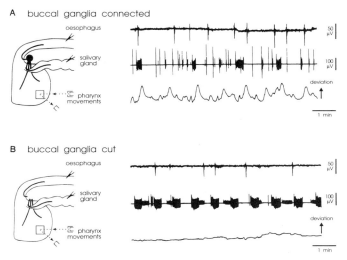

Figure 1. Buccal ganglia of *Helix pomatia* and their control of oesophagus, salivary glands and pharynx. A and B: activities with buccal ganglia connected (A) and cut (B), respectively. Schematical drawings give the experimental conditions. Field potential recordings of motor activities of oesophagus and salivary glands (upper two lines); recording of pharyngeal movements via light beam and photo diode (lower line).

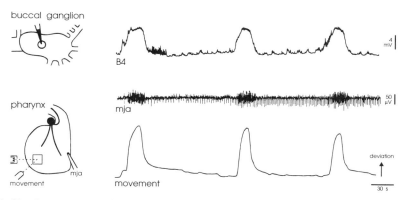

Figure 2. Simultaneous recordings of membrane potential of the motorneuron B4, of field potentials of anterior jugalis muscle (mja) and of pharynx movements in *Helix pomatia*. Movements were recorded via light beam and photo diode. Schematical drawings give the experimental conditions.

differentiated from each other by the differential shape of their depolarizations (Fig. 3). Synergistic neurons displayed depolarizations of similar time course (Fig. 3 A, B) and they were electrically coupled in part whereas in antagonistic neurons excitations were phase shifted to each other (Fig. 3 D and E).

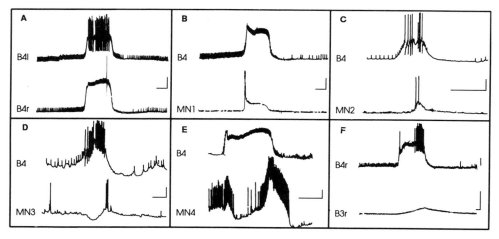

Figure 3. Simultaneous fluctuations of membrane potential in neurons of the buccal ganglia of *Helix pomatia*. A: neurons B4 of left (l) and right (r) buccal ganglion, B through F: fluctuations of membrane potential in a neuron B4 in comparison to the fluctuations in neuron B3 and in different motorneurons (MN 1 through MN 4). Pen recordings. calibration bars: 10 s; 10 mV.

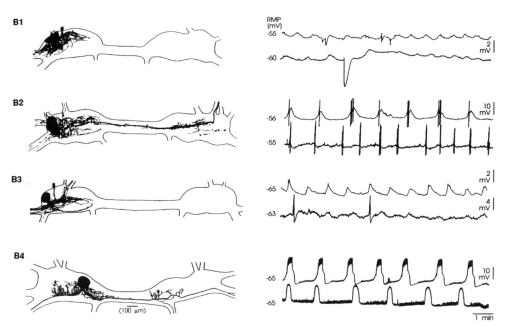

Figure 4. The identified giant neurons B1 through B4 in the buccal ganglia of *Helix pomatia*. left side: neurons drawn into the contour of the buccal ganglia after intracellular staining with cobalt-lysine. right side: typical examples of spontaneous activity of neurons B1 through B4. RMP: resting membrane potential. Action potentials are truncated.

The buccal ganglia of *Helix pomatia* contain four identified neuronal individuals which can be identified visually and which show identical properties from one animal to the next. The neurons are shown in Fig. 4 drawn into the contour of the ganglia. Each ganglion contains one set of the identified neurons. The typical spontaneous bioelectric activity of the neurons are shown in the right part of Fig. 4. Neuron B1 is a "power switch" for peristaltic activity of the oesophagus and stomach, neuron B2 activates salivation of the salivary glands (Altrup, 1987), neuron B3 can adapt the kidney to food intake (Altrup et al.1990), and neuron B4 activates monosynaptically 11 muscles of the pharynx (Peters and Altrup, 1984).

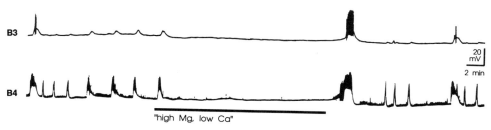

Figure 5. Effects of an increased magnesium (from 0 to 9 mmol/l) and decreased calcium ion concentration (from 9 to 2 mmol/l) in the bath fluid ("high Mg, low Ca") on spontaneous fluctuations of membrane potential of neurons in the buccal ganglia of *Helix pomatia*. B3 and B4: membrane potential of neurons B3 and B4. Action potentials are truncated.

Physiologic oscillations of bioelectric activity

The main functional role of the buccal ganglia consists in driving the muscles of the pharynx during food intake. The respective motor activity is oscillatory (biting, swallowing) with a low rate of 1/10 s to 1/10 min. The pattern is generated in the buccal ganglia and it is under modulatory control of the cerebral ganglia.

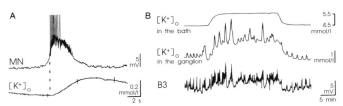

Figure 6. Relations between neuronal activities and K^+-concentration in the extracellular space in the buccal ganglia of *Helix pomatia*. K^+ was measured by ion selective microelectrodes. A: Activity of a motorneuron (MN) and intraganglionic K^+-concentration. Action potentials of MN are truncated. B: Effects of an increase in K^+- concentration in the bath on intraganglionic K^+-concentration and membrane potential of the neuron B3. Pen recordings.

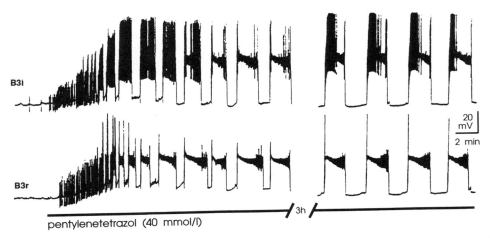

Figure 7. Effects of the epileptogenic drug pentylenetetrazol (40 mmol/l) on membrane potential of two neurons of the buccal ganglia of *Helix pomatia*. B3l, B3r: neurons B3 of left and right ganglion. Simultaneous pen recordings, interrupted for 3 hours.

Intracellular recordings from any studied neuron in the buccal ganglia showed these oscillations of membrane potential. Since oscillations were about simultaneous in all recorded neurons (cf. Fig. 3) they can be recorded as field potentials from the surface of the ganglion or from the neuropil.

The mechanisms underlying the oscillations have been studied in several series of experiments. A blockade of transmitter release by application of a "high Mg-low Ca" solution abolished the oscillations reversibly (Fig. 5). In motorneurons such as neuron B4, oscillations are induced by an increase in membrane conductance, since an increase in membrane potential concomittantly raised amplitudes of oscillations (Altrup and Speckmann, 1982). In other efferent neurons such as neurons B1 and B3, the oscillations result from increases in extracellular potassium concentration. Evidence for this comes from the following observations. (1) Increasing the membrane potential of neuron B3 or B1 did not affect amplitudes of oscillations. Thus, oscillations do not appear to result from changes in membrane conductance. (2) Membrane potential of neurons B1 and B3 followed the extracellular potassium concentration which in turn followed the motorneuronal activities (Fig. 6A). (3) Increases in extracellular potassium concentration (Fig. 6B) achieved by an elevation of K^+ in the bath solution were followed by corresponding shifts in the membrane potential of neurons B1 and B3 as in cases of spontaneous K^+ changes.

As a whole, the spontaneous bioelectrical oscillations result from an intraganglionic network of "function generating" neurons, i.e. the neuronal premotor network generating the motor program for food intake. Beside transmitter induced membrane potential oscillations, there are also oscillations induced by activity dependent increases in extracellular potassium concentrations.

Epileptic oscillations of bioelectric activity

Induction of epileptic activity. Epileptic activity can be induced by application of an epileptogenic drug. In the example in Fig. 7 pentylenetetrazol was applied during the simultaneous recording of both neurons B3. When the slow continuous flow of "Ringer" solution contained the epileptogenic drug, the neurons developed the typical epileptic depolarizations (paroxysmal depolarization shift, PDS; Ajmone-Marsan, 1961; Goldensohn and Purpura, 1963). PDS became synchronized within several minutes and then stayed synchronized for hours. The synchronized epileptic activity can also be recorded as field potentials from the surface of the ganglia or from the neuropil.

Properties of epileptic depolarizations. The epileptic oscillations of membrane potential showed differential properties when compared with the physiologic oscillations. (1) As can be seen in Fig. 8 A, the application of a "high Mg-low Ca" solution did not abolish the epileptic activity, although there are changes in shape of PDS. (2) The hyperpolarization of an epileptically active neuron showed typical changes of potentials. With moderate hyperpolarizations, frequency of occurrence of PDS decreased whereas amplitude of PDS increased. The increase was, however, relative since about the same level was reached by the plateau of the epileptic depolarization starting from an increased resting potential (cf. Witte et al., 1985a,b). With increases of membrane potential to above ca. -80 mV, epileptic depolarizations were abruptly blocked and they did not reappear with higher membrane potentials. (3) When the "Ringer" solution additionally contained a calcium antagonist, the epileptic depolarizations were reversibly depressed (Fig. 8 B).

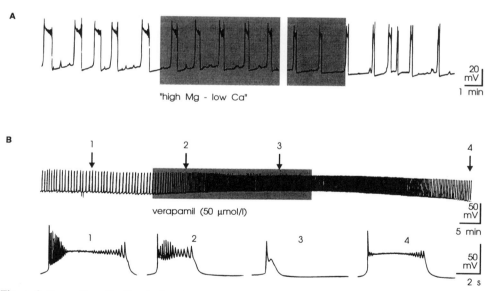

Figure 8. Properties of epileptic depolarizations (neuron B3 of the buccal ganglia of *Helix pomatia*). A: application of an increased magnesium (from 0 to 9 mmol/l) and decreased calcium ion concentration (from 9 to 2 mmol/l) in the bath fluid ("high Mg, low Ca"). Application is marked by the shaded area. Recording was interrupted for 30 min. B: application of the organic calcium antagonist verapamil (50 μmol/l, shaded area). Membrane potential of the neuron is shown with extended time scale in the lower recording. Tracings are related to each other by numbers.

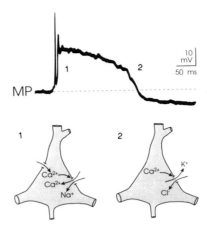

Figure 9. Calcium hypothesis of epileptic discharges. Upper recording: epileptic depolarization of a neuron. MP: membrane potential, 1, 2: depolarizing (1) and repolarizing phase of epileptic discharge (2). Lower drawings: membrane currents during depolarization, a calcium ion influx activates an unspecific cation current (1), and during repolarization, an increased intracellular calcium ion concentration activates potassium and chloride ion currents (2).

Synchronization of epileptic activity in neurons of the buccal ganglia primarily resulted from electrical coupling within the networks of neurons (Madeja et al., 1989). The electrical coupling signal can be separated when one of the neurons was hyperpolarized to above -80 mV. Several observations have led to the assumption of a synchronization via electrical synapses. (1) Epileptic depolarizations and their synchronization were not blocked by the "high Mg, low Ca" solution, such that "chemical" excitatory post-synaptic potentials could only play minor roles. (2) PDS were furthermore abruptly blocked with hyperpolarization. (3) Epileptic depolarizations do not result directly from changes in extracellular potassium concentration since these followed epileptic potentials. (4) Shape of the coupling signal corresponded to an electrically transmitted PDS of a pre-switched neuron and not to the action potentials in the beginning of a PDS. (5) Those neurons in the buccal ganglia became synchronized which have been demonstrated to be electrically coupled.

As a whole, the observations are in line with an extrasynaptic (non-synaptic) origin of the epileptic depolarizations and a primary synchronization via electrical synapses. The extrasynaptic origin is strongly supported by the observation that an isolated neuron, which is mechanically separated from the rest of the ganglion, can generate epileptic activity (Speckmann and Caspers, 1978).

Calcium hypothesis of epileptic activity. Several lines of evidence have led to the calcium hypothesis of epileptic activity. The hypothesis is schematically shown in Fig. 9. The steep depolarization of the epileptic potential which is mostly superimposed by action potentials, comes from the opening of voltage sensitive calcium channels (Fig. 9-1). The increase in intracellular calcium-concentration activates un-

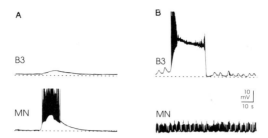

Figure 10. Comparison of spontaneous fluctuations of membrane potential of neuron B3 and of a motorneuron (MN) during non-epileptic condition (A) and during epileptic condition (B) in the buccal ganglia of *Helix pomatia*. Epileptic condition established by bath application of the epileptogenic drug etomidate (0.5 mmol/l). Action potentials are truncated.

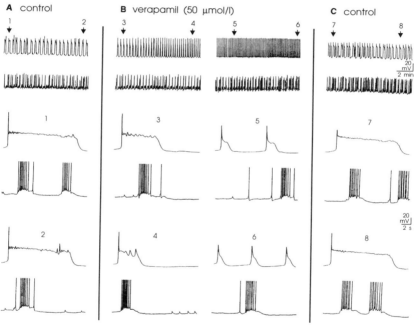

Figure 11. Effects of application of the calciumantagonist verapamil (50 µmol/l) on activities of neurons B3 and B4 in the buccal ganglia of *Helix pomatia* during epileptic conditions induced by bath application of the epileptogenic drug etomidate (0.5 mmol/l). Uppermost two recordings are slow pen recordings of neuron B3 (1st line) and of neuron B4 (2nd line). The lower lines show oscillographic recordings. The tracings are related to each other by numbers.

specific cation-channels which maintain the neuron in the depolarized plateau. The repolarization is then based on calcium-sensitive potassium and chloride channels which lead the membrane potential to the resting level (Fig. 9-2). The mechanisms of the "epileptic influence" on calcium channels of the membrane are not known at present. It is possible that the basic epileptic process consists in a decrease of threshold of voltage sensitive calcium channels. Such a decrease of threshold occurs when the concentration of an epileptogenic drug is increased (cf. Fig. 12).

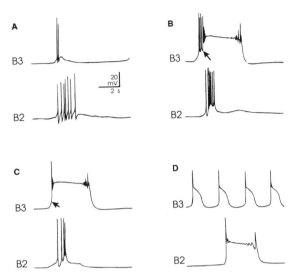

Figure 12. Membrane potential of neurons B3 and B2 in the buccal ganglia of *Helix pomatia* during application of different concentrations of the epileptogenic drug etomidate (A: 0.2; B: 0.5; C: 0.6; D: 0.8 mmol/l). Simultaneous oscillographic recordings. Arrows in B and C mark the threshold potential of epileptic depolarizations.

Relations between physiologic and epileptic oscillations

Principal independence of oscillations. Physiologic and epileptic oscillations were principally independent from each other. This can be derived from Figs. 10 and 11. In the example in Fig. 10, neuron B3 and a motorneuron were recorded simultaneously. Neuron B3 easily generates epileptic activity when treated with epileptogenic drugs whereas many efferent motorneurons are relatively insensitive. Fig. 10 A shows typical physiologic oscillations, which have small amplitudes in neuron B3 and high amplitudes in motorneurons. As described above they are evoked in neuron B3 by the extracellular potassium concentration and in the motorneurons by transmitter release. With application of the epileptogenic drug, the neuron B3 receiving little physiologic input developed epileptic oscillations and the motorneuron, which was strongly activated under physiologic conditions, showed a more regular activity and no epileptic oscillations. There is no indication that increased excitatory activity facilitates synchronization and thus leads to epileptic oscillations. After removal of the epileptogenic drug, the physiologic oscillations gradually reappeared.

In Fig. 11 a further example of action of calciumantagonists is shown (s. Fig. 8 B). In the control phase (Fig. 11 A), only the sensitive neuron B3 displayed epileptic oscillations whereas neuron B4 showed physiologic oscillations with increased frequency. They were asynchronous to those in neuron B3. With application of the calciumantagonist verapamil, only the epileptic oscillations disappeared and the physiologic ones remained about unaffected. With washing of the calciumantagonist, the epileptic oscillations reappeared in neuron B3.

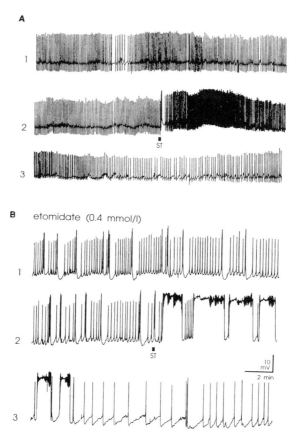

Figure 13. Effects of electrical stimulation (ST) of cerebrobuccal connective on membrane potential of neuron B3 in the buccal ganglia of *Helix pomatia* under non-epileptic (A) and epileptic conditions (B). Continuous recordings (1 to 3). ST: 20 stimuli, 1.5 ms each, in 4 s, double threshold intensity.

Thus, as a whole the epileptic activities do not result from synchronized physiologic activities. Whereas physiologic activities result from complex neuronal interactions in a network of neurons, epileptic activities are generated by different neurons which are rather poorly integrated synaptically into the functional network.

Physiologic oscillations triggering epileptic ones. It is shown in Figs. 12 and 13, that physiologic oscillations can trigger epileptic ones. The example in Fig. 12 A shows transmitter-induced depolarizations appearing simultaneously in neurons B2 and B3. At low concentrations of the epileptogenic drug (etomidate: 0.5 mmol/l), the synaptic inputs are seen to occasionally surpass the threshold of an epileptic depolarization (arrow in Fig. 12B). With increasing concentration of the epileptogenic drug (Fig. 12 C, etomidate: 0.6 mmol/l), threshold of epileptic depolarization decreased (arrow) and in the high concentration range (Fig. 12 D, etomidate: 0.8 mmol/l), neuron B2 started to generate epileptic activity whereas the sensitive neuron B3 became independent from synaptic inputs. The "pacemaker" potential at high epileptogenicity could result from an interplay between the calcium sensitive potassium/chloride channels active at the end of the PDS and the voltage sensitive calcium channels the threshold for opening of which is reduced to resting levels of membrane potential.

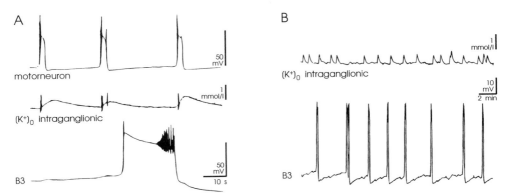

Figure 14. Relations between neuronal activities and K^+-concentration in the extracellular space in the buccal ganglia of *Helix pomatia* during epileptic conditions evoked by etomidate (0.5 mmol/l). K^+ was measured by ion selective microelectrodes. A: Activity of a motorneuron (MN), neuron B3 and intraganglionic K^+-concentration. Action potentials are truncated. B: Recordings of K^+-concentration in the ganglion and epileptic activity of neuron B3 at lower resolution of time base.

Results corresponding to the above observations were obtained also with the electrical stimulation of a pharyngeal nerve. The stimulation of any ipsilateral pharyngeal nerve evoked synaptic EPSP in neuron B3 (Altrup and Speckmann, 1982), which in turn can evoke a PDS when an epileptogenic drug is supplied. With long lasting trains of stimuli the probability of occurrence of epileptic oscillations can be increased.

With the stimulation of a cerebrobuccal connective long lasting modulatory activity appears in the buccal ganglia under non-epileptic conditions. As can be seen in Fig. 13 A, a short train of stimuli which lasted a few seconds induced a long lasting modulation of ganglionic activities for up to about 1 hour. After a period of activation which lasted about 30 min, the neuronal activity was transiently depressed. When the same train of stimuli was applied with a subthreshold concentration of an epileptogenic drug (Fig. 13 B), neuronal activities could shift from an interictal-like activity to an ictal-like activity during the period of activation.

Epileptic oscillations triggering physiologic ones. The epileptic oscillations induced by an epileptogenic drug consisted of synchronized neuronal activities which in turn could evoke "physiologic" activities. Examples are given in Figs. 14 and 15. In Fig. 14 A two neurons and the intraganglionic potassium concentration were recorded simultaneously. It can be seen that the epileptic activity in the motor neuron was followed by increases in extracellular potassium concentrations which in turn are reflected in the membrane potential of neuron B3. However, the increases in K^+ and membrane potential did not necessarily trigger PDS in neuron B3 (Fig. 14 B). In Fig. 15 neurons B1 and B3 are recorded simultaneously. The neurons, the dendrites of which occupy in part the same regions of the neuropil, are not coupled in the beginning of epileptic activity (Fig. 15 A). After 4 hours (Fig. 15 B), there was a marked coupling which re-decreased during the further epileptic activity (Fig. 15 C). This coupling can also be based on extracellular potassium concentration or on sprouting phenomena (Altrup and Speckmann, 1988). The mechanism of these alterations are unknown.

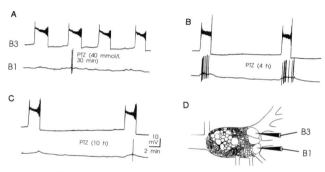

Figure 15. Changes of coupling between two neurons depending upon duration of epileptic activity. Membrane potential of neurons B3 and B1 in the buccal ganglia of *Helix pomatia* were simultaneously recorded during epileptic conditions (pentylenetetrazol, PTZ, 40 mmol/l) lasting 30 min (A), 4 h (B) and 10 h (C). Schematic drawing in C shows one buccal ganglion with superficially located neuronal somata. Neurons B3 and B1 are punctated by microelectrodes.

DISCUSSION

Buccal ganglia as model nervous system. Experimental observations have shown that the buccal ganglia of the snail are useful model nervous systems for the study of basic mechanisms in neurophysiology and in epileptology. When compared with vertebrate nervous systems, this model offers several advantages among which the presence of giant neurons is of special value. The giant neurons are cell individuals showing the same properties from one animal to the next. Thus, detailed information on the same neuron can be collected using different techniques. The technical advantages concerning the study of this small nervous system have led to knowledge about the functional meaning of neuronal activities.

One problem using invertebrate preparations in experimental medicine is expressed in the question whether or not the results can be transferred to the human nervous system. The answer to this depends on whether principles or details are to be transferred. Many studies have shown that principles as well as many details are conservatively maintained through phylogenesis e.g. the ionic bases of cell excitation, the use of the same neurotransmitters and learning mechanisms (cf. Kandel, 1976). Thus, in principle the nervous system of invertebrates function like vertebrate ones. It is consequently not amazing that in both the nervous system of man and the buccal ganglia of the snail the same drugs evoke epileptic activity (Fehér et al., 1988; Altrup et al., 1991; Madeja et al., 1989; Sugaya et al., 1973; Rayport and Kandel, 1981) or act antiepileptically (Altrup et al., 1992; Sugaya et al., 1987; Walden et al., 1993).

The functional role of "electrical" synapses in nervous systems (Sloper, 1972) is not fully understood. In invertebrates these synapses are often found to connect synergistic neurons. Thus, neuron B4 in the buccal ganglia of Helix pomatia is electrically coupled to other motor neurons which receive about the same transmitter-induced synaptic input from the pre-switched network and which supply monosynaptically about the same peripheral muscles (Peters and Altrup, 1984). Since action potentials are generally not synchronized in electrically coupled networks (cf. Altrup, 1987), the functional meaning of "electrical" synapses may be an enlargement of the field of the "chemical" synaptic inputs. This could explain why neurons of similar properties are electrically coupled. Furtheron, such electrically coupled networks can strongly activate the following plane of synaptic processing.

It is well-known that electric contacts of neurons play an important role during ontogeny of the nervous system (cf. Hadley et al., 1983). Enhanced electrical contacts could contribute to the higher vulnerability of premature nervous tissues to epileptogenic processes when compared to adult tissues. Furtheron, during regeneration neurons have often found to form "electrical" synapses. In the snail this process leads to a strong increase in electrical coupling during the first day of regeneration and a smaller enduring coupling (Bulloch and Kater, 1982). Such processes can result in an increased sensitivity against epileptic influences.

Nature of physiologic oscillations. The physiologic oscillations consist of slow changes in membrane polarization which are composed of post-synaptic potentials (cf. Altrup and Speckmann, 1982; Rose and Benjamin, 1981). The shape of the potentials in different neurons determine the peripheral motor pattern i.e. the differential activation of the peripheral muscles. To this aim the motor neurons get synaptic inputs from a complex pre-switched neuronal network (cf. Benjamin and Elliott, 1989; Peters and Altrup, 1984). As with the sensory receptor potentials at the input side of the nervous system, the exact position of action potentials in the course of time was not achieved. Action potentials rather appeared according to a probability the time course of which was given by the complex and composed synaptic depolarizations (cf. Stein, 1970).

Nature of epileptic oscillations. The results obtained so far demonstrate that epileptic oscillations showed properties not in line with the widespread idea that epileptic depolarizations result from enhanced physiologic oscillations. In contrast, epileptic oscillations are primarily evoked by extra-synaptic mechanisms, i.e. mainly via voltage sensitive calcium channels. A consequence of the described results is that synchronization in the nervous system is secondary to the appearance of epileptic depolarizations. Electrical contacts between neurons are a favorite means for primary synchronization since during each PDS of one of the neurons the probability of occurrence of a PDS in all the electrically coupled neurons is increased leading to synchronization. A secondary synchronization can be induced easily by the electrically coupled network since electrical contacts were mostly found between synergistic neurons. Thus, the following synaptic level of processing of information in the nervous system will receive maximal synaptic inputs. However, every means which simultaneously depolarizes neurons showing "epileptically altered calcium channels" can evoke synchronized epileptic activity.

REFERENCES

Ajmone-Marsan, C., 1961. Electrographic aspects of "epileptic" neuronal aggregates. *Epilepsia*, 2, 22-38.

Altrup, U., 1987. Inputs and outputs of giant neurons B1 and B2 in the buccal ganglia of Helix pomatia: an electrophysiological and morphological study. *Brain Res.*, 414, 271-284.

Altrup, U., Lehmenkühler, A., Madeja, M. and Speckmann, E.-J., 1990. Morphology and function of the identified neuron B3 in the buccal ganglia of Helix pomatia. *Comp. Biochem. Physiol.*, 97A, 65-74.

Altrup, U., Lehmenkühler, A. and Speckmann, E.-J., 1991. Effects of the hypnotic drug etomidate in a model nervous system (buccal ganglia, Helix pomatia). *Comp. Biochem. Physiol.* 99C, 579-587.

Altrup, U., Gerlach, G., Reith, H., Said, M.N. and Speckmann, E.-J., 1992. Effects of valproate in a model nervous system (buccal ganglia of Helix pomatia): I. Antiepileptic actions. *Epilepsia* 33, 743-752.

Altrup, U. and Peters, M., 1982. Procedure of intracellular staining of neurons in the snail Helix pomatia. *J. Neurosci. Meth.*, 5, 161-165.

Altrup, U. and Speckmann, E.-J., 1982. Responses of identified neurons in the buccal ganglia of Helix pomatia to stimulation of ganglionic nerves. *Comp. Biochem. Physiol.*, 72A, 643-657.

Altrup, U. and Speckmann, E.-J., 1988. Epileptic discharges induced by pentylenetetrazol: Changes of shape of dendrites. *Brain Res.*, 456, 401-405.

Benjamin, P.R. and Elliott, C.J.H. 1989. Snail feeding oscillator: the central pattern generator and its control by modulatory interneurones. In: *Cellular and neuronal oscillators*, edited by Jacklet, J. Marcel Dekker, New York: p. 173-214.

Bulloch, A.G.M. and Dorsett, D.A., 1979. The functional morphology and motor innervation of the buccal mass of Tritonia hombergi. *J. Exp. Biol.*, 79, 23-40.

Bulloch, A.G.M. and Kater, S.B. 1982. Neurite outgrowth and selection of new electrical connections by adult Helisoma neurons. *J. Neurophysiol.*, 48, 569-583.

Fehér, O., Erdélyi, L. and Papp, A., 1988. The effect of pentylenetetrazol on the metacerebral neuron of Helix pomatia. *Gen. Physiol. Biophys.*, 7, 506-516.

Fentress, J.C., 1976. *Simpler Networks and Behavior*. Sinauer Associates, Sunderland.

Goldensohn, E.S. and Purpura, D.P., 1963. Intracellular potentials of cortical neurons during focal epileptogenic discharges. *Science*, 139, 840-842.

Hadley, R.D., Kater, S.B. and Cohan, C.S., 1983. Electrical synapse formation depends on interaction of mutually growing neurites. *Science* 221, 466-468.

Kandel, E.R. 1976. *Cellular basis of behavior: an introduction to behavioral neurobiology*, Freeman Pub., San Francisco.

Madeja, M., Altrup, U. and Speckmann, E.-J., 1989. Synchronization of epileptic discharges: temporal coupling of paroxysmal depolarizations in the buccal ganglia of Helix pomatia. *Comp. Biochem. Physiol.*, 94C, 585-590.

Peters, M. and Altrup, U., 1984. Motor organization in pharynx of Helix pomatia. *J. Neurophysiol.* 52, 389-409.

Rayport, S.G. and Kandel, E.R., 1981. Epileptogenic agents enhance transmission at an identified weak electrical synapse in Aplysia. *Science* 213, 462-464.

Rose, R.M. and Benjamin, P.R., 1981. Interneuronal control of feeding in the pond snail Lymnaea stagnalis. I. Initiation of feeding cycles by a single buccal interneurone. *J. exp. Biol.* 92, 187-201.

Siegler, M.V.S., 1977. Motorneuron coordination and sensory modulation in the feeding system of the mollusc, Pleurobranchea californica. *J. Exp. Biol.*, 71, 27-48.

Sloper, J.J., 1972. Gap junctions between dendrites in the primate neocortex. *Brain Res.*, 44, 641-646.

Speckmann, E.-J. and Caspers, H. 1978. Effects of pentylenetetrazol on isolated snail and mammalian neurons. In: *Abnormal Neuronal Discharges*, edited by Chalazonitis, N. and Boisson, M. Raven Press, New York: p. 165-176.

Stein, R.B., 1970. The role in spike trains in transmitting and distorting sensory signals. In: *The Neurosciences, 2nd Study Program*, edited by Schmitt, F.O., Quarton, G.C., Melnechuk, T. and Adelman, G. The Rockefeller University Press, New York.

Sugaya, A., Sugaya, E. and Tsujitani, M., 1973. Pentylenetetrazol-induced intracellular potential changes of the neuron of the japanese land snail Euhadra peliomphala. *Jap. J. Physiol.*, 23, 261-274.

Sugaya, E., Onozuka, M., Furuichi, H., Kishii, K., Imai, S. and Sugaya, A., 1987. Phenytoin effects on pentylenetetrazol-induced intracellular calcium related abnormal phenomena. In: *Advances in Epileptology*, edited by Wolf, P., Dam, M., Janz, D. and Dreifuss, F.E. Raven Press, New York: p. 507-510.

Walden, J., Altrup, U., Reith, H. and Speckmann, E.-J., 1993. Effects of valproate on early and late potassium currents of single neurons. *Europ. Neuropsychopharmacol.*, 3, 137-141.

Witte, O.W., Speckmann, E.-J. and Walden, J., 1985a. Acetylcholine responses of identified neurons in Helix pomatia. I. Interactions between acetylcholine-induced and potential-dependent membrane conductances. Comp. Biochem. Physiol. 80C, 15-23.

Witte, O.W., Speckmann, E.-J. and Walden, J., 1985b. Acetylcholine responses of identified neurons in Helix pomatia. III. Ionic composition of the depolarizing currents induced by acetylcholine. Comp. Biochem. Physiol. 80C, 37-45.

CORTICO-HIPPOCAMPAL INTERPLAY: SYNOPSIS OF A THEORY

Robert Miller

Department of Anatomy and Structural Biology
University of Otago Medical School
Dunedin
New Zealand

This paper represents a synopsis of a more detailed argument which was published in 1991 in monograph form (Miller 1991; see also Miller 1989). That monograph developed a theory of the interaction between hippocampus and neocortex, included a great deal of anatomical and behavioural detail, and focused on the electrophysiology of the hippocampal rhythmic slow activity, otherwise known as the theta rhythm. The present paper aims to avoid most of the detail, and rather to paint the overall concept with broad brush strokes, identifying the main premises of the argument, the main thrust of the conclusions, and the implications for further work.

The first premise for this theory is the idea that **contexts for information processing are dealt with by the brain separately from detailed items**. This idea became prominent in the late 1970s with the publication by O'Keefe and Nadel (1978) of the monograph "The hippocampus as a cognitive map". Right at the beginning of that work, the subjects to be considered are announced as not only the hippocampus but also "representation of space", and "context dependent memory". The idea that space is an innately-known context is advocated, and traced back to some of the ideas in Immanuel Kant's book "Critique of Pure Reason". However, another way of reading Kant's arguments admits the need to separate contextual from other information, but does not give space (or any other context) the status of innate (i.e. genetically specified) knowledge. This alternative reading of Kant may be put in the following way: In deductive reasoning, one needs more than the overt premises of the argument to reach the apparent conclusion. One needs in addition a frame of reference, a statement of ground rules, in other words a context. This realization was the first seed which grew into the theory of cortico-hippocampal interplay.

One can make a similar statement about the ideas on learning theory which were developed in the first half of the present century. The earliest learning theorists appeared to believe that if an animal had been presented with appropriate conjunctions of stimulus, response (and reinforcement in some cases), then subsequently input would dictate output unconditionally. Later work showed that this was wrong: There was required in addition, that the appropriate dimension, context or schema for attention be activated. Selective attention can then be viewed as the process by which we select the right dimension or context for mental operations.

If we translate these ideas to the anatomy of the cerebral cortex, we can see a similar problem with the connectivity of that structure. Every cortical pyramidal cell gives and receives several thousand connections. If we consider a hypothetical pathway from sensory regions of cortex, where a stimulus is represented, to the motor cortex where a corresponding response is to be represented, a sequence of several synaptic relays is required to connect the two. At each stage convergence and divergence is likely to occur, to judge from the anatomy. Therefore neural activity in input lines cannot dictate that in output lines unconditionally. An additional mechanism is needed which can differentially enhance transmission at a small subset of the many possible relays involved. The need, on anatomical grounds, for such differential enhancement may correspond to the need for context representations in mental processes. The differential enhancement of transmission in a subset of the possible pathways may be controlled by the prevailing context for information processing.

Much of the reasoning from philosophy and from learning theory, mentioned above, was concerned with representation of space. Moreover, O'Keefe and Nadel were struck by the fact that many of the behavioural impairments in animals with hippocampal damage were in spatial behaviour. This then led them to perform experiments in which they were able to identify single cells in the hippocampus, which, in free moving animals, responded when the animal was in a particular place in its environment. This result has now been confirmed and extended in a number of laboratories. "Place" appears to be defined by coordinated representation of a variety of stimuli in the environment, which may be widely distributed spatially, employing a variety of sensory modalities, and possibly also motor information. O'Keefe and Nadel regarded *place* as the archetype of a context, such that all contextual representations, even for human thought, are formally similar to Euclidean space. However, if one takes the alternative reading of Kant's ideas, it is also possible for "context" to have a more abstract meaning, and that is the way the term is used here. Despite this modification of O'Keefe and Nadel's ideas, the hippocampus is still seen as the focus for representation of contexts. Representation of "place" thus is treated as a special case of the broader concept of "context".

$$\boxed{\Sigma_i \rightleftharpoons C}$$

Whichever way one defines a context, it is clear that context representations are stable over longer periods of time than the detailed items, which may be quite transient. Moreover, it is implicit that **there is a two-way interaction between contextual information and the smaller details.** On the one hand, many details, some of them quite transient, are needed to establish which context is to be used; on the other, the context which is in use helps to resolve the detail, especially in circumstances where some of the detail is ambiguous. This two-way interaction is represented symbolically below.

$$\Sigma_I \rightleftharpoons C$$

Obviously, in such a scheme, the hippocampus must get its input from somewhere. The likelihood is that multimodal information of such detail as is needed to define a context must be relayed from the larger part of the cortex, the so-called "neocortex" or "isocortex". The traditional view has been that hippocampus and isocortex are linked by the circuit of Papez via the anterior thalamus. However, from late 1970s connectionist studies have provided more and more evidence of direct links in both directions between hippocampus and isocortex, mainly via relays in the entorhinal cortex. Widespread areas of isocortex are involved, especially the association areas therein, and particularly the frontal areas, most remote from the hippocampus. Thus **one has the connections for a two-way interaction between the hippocampus and the neocortex.** This, then, could be the physical basis for the two way interaction between the broad contexts, and the detailed items.

What about the mechanism, the actual physiology of such an interaction? The mechanism envisaged is built upon two assumptions, the first about the theta rhythm, the second about axonal conduction delays.

The **hippocampal theta rhythm** (otherwise known as "rhythmic slow activity") seems to have something to do with the hippocampus's use of contexts. This rhythm has a frequency of 7-12 Hz in rats, and rather less in cats (3-8.5 Hz). Not very much is known of its properties in primates. The corresponding periods for this rhythm would be 83-143 msec in rats, and 118-330 msec in cats. In some species, especially the cat, the circumstances in which the hippocampus generates its theta rhythms are times when learning is occurring, involving focused attention and alertness and presumably the capacity to integrate small details of information into a bigger picture. A typical scenario would be a cat stalking its prey. In rodents such as the rat much has been written about the relationship of the hippocampal rhythmic slow activity to simple movements, especially locomotion. However, it should not be forgotten that the rat, in its natural habitat, is a territorial animal which probably does most of its learning by regular exploration of its rather large territory, presumably in a state of focused attention and alertness similar to that in which the cat hippocampus shows this rhythm. The review of the relation between behaviour and hippocampal theta activity (Miller 1991, Chapter 4) suggested that the correlation with locomotor activity, though important in some species, was full of exceptions, even for the rat, and did not fit very well at all in larger-brained animals. The conclusion was therefore reached that the correlation of theta activity with locomotor activity, when it occurs, is a reflection of a more fundamental correlation with some phases of learning. This suggests the hypothesis that the hippocampal theta rhythm is somehow involved in integration of details into an overall picture, during these phases of learning.

The other physiological premise from which the hypothesis of cortico-hippocampal interplay grew was related to the **properties of axons.** It is usually assumed, by those who attempt to understand or model the dynamics of large collections of nerve cells, that conduction along a single cortico-cortical axon cannot possibly contribute a delay of more than about 10 msec to signal transmission.

However, this assumption does not agree with available evidence from antidromic conduction studies in single neurones. In the early 1970s I undertook some experiments in cats, in which I measured the latencies of antidromic responses of single cortico-cortical axons (Miller, 1975). A proportion of those latencies were quite long, the longest I documented being 44 msec (for a conduction distance of about 10 mm). Swadlow and associates, working mainly in rabbits, have also documented cortico-cortical antidromic conduction times of over 30 msec in some axons (e.g. Swadlow, 1974; Swadlow and Weyand, 1981). Such results are not very well known by those who study massed electrical activity of the cerebral cortex. Moreover even this evidence (and evidence using any other method of assessing axonal conduction delays) is subject to a variety of biases, all of which underestimate or ignore the contribution of conduction delays arising in the finer calibre cortico-cortical axons. (This topic is discussed in Miller, 1993: This conference). Indeed, if one considers conduction in cortico-cortical axons in large brains, such as the human brain, it seems (to this author) entirely possible for there to be conduction delays in single axons, from soma to synapse, of 100 msec, or even 200 msec. However, there is as yet no empirical proof of this conjecture. It is very important to obtain unbiased estimates of the spectrum of axonal conduction delays within the cerebral hemispheres, when we are considering oscillatory phenomena in the brain, and the dynamics of the EEG. Swadlow (1991) considers that conduction time across the corpus callosum in rabbits can be 20-60 msec in a substantial proportion of axons. The distance between hippocampus and prefrontal cortex is longer than that for interhemispheric transmission via the corpus callosum, and the conduction delay can be expected to be longer in proportion.

Summing up, there are then three key ideas to be combined: (i) It is plausible to suggest some sort of interplay takes place between hippocampus and cortex, which might correspond to the interplay between stable contextual information and the transient detailed items "embedded" within each context. (ii) Rhythmic activity at theta frequency might be closely involved in this interplay, at times when contexts are being represented; (iii) Axonal conduction delays between hippocampus and sites in the neocortex could be quite long, and under some circumstances can be in the same range as the cycle time of the theta rhythm.

Pulling these three themes together, one can suggest that there might be routes for axonal conduction from hippocampus to association cortex, especially the frontal cortex, and back again to the hippocampus, which have conduction times which match the period of theta oscillations. Obviously there would be many other routes which did not have this property. But if one could select those pathways whose conduction time corresponded to the theta period, one would have a basis for interplay between hippocampus and cortex involving resonance between two structures at the theta frequency. This is illustrated in figure 1. The term **phase-locked loops** borrowed from engineering, is a short hand with which to describe this neural resonance

The approximate values for conduction time needed to make this scheme work could be as follows, taking values appropriate to the rat:- Hippocampus to prefrontal cortex: 50 msec; prefrontal cortex back to hippocampus: another 50 msec; delays within the trisynaptic hippocampal circuit: 40 msec. There may be additional links giving the circuit a few more synaptic relays, and the possibility of longer loop times, though it is not assumed that there are large numbers of synapses in the loop.

The total loop time could thus easily be of the order of 150 msec, quite similar to the theta period in the rat.

It has to be stressed that the axonal pathways which allow such resonance would be a careful selection from the repertoire of pathways which exists in anatomical terms. It is envisaged that the selection is made by strengthening of selected synaptic junctions, utilizing the mechanism which is now widely investigated, not only in the hippocampus, but also in the neocortex. This mechanism, first suggested by Hebb (1949), is that synapses are strengthened when activity on the presynaptic side of a junction coincides with postsynaptic activation. In the case of the resonant circuits between neocortex and hippocampus, one may envisage that there are a variety of synapses in the hippocampus relaying rhythmic signals back from the neocortex. But only those will be strengthened in which the peak of activity in the presynaptic fibre coincides with the peak of the activity generated by the theta rhythm in the hippocampus. These in turn are the synapses connected to axonal loops whose total loop time would be similar to that of the concurrent theta rhythm. That is the essence of the hypothesized mechanism, but there are many details of the dynamics involved which have yet to be worked out, which will require both experimental and theoretical methods.

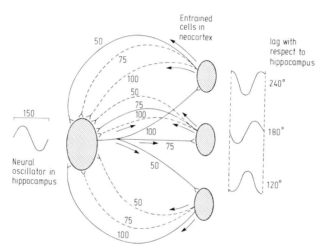

Figure 1. Scheme for self-organization of phase-locked loops. The master oscillator (hippocampus) has afferent and efferent links with a variety of neocortical loci. All links include lines with a range of delays. [Conduction delays (ms) for each axonal link are shown by **numbers** adjacent to each axon.] By strengthening of selected synaptic connections of axons with appropriate delays (**solid lines**, depicted as having total loop times of 150 ms), the neocortical loci can become entrained to the hippocampal rhythm. Different neocortical loci will have different phase relation with the hippocampal rhythm (**right**). (Reprinted from *Psychobiology* vol. 17, pp. 115-128, with permission of the Psychonomic Society, Inc.)

With that as the basic concept, there are some reasons for subdividing theta activity into two modes. These can be referred to as "idling theta" and "entrained theta" by analogy with the workings of a motor car. In the "idling" mode, the hippocampus is set into rhythmic activity as a result of rhythmic signals relayed from the medial septal nucleus, but in this mode resonance with the neocortex is not established. One can liken this activity to that of a car which is idling, with the clutch not engaged. At other times however, the hippocampo-neocortical loops are activated and resonance between hippocampus and neocortex is entrained. It is as if the clutch has been engaged, and the mechanism can do what it, as a whole, is designed to do.

Some of the evidence which led to this supposition came from reports of the theta activity in normal rats: This rhythm may occur at somewhat lower frequencies when the animal is not engaged in locomotion; and then becomes very vulnerable to the action of anticholinergic drugs. The higher frequency theta seen during locomotion (in rat) is vulnerable to low doses of anaesthetics such as urethane. Other evidence comes from the study of the effects of lesions disconnecting the hippocampus from the neocortex. Normally, walking in rats is always accompanied by theta activity in the higher frequency range. After disconnecting lesions it is not, as though the behavioural circumstances which normally promote resonant activity can no longer do so. When atropine is given to a normal rat during locomotion, theta activity survives. Hypothetically, the rhythmic signals from the septum are attenuated but the resonance between hippocampus and neocortex keeps the rhythm going to some extent. But in the lesioned rats, atropine completely abolishes all theta activity. Thus two manifestations of theta activity can be distinguished on grounds of frequency, behavioural correlates, pharmacology and effects of lesions. It has been suggested (e.g.Vanderwolf et al, 1978) that this division corresponds to two independent groups of pace-maker neurones in the medial septal nucleus. However, most of the evidence better fits the division between "idling" and "neocortical entrainment", as described above, with only a single pacemaker being involved (see Miller, 1991).

One important question is raised about the hypothesis of hippocampo-neocortical resonance: There is very little evidence that the neocortex can carry a theta rhythm, yet it should do so if the hypothesis of resonant interaction between neocortex and hippocampus is correct. However, the failure to detect an obvious theta rhythm in most of the neocortex, may arise for technical reasons. The theory predicts that each neurone in the neocortex involved in resonance should have a regular phase relationship with the rhythm in the hippocampus; but the actual phase relationship could vary widely from one such neocortical neurone to another. Since the EEG or electrocorticogram is usually recorded with quite large electrodes, it is likely that activity will be averaged from many sources, hypothetically with different phase relationships to the hippocampal rhythm. This may be the reason why no macroscopic rhythm is usually detected. In addition it is possible that even if one had the resolution to record the rhythm from a single neurone, the neocortical theta rhythm would not be easily seen in a single trace, but would require statistical techniques such as phase-locked averaging.

There are a number of **predictions** which follow from the proposed mechanism of resonance:

(i) With appropriate techniques, the neocortex should show theta activity, limited to the times when the hippocampus also carries this rhythm (though not necessarily at all of those times).

(ii) When it does so, it should have the same frequency as the hippocampal rhythm.

(iii) When it does so it should also have a regular phase relationship with the hippocampal rhythm. Indeed if one traces the various stages in the proposed loop (i.e. dentate gyrus, CA3, CA1, deep layers of entorhinal cortex, resonant nodes diffusely scattered through the neocortex, superficial layers of entorhinal cortex, and return to the dentate gyrus), there should be a steady progression of phase, amounting in total to 360 degrees of phase angle.

(iv) If these concomitant rhythms really are coordinated by a form of resonance, one would expect that in the hundreds of milliseconds after the neocortex picks up the rhythm from the hippocampus, the frequency spectrum of both structures should become progressively focused on a narrower frequency range. Likewise, over a series of learning trials (and comparing equivalent epochs in each trial) one may see narrowing of the frequency spectrum on specific frequencies.

Some data is already available concerning these predictions. Occasionally the neocortex can show macroscopic theta activity (especially the so-called frontal midline theta of Mizuki [1980]). Recently Prof Erol Basar and co-workers have produced new evidence that the neocortex in humans can carry theta activity. This involved filtering of event related potentials to reveal different frequency components. Increased theta activity occurred as an event-related response, with sharpening of the dominant peak, notable particularly in the theta frequency range (Basar-Eroglu et al, 1992). In cats an event-related enhancement of theta frequency rhythms has also been observed in the hippocampus (Basar et al, 1992). Simultaneous recordings of such event-related phenomena in both hippocampus and neocortex have not yet been reported, so it is not possible to evaluate the prediction that there be identical frequencies in, and consistent phase relationship between the two structures.

There is also some suggestive evidence from Buzsaki (1985), that within the hippocampal formation in the free-moving rat, there is regular phase sequence through the trisynaptic pathway through the hippocampus. Moreover Leung (1984) has provided evidence in rats that during REM sleep or waking there is a constant time-lead of theta activity in the dentate gyrus/hippocampal fissure compared with field CA1, which is efferent to the dentate gyrus. Thus phase angle difference between these two regions changes as theta frequency changes. Interestingly field CA3 of Ammon's horn appears not to generate a macroscopic theta rhythm in the urethane-treated rat, though the dentate gyrus and field CA1 do so. It is only in the free moving rat, when, hypothetically, the resonance phenomenon should be occurring, that the overt rhythms are sustained, with regular time delay, through each step of this trisynaptic pathway.

In addition, there is already some evidence for focusing of theta rhythms on narrower frequency bands as learning proceeds (reviewed in Miller, 1991). At the time of most rapid learning, hippocampal theta activity becomes most highly rhythmic. The prediction should also apply to neocortical activity.

The above paragraphs have outlined a mechanism (and empirical predictions derived from it) for a form of resonance allowing rhythmic interplay between hippocampus (the "master oscillator") and the neocortex (entrained to the master oscillator). However, there is the additional question of whether this mechanism can actually accomplish the information processing task which led to the development of this idea. This is still largely conjecture, and is a much more difficult question to answer.

In part this question is a subject for experimentalists, particularly for those who study change of theta activity during the course of registration of information during learning. Some evidence of this nature exists, in the older literature especially in the work of the late E.Grastyan and of W.R.Adey. Another experimental test is suggested by an experiment performed by E.R.John et al (1969). The experiment produced a situation in which the experimental cats could interpret a single visual stimulus in two different ways (as indicated by the cat's behaviour). Consistent differences were found in event-related neocortical waveforms, according to how the cat interpreted the self-same stimulus. It seems likely that the differences in "interpretation" were dependent on "context" (in the sense used in the present paper). Therefore, if that experiment were repeated in cats with hippocampal lesions, one would predict that context-dependent difference in cortical waveforms produced by a single stimulus should not occur, or the degree of difference should be reduced.

Apart from such experimental work, the evaluation of the information processing aspects of this hypothesis of neocortico-hippocampal interplay is a task for the future, in which theoreticians, especially mathematicians and modelers of neural networks have a key role to play. The core of the hypothesis, as far as information processing is concerned, is the prediction that such phase-locked loops of neural activity can actually allow representations of contexts to appear, when initially there was only the uncoordinated representation of items. In addition it would be expected that when a contextual representation has been formed in this way, it can help to resolve ambiguity of representation of some of the less-well encoded details, and thus make large-scale neocortical functioning more effective.

I am well aware that the hypothesis, however simple in outline, is actually very complex in detail. Because of this, my own methods (based mainly on careful scrutiny of published literature) are not precise enough to provide the definitive arguments that the scheme would actually work as it is supposed to do. What is needed then is a thorough simulation of the several structures in interaction. Undoubtedly this would involve some simplification, and mathematical analysis of the system may suggest ways in which this can be done. But, given this it may nevertheless be possible to build into the simulation real quantitative physiological data, and to express the results of the simulation in equally real terms.

Acknowledgment

The author is supported by the Health Research Council of New Zealand.

REFERENCES

Basar, E., Basar-Eroglu, C., Parnefjord, R., Rahn, E., Schuermann, M., 1992, Evoked potentials: ensembles of brain induced rhythmicities in the alpha, theta and gamma ranges. in: "Induced rhythms in the brain", E. Basar and T.H. Bullock eds., Birkhauser, Boston, Basel, Berlin.

Basar-Eroglu, C., Basar, E., Demiralp, T. and Schuermann, M., 1992, P300-response: possible psychophysiological correlates in delta and theta frequency channels. A review. *Int. J. Psychophysiol.*, 13: 161.

Buzsaki, G., 1985, Electroanatomy of the hippocampal slow activity (RSA) in the behaving rat, in: "Electrical activity of the archicortex", G. Buzsaki, C.H. Vanderwolf eds., Akademiai Kiado, Budapest.

John, E.R., Shimococki, M., and Bartlett, F., 1969, Neural readout from memory during generalization, *Science*, 164: 1534.

Hebb, D.O.,1949, The organization of behavior, John Wiley, New York.

Leung, L.-W.S.,1984, Theta rhythm during REM sleep and waking: correlations between power, phase and frequency. *Electroencephalogr.Clin.Neurophysiol.*, 58:457.

Miller, R., 1975, Distribution and properties of commissural and other neurons in cat sensorimotor cortex. *J.Comp.Neurol.*, 164: 361.

Miller, R., 1989, Cortico-hippocampal interplay: self-organizing phase-locked loops for indexing memory. *Psychobiology*, 17: 115.

Miller, R., 1991, Cortico-hippocampal interplay and the representation of contexts in the brain., Studies of Brain Function Vol. 17., Springer, Berlin, Heidelberg, New York.

Miller, R., 1993, What is the contribution of axonal conduction delay to temporal structure in brain dynamics? Symposium: Oscillatory event related brain dynamics (This volume).

Mizuki, Y., Masotoshi, T., Isozaki, H., Nishijina, H., and Inanaga, K., 1980, Periodic appearance of theta rhythm in the frontal midline area during performance of a mental task, *Electroencephalogr.Clin.Neurophysiol.* 49:345.

O'Keefe, J., and Nadel, L.,1978, The hippocampus as a cognitive map, Clarendon Press, Oxford.

Swadlow, H.A.,1974, Properties of antidromically activated callosal neurons and neurons responsive to callosal input in rabbit binocular cortex. *Exp. Neurol.*, 43, 424.

Swadlow, H.A., and Weyand, T.G., 1981, Efferent systems of the rabbit visual cortex: laminar distribution of the cells of origin, axonal conduction velocities and identification of axonal branches. *J.Comp.Neurol.*, 203:799.

Swadlow, H.A., 1991, Efferent neurons and suspected interneurons in second somatosensory cortex of awake rabbit: receptive fields and axonal properties. *J.Neurophysiol.*, 66:1392.

Vanderwolf, C.H., Kramis, R., and Robinson, T.E.,1978, Hippocampal electrical activity during waking behavior and sleep: analyses using centrally acting drugs, *in*: Functions of the septo-hippocampal system, CIBA Symposium, 58, Elsevier, Amsterdam.

WHAT IS THE CONTRIBUTION OF AXONAL CONDUCTION DELAY TO TEMPORAL STRUCTURE IN BRAIN DYNAMICS?

Robert Miller

Department of Anatomy and Structural Biology
University of Otago Medical School
Dunedin
New Zealand

A variety of experimental approaches to forebrain dynamics have revealed consistent correlations between signals in single neurones occurring at different times. Some of the temporal correlations involve action potentials separated in time by up to several hundred milliseconds. Abeles (1982) for instance shows such temporal dependencies in three-way correlations of neuronal spike trains. Villa and Abeles (1990) have demonstrated highly consistent temporal patterning amongst simultaneously recorded thalamic neurones. One of their figures, shown below (Figure 1), illustrates the precise dependencies of the firing of one neurone on that of others occurring several hundred milliseconds apart. Wright and Sergejew (1991) carried out cross-correlation between EEG signals at different loci in the cerebral cortex, and derived a propagation velocity for EEG of 0.1-0.29 m/sec. This clearly implies consistent correlation of signals which are temporally separated by quite long intervals (compared with the usual time scale of electrophysiologists).

The question arises: What components of neural tissue can give rise to temporal delays of this long duration which are also so precise? Axonal conduction delays are very precisely timed, but it is usually thought that they cannot give rise to delays more than perhaps 10 msec. Therefore the idea that axonal conduction time can play a major part in these delays is usually dismissed quickly. On the contrary, in this short paper, I want to argue that axonal conduction delays play a much more important part in these time dependencies than is generally recognized. Three simple points will be made. Firstly the usual assumptions made about axonal conduction time will be documented. Secondly, some of the actual evidence on this subject will be referred to. This suggests that axonal conduction times may be substantially longer than is usually assumed. Finally, a variety of biases will be listed, which apply

Oscillatory Event-Related Brain Dynamics, Edited
by C. Pantev, Plenum Press, New York, 1994

to the actual evidence, and make it likely than the usual assumptions are even further from the truth than the available evidence would suggest, if taken at face value.

Assumptions about conduction delays in cortico-cortical axons

A few statements are quoted below about what is usually assumed by those who study brain dynamics:

Abeles (1982) writes: "Time dependencies of tens or even hundreds of milliseconds must involve a large number of synapses. If one neuron can affect another through dozens of interneurons we must have a chain of interneurons which transmits with almost no failure"

Gassanov et al (1985) write: "Most of the cross-interval histograms between neighbouring neuron pairs had peaks with 1-2 ms latencies . . but direct connections are more difficult to account for cross interval histograms between neurons of different cortical regions with peaks of quite long latencies, 10-15 msec and even 60 msec."

Tank and Hopfield (1987) consider it unlikely that delays longer than 10 msec can be explained on the basis of axonal conduction.

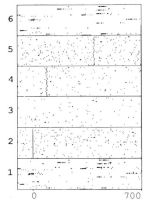

Figure 1: Raster displays from six neurones simultaneously recorded from thalamus of a cat anaesthetised by 80% nitrous oxide/20% oxygen. The separate sets of traces from the six neurones have been rearranged so that a subset of the action potentials in neurone 2 are aligned, and traces of the other neurones recorded at the same time given corresponding temporal positions. Note the remarkable display of correlated firing in neurones 4 and 5, at intervals several hundred milliseconds after that of the aligned spikes in neurone 2. Reprinted from Villa and Abeles (1990), *Brain Research* 509: 157-165, with permission from Elsevier Press.

Nunez (1989) builds a theory for the generation of the human EEG on the assumption that "the bulk of corticocortical action potential velocities . . lie in the 6-9 m/sec range".

Wright and Sergejew (1991) compute an estimate of wave velocity for the EEG of 0.1-0.29 m/sec. They consider that this "is much lower than the velocity of conduction in myelinated or unmyelinated axons and indicates that wave conduction probably depends upon comparatively short range fibre connections, thus introducing many synaptic delays".

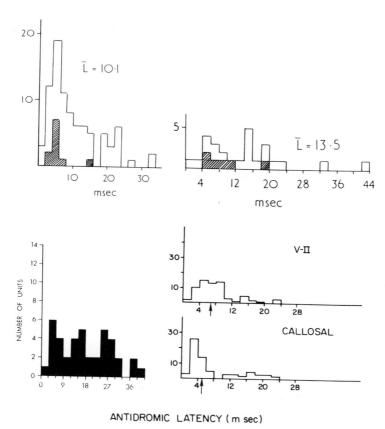

Figure 2: Composite diagram, from various sources, showing latency histograms for antidromic responses in ipsilateral cortico-cortical and callosally projecting single neurones. **Upper left:** Callosal neurones in primary somatosensory cortex of cat (from Miller, 1975); **Upper right:** Neurones in primary somatosensory cortex of cat projecting to second somatosensory cortex (from Miller, 1975); **Lower left:** Callosal neurones in rabbit visual cortex (from Swadlow, 1974); **Lower right:** Neurones in rabbit visual area V-I projecting to ipsilateral area V-II (above) and across the callosum (below) (from Swadlow and Weyand, 1981). Reprinted with permission from *Journal of Comparative Neurology* [Miller, 1975; Swadlow and Weyand, 1981] and from *Experimental neurology* [Swadlow, 1974].)

Histograms of conduction delays

The most precise method of assessing the spectrum of conduction delays possible in cortico-cortical axons involves recording of antidromic single unit responses. Glass micropipettes have an obvious bias for the large sized neurones, which probably have the largest calibre axons and highest conduction velocity. A few studies avoid this source of bias, by using tungsten, or tungsten-in-glass microelectrodes.

Some of such results are illustrated above (Figure 2). On the basis of such measurements Swadlow suggests that the estimated interhemispheric conduction time for the slower 40% of callosal axons is 20-60 msec. However, he is equivocal about whether such axons are capable of faithfully driving post-synaptic neurones.

Biases on all present estimates of axonal conduction time in cortico-cortical axons

(i) Recording techniques may have a sampling bias in favour of larger neurones, which are also likely to have larger-calibre (and faster conducting) axons.

(ii) Even with tungsten microelectrodes the smallest extracellular unit potentials, produced by the smaller cell bodies, may produce a signal which is so small that it is indistinguishable from noise in a single trace. Signal averaging has seldom been used to improve the detection of very tiny single unit potentials. These small cell bodies, whose response are usually undetected, are likely to be those which project the finer calibre axons with lowest conduction velocities.

(iii) Antidromic stimulation may ignore the terminal arborization of an axon, where calibre is likely to be smallest, and conduction velocity slowest.

(iv) If antidromic stimulation is actually within the arborization, only the most rapidly conducting pathway in that arborization may be detected, the others being canceled by collision.

(v) Conduction delay from soma to synapse may be underestimated because the full conduction distance is not taken into consideration. (Antidromic single unit responses have generally not been sought from those cells which, on the basis of anatomical connection studies, would be expected to have the longest trajectories.)

(vi) Almost all antidromic stimulation studies with tungsten microelectrodes have been conducted on smaller-brained animals (rabbit, cat) and seldom on primate species.

(vii) Latency measurements based on orthodromic responses to electrical stimulation are also biased in favour of rapidly conducting connections because only such connections will have sufficient synchrony at the post-synaptic site to summate and produce a suprathreshold postsynaptic response.

CONCLUSIONS

All published empirical information about axonal conduction delays in the forebrain is biased in favour of the rapidly conducting axons. The degree of bias may be very large. In the human brain it may represent an underestimate of the functionally important conduction time by a factor of ten or even twenty. Thus it is probable that significant numbers of cortico-cortical axons exist in the human hemispheres which impose conduction delays between soma and synapse of up to 100 ms. or even 200 ms.

Acknowledgment

The author is supported by the Health Research Council of New Zealand.

REFERENCES

Abeles, M., 1982, Local cortical circuits. Studies in brain function No 6, Springer, Berlin, Heidelberg, New York.

Gassanov, V.G., Merzhanova,G.Kh., and Galashina,A.G., 1985, Interneuronal relations within and between cortical areas during conditioning in cats, *Behav. Brain Res.*, 15: 137.

Miller, R., 1975, Distribution and properties of commissural and other neurons in cat sensorimotor cortex. *J.Comp.Neurol.*, 164: 361.

Nunez, P., 1989, Electric fields of the brain: the neurophysics of the EEG. Oxford University Press, New York.

Swadlow, H.A.,1974, Properties of antidromically activated callosal neurons and neurons responsive to callosal input in rabbit binocular cortex. *Exp. Neurol.*, 43: 424.

Swadlow, H.A., and Weyand, T.G., 1981, Efferent systems of the rabbit visual cortex: laminar distribution of the cells of origin, axonal conduction velocities and identification of axonal branches. *J.Comp.Neurol.*, 203:799.

Swadlow, H.A., 1991, Efferent neurons and suspected interneurons in second somatosensory cortex of awake rabbit: receptive fields and axonal properties. *J.Neurophysiol.*, 66:1392.

Tank, D.W., and Hopfield, J.J.,1987, Neural computation by concentrating information in time, *Proc.Natl Acad. Sci*, 84: 1896.

Villa, A.E. and Abeles, M., 1990, Evidence for spatiotemporal firing patterns within the auditory thalamus of the cat, *Brain Res.*, 509: 325.

Wright,J.J. and Sergejew,A., 1991, Radial coherence, wave velocity and damping of electrocortical waves. *Electroencephalogr.Clin.Neurophysiol.* 79:403-412.

COHERENT ASSEMBLY DYNAMICS IN THE CORTEX: MULTI-NEURON RECORDINGS, NETWORK SIMULATIONS AND ANATOMICAL CONSIDERATIONS

Ad Aertsen,[1] Michael Erb,[2] Günther Palm,[3] and Almut Schüz[4]

[1]Institut für Neuroinformatik, Ruhr-Universität, Bochum, Germany
[2]Abt. Biophysik, Philipps-Universität, Marburg, Germany
[3]Abt. Neuroinformatik, Universität Ulm, Ulm, Germany
[4]Max-Planck-Insitut für Biologische Kybernetik, Tübingen, Germany

INTRODUCTION

The anatomical structure of the neo-cortex and, particularly, its massive connectivity strongly suggest that the functional organization of this part of the brain is based upon interactions within and among groups of cells (Braitenberg and Schüz, 1991). This observation has prompted neurobiologists as early as Sherrington (1941) and Hebb (1949) (see also James, 1890) to speculate that cortical neurons do not act in isolation, but rather that they organize into *cell assemblies* for various computational tasks (see Gerstein et al., 1989, for a review of different definitions of the concept of cell assembly). One operational definition for the cell assembly has been particularly influential: near-simultaneity or some other specific timing relation in the firing of the participating neurons. As, for instance, elaborated in the concept of the '*synfire chain*' (Abeles, 1982, 1991), the synaptic influence of multiple neurons converging onto others in the cortical network is much stronger if they fire in (near-) coincidence. Thus, *temporal coherence* or *synchronous firing*, postulated as a mechanism for perceptual integration (Hebb, 1949), would in fact be directly available to the brain as a potential neural code (Perkel and Bullock, 1968; Gerstein and Michalski, 1981; Johannesma et al., 1986).

Inspired by these ideas, there has been a considerable interest recently in the simultaneous observation of the activities of many separate neurons, preferably in awake, behaving animals. These 'multi-neuron activities' are then analyzed for possible signs of (dynamic) interactions within and between the hypothetical cell assemblies. In this paper we review some of the evidence produced in these experiments, and compare these findings with the results of a theoretical study on the

Oscillatory Event-Related Brain Dynamics, Edited
by C. Pantev, Plenum Press, New York, 1994

activity dynamics in biologically inspired neural network models. We will discuss how our findings fit with the cortical neuroanatomy, and speculate on the consequences for the generation of cell assemblies in sensory and non-sensory cortical areas by means of Hebbian learning.

DYNAMICS OF FUNCTIONAL COUPLING BETWEEN CORTICAL NEURONS

The conventional method to analyse neural interactions is to cross-correlate the spike trains from neurons that were recorded simultaneously under some appropriate stimulus and/or behavioral conditions (Perkel et al., 1967; Aertsen and Gerstein, 1985). Following Moore et al. (1966), peaks and troughs in the cross-correlograms, after comparison with appropriate control measurements, are characterized by parameters describing their shape (symmetry, width, sign) and delay, and their possible dependence on stimulus or behavioral features. On the basis of such descriptors, the type and temporal acuity of the interaction exhibited by the neurons are then interpreted in terms of their 'functional coupling'. Recent developments in analysis methodology have considerably expanded the scope of these studies. It is now possible to examine the cooperativity in larger groups of neurons (Gerstein et al., 1985; Gerstein and Aertsen, 1985; Aertsen et al., 1987), to study the fine-temporal properties of firing correlation between two or three neurons (Aertsen et al., 1989; Palm et al., 1988), and to demonstrate the existence of accurate spatio-temporal firing patterns in the activity of single and multiple neurons (Dayhoff and Gerstein, 1983; Abeles and Gerstein, 1988; Abeles et al., 1993a, b).

Particularly, the *Joint-PSTH* (Aertsen et al., 1989), a temporal decomposition matrix of the ordinary cross-correlogram, was designed to highlight the detailed time structure of firing correlation among two neurons, and its possible time-locking to a third event, such as a stimulus or a behavioral event. Appropriate normalization of the Joint-PSTH enables us to distinguish between contributions due to stimulus- or behavior-induced modulations of the individual neuron firing rates, and those from interneuronal spike correlations, and to evaluate these differences for statistical significance (Palm et al., 1988). This normalization essentially scales the cross-covariance between the firings of two neurons for the associated pair of auto-covariances of the individual neurons. This is accomplished by subtracting from each bin in the Joint-PSTH matrix the product of the two corresponding bins in the individual PST histograms, and dividing the result by the product of the standard deviations of these same PST bins. Subsequent summations over appropriate diagonal and para-diagonal bins in the matrix plane give us the appropriately normalized, 'neuronal interactions only', PST coincidence histogram and cross-correlogram. Thus, we can examine fast, stimulus-related changes in the interactions between two observed neurons. (For a more detailed account and mathematical formulations we refer to Aertsen et al., 1989; Palm et al., 1988).

The salient result of such direct assembly observation is that the functional coupling among cortical neurons is *context-dependent* and *dynamic* on several different time scales. This is illustrated in Fig. 1 for an example drawn from cat visual cortex recordings, made in the Krüger laboratory (Krüger, 1982); similar results have been reported for other cortical areas (Aertsen and Gerstein, 1991; Aertsen et al., 1991; Vaadia et al., 1991; Vaadia and Aertsen, 1992).

Fig. 1 shows the Joint-PSTH of two neurons in cat Area 17, recorded simultaneously during presentation of a bar stimulus, moving back and forth over a series of trials (n=22) of 4 seconds each. The movement was perpendicular to the bar orientation and proceeded in opposite directions from 0-1.8 and 2.2-4 seconds, respectively; inbetween, the stimulus was stationary (1.8-2.2 s). The two panels show the 'raw' (Fig. 1A) and the normalized Joint PSTH (Fig. 1B), together with the associated marginal distributions for the selected two neurons. Within each panel, the individual PST histograms are shown along the horizontal and vertical axes. The density of coincident firing is indicated in the Joint PSTH matrix, according to the grey code above the matrix. At the right of each panel are respectively the PST coincidence histogram (along the diagonal rising to right) and the conventional cross-correlogram (perpendicular to the diagonal and descending to right). The PST coincidence histogram measures the counts near the diagonal of the Joint matrix and represents the time-locked average of near-coincident firings of the two neurons in the same sense as the ordinary PST histogram represents the stimulus time-locked average of the individual neuron's firings.

In the 'raw' JPSTH panel (Fig. 1A) we note that the individual PST histograms show a strong stimulus-locked increase in firing for both neurons as the stimulus enters their receptive fields, with a clear preference for the first direction of motion. These are direction (as well as orientation) sensitive neurons. Similarly, the 'raw' Joint-PSTH matrix shows considerable hills at locations matching the PST-peaks, and corresponding peaks are visible in the PST-coincidence histogram. In other words, individual, joint, and near-coincidence firings all are increased during portions of both directions of movement, but more so in the first direction. We refer to this time-dependent co-variation of the firing rates as 'rate coherence'. This type of activity correlation is to be distinguished from 'event coherence', which expresses the degree of firing synchrony at the level of individual spike occurrences (Neven and Aertsen, 1992). This latter form of correlation is measured by the normalized Joint-PSTH (Fig. 1B), i.e. the correlation after 'removing' the contribution by stimulus-locked modulations of the individual firing rates. Note that after such normalization the time-averaged crosscorrelogram, unlike its 'raw' counterpart in Fig. 1A, shows a distinct peak, straddling the origin. This clearly points at the presence of spike synchronization which exceeds the mere co-variation of firing rates. The time course of this spike correlation is captured by the normalized PST-coincidence histogram along the diagonal. During most of the response to the *second* direction of motion (roughly, between 3000 and 4000 ms), there is a positive level of near-coincident firing. This level is not constant, though, but rather follows the time course of the individual neurons' firing rates. Note, however, the brief, but strong burst of excess co-firing around 2500 ms, i.e. just when the responses to the second direction of motion start to develop. The situation is much more complicated during the *first*, and opposite, direction of movement (roughly, between 0 and 2000 ms). Here we also observe a varying level of correlated firing, but its time course is altogether different from the individual neurons' responses. After an initial, large positive peak, coinciding with the rising phase of the individual responses, the correlation breaks down to zero and even becomes negative (i.e. there is a lower than expected amount of near-coincidence firing) as the individual firing rates reach high values.

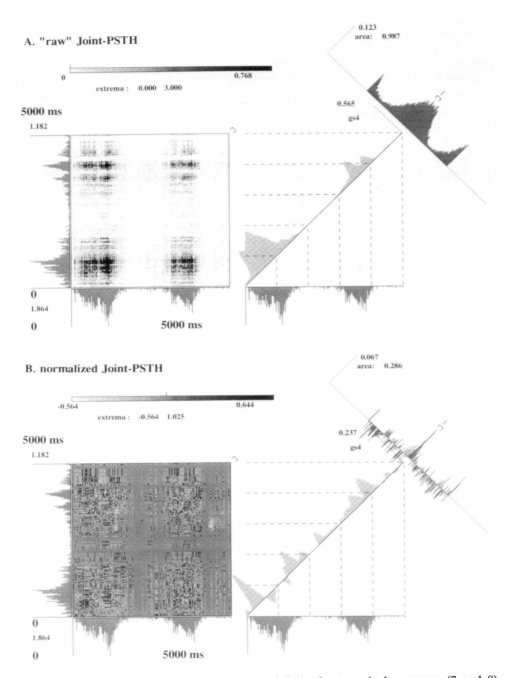

Figure 1. Stimulus-locked dynamic correlation of firing for two single neurons (7 and 8), recorded simultaneously from the cat visual cortex during repeated presentation of a moving bar

This stage of zero or negative correlation lasts for most of the duration of the peak response rates, and only returns to positive values when the individual responses have largely decayed. (A similar case of coupling among visual cortex neurons with even more pronounced sign-reversal is described in Gerstein and Aertsen, 1985.)

In the example shown here, the switching from a condition of co-firing to one of anti-co-firing seems to be associated, at least roughly, with the transitions from low to high firing rate of the neurons involved (compare the time course of the PST-histograms with that of the normalized PST-coincidence histogram). This pattern is, in fact, very suggestive of a non-monotonic relation between the degree of spike synchrony and the amplitude of the single neuron responses. The correlation is positive when firing is weak to moderately strong (i.e. during the rising and falling stages of the first response and throughout most of the second response), whereas it becomes zero or even negative when the firing rates are high. For the first (and stronger) response, this non-monotonicity causes the interaction to behave essentially like a 'transient' version of the individual responses. Thus, it has essentially the quality of a 'time derivative' (or, rather, its absolute value), 'signalling' the changes (rise and fall) of the responses, rather than their magnitude. Since the second response does not reach such high values, the non-monotonicity does not take such strong effect and, correspondingly, the individual response waveforms are more closely reflected in the time course of the interaction (although some of the above described 'transiency' is also present here).

Summarizing, the example in Fig. 1 clearly demonstrates that near-coincidence firing can be strongly modulated by the stimulus presentation. These two neurons are repeatedly switching from a condition of co-incident firing to a stimulus-related period of zero correlation or even anti-coincidence. We have observed and reported similar phenomena in other recordings, both from visual cortex (Gerstein and Aertsen, 1985) and other cortical areas, such as the prefrontal cortex (Aertsen et al., 1991; Vaadia et al., 1991; Vaadia and Aertsen, 1992). (A review on experimental results, based on recordings from a variety of regions in the CNS of different animal species, made in several laboratories, is currently in progress; Gerstein and Aertsen, in preparation.) These findings demonstrate that cortical neurons may exhibit rapid modulations of discharge synchronization that are related to

---

stimulus. The upper panel (A) shows the 'raw' Joint-PST histogram; the lower panel (B) shows the Joint-PST histogram after normalization for stimulus-induced nonstationarities in the single neuron firing rates. The display format of both panels is the same. The left-hand half of the panel shows the Joint-PSTH matrix and the two ordinary PST histograms along its x- and y-axes (binwidth: 25 ms). Values in the Joint-PSTH matrix are coded in grey as indicated in the bar above the matrix. The tic mark above the grey bar corresponds to the value zero. All counts were divided by the number of stimulus presentations. The right-hand half of the panel shows the PST coincidence histogram (running along the diagonal from lower left to upper right) and the ordinary cross-correlation histogram (perpendicular to the diagonal, and running from upper left to lower right). The PST coincidence histogram was smoothed using a gaussian with sigma of four bins; this particular value (gs4), as well as the location and width of the selected diagonal band are indicated in each panel. The position of true coincidence (zero delay) in the cross-correlogram coincides with the intersection point of the PST coincidence histogram and the cross-correlogram; it is indicated by a tic mark above the diagonal band marker above the correlogram. Numbers of spikes: 1420 (neuron 7, x-axis), 788 (neuron 8, y-axis), recorded during 22 stimulus trials of 5 seconds each.

stimulus context and behavioral state. Such modulations -which could not be inferred from single neuron observations- may switch the neurons' firing behavior from being mutually incoherent into a particular coherent state of joint synchrony, or, alternatively, from one particular pattern of mutual coherence into a different one. Each such pattern may last for only a few tens to hundreds of milliseconds. Finally, the observed modulations in synchronous firing may be, but need not be, associated with changes in either of the neurons' individual firing rates. In fact, the spike correlation may follow a time course that deviates appreciably from that of the firing rates of either of the individual neurons. In general, rate coherence and event coherence present two instances of firing synchrony at different levels of temporal acuity. At a course level of time resolution the correlation of firing is governed by the co-variation of the firing rates (*rate coherence*), at a more fine-grained level of time resolution, the detailed spike correlation (*event coherence*) becomes the dominant term (Neven and Aertsen, 1992). In any specific case, these different types of correlation may be, but need not be, correlated. These various phenomena appear to be robust across different regions in the brain, and across a variety of animal species. We conclude that dynamic cooperativity as expressed in the different types of firing correlations presents an emergent property of neuronal assembly organization in the brain.

FUNCTIONAL COUPLING IN MODEL NEURAL NETWORKS

These findings in cortical multi-neuron recordings suggest that the usual concept of neurons with static interconnections of fixed or only slowly changing efficacy (during learning, for example) is no longer appropriate. Instead, one should distinguish between *structural* (or anatomical) *connectivity* on the one hand and *functional coupling* (or effective connectivity) on the other (Aertsen and Gerstein, 1991; Aertsen and Preissl, 1991). Whereas the former can presumably be described as (quasi) stationary, the latter may be highly dynamic and context-sensitive. These findings raise a two-fold question: what is the nature of the underlying mechanisms, and what are the functional implications? In a number of theoretical studies we have sought to elucidate these issues. Several different mechanisms may be invoked to mediate the transition from static, anatomic connectivity to dynamic, functional coupling. On the one hand, the underlying mechanism may be *local*, as in von der Malsburg's proposal of rapid modulation of synaptic efficacy (von der Malsburg, 1981, 1986). On the other hand, more *global* network effects might be involved, as for instance in Sejnowski's notion of the 'skeleton filter' (Sejnowski, 1981).

In order to obtain more insight into these different alternatives, we investigated the activity dynamics in various physiology-oriented networks of model neurons, both of the spiking and of the analog type (Aertsen and Preissl, 1991; Erb, 1992; Erb and Aertsen, 1992; Aertsen et al., 1994). These networks were designed to capture the principal features of real cortical networks, both in terms of anatomy and of physiology. Of special importance in this context is the physiological constraint of 'sparse' firing: the activity in the network is required to be of low firing rate, such as typically observed in the neocortex (Abeles et al, 1990). As it turned out, maintaining stability in such a feedback system of sparsely firing neurons is by no means a trivial problem (Erb and Aertsen, 1992). In fact, stable solutions could only be reached for confined 'islands' in parameter space. Outside this restricted range, the network activity either died, exploded or developed strong, coherent oscillations, with frequency and amplitude dynamics governed predominantly by the inhibition parameters, but largely independent of (1) the network architecture (uniform, random or structured), (2) the spiking or analog nature of the neural activity, and, albeit to a lesser extent, (3) the linear or nonlinear nature of the neural threshold function. In the present report we focus on the issues of stimulus-dependence and the dynamics of functional coupling. Hence, we took great care to select parameter values such that the network was operating in a regime of stable, sparse firing under non-stimulus conditions (low rate 'spontaneous' activity).

We investigated the behavior of a feedback neural network of 100 spiking model neurons with fixed synaptic connections (a detailed description of the network model and the corresponding equations is given in Erb and Aertsen, 1992; Aertsen et al., 1994). Briefly, the model neurons are connected by excitatory synapses, inspired by neuroanatomical findings that about 90% of the cortical synapses are of this type (Braitenberg and Schüz, 1991). The synapses are modeled as lowpass filters with delayed response, transforming the incoming spike activity into EPSP's. These are summated linearly to yield the instantaneous value of the membrane potential at the cell body. The probability of spike generation is described by a sigmoid function of this membrane potential; this probability, in turn, is modulated by a refractory mechanism, driven by the recent spike history of the neuron. The final spike output is obtained by using the firing probability as the instantaneous rate of a stochastic event generator. In addition to this basic network of spiking neurons (the 'pyramidal cells'), the model comprises a global inhibitory mechanism for activity regulation, consisting of two parallel, linear branches: a fast one, with a time constant similar to that of the excitatory synapses (5 ms), and a slow one with a considerably larger time constant (200 ms). The fast branch mimics the inhibitory action of the stellate cells, the slow one serves to regulate the overall activity in the network towards a preset value (threshold control; Braitenberg, 1978b; Palm, 1982). In the experiments reported here, parameter values were chosen such that they conform to physiological observations, while ensuring stability under sparse firing conditions, with an overall network activity in the order of 10-20 spikes per neuron per second on average.

For the present simulations we used a network with fixed connectivity, explored in a preceding study of associative memory and the performance of different learning rules (Erb, 1985,; Palm, 1986, 1987). In particular, we used a connectivity matrix in which were embedded the memory traces of a set of 10 sparsely occupied, randomly distributed input patterns. The resulting connectivity matrix, after reordering the neurons such that the input patterns become more or less segregated, is characterized by a clear block-like structure along the diagonal (Aertsen et al., 1994). Each of these compact blocks (containing on the average some 7 neurons) denotes a heavily interconnected 'cell assembly', preferentially associated with one of the input patterns. Clearly, the segregation is not complete, since the input patterns are random, not orthogonal. Hence, associations do overlap, resulting in (usually weaker) connections between the assemblies. On average, each assembly involves some 10 neurons, having an overlap of 2 to 3 neurons with one or more of the other assemblies.

STIMULUS-DEPENDENT COUPLING IN A NETWORK OF SPIKING MODEL NEURONS

In a series of simulations, we investigated the flow of activity in the network upon presentation of a continuously varying stimulus. These 'experiments' were designed to mimic the physiological situation. By adopting this approach, we could apply the same analytical tools as described for the cortical recordings and, thus, directly confront the outcome of the physiological experiments with that of the theoretical investigation. We refer to such experiments on theoretical constructs as 'in virtu' recordings, to distinguish them from their *in vivo* and *in vitro* physiological counterparts (Aertsen et al., 1994).

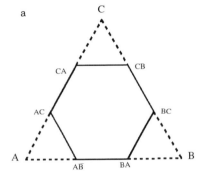

b		A	B	C
	SE1	9	4	2
	SE2	8	7	10
	SE3	4	2	10
	SE4	9	8	7
	SE5	9	2	4
	SE6	8	10	7
	SE7	4	10	2
	SE8	9	7	8

Figure 2. Composition of the stimulus ensembles, used to test the model neural network. The symbols A, B and C denote three different input patterns, selected from the stored set of ten patterns; the hexagon symbolizes the stimulus trajectory. Further explanation in Text.

The stimulus protocol, based on the originally stored input patterns, is schematically illustrated in Fig. 2. The symbols A, B and C denote any triplet from the set of stored patterns; the hexagon symbolizes the path taken by the stimulus. This path was designed to investigate the switching behavior of the network to varying mixtures of stored input patterns. Thus, we avoided the attractors themselves and concentrated our attention on the boundary regions. For instance, at any point along the branch AB-BA, the stimulus consists of a superposition of the two patterns A and B, with weights linearly varying from a ratio 2:1 to 1:2. One such branch has a duration of 100 ms, sampled in steps of 1 ms. Thus, one 'round-trip' along the hexagon takes 600 ms. By making different selections for the triplet (A,B,C), we generated four different stimulus ensembles; by additional variation of the direction along the hexagon ('clockwise' or 'anti-clockwise'), we arrived at a total of eight different stimulus ensembles (SE1 to SE4, and SE5 to SE8; Fig. 2b).

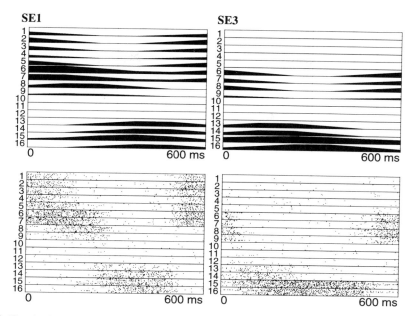

Figure 3. '*In virtu*' recordings from 16 out of 100 neurons in the model network upon presentation of two different test stimuli (SE1 and SE3; Fig. 2). The top panels show the inputs to each of these 16 neurons, the bottom panels show the spike trains these neurons generate in response to 10 consecutive stimulus presentations.

The response of the network to these various stimuli was investigated by recording the spike activity of a subgroup of 16 of the 100 model neurons, in much the same way a physiological multi-neuron recording would pick up a fraction of the entire population of neurons. These 16 neurons were chosen such that each of them belonged to one or two assemblies, associated with the stored input patterns used for generating the stimuli. The 'synaptic' connectivity matrix for this subset of 16 neurons is shown at the bottom left of Fig. 4. Notice the block-like connectivity structure of the various cell assemblies and the interconnections between them. This is outlined in the simplified membership diagram to the right, depicting the three dominant assemblies (numbered 1, 2, 3) within the group of 16. Note, however, that the actual connectivity matrixes more complex, and that more than three, partly overlapping, assemblies can be discerned.

Fig. 3 shows the temporal variation of the input to the selected 16 neurons for two different stimuli (SE1 and SE3; top panels) and the associated spike activities, collected over 10 trials of 600 ms each (bottom panels). Notice the qualitative similarity of this picture with the results of physiological recordings and, particularly, how the various neurons reflect the time course of their input in the temporal modulation of their (generally low) firing rates. We made simulations like these for each of the eight different stimulus ensembles SE1 to SE8. The multi-neuron spike trains 'recorded' during these simulations were first analysed for neural interactions using the 'network correlation matrix' algorithm (Aertsen et al., 1987, 1994). Such a network correlation matrix provides a global measure of the spike coherences among the entire set of recorded neurons in a single picture. This is accomplished by collapsing the diagonal trace in the normalized Joint-PSTH for each possible pair among the 16 neurons into a single number, the time-averaged spike coherence for that particular pair. These numbers are then arranged into the form of a 16x16 matrix, comparable in format to the synaptic connectivity matrix at the bottom left of Fig. 4. The results of this procedure, representing the time-averaged functional coupling among the 16 neurons for each of the 8 different stimuli, are shown in the top two rows of Fig. 4.

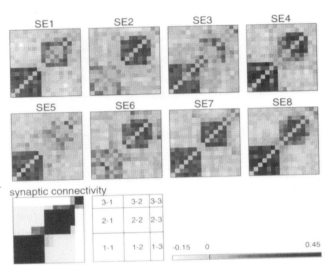

Figure 4. Network correlation matrices (top eight panels), representing the functional coupling among a subset of 16 neurons in the model network in response to eight different test stimuli (SE1 to SE8; Fig. 2). All stimuli were presented at equal input strength of 0.25. The synaptic connectivity matrix for these 16 neurons is shown in the bottom panel (left), together with a simplified diagram of the three dominant cell assemblies (1, 2, 3), present in this group of 16 neurons (right). The strength of coupling is coded in grey according to the grey bar (bottom right).

Two interesting comparisons can be made at this point. First, the network correlation matrices for these eight different stimulus ensembles reveal distinct stimulus-dependent differences in the degree of coupling among the neurons. Different subgroups of neurons, defined by a strong and approximately uniform degree of coupling within a group, more or less clearly pop out, depending on the stimulus used. Such results are highly reminiscent of the stimulus-dependent coupling matrices we observed in physiological recordings (Aertsen et al., 1987, 1994). In addition, we can also compare the network correlation matrices with the synaptic connectivity matrix in the bottom panel. Note, however, that this second comparison is usually not possible in the physiological context and, thus, presents a distinct advantage of such 'in virtu' experiments. In particular, it demonstrates how, depending on the stimulus context, different portions of the underlying synaptic network express themselves in the correlation matrix, whereas others, in spite of their strong and persistent synaptic connectivity, do not show up at all. For instance, the stimulus ensemble SE3 mainly reveals coupling among the first assembly (neurons 1 to 7), stimulus ensemble SE2 predominantly shows interactions in the second group (neurons 8 to 13), whereas both the first and second assembly (or, rather, different portions of the latter) are manifest with stimulus ensembles SE1 and SE4. None of these four stimuli is apparently able to reveal the third assembly. Observe also how the time-reversed stimulus ensembles SE5 and SE7 give rise to distinct differences as compared to their counterparts SE1 and SE3; such hysteresis effects are less visible for the stimuli SE6 and SE8.

In a related series of experiments we also studied the effect of changing the amplitude of the stimuli (Aertsen et al, 1994). The implicit assumption here is, of course, that such changes of input strength are monotonically related to changes in the intensity (or contrast, or some other appropriate parameter) of sensory stimuli in physiological experiments. The principal result of those experiments was that the correlation strengths between the various members of an assembly remain uniform and show identical dependence on input amplitude. Generally, the dependence is non-monotonic, with a maximum at some intermediate value of stimulus intensity, and consistent fall-off at larger amplitudes. This suggests that there exists a restricted range of intermediate amplitude values in which the assembly organization can unfold most clearly. Very weak or very strong input strenghts either do not address the network adequately or enslave it to such an extent that no room is left for the assemblies to organize. In addition, there are also clear differences in the amplitudes and the precise form of the intensity dependence, both when different assemblies are compared for the same stimulus and for each single assembly compared across different stimuli (Aertsen et al, 1994).

This differential dependence on the composition as well as on the strength of the input stimuli demonstrates how the network may drastically change its outward appearance, depending on the stimuli used to investigate its performance. Stated in more functional terms, it explains how the network reorganizes the topology of its functional coupling, depending on the computational task it is faced with.

DYNAMICS OF FUNCTIONAL COUPLING IN A MODEL NEURAL NETWORK

After this global analysis of network coupling, we now turn to the dynamics of coupling between individual model neurons using Joint-PSTH analysis. We will consider the interactions both within and across different cell assemblies.

In order to obtain reliable statistics, we presented the stimulus ensemble SE1 to the network over 250 consecutive trials. Fig. 5 shows the normalized Joint-PSTHs of two selected pairs of model neurons. The normalized Joint-PSTH for the neuron pair (6,7), both taken from a single cell assembly (group 1), are shown in Fig. 5A. In contrast, Fig. 5B shows the normalized Joint-PSTH for the neuron pair (6,8), taken from two different cell assemblies (group 1: neuron 6, and group 2: neuron 8). Both Joint-PSTHs are displayed using the same scales, in order to enable comparison.

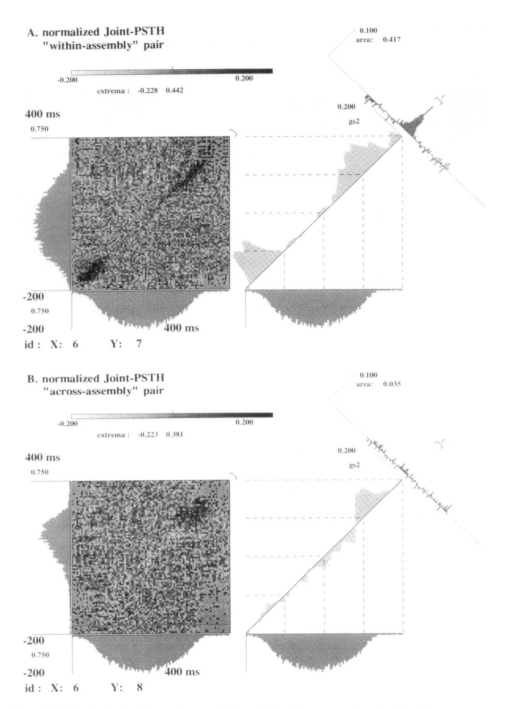

Figure 5. Stimulus-locked dynamic correlation of firing between selected pairs of neurons from the model network during repeated presentation of the test stimulus SE1. Normalized Joint-PST histograms were determined for a 'within-assembly' pair (A) and for an 'across-assembly' pair (B). The within-assembly pair consists of the neurons (6,7), both members of a single cell as-

Inspection of the normalized Joint-PSTH for the within-assembly pair (Fig 5A) reveals that the detailed time course of the spike correlation between the two model neurons is very similar to that observed between the cortical neurons in Fig. 1B. Particularly, the normalized PST-coincidence histogram along the diagonal of Fig. 5A reveals positive correlation in the rising and falling phases of the responses, while the correlation breaks down to zero inbetween, and stays down for most of the duration of the peak response rates. Comparison with further pairs taken from a single assembly confirmed that the behavior exhibited by these two neurons is representative for the interactions taking place within a model assembly. The situation is quite different, however, for the interactions between model neurons that belong to different assemblies, as is illustrated for the across-assembly pair (6,8) in Fig. 5B. Observe that in this case the time-averaged normalized correlation (top right) is essentially zero. This explains the zero coupling among such across-assembly neurons in the network correlation matrix (Fig. 4, SE1), as this latter matrix essentially represents the time-averaged spike correlations. Interestingly, however, this absence of time-averaged correlation does not imply that the coupling between the two neurons is constantly zero throughout the stimulus trial. Rather, the normalized PST-coincidence histogram along the diagonal exhibits an alternating trajectory. It passes through a prolonged, shallow negativity during most of the elevated single neuron responses, before rising to a brief burst of positive spike correlation as the responses have almost decayed down to background level. This result once more demonstrates the dangers of taking the usual, time-averaged crosscorrelogram as evidence for the presence or absence of neural interactions. Contributions of opposite sign may cancel in the time integral and, hence, leave no trace in the correlogram (see also Aertsen and Gerstein, 1991). More interestingly, though, are the functional implications of this type of interaction. Apparently the two model assemblies compete with each other during most of the time they are both strongly activated by the stimulus. This competition is presumably mediated by the global inhibition in our model network. It only ceases to exist, or, rather, it is overridden by the positive cross-coupling between the assemblies, after the firing rates have decayed to the extent that the simultanous co-existence of two active assemblies no longer poses a challenge to the activity regulation by the slow inhibition.

Summarizing, these simulations demonstrate that a network with fixed synaptic connectivity can exhibit strong and extremely dynamic context-dependence of functional coupling, expressed in the neuronal spike correlations. Thus, considerable and rapid changes in coupling may occur, without any associated changes in local connectivity. In contrast, our findings rather point at a more global mechanism, in which the instantaneous degree of coupling among the neurons is controlled by the stimulus-dependent flow of activity in the entire feedback network.

---

sembly (group 1; Fig. 4). The 'across-assembly' pair consists of the neurons (6,8), taken from different cell assemblies (groups 1 and 2; cf. Fig. 4). Observe the similarity of the correlation dynamics between the visual cortex neurons (Fig. 1B) and those between the model neurons 6,7 (Fig. 5A). Observe also that quite different coupling dynamics are exhibited by the across-assembly pair (6,8) (Fig. 5B). The format of these Figures is the same as in Fig. 1. Numbers of spikes: 9244 (neuron 6), 8426 (neuron 7), 4136 (neuron 8), recorded during 250 stimulus trials of 600 ms each.

Such global mechanisms do not have to invoke rapid modulations of synaptic efficacy, although they obviously do not exclude them as an additional mechanism either. Further experimental evidence is required to clear this issue.

In a related model study (Boven and Aertsen, 1989; Aertsen and Preissl, 1991) we found evidence for the hypothesis that the principal cause of these various dynamic effects may reside in modulations of the background activity projecting onto the observed neurons. Due to the subthreshold nature of cortical EPSP's, action potentials from a single neuron will not suffice to make a receiver neuron fire. For this they need the coincident arrival of spikes at the other synapses of that neuron. These are provided by the diffuse background activity coming from the remainder of the network. The pool activity thus provides a background level, which, depending on its magnitude, will make activity from the driver neuron more (or less) viable in eliciting activity from its target neuron. Hence, the background activity functions as a control parameter, determining the operating point of the receiver neuron, and, thereby, the efficacy of the otherwise subthreshold connections. Due to the stochastic and pulse-like nature of the pool activity arriving at the target neuron, the background level of the membrane potential will fluctuate and, hence, occasionally reach threshold, causing the neuron to fire. Since these 'accidental' firings do not require coincident arrival of spikes from the driver neuron, they do not contribute to the efficacy of the connection. On the contrary, they are counterproductive since they cause spikes from the driver neuron to arrive more often at its target neuron when this is in a non-responsive state ('refractoriness'). The size of the membrane potential fluctuations and, hence, the amount of occasional firing is proportional to the pool coupling. This implies that the smaller this pool coupling, the closer one can regulate the operating point of the target neuron towards being liable to fire upon arrival of spikes from the driver neuron, yet without the danger of eliciting too many accidental firings. Rather than a 'noisy' control parameter, one thus obtains a smooth, and thereby more influential regulation of synaptic efficacy. Moreover, it could be shown that very similar and equally rapid effects can be obtained by dynamically manipulating the internal correlation structure of the background activity, while keeping the magnitude of its firing rate constant (Gerstein et al., 1989; Bedenbaugh et al., 1988, 1990).

We conclude that, even with the highly simplifying assumptions in this model, we have captured the essence of a nonlinear mechanism by which 'diffuse' background activity may govern the dynamic linking of neurons into functional groups. It is based on activity variations in the entire network, being 'projected down' onto the connections among the recipient neurons. As a result, rapid changes in activity in the entire network induce equally fast, nonlinearly distorted modulations of synaptic efficacy. These, in turn, are expressed in rapid modulations of spike correlation in the cortical activity, as can be observed in multi-neuron recordings. Thus, many of the features we encountered in our study of interactions among cortical neurons were reproduced in these simulations in remarkable detail. This is all the more surprising, since our model network was not designed with that particular purpose in mind. Rather, it was inspired by much more general considerations, originating from statistical neuroanatomy and basic physiology. We conclude that the observed consistency with the physiological interactions clearly speaks in favor of the biological plausibility of this type of neural network model. Moreover, it lends support to our thesis that most, if not all, of the dynamic context-dependence of cortical coupling can, in fact, be attributed to the global mechanisms embedded in such probabilistic feedback networks.

HOW DO THE PHYSIOLOGICAL OBSERVATIONS FIT WITH CORTICAL NEUROANATOMY?

We have shown how a simple model of 100 excitatorily coupled spiking neurons (with inhibitory control of the overall activity) can reproduce many of the phenomena concerning neural interactions that are observed in real brains. But clearly this model is highly unrealistic, anatomically, for two reasons:
- 100 is a much too small number of neurons, and
- the real connectivity among tens or hundreds of thousands of neurons is far from full, i.e. the chance of any two neurons having a synaptical connection is much lower than 1.

The question now can be raised, whether the same phenomena can also be observed in more realistic models of cortical connectivity. Indeed, simulations with larger networks and different degrees of average connectivity show that effects of stimulus-dependent synchronisation, leading to dynamic variations of functional coupling are typical for low connectivity networks of spiking neurons (Wennekers and Palm, 1994), in particular if the connectivity is structured in a way that corresponds to a superposition of more highly interconnected subgroups (assemblies). Also, it is feasible to establish relatively stable firing patterns of low average firing rate in such networks with low average connectivity. If the average activity and/or the average connectivity is too high, then too many neurons in the network will get synchronized. This typically results in an unnatural exactness in the firing of almost all observed neurons (neurons firing like marching troops), that can be seen in many simulations of event-coherence binding (Glünder and Nischwitz, 1993; Deppisch et al., 1987; Mirollo and Strogatz, 1990) and that cannot easily allow for varying activation and interaction patterns. Thus, the observed effects of neural interaction are indeed compatible with large networks of low average connectivity holding a number of more highly interconnected subgroups, which we like to call cell assemblies.

In accordance with Hebb's original ideas (Hebb, 1949), it is today widely believed that such assemblies are indeed present in the cortical network, where they have presumably been formed by Hebbian learning (e.g. Gerstein et al., 1989). There is, however, a crucial problem related to this idea: Is the neural excitation provided by external input in a network with low connectivity large enough to produce such assemblies? In other words: Are there sufficiently many excitatory synapses converging on each neuron in the assembly-to-be that will be strengthened (as the Hebb rule stipulates) due to a coincidence of pre- and postsynaptic activity?

If we can answer this question positively, then we have an extremely plausible model for the observed physiological effects, that would also be consistent with the known cortical anatomy. Moreover, we can then hope to obtain a refined picture of the local (areal) structure of realistic cortical cell assemblies, both in space and in time.

THE GENERATION OF ASSEMBLIES IN SENSORY AND NONSENSORY AREAS BY MEANS OF HEBBIAN LEARNING

We will now enter into some more elaborated speculations, concerning the size and appearance of cell assemblies, which are based on informed guesses about excitation thresholds and contact probabilities of cortical neurons, based on anatomical observations (Braitenberg, 1978; Schüz and Palm, 1989; Braitenberg and Schüz, 1991). The basic idea is that a typical assembly can be thought of as a collection of local sub-assemblies, where each local sub-assembly is confined within the range which can be reached by a typical pyramidal cell, e.g. a cortical area of about 1 mm x 1 mm in the mouse (Braitenberg and Schüz, 1991) or somewhat more in larger animals (Creutzfeldt et al., 1977; Gilbert and Wiesel, 1983; Martin, 1984; DeFelipe et al., 1986), let us say 2 mm x 2 mm. Based on this idea we estimate the number of excitatory cortico-cortical and intracortical synapses which a typical neuron in the assembly receives from the assembly. We assume that these synapses are Hebb-synapses and we postulate that there should be at least about 5 to 8 of them converging onto each neuron in the assembly. This number is based on estimates on the excitation threshold by Abeles (1991) and on the idea that a reasonable proportion of these excitatory synapses could be provided by nonspecific inputs. Our further considerations are related to, but different from the recent calculations of Wickelgren (1992) and Miller (1994).

With respect to timing we assume that within 2 mm of cortical distance a timing accuracy of up to 4 msec can be kept (cf. Nelken, 1992). Thus, we base our analysis on a time window of 4 msec. The following analysis may seem a little naive, since there are much more refined methods of calculation demonstrated for example by Abeles (1991), but such calculation would yield roughly the same results.

Before we enter into the detailed calculations, let us elaborate on some of the anatomical facts which play a role in this context. As mentioned above, the cortex in no way reaches the degree of connectivity of a fully connected network. In the mouse, the average neuron is connected to only about 1/2000 of the neurons in the neocortex. On the other hand, the largest possible connectivity (within the given anatomical constraints) is nearly reached, as indicated by the numbers in Fig. 6. In the cortical neuropil the processes of the different neurons are intermingled to such a degree that an individual pyramidal cell cannot but connect via nearly each of its synapses to a different postsynaptic neighbour. Thus, the activity of an individual pyramidal cell cannot be assumed to activate any of those neighbours on its own, unless other neurons within the same region (and converging onto some of the same neurons) happen to be activated at the same time. In primary sensory areas, the occurrence of this condition may be favoured by two anatomical properties:

1) the relative density of the ramifications of sensory input fibers (Braitenberg, 1978 b; Martin, 1984) and
2) the tendency for an ordered organization of these, such as the topographical organization of the input to the primary visual cortex.

With respect to intra-cortical and cortico-cortical fibers, the anatomical answer to the weakness of individual synaptic connections between individual neurons may be the well-described phenomenon of patchy projections. In contrast to the diffuse connections of the individual neuron, groups of neurons tend to stay together when they project from one region to another (e.g. Bullier et al., 1984; Pandya and Yeterian, 1985; Van Essen et al., 1986; Gilbert and Wiesel, 1989; Amir et al., 1993; Malach et al., 1993). By this they may have a chance to activate a new set of neurons there. Thus, the combination of the diffuseness of connections with the patchiness of projections may be the anatomical solution to the dilemma of achieving a high degree of connectivity on the one hand and to get a message across on the other.

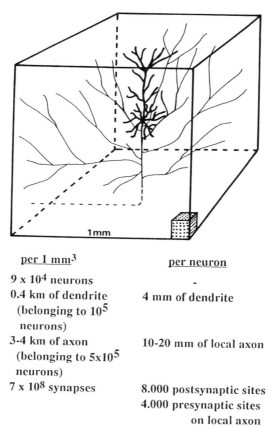

per 1 mm^3	per neuron
9 x 10^4 neurons	-
0.4 km of dendrite (belonging to 10^5 neurons)	4 mm of dendrite
3-4 km of axon (belonging to 5x10^5 neurons)	10-20 mm of local axon
7 x 10^8 synapses	8.000 postsynaptic sites
	4.000 presynaptic sites
	on local axon

Figure 6. Neighbourhood relationships of a pyramidal cell. The size of the cube is chosen such as to just enclose all of the local axonal ramifications of this neuron which, in the mouse, corresponds to about 1 mm^3. The dots in the small cube in the lower right corner indicate the density of neurons. Below on the right, some measures for the average pyramidal cell in the mouse cortex are listed. On the left, the corresponding total numbers of neurons, neuronal processes and synapses in 1 mm^3 are given. Modified from Schüz and Braitenberg (1994).

First let us consider a local assembly that is situated in a primary sensory area of the cortex like V1. Such an assembly can be formed by Hebbian strengthening of sufficiently many synapses between the neurons of the assembly. This can easily happen if the neurons in the assembly are stimulated together often enough by the sensory input.

Once the assembly is formed, each neuron should receive coincident input through at least a certain number s of strengthened synapses from the others. In each time window (of 4 msec) only a certain percentage p of the neurons in the assembly are active. We have to assume a certain time window for firing coincidence that is effective in firing the next neuron, let us again say 4 msec in view of the typical duration of EPSPs. If the assembly size is A, then on average pA neurons were active within the last 4 msec, thus the probability p' that a neuron receives at least s strengthened synaptic inputs can be estimated from a binomial distribution (as in Miller, 1994) or simply from a Poisson distribution with mean $pApl$ - where pl is the probability of a local connection between two neurons within 2 millimeter distance. This probability p' should be larger than p.

Apart from the time windows we also have to specify the numbers p, pl and s. Neurons that are active, fire at a rate which can vary greatly, but we assume here a value of 50 spikes per second for a typical cortical neuron. Thus for a 4 msec window we have $p = 1/5$. The excitation threshold of cortical neurons is assumed to be around 8 to 10 EPSPs within about 3-4 msec. Since we assume that our neurons already receives some baseline activation through the sensory stimulation and also through non-specific thalamic activation we will work with $s = 4$. The value for pl can be estimated to be around 1/10 (Braitenberg, 1978b), but it depends strongly on the distance between the two neurons. To be on the safe side, we assume $pl=1/20$.

With these numbers the condition $p' > p$ is satisfied for $A > 230$. Thus we can assume that purely local assemblies in sensory areas should have about 250 neurons, which are all in an elevated state of activity when the assembly is activated, but not all active at exactly the same times, the 'event coherence probability' p being about $p = 1/5$.

Next we want to consider global assemblies encompassing several areas. To bind neurons in different areas together, we need cortico-cortical connections. Between sensory areas that are cortico-cortically connected, this is easy once the sub-assemblies have been formed. In addition, it would be good to have a special representation for the co-occurence of neural events in several sensory areas. This could be done by recruiting representative neurons in other non-sensory areas of the cortex. This idea is referred to as 'chunking' in the psychological literature. In particular Wickelgren (1992) has analysed this situation in the context of cell assemblies. The problem with this idea is that there is no sensory stimulation in such an association area and, hence, we may only assume an unspecific thalamic excitation.

Let us assume that only $s = 6$ additional non-local synapses are sufficient to fire a typical neuron in an 'attentive' (pre-excited) association area. Even with this assumption it turns out to be hard to recruit sufficiently many neurons in such an area. We need an additional source of excitation. This source would be even more efficient if it would produce coincident excitation within our 4 msec time window.

We share with Robert Miller (1991, 1994) the idea that the hippocampus may be such a source of excitation (in particular to cortical association areas). Thus the hippocampus is thought of as sending out packets of coincident activation which are propagated from area to area across the cortex. We assume that about h neurons in the hippocampus being simultaneously activated by the theta-rhythm send axons to a typical cortical (association) area. They have a chance p_g to connect to one arbitrarily chosen neuron in that area. The cortico-cortical projection neurons are assumed to have roughly the same chance p_g of making a global connection, but only a fraction f of cortical cells takes part in the necessary cortico-cortical projection.

Thus the probability p' that a neuron in a non-sensory area gets at least s synaptic inputs via hippocampal and/or cortico-cortical projections from a other areas can again be calculated from a Poisson distribution, this time with mean $(pAa \cdot f + h)p_g$. In this formula we take $p_g = p_l = 1/20$ and $a \cdot f = 1/12$. We assume that the number of neurons which provide a cortico-cortical association fiber ranges between 1/5 and 4/5 and that there should be roughly as many projections converging on an area as diverging from an area. This number is 4-5 in the mouse (Greilich, 1984; Palm, 1986) and we should assume that at least 2-3 sensory subassemblies give input to our non-sensory population. Thus $a \cdot f$ may be between 1/8 and 1/16.

There should be about $p \cdot A$ neurons recruited within one time window in order to have a similar effect as a subassembly in a sensory area. Thus we require that $p' > (A/N)p$. The number N of neurons below 4 mm^2 of cortex is between $2 \cdot 10^5$ and $4 \cdot 10^5$ (Rockel et al., 1980; Beaulieu and Colonnier, 1989; Schüz and Palm, 1989). This requirement is satisfied for $h > 15$.

Thus the hippocampus can indeed provide the necessary input for recruiting chunking neurons in a cortical association area, if it sends out roughly $h = 15$ axons per 4 mm^2, i.e. 4 axons per mm^2 of non-sensory cortical areas. If all non-sensory areas would receive such a hippocampal input, then about $4 \cdot 10^5$ cells in the hippocampus would have to be activated coherently by the theta-rhythm. (Even if all cortical areas, sensory and non-sensory, were activated, this number would be about $6 \cdot 10^5$, but it seems that the input from the hippocampal formation to the cortex is mainly restricted to association areas (Miller, 1991).) These numbers are not unreasonable for the hippocampal formation (see also Miller, 1991).

So we might consider this 'injection' of coherent activity packets into the cortex as one possible function of the hippocampus. As discussed in more detail by Miller (1991, 1994) and Kuroda (1991), this idea is well compatible with many experimental and clinical observations. For example anterograde amnesia, which is typically induced by hippocampal lesions in humans, is a loss of the ability to store new information which is typically complex and multi-modal and thus might require the recruitment of 'chunking' subassemblies in cortical association areas.

A note of caution may be appropriate at this point. Since all the empirical estimates assumed in this section about neural excitation threshold, intracortical connection probabilities and the like, are quite uncertain and hard to obtain, we should analyze the stability of the resulting picture with respect to these parameters. Table 1 serves essentially this purpose. We observe that by far the most crucial parameter is the excitation threshold of a typical cortical pyramidal cell, reflected by our parameter s. Thus it would be extremely important to know more about the realistic value for s in various cortical areas.

The emerging picture of cortical cell assemblies based on the above calculations is also very interesting. Cell assemblies may vary greatly in size, depending on the number of areal sub-assemblies that they comprise. Up to now there is no easy way of estimating this number of sub-assemblies. Earlier calculations based on their information content (Palm, 1986) hint in the direction that a total assembly should have somewhere around 10^4 neurons with a working range from a few thousand to several tenthousand. A sensory subassembly should normally have at least around 250 neurons, which are not all firing synchroneously, whereas a typical non-sensory sub-assembly should have at least around 50 neurons, which fire synchroneously. It seems plausible that each assembly should have at least one sensory sub-assembly and that large assemblies should have a reasonable fraction of sensory sub-assemblies. Hence, the total number of sub-assemblies should be somewhere between 20 and 1000. This is still a small fraction of the about 40.000 local patches of 4 mm^2 size that would cover a human cortex. In large cortical areas (many mm^2) it is possible that several local subassemblies of about 2-4 mm^2 size belong to the same assembly. In sensory areas this will probably be the usual case.

Table 1. Possible parameter configurations for the parameters p, pl, s, A, s' and h. All parameters are described in the text. s is the 'threshold' value assumed for a sensory cortical area, s' the same 'threshold' value assumed for a non-sensory cortical area. The two values for h (last two columns) correspond to the assumptions $a \cdot f = 1/12$ and $a \cdot f = 1/6$, respectively,

p	pl	s	A	s'	h	
0.2	0.05	4	230	6	15	11
0.1	0.05	4	360	6	15	12
0.2	0.05	5	320	7	23	18
0.1	0.05	5	500	7	22	18
0.2	0.1	4	120	6	8	6
0.1	0.1	4	190	6	8	7
0.2	0.1	4	120	7	12	10
0.1	0.1	4	190	7	12	10

In sensory areas the average activity should be higher than in non-sensory areas. In contrast, event coherence (as opposed to rate coherence) should be more strongly expressed in non-sensory association areas. The higher activity level in sensory areas gives rise to an interesting possibility of information processing that has been discussed in the recent literature: In addition to rough coincidence of activation (time windows of 100 msec), termed rate coherence, the cortex may make use of fine coherence (time windows of a few msec as discussed in this paper), termed event coherence, to articulate some sub-structure in the local subassemblies (Neven and Aertsen, 1992; Aertsen and Arndt, 1993). This should be possible in all areas where the subassemblies are somewhat larger (at least 50 neurons per mm^2), mainly in the sensory areas. These coherently firing subgroups of sensory assemblies could be bound by coincident activation through cortico-cortical projections to different chunking assemblies in non-sensory areas. In this way the global structure of multi-areal assemblies provides a natural mechanism to 'read out' an interpretation of the sensory substructures. The global structure itself can be formed by Hebbian learning in cortico-cortical projections, provided that during learning the non-sensory neurons get enough coincident activation from some external source - presumably the hippocampus (for an alternative scenario, involving the basal ganglia, see Plenz and Aertsen, 1993).

CONCLUSIONS

We have demonstrated that the experimentally observable interactions between cortical neurons can vary dynamically on a time scale of ten milliseconds, depending on the stimulus and the behavioral context. By means of 'in virtu' recordings from simulated networks of spiking neurons we have shown that these observations do not need fast local mechanisms of synaptic modification, but can be observed in networks with fixed synaptic efficacies.

In fact, all experimental results can be explained by the influence of global network activity (i.e. of many experimentally unobserved neurons) on the functional coupling between the observed neurons embedded in such networks. The structural properties of the network that enhance the occurrence of dynamic functional coupling are the following:

1) sparse connectivity in a very large network
2) low average activity
3) structured connectivity that favours excitatory connections within a number of subgroups of neurons (assemblies)
4) a spike-generating mechanism that requires several coincident (or near-coincident) inputs to produce a spike.

These properties are well compatible with the anatomy and physiology of the cerebral cortex and with the idea of Hebbian learning of cell assemblies. Accordingly, we developed a picture of the most plausible size and appearance of typical cell assemblies that is compatible with the known sparseness of the intra-cortical and cortico-cortical connectivity. Such an assembly has to be viewed as a rather complex spatio-temporal structure that consists of many synchronously firing local subgroups, each of about 50 neurons. These subgroups activate each other in a cyclic fashion, such that only a fraction of about 1/10 to 1/5 of the assembly is active at any 'moment' (i.e. within a 2-4 msec time-window). This temporal structure allows in a natural way to 'parse' larger sub-assemblies in sensory cortical areas into subsets of different semantical binding as represented in the non-sensory cortical areas. This refined concept of cell assemblies is therefore a natural mechanism for 'solving' the 'binding problem', and is fully consistent with the requirements as specified by Neven and Aertsen (1992) and others, as well as with the known physiological mechanism and anatomical constraints.

Acknowledgements

This paper summarizes results from a number of ongoing collaborations. Contributions by and discussions with Moshe Abeles, Valentino Braitenberg, George Gerstein, Jürgen Krüger, Eilon Vaadia and their colleagues are gratefully acknowledged. Multi-neuron spike train data were kindly made available by Jürgen Krüger. This reseach was supported by the Max-Planck-Institute for Biological Cybernetics (Tübingen), the Neurobiology Program of the "Bundesministerium für Forschung und Technologie" (BMFT), The German-Israeli Foundation for Research and Development (GIF), the Volkswagen Foundation, and the Neural Network Program of the "Deutsche Forschungsgemeinschaft" (DFG).

REFERENCES

Abeles, M., 1982, "Local Cortical Circuits. An Electrophysiological Study", Springer, Berlin

Abeles, M., 1991, "Corticonics. Neural Circuits in the Cerebral Cortex", Cambridge University Press, Cambridge, UK

Abeles, M., and Gerstein, G.L., 1988, Detecting spatiotemporal firing patterns among simultaneously recorded single neurons, *J. Neurophysiol.* 60: 909-924

Abeles, M., Vaadia, E., and Bergman, H., 1990, Firing patterns of single units in the prefrontal cortex and neural network models, *Network* 1: 13-35

Abeles, M., Bergman, H., Margalit, E., and Vaadia, E., 1993a, Spatio-temporal firing patterns in the frontal cortex of behaving monkeys, *J. Neurophysiol.* 70: 1629 - 1643

Abeles, M., Prut, Y., Bergman, H., Vaadia, E., and Aertsen, A., 1993b, Integration, synchronicity and periodicity, *in:* "Brain Theory: Spatio-Temporal Aspects of Brain Function", A. Aertsen, ed., Elsevier Science Publ., Amsterdam, New York

Aertsen, A.M.H.J., and Gerstein, G.L., 1985, Evaluation of neuronal connectivity: sensitivity of cross correlation, *Brain Res.* 340: 341-354

Aertsen, A., Bonhoeffer, T., and Krüger, J., 1987, Coherent activity in neuronal populations: analysis and interpretation, *in:* "Physics of Cognitive Processes", E.R. Caianiello, ed., World Scientific Publishing, Singapore

Aertsen, A.M.H.J., Gerstein, G.L., Habib, M.K., and Palm, G., 1989, Dynamics of neuronal firing correlation: modulation of "effective connectivity", *J. Neurophysiol.* 61: 900-917

Aertsen, A.M.H.J., and Gerstein, G.L., 1991, Dynamic aspects of neuronal cooperativity: fast stimulus-locked modulations of 'effective connectivity', *in:* "Neuronal Cooperativity", J. Krüger, ed., Springer, Berlin

Aertsen, A., and Preissl, H., 1991, Dynamics of activity and connectivity in physiological neuronal networks, *in:* "Nonlinear Dynamics and Neuronal Networks", H. Schuster, ed., VCH Verlag, Weinheim

Aertsen, A., Vaadia, E., Abeles, M., Ahissar, E., Bergman, H., Karmon, B., Lavner, Y., Margalit, E., Nelken, I., and Rotter, S., 1991, Neural interactions in the frontal cortex of a behaving monkey: Signs of dependence on stimulus context and behavioral state, *J. f. Hirnforschung* 32: 735-743

Aertsen, A., and Arndt, M., 1993, Response synchronization in the visual cortex. *Curr. Opinion Neurobiol.* 3: 586-594

Aertsen, A., Erb, M., and Palm, G., (1994) Dynamics of functional coupling in the cerebral cortex: An attempt at a model-based interpretation. *Physica D* (in press)

Amir, Y., Harel, M., and Malach, R., 1993, Cortical hierarchy reflected in the organization of intrinsic connections in macaque monkey visual cortex, *J. Comp. Neurol.* 334: 19-46

Beaulieu, C., and Colonnier, M., (1989) Number of neurons in individual laminae of areas 3B, 4, and 6a of the cat cerebral cortex: a comparison with major visual areas, *J. Comp. Neurol.* 279:228-234

Bedenbaugh, P.H., Gerstein, G.L., Boven, K.-H., and Aertsen, A.M.H.J., 1988, The meaning of stimulus dependent changes in cross correlation between neuronal spike trains, *Soc. Neurosci. Abstr.* 14: 651

Bedenbaugh, P.H., Gerstein, G.L., and Aertsen, A.M.H.J., 1990, Dynamic convergence in neural assemblies, *Soc. Neurosci. Abstr.* 16: 1224

Boven, K.-H., Aertsen, A., 1989, Dynamics of activity in neuronal networks give rise to fast modulations of functional connectivity, *in:* "Parallel Processing in Neural Systems and Computers", R. Eckmiller et al., eds., Elsevier Science Publishers

Braitenberg, V., 1978 a, Cortical architectonics: general and areal, *in:* "Architectonics of the Cerebral Cortex", M.A.B. Brazier, H. Petsche, eds., Raven Press, New York, pp 443-465

Braitenberg, V., 1978b, Cell assemblies in the cerebral cortex *in:* "Theoretical Approaches to Complex Systems", P. Heim, G. Palm, eds., Springer, Berlin Heidelberg New York, pp 443-465

Braitenberg, V., Schüz, A., 1991, Anatomy of the cortex, Statistics and geometry. Springer, Berlin Heidelberg

Bullier, J., Kennedy, H., and Salinger, W., 1984, Branching and laminar origin of projections between visual cortical areas of the cat, *J. Comp. Neurol.* 228:329-341

Creutzfeldt, O.D., Garey, L.J., Kuroda, R., and Wolff, J.-R., (1977) The distribution of degenerating axons after small lesions in the intact and isolated visual cortex of the cat, *Exp. Brain Res.* 27:419-440

Dayhoff, J.E., and Gerstein, G.L., 1983, Favored patterns in spike trains. I. Detection, II. Application, *J. Neurophysiol.* 49: 1334-1348, 1349-1363

DeFelipe, J., Conley, M., Jones, E.G., (1986), Long-range focal collateralization of axons arising from corticocortical cells in monkey sensory-motor cortex, *J. Neurosci.* 6 (12): 3749-3766

Deppisch, J., Bauer, H.U., Schicken, T., König, P., Pawelzik, K., and Geisel, T., 1987, Alternating oscillatory and stochastic states in a network of spiking neurons, *Network* 4: 243-257

Erb, M., 1985, Computersimulation eines assoziativen Speichers als Modell für die Funktion der Großhirnrinde (in German), MSc Thesis, Eberhard-Karls-Universität Tübingen, FRG

Erb, M., 1992, Simulation neuronaler Netze: Stabilität, Plastizität und Konnektivität (in German) PhD Thesis, Eberhard-Karls-Universität Tübingen, FRG

Erb, M., Aertsen, A., 1992, Dynamics of activity in biology-oriented neural network models: stability at low firing rates, *in:* "Information Processing in the Cortex: Experiments and Theory", A. Aertsen, V. Braitenberg, eds., Springer, Berlin

Gerstein, G.L., and Michalski, A., 1981, Firing synchrony in a neural group: putative sensory code, *in:* "Neural Communication and Control", G. Szekely, E. Labos, S. Damjanovich, S., eds., Adv Physiol Sci Vol 30, Pergamon Press, Akademiai Kiado, Budapest

Gerstein, G., and Aertsen, A., 1985, Representation of cooperative firing activity among simultaneously recorded neurons, *J. Neurophysiol.* 54: 1513-1527

Gerstein, G., Perkel, D., and Dayhoff, J., 1985, Cooperative firing activity in simultaneously recorded populations of neurons: detection and measurement, *J. Neurosci.* 5: 881-889

Gerstein, G.L., Bedenbaugh, P., and Aertsen, A.M.H.J., 1989, Neuronal Assemblies. *IEEE Trans Biomed Engineering* 36: 4-14

Gilbert, C.D., Wiesel, T.N., (1983), Clustered intrinsic connections in cat visual cortex, *J. Neurosci.* 3:1116-1133

Gilbert, C.D., and Wiesel, T.N., 1989, Columnar specificity of intrinsic horizontal and coritocortical connections in cat visual cortex. *J. Neurosci.* 9 (7): 2432-2442

Glünder, H., Nischwitz, A., 1993, On spike synchronization, *in:* "Brain Theory",. A. Aertsen, ed., Elsevier

Greilich, H., (1984), "Quantitative Analyse der cortico-corticalen Fernverbidungen bei der Maus", Dissertation at the Faculty of Biology at the University of Tübingen

Hebb, D., 1949, "The Organization of Behavior. A Neuropsychological Theory", Wiley, New York

James, W., 1890, Psychology (Briefer Course), *in:* "Neurocomputing", J.A. Andersen, E. Rosenfeld, eds., 1989, MIT Press, Cambridge, MA

Johannesma, P., Aertsen, A., van den Boogaard, H., Eggermont, J., and Epping, W., 1986, From synchrony to harmony: Ideas on the function of neural assemblies and on the interpretation of neural synchrony, *in:* "Brain Theory", G. Palm, A. Aertsen, eds., Springer, Berlin

Krüger, J., 1982, A 12-fold microelectrode for recording from vertically aligned cortical neurones, *J. Neurosc. Meth.* 6: 347-350

Kuroda, Y., 1991, Rapid consolidation of associations between multimodal memories by intrahippocampal connections through corticohippocampal closed circuits, *Concepts Neurosci.* 2: 221-228

Malach, R., Amir, Y., Harel, M., and Grinvald, A., 1993, Relationship between intrinsic connections and functional architecture revealed by optical imaging and in vitro targeted biocytin injections in primate striate cortex, Proc. Natl. Acad. Sci. USA, Vol. 90:10469-10473

Martin, K.A.C., (1984), Neuronal circuits in cat striate cortex, *in:* "Cerebral Cortex, Vol. 2, Functional Properties of Cortical Cells", E.G. Jones, A. Peters, eds., Plenum Press, New York, London

Miller, R., 1991, "Cortico-Hippocampal Interplay", Springer, Berlin

Miller, R., 1994, An interpretation, based on cell assembly theory, of the psychological impairments following lesions of the hippocampus and related structures, *in:* "The Memory System of the Real Brain", J. Delacour, ed., World Scientific, Singapore

Mirollo, R.E., and Strogatz, S.H., 1990, Synchronization of pulse-coupled biological oscillators, *SIAM J. Appl. Math.* 50 No. 6: 1645-1662

Moore, G.P., Perkel, D.H., and Segundo, J.P., 1966, Statistical analysis and functional interpretation of neuronal spike data, *Ann. Rev. Physiol.* 28: 493-522

Nelken, I., ,1992, A probabilistic approach to the analysis of propagation delays in large cortical axonal trees, *in:* "Information Processing in the Cortex: Experiments and Theory", A. Aertsen, V. Braitenberg, eds., Springer, Berlin

Neven, H., and Aertsen, A., 1992, Rate coherence and event coherence in the visual cortex: a neuronal model of object recognition. *Biol. Cybern.* 67: 309-322

Palm, G., 1980, On associative memory, *Biol. Cybern.* 36: 19-31

Palm, G., 1982, "Neural Assemblies. An Alternative Approach to Artificial Intelligence", Springer, Berlin

Palm, G., 1986, Associative networks and cell assemblies, *in:* "Brain Theory", G. Palm, A. Aertsen, eds., Springer, Berlin

Palm, G., 1987, On associative memories, *in:* "Physics of Cognitive Processes", E.R. Caianiello ed., World Scientific Publishing, Singapore

Palm, G., 1991, Memory capacities of local rules for synaptic modification, *Concepts Neurosci.* 2: 97-128

Palm, G., Aertsen, A.M.H.J., and Gerstein, G.L., 1988, On the significance of correlations among neuronal spike trains, *Biol. Cybern.* 59: 1-11

Pandya, D.N., Yeterian, E.H., 1985, Architecture and connections of cortical association areas, *in:* "Cerebral Cortex, Vol. 4, Association and Auditory Cortices", A. Peters, E.G. Jones, eds., Plenum Press, New York, London

Perkel, D.H., Gerstein, G.L., and Moore, G.P., 1967, Neuronal spike trains and stochastic point processes. II. Simultaneous spike trains, *Biophys. J.* 7: 419-440

Perkel, D.H., and Bullock, T.H., 1968, Neural coding, *Neurosci. Res. Progr. Bull.* 6

Plenz, D., and Aertsen, A., 1993, The basal ganglia: minimal coherence detection on cortical activity distributions, *in:* "The Basal Ganglia IV. New Ideas and Data on Structure and Function", G. Percheron, J.S. McKenzie, J. Féger, eds., Plenum Press, New York (in press)

Rockel, A.J., Hiorns, R.W., and Powell, T.P.S., (1980), The basic uniformity in structure of the neocortex, *Brain* 103:221-244

Schüz, A., and Palm, G., 1989, Densities of neurons and synapses in the cerebral cortex of the mouse, *J. Comp. Neurol.* 286: 442-455

Schüz, A., and Braitenberg, V., 1994, Constraints to a random plan of cortical connectivity, in: "Structural and Functional Organization of the Neocortex. A Symposium in the Memory of Otto D. Creutzfeldt", A. Albowitz, K., Albus, U. Kuhnt, H.C. Nothdurft, P. Wahle, eds., Springer, Berlin, Heidelberg (in press)

Sejnowski, T.J., 1981, Skeleton filters in the brain, *in:* "Parallel Models of Associative Memory", G.E. Hinton, J.A. Anderson, eds., Lawrence Erlbaum Assoc Publ, Hillsdale

Sherrington, C., 1941, "Man on his Nature. The Gifford Lectures, Edinburgh 1937-38", Cambridge University Press, Cambridge, UK

Vaadia, E., Ahissar, E., Bergman, H., and Lavner, Y., 1991, Correlated activity of neurons: a neural code for higher brain functions? *in:* "Neuronal Cooperativity", J. Krüger, ed., Springer, Berlin

Vaadia, E., and Aertsen, A., 1992, Coding and computation in the cortex: single-neuron activity and cooperative phenomena, *in:* "Information Processing in the Cortex: Experiments and Theory", A. Aertsen, V. Braitenberg, eds., Springer, Berlin

van Essen, D.C., Newsome, W.T., Maunsell, J.H.R., and Bixby, J.L., 1986, The projections from striate cortex (V1) to areas V2 and V3 in the macaque monkey: asymmetries, areal boundaries, and patchy connections, *J.Comp. Neurol.* 244:451-480

von der Malsburg, C., 1981, The correlation theory of brain function, *Internal report 81-2,* Max-Planck-Institute for Biophysical Chemistry, Göttingen (FRG)

von der Malsburg, C., 1986, Am I thinking assemblies? in: "Brain Theory", G. Palm, A. Aertsen, eds., Springer, Berlin

Wennekers, Th., and Palm, G, 1994, in preparation

Wickelgren, W.A., 1992, Webs, cell assemblies, and chunking in neural nets, *Concepts Neurosci.* 3: 1-53

TEMPORAL ASPECTS OF INFORMATION PROCESSING IN AREAS V1 AND V2 OF THE MACAQUE MONKEY

L.G. Nowak, M.H.J. Munk, N. Chounlamountri, J. Bullier

Cerveau et Vision
INSERM Unité 371
18 avenue du Doyen Lépine
69500 Bron/Lyon France

INTRODUCTION

In mammals, the neocortex devoted to each sensory modality is subdivided into a number of functional areas. Each of these areas contains a more or less complete representation of the sensory surface and is connected to numerous other cortical areas of the same modality. As a consequence, a sensory stimulus activates neurons in several of these interconnected cortical areas. One of the present challenges of sensory physiology is to discover how information concerning a given object in the external world is integrated between its representations in the different cortical areas. In particular, it is important to associate a common tag to the neurons coding for the same stimulus and which are distributed in different cortical areas. Early theoretical studies (Milner, 1974; von der Malsburg and Schneider, 1986; Abeles, 1982) suggested that neurons activated by the same stimulus could be bound together by synchronization of the time of occurence of their action potentials.

A number of recent studies in cat visual cortex have demonstrated the presence of synchronization of firing between neurons of different areas associated with the same stimulus. Thus, Engel and collaborators showed that neurons in the suprasylvian sulcus (area PMLS) and area 17 oscillate at the same frequency with zero phase lag when activated by the same stimulus (Engel et al., 1991a). Synchronous oscillations in the gamma-band range (around 40 Hz) have also been reported between neurons of opposite cortical hemispheres (Engel et a., 1991b). A similar synchronous oscillation in the local field potentials of areas 17 and 18 has been reported by Eckhorn and his collaborators (Eckhorn et al. 1988). Recording from single units in areas 17 and 18, we found that some neurons synchronize their firing if their respective receptive field properties fulfil certain conditions (Nelson et al 1992). Thus, neu-

Oscillatory Event-Related Brain Dynamics, Edited
by C. Pantev, Plenum Press, New York, 1994

rons with overlapping receptive fields and similar orientation selectivities tend to show synchronized firing with a temporal jitter of a few milliseconds. This corresponds to sharp peaks centered on the origin of time in the cross-correlograms computed from the activity of the two neurons studied. Neurons with receptive fields distant by less than 10 degrees can display a different type of synchronization characterized by a jitter of a few tens of milliseconds. Finally, pairs of neurons often show a very loose type of synchronization with a jitter of several hundred milliseconds, regardless of their receptive field separations and properties. Similar results have been found when cross-correlograms were computed between areas 17 of opposite hemispheres (Bullier et al 1993a).

All the studies reviewed above concern the synchronization of firing between neurons located in different cortical areas of the cat visual cortex. It is known that the cat visual system is organized in a way similar to many other mammals but different from primates (Bullier et al 1993b). In the cat, the lateral geniculate nucleus (LGN), which relays the visual information from the retina to the cortex, contains a large number of relay cells projecting to practically all the visual cortical areas known in this species (Bullier 1986; Dreher 1986). Furthermore, lesion or inactivation of area 17, the main target of the LGN, leads to little change in the responses of neurons in other cortical areas (Dreher and Cottee, 1975; Sherk 1978; Spear and Baumann 1979; Casanova et al 1992). Because of such widely divergent cortical projections of the LGN and the independence of the responses of neurons in different cortical areas, it is usually concluded that the cat visual system is organized in parallel. Neurons in different areas of the cat visual cortex can be activated at the same time by the diverging thalamic afferents which could therefore play an important role in the synchronization of neurons in different areas.

The organization of the primate visual system is different : the relay cells of the LGN project almost exclusively to area 17 (Yukie and Iwaï 1981; Benevento and Yoshida 1981; Bullier and Kennedy 1983). When area 17 is inactivated or lesioned, other cortical areas are silenced or their activity strongly reduced (Bullier et al 1993b). The familiar expression of a bottleneck associated to the role of area 17 in the transfer of visual information is a good illustration of this organization. Because of this, the primate visual system is thought to be organized serially or hierarchically, with a series of processing stages corresponding to the successive cortical areas. As a consequence, one would expect neurons in areas located beyond area 17 to be activated later than those in area 17. It appears therefore unlikely that neurons of area 17 and extrastriate cortex in the monkey could be synchronized. This is an important issue since it questions the generality of the phenomenon of synchronization of neurons belonging to different cortical areas and its possible role in perception.

We have tested whether the different organizations of primate and non-primate visual systems lead to differences in the synchronization by recording in area 17 (also called V1) and in the adjacent area V2 (Felleman and Van Essen 1991). These two areas clearly belong to different stages of processing of visual information since inactivating V1 leads to a complete silence of neurons in V2 (Schiller and Malpeli 1977; Girard and Bullier 1989) and the small LGN input to V2 comes principally from regions of the LGN which do not receive retinal input (Bullier and Kennedy 1983).

We have investigated the temporal relationship between the activities in areas V1 and V2 by two methods 1) Measuring the latencies of the responses of cortical neurons in areas V1 and V2 to flashes of light to determine whether the responses of neurons in V2 lag behind those of neurons in V1. 2) Calculating cross-correlations between the activities of neurons in areas V1 and V2 to see whether they reveal synchrony of firing or patterns characteristic of successive processing stages, as expected from the anatomical organization.

RESPONSE LATENCIES TO FLASH STIMULI

Figure 1 presents examples of responses to flash in neurons of V1 and V2 in macaque monkeys paralysed and anaesthetised with nitrous oxide supplemented with fentanyl. The left part of the post-stimulus time histograms contains the activity of the cell before the flash which is turned ON at time 0. The stimulus consisted of a spot of light 1 to 2 degree wide centered on the receptive fields of the neurons in V1 and V2 which were simultaneously recorded. Responses to light ON and light OFF were recorded.

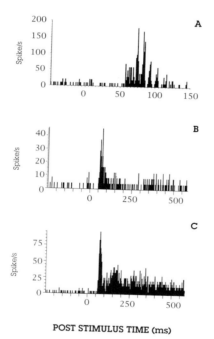

POST STIMULUS TIME (ms)

Figure 1. Examples of responses of neurons in V1 (A and B) and V2 (C) to a small spot of light centered in the receptive field and flashed ON at time 0. Note the high-frequency (about 100 Hz) oscillatory response in A, the transient response in B and the transient component followed by a sustained response in C. The time scales are different in A and B-C.

The average rate of spontaneous activity and its variability was estimated from the firing of the cell before time 0 (assuming a Poisson distribution) and a threshold value for estimating the beginning of the response (p<0.01 for 3 successive bins) was determined following the method of Maunsell and Gibson (1992). The latency of the cell was taken as the time at which the level of activity of the cell crossed that threshold.

In figure 1A is presented an example of an oscillatory response to light ON of a V1 neuron. Such high-frequency oscillatory responses were recorded almost exclusively in layers 4Calpha, 4C beta and 4B of area V1, confirming the findings of Maunsell and Gibson (1992).

Figures 1B and 1C illustrate examples of ON responses for a neuron in V1 (Fig. 1B) and another in V2 (Fig. 1C; note the different time scales in Figs. 1A and 1B-C). In most cases, the responses of V1 and V2 neurons consisted of an early transient response, which facilitated the estimation of the latency, followed by a sustained component.

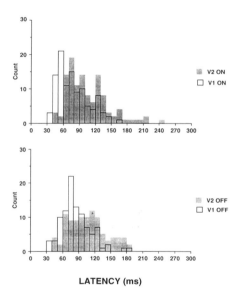

LATENCY (ms)

Figure 2 : Frequency distributions of latencies in areas V1 and V2 to onset (ON, upper histograms) or offset (OFF, lower histograms) of a small spot of light. Bin width is 10 ms wide.

We have recorded from 106 sites in V1 and 115 sites in V2. The frequency distribution histograms of the latencies of neurons in V1 and V2 to onset and offset of the flash are presented in Figure 2. It is clear that the latencies of V2 neurons lag by about 10-20 ms behind those of neurons in V1, as expected given that information has to flow through V1 before reaching V2 (see also Raiguel et al 1989). The histograms in Fig. 2 also indicate that there is a substantial range of latencies (60 to 150 ms) for which the latencies in the two areas overlap substantially. During that extended period, therefore, processing is done simultaneously in the two areas. Therefore, one cannot conclude that the two areas work in strictly serial fashion, with V1 processing information before transferring it to V2.

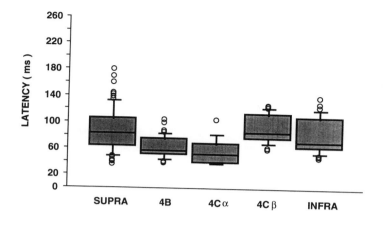

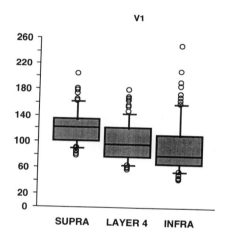

Figure 3 : Summary of latency distributions in different layers of areas V1 and V2. Bars indicate different percentile levels. The lowest small bar represents the 10% percentile. The box corresponds to the population enclosed between the 25 and 75% percentiles, the 50% percentile (median) is indicated by the elongated bar contained in the box. The highest small bar corresponds to the 90% percentile. Data points not enclosed in the percentiles ranges are shown as open dots. Infra : infragranular layers 5 and 6. Supra : supragranular layers 2 and 3. Most of the shortest latencies are found in layers 4C alpha and 4B. Latencies do not differ significantly in layer 4C beta, infragranular and supragranular layers of V1. The shortest latencies in area V2 are observed in infragranular layers. Latencies in infragranular layers of V2 do not differ significantly from those found in supra and infragranular layers in V1.

The anatomical organization of the information transfer through V1 and V2 is relatively well known (Lund 1988). Axons from the LGN terminate mostly in layers 4C alpha and 4C beta of V1 (Hubel and Wiesel 1972; Lund and Boothe 1975; Bullier and Henry 1980); cells in these layers then project to layer 4B and the supragranular layers 2 and 3 (Fitzpatrick et al 1985) from which axons project to the infragranular layers 5 and 6 (Blasdel et al 1985). The information is transferred from V1 to V2 mostly by axons originating in neurons of the supragranular layers of V1 (Kennedy and Bullier 1985) and terminating mainly in layer 4 and the lower part of layer 3 of area V2 (Rockland and Pandya 1979; Lund et al. 1981). Neurons in layers 4 and lower 3, in turn, project to supragranular layers of V2 (Lund et al 1981).

These morphological results suggest that there are successive stages of information transfer through areas V1 and V2. The first stage would correspond to layers 4C alpha and beta of area V1, the next one to layer 4B and layers 2-3, followed by layers 5-6 of V1. Information is sent from supragranular layers of V1 to layer 4 of V2 which constitutes a later stage followed by the supragranular layers of V2. Latencies of action potentials or slow potentials evoked by electrical stimulation of the afferent pathways confirm the presence of these successive stages (Bullier and Henry 1980; Mitzdorf and Singer 1979). One therefore expects that, following stimulation by light flashes, progressively longer latencies would also be observed at each of these successive stages.

The results presented in figure 3 show the different percentiles of the latency distributions in different cortical laminae of areas V1 and V2 . In area V1, two groups of layers can be distinguished : layers 4B and 4C alpha with short latencies, and layers 4C beta and the infragranular and supragranular layers with longer latencies. The latency differences between these two groups are highly significant.

The earliest latencies in V1 are observed in layers 4Calpha and 4B, which relay the information from the magnocellular layers of the LGN (Hubel and Wiesel 1972; Lund 1973; Bullier and Henry 1980). Such an early activation is consistent with the results of Marocco (1976), who showed that the LGN neurons that respond transiently to flashes and are concentrated in the magnocellular layers (Dreher et al 1976), have shorter latencies than neurons with sustained responses which tend to be concentrated in the parvocellular layers. The similarity in latency difference between transient and sustained cells in the LGN and between neurons in layers 4Calpha and 4Cbeta in cortex suggests that the early responses in layers 4Calpha and 4B reflect their role in relaying information from the magnocellular LGN layers. This interpretation is also consistent with the findings of Maunsell and Gibson (1992) who showed that the earliest latencies in V1 disappear after lesions of the magnocellular layers of the LGN. Thus, the major difference in latency between layers 4C alpha/4B and 4C beta most likely reflects latency differences in LGN neurons that probably originate in the retina. Evidence for successive stages of processing within area V1 is lacking in figure 3. There is no significant difference between the latencies in layers 4C beta and in the supragranular and infragranular layers, despite the presence of a strong thalamic input in layer 4Cbeta. Also, in supragranular and infragranular layers one can see latencies as short as in layer 4C alpha (Fig. 3). This suggests that, at least for the magnocellular pathway, the cortical integration time is very short (1.5 ms in Bullier and Henry 1980).

The results in V1 therefore do not support a step-by-step model of information processing as expected from the anatomy and the latencies of evoked potential to electrical stimulation. Instead, it appears that there are two parallel pathways, a fast and a slow, and that, within each pathway, spikes are emitted almost simultaneously in the different laminae.

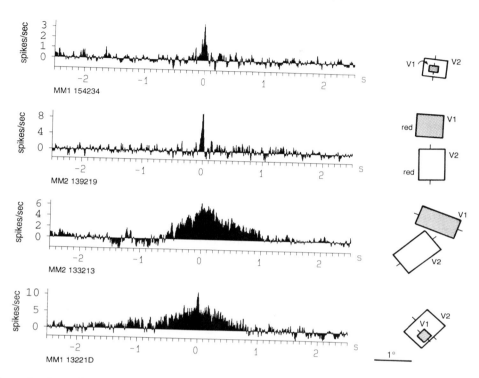

Figure 4 : Examples of cross-correlation histograms (CCHs) computed from the activity of neurons in areas V1 and V2. The CCHs are difference CCHs after subtraction of the shift predictor (see text). All CCHs use V1 neurons as reference neurons. The upper two histograms are examples of narrow CCH peaks, the third CCHs is an example of a broad peak while the lowest CCH corresponds to a combination of narrow and broad peaks. On the right are represented the receptive fields of neurons in V1 and V2. The small bars on each side of the receptive field indicate the optimal orientation of the neuron.

The latencies in layer 4 of area V2 are longer than those in the supragranular layers of V1, as expected from the interconnections between these two neuronal groups. What is more surprising is that the latencies of infragranular layers of V2 are not significantly different from those in supra- or infragranular layers of V1, which constitute their source of activation. Latencies in V2 infragranular layers are also shorter than those of layer 4 neurons, which are supposed to be the main targets of the V1 afferents. It is known that neurons of the infragranular layers of area V2 have apical dendrites that bear spines on their portion crossing layer 4 (Lund et al 1981). Input from V1 terminating in layer 4 of V2 could therefore activate directly neurons in V2 infragranular layers. The early latencies of these neurons could be due to a selective activation by afferents from layer 4B of V1 which contains neurons with short latencies and which project to V2 (Kennedy and Bullier 1985). The longer latencies observed in layer 4 of V2 could be explained if neurons in that layer were selectively activated by neurons in the supragranular layers of V1.

The latency similarity of neurons in the infragranular layers in areas V1 and V2 is interesting since neurons in layer 5 of both areas have common subcortical targets such as the pulvinar or the superior colliculus (Trojanowski and Jacobson 1977; Lund et al 1981; Fries 1984). It may be important, for a correct integration of information coming from both areas, that the neurons converging on a common site be activated at the same time. We observed short latencies in many layer 6 neurons of area V2 (not shown here). This layer sends a heavy feedback projection to area V1 (Kennedy and Bullier 1985). This suggest that neurons in this layer may be particularly important for the early control of responses of neurons in area V1.

The long latencies observed in the supragranular layers of area V2 (Fig. 3) are difficult to interpret at the moment because they have been measured in a few penetrations. It is known that the supragranular layers of V2 contain different subsystems revealed by cytochrome oxidase staining and relaying in parallel information from the magno and parvo-cellular streams (Hubel and Livingstone 1987). Recently, we combined cytochrome-oxidase histochemistry and latency measurements in V2 supragranular layers. We found neurons with short latencies in thick cytochrome oxidase bands which project to area MT and relay the magnocellular layer activity coming from layer 4B of V1 (Livingstone and Hubel 1987). This suggests that the results presented in figure 3 were obtained in penetrations that undersampled the thick cytochrome oxidase bands.

In conclusion, latency measurements in areas V1 and V2 do not simply reflect the sequential order of processing expected from the anatomical results. It appears that, in area V2 as well as in V1, much of the processing occurs in parallel in a fast magno-dominated pathway and in a slower parvo- pathway. The latency study also reveals that many neurons in areas V1 and V2, particularly in the infragranular layers, are activated simultaneously and are therefore likely to exert reciprocal interactions.

CROSS-CORRELATION BETWEEN NEURONS IN AREAS V1 AND V2

280 pairs of single and multi units have been recorded in areas V1 and V2 of cynomolgus monkeys paralysed and anaesthetised with nitrous oxide supplemented with Fentanyl, Halothane or Isofluorene. Cross-correlograms were computed during visual stimulation with moving bars or small spots flashed ON and OFF. The amount of correlation due to common activation by the visual stimulus was estimated by computing a cross correlogram after shifting one of the spike trains by one stimulus period. The resulting cross correlogram, called the shift predictor, was subtracted from the original cross-correlogram to give the difference cross-correlogram (Perkel et al 1967). Figure 4 illustrates representative examples of difference cross-correlogram histograms (CCHs) computed between single neurons in V1 and V2 together with the corresponding receptive fields. The upper histogram shows a sharp peak in the CCH computed from neurons with overlapping receptive fields and same optimal orientation. The second CCH has been calculated from the responses of two units with non-overlapping receptive fields but same color selectivity. The third ex-

ample is a broad peak in the crosscorrelogram calculated from the responses of two neurons with non-overlapping receptive fields and different optimal orientations. The lowest histogram corresponds to two neurons with overlapping receptive fields and same optimal orientation. The CCH contains a combination of a broad and a narrow peak. In addition, in about 10% of the cross-correlograms, we observed oscillations with frequencies in the alpha range.

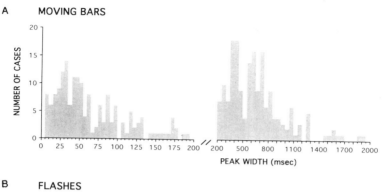

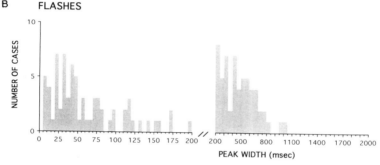

Figure 5 : Frequency distribution histograms of the width at half-height of the CCH peaks computed with moving bars (A) or flashing bars (B) as stimuli. Note that in both cases, the distribution of peak width is bipartite with narrow (less than 200 ms) and broad CCHs (wider than 200 ms).

The presence of narrow and broad peaks in these examples suggest that between monkey areas V1 and V2, as in the cat visual system, different types of cross-correlograms can be distinguished on the basis of the width of the peaks. To test this we measured the width at half height of the cross-correlogram peaks recorded between V1 and V2 neurons. The resulting distributions of peak width are presented in figure 5. The upper part of the figure corresponds to CCHs computed while the neurons were stimulated with moving bars while the lower histograms correspond to cases with flash stimulation. It is clear that the two distributions are bipartite. The class of broad correlograms with widths larger than 200 ms appears similar to the class of the broadest peaks recorded in the cat (Nelson et al., 1992). There is a second group of CCH peaks with widths smaller than 200 ms. The presence of narrow and broad types of peaks in the CCHs recorded between neurons in V1 and V2 is interesting since similar observations have been made in every crosscorrelation study between cortical neurons, within a cortical area and between two different areas, in anaesthetised and in behaving animals, in visual and frontal cortex (Nelson et al 1992; Krüger and Aiple 1988; Gochin et al 1991; Vaadia et al., 1991). The temporal interaction between neurons in areas V1 and V2 therefore obeys general principles of information processing by the cortex.

In order to determine how information is transferred between neurons of areas V1 and V2, one can examine the position of the peak of the CCHs. A cross-correlogram computed between two neurons, A and B, records the firing rate of neuron B as a function of the time before or after a spike is emitted in neuron A. (Abeles, 1982). When the CCH is flat, it makes no difference to the activity of neuron B whether a spike has been emitted or not in neuron A. These two neurons are said to fire independently. In certain cases, the CCH computed between the activities of neurons A and B contains a sharp peak displaced with respect to the origin of time (e.g. Levick et al. 1972; Mastronarde, 1987; Tanaka, 1983). This is interpreted as meaning that neuron A activates neuron B. Finally, in many CCHs recorded between neurons of different areas in the cat (Nelson et al 1992; Engel et al 1991a,b; Bullier et al 1993a), the peak is centered on or close to the origin of the abscissa. This means that the maximum increase in firing rate of neuron B is synchronized with the firing of a spike in neuron A. This is interpreted as synchronization of activity between neurons A and B.

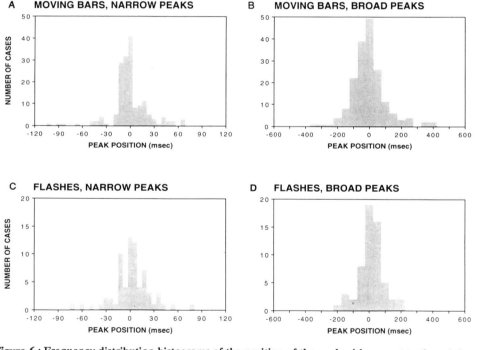

Figure 6 : Frequency distribution histograms of the position of the peak with respect to the origin of the abscissa. Negative values of peak position correspond to activity in V2 neurons leading that in V1 neurons. Positive values correspond to the opposite. Value of 0 correspond to synchrony. Bin width for narrow peaks (A and C) : 5 ms. Bin width for broad peaks (B and C): 50 ms.

In order to test for the presence of serial connections between areas V1 and V2, we have searched for the presence of direct drive in the CCHs computed between areas V1 and V2. This was done by measuring the position of the CCH peak with respect to the origin of the abscissa for narrow and broad types of coupling with moving bars or flashes as stimuli. The resulting frequency distributions are presented in figure 6. A negative value of the peak position in figure 6 corresponds to a situation in which the coupled spikes in a V2 neuron lead the corresponding spikes in the V1 neuron whereas positive values correspond to spikes in V2 following those of V1. In all cases, the most often observed latency is zero meaning that units in V1 and V2 are most likely to fire in synchrony. The variability of the CCH peak position is high (30 ms for narrow, 200 ms for broad peaks), but smaller than the widths of the peaks (Fig. 5). As a consequence, when the peak position is displaced, in most cases the origin of time is located very close to the CCH peak. Thus practically all CCHs contain the origin of time in the upper portion of the peak. In only a few occasions, we observed a peak which was clearly displaced from the origin of time which could be a clear sign of serial activation, as observed in the retino-geniculate or geniculo-cortical connections (Levick et al 1972; Tanaka 1983; Mastronarde 1987). Signs of serial activation in both directions (V1 driving V2 or the opposite) were observed.

When moving bars are used for stimulation, there are more negative than positive peak positions (Fig. 6A, B). This means that more units in V2 give spikes which precede, rather than follow, the coupled spikes in the corresponding V1 neuron. This result is surprising, given the serial organization of V1 and V2 and the longer latencies of V2 neurons (Fig. 2). This can probably be explained by the fact that the receptive fields of neurons in V2 are larger than those in V1. Because of this, V2 neurons have a tendency to be activated before the V1 neurons by a moving bar. The coupled spikes, which are recorded by the CCH, constitute only a small proportion (10-20%) of the total number of spikes given by the two neurons in response to the moving bar. Despite this, it is likely that the early activation of the V2 neurons influences the functional interaction between the V1 and V2 neurons revealed by the CCHs. This tendency for V2 spikes to lead the corresponding coupled activity in V1 disappears with the use of flashes (Fig. 6C, D). In this condition, there are approximately the same number of V1 neurons that give spikes which precede those of V2 neurons as the opposite. In other words, when stimulated by flashes, areas V1 and V2 are synchronized.

CONCLUSION

Our results do not support the step-by-step model of information processing suggested by the anatomy. This would lead to successively longer latencies for different anatomical stages. It is clear that such is not the case within each area. Furthermore, the observation that latencies are similar in infragranular layers of V1 and V2 suggests that much processing in these two areas occurs simultaneously.

Measurement of the latencies of neurons after light flashes and cross-correlation are different methods for testing the rank of different neurons. In the case of flashing stimuli, the latency is given by the first spikes elicited by the stimulus. Thus, differences in latencies correspond to differences in timing of the first spike. Cross-correlograms on the other hand, are based on a large number of spikes elicited during the response. They also reveal the statistical dependency of spikes emitted by one neuron on the spikes of another neuron. Despite these differences, the results of cross-correlation experiments also lead to the same conclusion. A step-by-step model of organization between areas V1 and V2 would lead to peaks displaced toward positive values in the cross-correlograms computed between V1 and V2 neurons. This was not the case and the symmetric distribution of the CCH peak position suggests that area V2 neurons are as likely to activate V1 neurons as the contrary. This indicates that the two areas work interactively and that they should be considered more as a pair of interacting units processing information in common than as successive stages processing information in a serial fashion. Such a model of a pair of interacting units is in keeping with the fact that areas V1 and V2 receive most of their afferents from common subcortical and cortical sources (Kennedy and Bullier 1985), with the notable exception of relay cells in the LGN which project only to V1.

Another important result of our studies is that, despite the different organisation of the visual pathways in the cat and monkey, neurons in different cortical areas of these two species tend to synchronize their firing. Synchronization of firing appears therefore to be a general mechanism of information processing between and within cortical areas. This gives support to the idea that synchronisation is important for cortical processing and perception (Singer 1993).

REFERENCES

Abeles, M., 1982, Quantification, smoothing, and confidence limits for single-units histograms, *J. Neurosci. Methods* 5: 317-325.

Abeles, M. 1982 "Local Cortical Circuits", Springer-Verlag, Berlin, Heidelberg, NewYork.

Benevento, L. A. and Yoshida, K., 1981, The afferent and efferent organization of the lateral geniculo-prestriate pathways in the macaque monkey, *J. Comp. Neurol.* 203: 455-474.

Blasdel, G.G., Lund, J.S. and Fitzpatrick, D., 1985, Intrinsic connections of macaque striate cortex: Axonal projections of cells outside lamina 4C. *J. Neurosci.* 5: 3350-3369.

Bullier, J. and Henry, G.H., 1980, Ordinal position and afferent input of neurons in monkey striate cortex, *J. Comp. Neurol.* 193 : 913-935.

Bullier, J. and Kennedy, H., 1983, Projection of the lateral geniculate nucleus onto cortical area V2 in the macaque monkey, *Exp. Brain Res.* 53: 168-172.

Bullier, J., 1986, Axonal bifurcation in the afferents to cortical areas of the visual system. In : "Visual Neuroscience", edited by J.D. Pettigrew, K.J. Sanderson and W.R. Levick. Cambridge Univ. Press, p. 239-259.

Bullier, J., Munk, M.H.J. and Nowak, L.G., 1993a, Corticocortical connections sustain interarea synchronization. *Concepts in Neurosciences* 4 : 101-117.

Bullier, J., Girard, P. and Salin, P.A., 1993b, The role of area 17 in the transfer of information to extrastriate visual cortex. In : "Primary Visual Cortex in Primates " "Cerebral Cortex vol. 10", edited by A. Peters and K. S. Rockland, Plenum Pub. Corp., chap. 7, p. 301-330.

Casanova, C., Michaud, Y., Morin, C., McKinley, P.A. and Molotchnikoff, S., 1992, Visual responsiveness and direction selectivity of cells in area 18 during local reversible inactivation of area 17 in cats, *Visual Neurosci.* 9: 581-593.

Dreher, B. and Cottee, L.J., 1975, Visual receptive-field properties of cells in area 18 of cat's cerebral cortex before and after lesions in area 17, *J. Neurophysiol.* 38 : 735-750.

Dreher, B., 1986, Thalamocortical and corticocortical interconnections in the cat visual system: relation to the mechanisms of information processing. In : "Visual Neuroscience", edited by J.D. Pettigrew, K.J. Sanderson and W.R. Levick. Cambridge Univ. Press, p. 239-259.

Dreher, B., Fukada, Y. and Rodieck, R.W., 1976, Identification, classification and anatomical segregation of cells with X-like and Y-like properties in the lateral geniculate nucleus of old-world primates, *J. Physiol.* 258 : 433-452.

Eckhorn, R., Bauer, R., Jordan, W., Brosch, M., Kruse, W., Munk, M. and Reitboeck, H.J., 1988, Coherent oscillations : a mechanism of feature linking in the visual cortex ?, *Biol. Cybern*. 60: 121-130.

Engel, A.K., König, P., Kreiter, A.K. and Singer, W., 1991b, Interhemispheric synchronization of oscillatory neuronal responses in cat visual cortex. *Science* 252 : 1177-1179.

Engel, A.K., Kreiter, A.K., König, P. and Singer, W., 1991a, Synchronization of oscillatory neuronal responses between striate and extrastriate visual cortical areas of the cat. *Proc. Natl. Acad. Sci. USA* 88 : 6048-6052.

Felleman, D.J. and Van Essen, D.C., 1991, Distributed hierarchical processing in the primate cerebral cortex, *Cerebral cortex* 1 : 1-47.

Fitzpatrick, D., Lund, J.S. and Blasdel, G.G., 1985, Intrinsic connections of macaque striate cortex. Afferent and efferent connections of lamina 4C.*J. Neurosci.* 5: 3329-3349.

Fries, W., 1984, Cortical projections to the superior colliculus in the macaque monkey : a retrograde study using horseradish peroxidase. *J. Comp. Neurol*. 230: 55-76.

Girard, P. and Bullier, J., 1989, Visual activity in area V2 during reversible inactivation of area 17 in the macaque monkey, *J. Neurophysiol.* 62 : 1287-1302.

Gochin, P.M., E.M. Miller, C.G. Gross and G.L. Gerstein, 1991, Functional interactions among neurons in inferior temporal cortex of the awake macaque. *Exp. Brain Res*. 84 : 505-516.

Hubel, D. H. and Livingstone, M. S., 1987, Segregation of form, color, and stereopsis in primate area 18. *J. Neurosci* 7: 3378-3415.

Hubel, D.H. and Wiesel, T.N., 1972 Laminar and columnar distribution of geniculo-cortical fibers in the monkey. *J. Comp. Neurol.* 146 : 421-450.

Kennedy, H. and Bullier, J., 1985, A double-labelling investigation of the afferent connectivity to cortical areas V1 and V2 of the macaque monkey, *J. Neuroscience* 5: 2815-2830 .

Krüger, J. and Aiple, F., 1988, Multimicroelectrode investigation of monkey striate cortex: spike train correlations in the infragranular layers. *J. Neurophysiol.* 60: 798-828.

Levick, W.R., Cleland, B.G. and Dubin, M.W., 1972, Lateral geniculate neurons of cat: retinal inputs and physiology. *Invest. Ophthalmol.* 11: 302-311.

Livingstone, M. S. and Hubel, D. H., 1987, Connections between layer 4b of area 17 and the thick cytochrome oxidase stripes of area 18 in the squirrel monkey. *J. Neurosci*. 7: 3371-3377.

Lund, J.S. and Boothe, R.G., 1975, Interlaminar connections and pyramidal neuron organisation in the visual cortex, area 17, of the macaque monkey. *J. Comp. Neurol.* 159 : 305-334.

Lund, J.S., 1988, Anatomical organization of macaque striate visual cortex. *Ann. Rev. Neurosci.* 11: 253-288.

Lund, J.S., Hendrickson, A.E., Ogren, M.P. and Tobin, E.A., 1981, Anatomical organization of primate visual cortex area V2, *J. Comp. Neurol.* 202: 19-45.

Marrocco, R.T., 1976, Sustained and transient cells in monkey lateral geniculate nucleus: conduction velocities and response properties, *J. Neurophysiol.* 39: 340-353.

Mastronarde, D.N., 1987, Two classes of single-input X-cells in cat lateral geniculate nucleus. II. Retinal inputs and the generation of receptive-field properties. *J. Neurophysiol.* 576 : 381-413.

Maunsell, J.H.R. and Gibson, J.R., 1992, Visual response latencies in striate cortex of the macaque monkey. *J. Neurophysiol.* 4 : 1332-1334 .

Milner, P.M., 1974, A model for visual shape recognition. *Psycholog. Review* 81:521-535.

Mitzdorf, U. and Singer, W., 1979, Excitatory synaptic ensemble properties in the visual cortex of the macaque monkey: a current source density analysis of electrically evoked potentials. *J. Comp. Neurol.* 187: 71-84.

Nelson, J.I., Salin, P.A., Munk, M.H.J., Arzi, M. and Bullier, J., 1992, Spatial and temporal coherence in corticocortical connections : a cross-correlation study in areas 17 and 18 in the cat. *Visual Neurosci.* 9 : 21-38.

Perkel, D.H., Gerstein, G.L. and Moore, G.P., 1967, Neuronal spike trains and stochastic point processes. II Simultaneous spike trains, *Biophys. J.* 7:419-441.

Raiguel, S.E., Lagae, L., Gulyas, B.,and Orban, G.A., 1989, Response latencies of visual cells in macaque areas V1, V2 and V5. *Brain Res.* 493: 155-159.

Rockland, K. S. and Pandya, D. N., 1979, Laminar origin and terminations of cortical connections of the occipital lobe in the rhesus monkey. *Brain Res*. 179: 3-20.

Schiller, P. H. and Malpeli, J. G., 1977, The effect of striate cortex cooling on area 18 cells in the monkey, *Brain Res.* 126: 366-369.

Sherk, H., 1978, Area 18 cell responses in cat during reversible inactivation of area 17. *J. Neurophysiol.* 41: 204-215.

Singer, W., 1992, Synchronization of cortical activity and its putative role in information processing and learning. *Ann. Rev. Physiol.* 55:349-74.

Spear, P.D. and Baumann, T.P., 1979, Effects of visual cortex removal on receptive-field properties of neurons in lateral suprasylvian visual area of the cat. *J. Neurophysiol.* 42: 31-56.

Tanaka, K., 1983, Cross-correlation analysis of geniculostriate neuronal relationships in cats. *J. Neurophysiol.* 49: 1303-1318.

Trojanowski, J.Q. and Jacobson, S., 1977, The morphology and laminar distribution of cortico-pulvinar neurons in the rhesus monkey. *Exp. Brain Res.* 28: 51-62.

Vaadia, E., Ahissar, E., Bergman, H. and Lavner, Y., 1991, Correlated activity of neurons: a neural code for higher brain functions ? In J. Krüger (ed.) "Neuronal Cooperativity", Springer-Verlag, Berlin, Heidelberg.

von der Malsburg, C. and Schneider, W., 1986, A neural cocktail-party processor, *Biol. Cybern.* 54:29-40.

Yukie, M. and Iwaï, E., 1981, Direct projection from the dorsal lateral geniculate nucleus to the prestriate cortex in macaque monkeys. *J. Comp. Neurol.* 201: 81-97.

OSCILLATIONS AND SYNCHRONY IN THE VISUAL CORTEX: EVIDENCE FOR THEIR FUNCTIONAL RELEVANCE

Pieter R. Roelfsema, Andreas K. Engel, Peter König, and Wolf Singer

Max-Planck-Institut für Hirnforschung
Deutschordenstr. 46,
60528 Frankfurt, Germany

INTRODUCTION: THE CONCEPT OF TEMPORAL CODING

The past 20 years of research on the visual cortex have revealed an organizational complexity that was largely unforeseen (Livingstone and Hubel, 1988; Felleman and Van Essen, 1991; Zeki, 1993). Electrophysiological studies on receptive field properties and visual maps as well as research on the pattern of afferent and efferent projections have inspired investigators to subdivide the visual cortex in ever smaller compartments each of which is characterized by distinct response properties and connections to other brain structures. As a result, more than 30 distinct visual areas have now been classified in the monkey visual system (Felleman and Van Essen, 1991), and presumably such areas are equally numerous in humans and other higher mammals. This parcellation is assumed to reflect some kind of functional specialization since neurons in each of these visual areas are, at least to some degree, selective for a characteristic subset of stimulus features. Thus, for instance, some areas contain predominantly colour-selective cells, while others primarily process information about the form of an object or its direction of motion in the visual field. As a consequence of this functional specialization, any object present in a visual scene will activate neurons in many cortical areas simultaneously. Accordingly, object representations in the visual system are likely to correspond to large and highly distributed assemblies of feature-detecting neurons (Hebb, 1949).

The parallel and distributed nature of visual processing naturally raises the question of how disseminated neuronal responses can be integrated into coherent representational states. This problem seems particularly intriguing since there is no single area in the cortex where all processing pathways would converge (Zeki, 1993). Moreover, this so-called "binding problem" is aggravated by the fact that natural scenes always comprise multiple objects that must be represented simultaneously.

Oscillatory Event-Related Brain Dynamics, Edited
by C. Pantev, Plenum Press, New York, 1994

Thus, any mechanism supposed to solve the binding problem must be able to selectively "tag" feature-selective neurons that code for the same object and to demarcate their responses from those of neurons activated by other objects in order to avoid the illusory conjunction of features (von der Malsburg, 1981). Figure 1 illustrates this situation schematically. If only one object is present in a visual scene the resultant neuronal activity pattern is unambiguous (Fig. 1A). However, if several objects activate feature-sensitive cells in a cortical area, it cannot be determined from the activity pattern as such which of the neurons belong to which particular object representation (Fig. 1B).

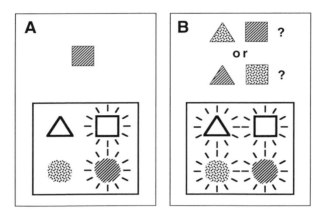

Figure 1. Schematic illustration of the binding problem. (A) As an example, we consider a network which contains only four hypothetical feature detectors: two "form detectors" and two "texture-sensitive" cells (bottom). If only one object is presented (top), the activity pattern in the network is unambiguous: the set of active cells can simply be taken as the object representation. (B) However, if several objects are presented (top) an ambiguous situation emerges: the actual conjunctions of features as they are present in the visual field cannot be inferred from the pattern of activated neurons (bottom).

A classical attempt to avoid this problem is displayed in Figure 2. As suggested by Barlow (1972) one might introduce neurons into the network which are selective in multiple stimulus dimensions. According to Barlow such neurons, which he termed "cardinal cells", should receive convergent input from the primary processing stages and, thus, acquire highly complex response properties (Fig. 2A). However, this model provides a solution to the binding problem only at enormous costs: it requires neurons which are selective for any possible combination of features. Since the number of neurons will grow exponentially with the number of feature dimensions this constitutes a serious limitation for the representation of any realistic visual environment. Another disadvantage of using highly selective neurons is the loss of representation flexibility, because every new object that is encountered will require the recruitment of a new dedicated cell.

An alternative solution to the binding problem has been proposed by Abeles (1982) and von der Malsburg (1981, 1986) and, in a preliminary form, also by Milner (1974). They suggested that neurons responding to the same object in the visual field might synchronize their discharges with a precision of a few milliseconds (Fig. 2B). In contrast, no synchronization should occur between cells encoding features of different objects.

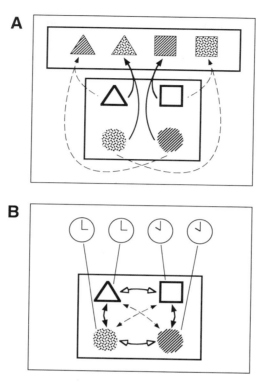

Figure 2. Potential solutions to the binding problem. (A) Object representation by "cardinal cells". The model assumes that object-specific cells are located at a higher level of processing which receive convergent input from lower-level feature detectors. If e.g. a stippled triangle and a hatched square have been presented, convergent connections are assumed to activate the appropriate set of cardinal cells (thick arrows). Note that for this coding scheme to work a complete set of cardinal cells is required for every module or "hypercolumn" in a retinotopic map. (B) Binding by a combinatorial temporal code. Since the model implies that feature conjunctions are expressed by correlated firing, it does not require high-level units which read the activity at more peripheral processing levels. In this scheme, a stippled triangle and a hatched square would be represented by synchronization of the appropriate feature detectors (clocks). Synchronizing connections are drawn with filled arrowheads. Unfilled arrowheads indicate desynchronizing connections between neurons coding for different values of the same feature dimension.

In psychophysical terms, this process of response synchronization would correspond to the binding of local object features into organized percepts as described e.g. by the Gestalt psychologists (Köhler, 1930). This concept complements and extends the classical notion of object representation by distributed neuronal assemblies (Hebb, 1949). As in the Hebb-model, representations are generated in a highly flexible and economic manner because a given neuron can, at different times, participate in a large number of different assemblies. Thus, new objects can readily be encoded as new patterns of activity over the same basic set of neurons which, in principle, only need to encode primitive object features. However, because it makes time available as an additional coding dimension the binding mechanism advocated by Abeles and von der Malsburg combines these advantages with the possibility of coactivating multiple object representations. Therefore, if actually existing in the brain such a mechanism would provide an attractive way of solving the binding problem.

EXPERIMENTAL EVIDENCE FOR A TEMPORAL CODE IN THE VISUAL SYSTEM

Indeed, recent experimental studies in cat and monkey visual system provide supportive evidence for the concept of a temporal code (for review, see Engel et al., 1992a; Singer, 1993). In cats, several studies have demonstrated that spatially separate cells within the primary visual area can synchronize their spike discharges in both anaesthetized and awake animals (Michalski et al., 1983; Ts'o et al., 1986; Eckhorn et al., 1988; Gray et al., 1989a, 1989b; Engel et al., 1990). In most cases, the recorded cells synchronize with zero phase lag, which holds even if the recording sites are separated by more than 7 mm (Gray et al., 1989a; Engel et al., 1990). Interestingly, correlated firing is not only found between cells with similar response properties but also, for instance, between cells with different preferred orientations. These findings agree with the proposal that synchrony might be the "glue" that binds distributed neuronal activity into coherent representations. Further experiments on anaesthetized cats have revealed that response synchronization can well extend beyond the borders of a single visual area. Thus, for instance, correlated firing has been observed between areas 17 and 18 (Eckhorn et al., 1988; Nelson et al., 1992) as well as between area 17 and the posteromedial area in the lateral suprasylvian sulcus (PMLS) (Engel et al., 1991a). These findings are of particular interest because, as mentioned already, different areas are assumed to process different aspects of visual objects (Livingstone and Hubel, 1988; Felleman and Van Essen, 1991) and, hence, interareal synchronization could mediate binding across feature maps which presumably is required for the complete representation of visual scenes. In addition, synchronization has been observed between the primary visual cortex of the two cerebral hemispheres (Engel et al., 1991b), in accordance with the requirement of feature binding across the midline of the visual field.

A key finding is that both within and across visual areas response synchronization depends critically on the stimulus configuration. We could recently demonstrate that spatially separate cells in cat visual cortex show strong synchronization only if they respond to the same visual stimulus. However, if responding to two independent stimuli, the cells fire in a less correlated manner or even without any fixed temporal relationship (Gray et al., 1989a; Engel et al., 1991a, 1991c). Figure 3 gives one example for this effect. These observations provide strong support for the hypothesis that correlated firing could provide a dynamic binding mechanism which permits the formation of assemblies in a flexible manner. In addition, they demonstrate that Gestalt criteria such as continuity or coherent motion, which have psychophysically been shown to determine scene segmentation, are important for the establishment of synchrony among neurons in the visual cortex.

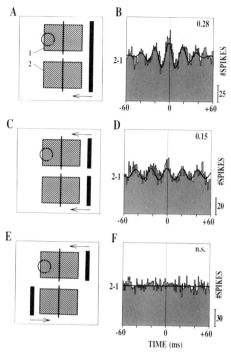

Figure 3. Synchronization in the visual cortex is influenced by stimulus coherence. Data from an experiment in which multiunit activity was recorded from two sites separated by 7 mm in the primary visual cortex of an anaesthetized cat. (A), (C), (E) Schematic plots of the receptive fields. Both cell groups preferred vertical orientations. The colinear arrangement of the fields allowed the comparison of three different stimulus paradigms: a long continuous light bar moving across both fields (A), two independent light bars moving in the same direction (C), and the same two bars moving in opposite directions (E). The circle represents the center of the visual field, and the thick line drawn across each receptive field indicates the preferred orientation. (B), (D), (F) The respective crosscorrelograms obtained with each stimulus paradigm. Using the long light bar, the two responses were synchronized as indicated by the strong modulation of the correlogram (B). If the continuity of the stimulus was interrupted, the synchronization became weaker (D), and it totally disappeared if the motion of the stimuli was incoherent (F). The graph superimposed on each of the correlograms represents a damped cosine function that was fitted to the data to assess the strength of the modulation. The number in the upper right corner indicates the "relative modulation amplitude", a measure of correlation strength that was determined by computing the ratio of the amplitude of the fitted function over its offset. Abbreviation: n.s., not significant. (Slightly modified from Engel et al., 1992a)

Similar evidence for response synchronization is available in the monkey visual system. In this species, synchronization of spatially separate neurons has been demonstrated within striate cortex (Ts'o and Gilbert, 1988; Krüger and Aiple, 1988) and within extrastriate regions such as the caudal superior temporal sulcus (Kreiter and Singer, 1992; Engel et al., 1992b) and the inferior temporal cortex (Gochin et al., 1991). In addition, correlated firing has been reported to occur between neurons of the primary and secondary visual area (Eckhorn et al., 1993a; Munk et al., 1993). Finally and most importantly, the stimulus-dependence of neuronal interactions has now also been confirmed in monkeys (Kreiter et al., 1992). Taken together, these studies suggest that a temporal binding mechanism might be a general operating principle in the visual system of higher vertebrates.

The findings on synchrony reviewed above raise the question of how temporal synchrony is mediated, in particular between remote neurons located in different areas or even hemispheres. There is now some evidence to suggest that synchrony is established at the cortical level and not mediated by a common driving input from subcortical structures. In the case of interhemispheric interactions it has been shown directly that cortico-cortical connections account for the observed interactions. If the corpus callosum is sectioned, response synchronization between the hemispheres disappears whereas synchronization within either hemisphere is preserved (Engel et al., 1991b). Further evidence for the role of intracortical connections comes from experiments on strabismic cats. In the primary visual cortex, squint induction early in development leads to a breakdown of binocularity (Hubel and Wiesel, 1965). The two sets of monocular cells tend to be clustered in columns driven by the left or the right eye. In a recent anatomical study of cats with divergent squint, Löwel and Singer (1992) demonstrated that only territories with the same ocular dominance are linked by tangential intracortical connections. In a subsequent crosscorrelation study we could show that in these animals response synchronization occurs also preferentially between neurons driven by the same eye (König et al., 1993). Evidently, this correlation of functional interactions and connection topology provides further support for the notion that synchrony is achieved by coupling at the cortical level.

FUNCTIONAL RELEVANCE OF SYNCHRONY: LESSONS FROM ABNORMAL VISION

Although the hypothesis of von der Malsburg offers an attractive conceptual scheme for understanding the integration of distributed neuronal responses, definitive evidence that the brain actually uses synchronization in exactly this way has not yet been obtained. Nonetheless, our study on cats with divergent squint provides the first hint that neuronal synchronization is indeed functionally relevant and related to the animal's perception and behaviour. Typically, humans and animals with a divergent strabismus develop a pattern of alternating fixation with the two eyes (for review, see Duke-Elder and Wybar, 1973; von Noorden, 1990). Usually, monocular vision is undisturbed in these subjects, but they show a striking inability of combining information into a single percept that arrives simultaneously through the two eyes. As mentioned above, this deficit is accompanied by a loss of synchronization between neuronal populations with different ocular dominance (König et al., 1993). This correspondence between a functional deficit and loss of neuronal interactions clearly argues for the importance of correlated neuronal firing in normal visual perception.

Further evidence for the functional relevance of synchrony among cortical neurons comes from a recent crosscorrelation study of cats with convergent squint

(Roelfsema et al., 1993, 1994). In contrast to the situation for subjects with a divergent strabismus, convergent squinters often use only one eye for fixation. The non-fixating eye then develops a syndrome of perceptual deficits called strabismic amblyopia (reviewed in Duke-Elder and Wybar, 1973; von Noorden, 1990). Symptoms of strabismic amblyopia include a reduced acuity of the affected eye, temporal instability and spatial distortions of the visual image, and the so-called crowding phenomenon: discrimination of details deteriorates further if other contours are nearby (Levi and Klein, 1985). Clearly, at least some of these deficits indicate a reduced capacity of integrating visual information and an impairment of feature binding. Interestingly, the neural correlate of strabismic amblyopia has not been identified so far. In contrast to effects that occur when amblyopia results from monocular deprivation (Hubel and Wiesel, 1970), cortical neurons driven by the amblyopic eye of squinting animals do not show a loss of responsiveness and have essentially normal response properties (Chino et al., 1983; Freeman and Tsumoto, 1983; Crewther and Crewther, 1990; Blakemore and Vital-Durand, 1992).

The results of our crosscorrelation study indicate that the perceptual deficits of subjects with strabismic amblyopia may rather be due to a disturbance of intracortical interactions (Roelfsema et al., 1993, 1994). In this study, we investigated neuronal responses and temporal correlations in the primary visual cortex of cats with convergent squint that had developed an amblyopia of the deviating eye. Prior to the physiological measurements, the visual deficit was verified by means of a behavioural test which revealed the animal's inability to see high spatial frequency gratings with the amblyopic eye. As illustrated in Figure 4 we observed highly significant differences in the synchronization behaviour of cells driven by the normal and the amblyopic eye, respectively. Responses to single moving bar stimuli recorded from neurons dominated by the amblyopic eye showed much weaker correlation than responses of neurons driven by the normal eye (Fig. 4A,B,E). In addition, synchronization between neurons dominated by different eyes was virtually absent (Fig. 4C,E), confirming the results that we had obtained in cats with divergent squint (König et al., 1993). However, the amplitudes of the responses to bar stimuli were very similar for neurons driven by the normal and amblyopic eye (Fig. 4D). Interestingly, both neuronal populations responded equally well even to high spatial frequency gratings which the cat had been unable to discriminate with the amblyopic eye during behavioural testing (data not shown). However, using such stimuli the synchrony between neurons dominated by the amblyopic eye was particularly impaired, whereas interactions were maintained between cells driven by the normal eye (Roelfsema et al., 1993, 1994).

These results suggest two important conclusions. First, they indicate that strabismic amblyopia is indeed accompanied by an impairment of intracortical interactions. Thus, they may provide a new explanation for some of the perceptual disturbances associated with this perceptual malfunction. The crowding phenomenon, for instance, which results from a pathological interference between spatially contiguous contours, would then appear as a consequence of the disruption of intracortical binding mechanisms. Indirectly, this conclusion corroborates the idea that neuronal synchrony is indeed employed for feature integration and assembly formation. Second, these data extend the results obtained in cats with divergent squint and provide further evidence that synchrony among cortical neurons is indeed functionally relevant. Our data show directly that neurons in the animal's brain are responding to stimuli that it had not been able to discriminate behaviourally. Given the finding of reduced synchrony, the most likely explanation for this apparent paradox is that the responses of individual feature detectors as such is not sufficient to influence the animal's perceptual decision and behaviour. Rather, what is required is their joint synchronous activity - the cells have to be linked into coherently active assemblies to be functionally effective.

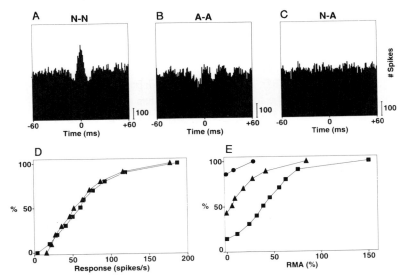

Figure 4. Response amplitudes and neuronal synchronization in the primary visual cortex of cats with strabismic ambylopia. The data shown here were recorded with bar-shaped stimuli. (A), (B), (C) Examples of crosscorrelograms between cells driven by the normal eye (N-N), by the amblyopic eye (A-A) and between cells dominated by different eyes (N-A). Note that temporal correlation is strong if both recording sites are driven by the normal eye. Synchronization is, on average, much weaker between cells dominated by the amblyopic eye and is in most cases negligible if the recording sites receive their input from different eyes. The relative modulation amplitudes (cf. Fig. 3) of the correlograms shown correspond to the 60th percentile of the respective distributions displayed in (E). Scale bars indicate the number of spikes. (D) Cumulative distributions of response amplitudes (evoked minus spontaneous activity) at sites dominated by the normal eye (squares, n=80) and by the amblyopic eye (triangles, n=65). For each graph, the ordinate represents the percentage of recording sites that had a lower response amplitude than the value indicated on the abscissa. Note the similiarity of the distributions. (E) Cumulative distributions of the relative modulation amplitudes (RMA) of crosscorrelograms for N-N pairs (squares, n=56), A-A pairs (triangles, n=38) and N-A pairs (circles, n=44). Note that the three graphs are clearly different (differences being highly significant). (Modified from Roelfsema et al., 1994).

THE POTENTIAL ROLE OF OSCILLATORY NEURONAL ACTIVITY

A striking observation made in many of the crosscorrelation studies described above is that the emergence of synchronized states in the cortex is frequently associated with oscillatory firing patterns of the respective neurons (Eckhorn et al., 1988; Gray et al., 1989a; Engel et al., 1990, 1991a, 1991b, 1991c; König et al., 1993; Kreiter and Singer, 1992). These temporal patterns, which induce troughs and additional peaks in correlograms (e.g. Fig. 3), were in the visual system first described by Gray and Singer (1987, 1989). They observed that, with appropriate stimulation, neighbouring neurons tend to engage in grouped discharges which recur at frequencies in the gamma range between 30 and 70 Hz. Figure 5 shows an example of such an oscillatory response recorded from the primary visual cortex of an anaesthetized cat. Similar oscillations have been observed in various extrastriate visual areas in the anaesthetized preparation (Eckhorn et al., 1988; Engel et al., 1991a) and in the striate cortex of awake cats (Gray et al., 1989b; Gray and Viana Di Prisco, 1993). In the monkey visual system, oscillatory activity in the gamma frequency range has been demonstrated in extrastriate motion-sensitive visual areas of both awake and anaesthetized animals (Kreiter and Singer, 1992; Engel et al., 1992b), in the striate cortex (Livingstone, 1991; Maunsell and Gibson, 1992; Eckhorn et al., 1993b) and in inferotemporal visual cortex (Nakamura et al., 1992), although one group has recently also reported negative evidence (Young et al., 1992).

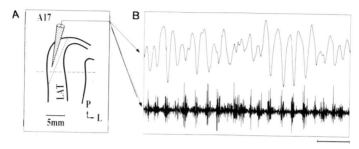

Figure 5. Example of an oscillatory response in cat visual cortex. (A) Schematic plot of the position of a recording electrode in area 17. (B) By filtering in different bandpasses, a local field potential (top, positivity upward) and multiunit spike activity (bottom) are extracted from the raw electrode signal. As a light bar is passed through the cells' receptive field a clear oscillatory response is observed in the field potential, indicating that the neurons of the recorded cluster have engaged in a coherent and rhythmic firing pattern. Note the variability of both amplitude and frequency in this time series. The multiunit activity shows directly that different neurons (as indicated by spikes of different sizes) synchronize their discharges into a burst-and-pause firing pattern. Note that the spike bursts are in-phase with the peak negativity of the field potential. Scale bar: 50 ms. Abbreviations: A17, area 17; LAT, lateral sulcus; P, posterior; L, lateral.

In the preceding sections, we have argued that synchronous firing of spatially separate neurons may provide the "glue" which dynamically binds distributed neurons into assemblies. This, then raises the question of what role the local oscillatory firing patterns might play as part of such a temporal coding scheme. Recently, this issue has been the subject of substantial controversy. Whereas one group has questioned the existence of fast induced oscillations altogether (Young et al., 1992) others have, while acknowledging this phenomenon, doubted its functional relevance and considered it as an epiphenomenal by-product of cortical processing (Ghose and Freeman, 1992; Tovée and Rolls, 1992). Presumably, this issue cannot be settled based on the experimental evidence which is currently available. However, in the following we will argue that an oscillatory temporal structure of neuronal responses may well be relevant for cortical information processing. With respect to its functional role, one possible assumption would be that the oscillatory temporal structure of a neuron's response encodes certain features of the visual input. However, this seems unlikely because the temporal modulation of visual cortical responses depends only to a minor extent on parameters of the stimulus such as its orientation (Gray et al., 1990). An alternative possibility is suggested by several arguments derived from simulation studies (König and Schillen, 1991; Schillen and König, 1991). The results of these studies indicate that an oscillatory modulation of neuronal responses may have crucial advantages for the establishment of synchrony between widely remote sites and thus, may be instrumental as a "carrier signal" for a temporal code in the brain (Engel et al., 1992a, 1992c, 1992d).

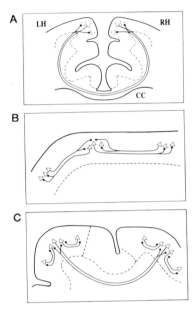

Figure 6. Potential advantages of an oscillatory temporal structure in neuronal responses. (A) Synchronization of neuronal discharges without phase lag despite considerable conduction delays in long-range cortico-cortical connections. (B) Synchronization of cell groups which are not directly coupled via intermediate neuronal groups. (C) Synchronization despite anisotropic delays, which may occur due to large variations in the length or myelination of coupling connections. Abbreviations: LH, left hemisphere; RH, right hemisphere; CC, corpus callosum.

As described in the preceding sections, both our experiments on interhemispheric interactions and on exotropic squinters provide evidence for the assumption that response synchronization is mediated by connections at the cortical level and not by synchronously driving subcortical input (Engel et al., 1991b; König et al., 1993). Interestingly, even interareal and interhemispheric interactions have consistently been observed to occur with near-zero phase lag (Eckhorn et al., 1988; Engel et al., 1991a, 1991b; Nelson et al., 1992), although the underlying connections are known to exhibit considerable transmission delays (e.g. Innocenti, 1980). As suggested by simulation studies (König and Schillen, 1991; Schillen and König, 1991), the establishment of synchrony without phase lag may be facilitated under these conditions if the respective neurons show oscillatory firing patterns (Fig. 6A). These models demonstrate that due to the recurrent temporal structure of such patterns reciprocally coupled neurons can entrain each other and improve synchrony within a few oscillatory bursts. Further advantages of oscillatory activity suggested by these models are that oscillating neurons can be synchronized via polysynaptic linkages without adding up of small phase-lags (Fig. 6B) and that synchrony in such networks is robust despite considerable variation of the conduction delays (Fig. 6C). If these predictions hold true, cortical long-range synchronization should be closely correlated with the occurrence of oscillatory activity. However, such a relation would not necessarily hold for synchrony between closely spaced cells, since these tend to be strongly coupled without major delays in the respective connections. Recently, we have obtained physiological evidence in support of this hypothesis (König et al., 1994).

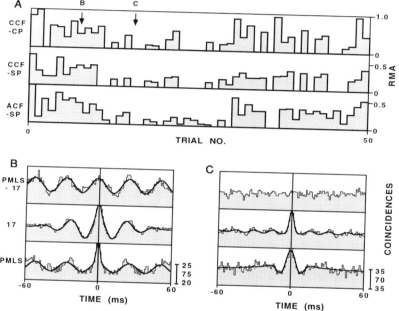

Figure 7. Single sweep analysis of long-range interactions. One of the recording sites was located in area 17, the other in area PMLS. The cells at the two recording sites had nonoverlapping receptive fields but similar orientation and velocity preferences. In this case, 50 trials had been recorded with the same stimulus. (A) Intertrial variability of relative modulation amplitudes (cf. Fig. 3) computed separately for the center peak (top) and the satellite peak (middle) of the crosscorrelogram, and of the arithmetic mean of the satellite peak amplitudes in the two respective autocorrelograms (bottom). Note the strong quantitative covariance of all three parameters. The arrows indicate two representative trials which are illustrated in Figures B and C. (B) Example of a trial where both cell groups showed narrow-banded oscillations, as indicated by the presence of multiple peaks and troughs in the autocorrelograms (middle and bottom). This is reflected in a strong modulation of the corresponding crosscorrelogram (top). (C) Example of a sweep where no temporal correlation was evident in the crosscorrelation function (top). The two autocorrelograms (middle and bottom) showed only a center peak without significant satellite peaks. As in Fig. 3, the thick continuous line superimposed on the correlograms represents the damped cosine function fitted to the data. Abbreviations: CCF, crosscorrelation function; ACF, autocorrelation function; CP, center peak; SP, satellite peak; RMA, relative modulation amplitude; 17, area 17.

We reanalyzed data from cat visual cortex on short- and long-range interactions within area 17 (the latter having interelectrode distances above 2 mm), as well as on interhemispheric interactions and on interactions between areas 17 and PMLS

(Engel et al., 1990; Engel et al., 1991a, 1991b). In these data sets we investigated the coincidence of synchrony and oscillations, as evidenced by a significant center peak in the respective crosscorrelogram and by significant satellite peaks in the auto- and crosscorrelograms, respectively. As a criterion for the presence of an oscillatory temporal structure, at least one significant satellite peak had to be present in auto- or crosscorrelograms on either side of the center peak (for details of the methods for data processing see Engel et al., 1990; König, 1994). This analysis revealed a close relationship between long-range synchronization and occurrence of oscillatory firing patterns. If synchronization was observed over large distances within striate cortex, between areas or across the hemispheres it was almost always accompanied by an oscillatory component in the auto- and crosscorrelograms. In contrast, significant short-range interactions occurred both with and without oscillatory modulation of the correlograms. In all cases, the observed oscillations had frequencies in the range between 30 and 70 Hz. In several cases exhibiting interhemispheric or interareal synchronization the relationship of oscillations and synchrony was, in addition, examined on a trial by trial basis. As shown for one of these cases in Figure 7, we examined the relationship between the amplitude of the center peak in the crosscorrelogram and that of the first satellite peak in both the auto- and crosscorrelation function of each trial. In all cases analyzed, these parameters were clearly correlated which indicates that, for long-range interactions in cat visual cortex, there is not only a qualitative relation between synchrony and oscillations, but also a quantitative covariance between the strength of synchronization and the degree of rhythmicity of the respective spike trains.

These data are compatible with the hypothesis that oscillatory firing patterns may facilitate the establishment of synchrony between widely separate neuronal populations in the brain (Engel et al., 1992a, 1992c, 1992d; Singer, 1993). If so, these discharge patterns may actually be a prerequisite for the binding of distributed neurons into assemblies. Clearly, however, the potential role of oscillatory firing patterns needs to be substantiated by further experiments. Based on the available data, which merely demonstrate a correlation of synchrony with the appearance of oscillatory firing patterns, we cannot rule out the possibility that synchrony is achieved by mechanisms which do not strictly require a band-limited temporal structure of neuronal responses.

CONCLUSIONS

In this chapter, we have reviewed experimental evidence for the notion of a temporal code. The available data support the hypothesis that correlated firing may functionally relevant in the brain for the binding of distributed neurons into coherently active assemblies. Moreover, the results suggest that oscillatory firing patterns with frequencies in the gamma range may play a crucial role as carrier signals for the establishment of synchrony. So far, we have restricted our discussion to evidence obtained in the visual system. However, a number of studies in non-visual sensory modalities and in the motor system indicate that synchrony and oscillatory activity may actually be quite ubiquitous in the nervous system (cf. also other chapters of this volume). Synchronization in the gamma frequency range is well known to occur in the olfactory bulb and entorhinal cortex of various species, where these phenomena have been related to the integration of odor information (e.g. Freeman, 1988). In the auditory cortex, synchronized gamma oscillations have been described by several groups (Galambos et al., 1981; Ribary et al., 1991; Pantev et al., 1991). Actually, with respect to this modality it has first been suggested that a temporal code may play an essential role in sensory segmentation (von der Malsburg and Schneider, 1986). In the somatosensory system, oscillatory activity has been described by Rougeul et al. (1979) and, more recently, by Ahissar and Vaadia (1990). In addition, the possibility of interactions in this frequency range across spatially separate sites has been demonstrated (Murthy and Fetz, 1992). Similar evidence is available for the motor system (Murthy and Fetz, 1992; Sanes and Donoghue, 1993). Based on these findings, it seems justified to generalize the results obtained in the visual cortex and to suggest that neuronal synchronization, in particular at frequencies in the gamma band, may be of general relevance for cortical processing. Remarkably, there is evidence to suggest that sensorimotor integration may also be achieved by a temporal code. Recently, synchronization between somatosensory and motor cortex has been observed (Murthy and Fetz, 1992), and there are indications that neurons in the visual and motor cortex can engage in synchronous activity (Sheer and Grandstaff, 1970; Bressler and Nakamura, 1993). Thus, temporal correlations may serve for the linkage of neurons encoding both sensory and motor aspects of behaviour. The specificity of such interactions might allow, for instance, the selective channeling of sensory information to different motor programs that are concurrently executed, which is required e.g. during bimanual manipulation under visual control. We assume that further studies of the synchronization between sensory cortices and the motor system may lead to valuable insights into how temporally coded sensory information might be used for suitable behavioural adjustments. Taken together, the results obtained in other cortical modalities corroborate the conclusions drawn from the studies on the visual system and make it tempting to speculate that temporal codes may be of general importance in the nervous system.

REFERENCES

Abeles, M., 1982, "Local Cortical Circuits", Springer, Berlin.

Ahissar, E., and Vaadia, E., 1990, Oscillatory activity of single units in a somatosensory cortex of an awake monkey and their possible role in texture analysis, *Proc. Natl. Acad. Sci. USA* 87:8935-8939.

Barlow, H.B., 1972, Single units and sensation: a neuron doctrine for perceptual psychology? *Perception* 1:371-394.

Blakemore, C., and Vital-Durand, F., 1992, Different neural origins for 'blur' amblyopia and strabismic amblyopia. *Ophthal. Physiol. Opt.* 12:83.

Bressler, S.L., and Nakamura, R., 1993, Inter-area synchronization in macaque neocortex during a visual pattern discrimination task, *in*: "Computation and Neural Systems", F.H. Eeckman and J.M. Bower, eds., Kluwer Academic, Boston.

Chino, Y.M., Shansky, M.S., Jankowski, W.L., and Banser, F.A., 1983, Effects of rearing kittens with convergent strabismus on development of receptive-field properties in striate cortex neurons. *J. Neurophysiol.* 50:265-286.

Crewther, D.P., and Crewther, S.G., 1990, Neural site of strabismic amblyopia in cats: spatial frequency deficit in primary cortical neurons. *Exp. Brain Res.* 79:615-622.

Duke-Elder, S., and Wybar, K., 1973, "System of Ophthalmology", Vol. VI, "Ocular Motility and Strabismus", Kimpton, London.

Eckhorn, R., Bauer, R., Jordan, W., Brosch, M., Kruse, W., Munk, M., and Reitboeck, H.J., 1988, Coherent oscillations: a mechanism for feature linking in the visual cortex? *Biol. Cybern.* 60:121-130.

Eckhorn, R., Frien, A., Bauer, R., Kehr, H., Woelbern, T., and Kruse, W., 1993a, High frequency (50-90 Hz) oscillations in visual cortical areas V1 and V2 of an awake monkey are phase-locked at zero delay, *Soc. Neurosci. Abstr.* 19:1574.

Eckhorn, R., Frien, A., Bauer, R., Woelbern, T., and Kehr, H., 1993b, High frequency (60-90 Hz) oscillations in primary visual cortex of awake monkey, *Neuroreport* 4:243-246.

Engel, A.K., König, P., Gray, C.M., and Singer, W., 1990, Stimulus-dependent neuronal oscillations in cat visual cortex: inter-columnar interaction as determined by cross-correlation analysis, *Eur. J. Neurosci.* 2:588-606.

Engel, A.K., Kreiter, A.K., König, P., and Singer, W., 1991a, Synchronization of oscillatory neuronal responses between striate and extrastriate visual cortical areas of the cat, *Proc. Natl. Acad. Sci. USA* 88:6048-6052.

Engel, A.K., König, P., Kreiter, A.K., and Singer, W., 1991b, Interhemispheric synchronization of oscillatory neuronal responses in cat visual cortex, *Science* 252:1177-1179.

Engel, A.K., König, P., and Singer, W., 1991c, Direct physiological evidence for scene segmentation by temporal coding, *Proc. Natl. Acad. Sci. USA* 88:9136-9140.

Engel, A.K., König, P., Kreiter, A.K., Schillen, T.B., and Singer, W., 1992a, Temporal coding in the visual cortex: new vistas on integration in the nervous system, *Trends Neurosci.* 15:218-226.

Engel, A.K., Kreiter, A.K., and Singer, W., 1992b, Oscillatory responses in the superior temporal sulcus of anaesthetized macaque monkeys, *Soc. Neurosci. Abstr.* 18:12.

Engel, A.K., König, P., and Schillen, T.B., 1992c, Why does the cortex oscillate? *Curr. Biol.* 2:332-334.

Engel, A.K., König, P., and Singer, W., 1992d, The functional nature of neuronal oscillations - reply, *Trends Neurosci.* 15:387-388.

Felleman, D.J., and Van Essen, D.C., 1991, Distributed hierarchical processing in the primate cerebral cortex, *Cerebral Cortex* 1:1-47.

Freeman, R.D., and Tsumoto, T., 1983, An electrophysiological comparison of convergent and divergent strabismus in the cat: electrical and visual activation of single cortical cells. *J. Neurophysiol.* 49:238-253.

Freeman, W.J. , 1988, Nonlinear neural dynamics in olfaction as a model for cognition, *in*: "Dynamics of Sensory and Cognitive Processing by the Brain", E. Basar, ed., Springer, Berlin.

Galambos, R., Makeig, S., and Talmachoff, P.T., 1981, A 40-Hz auditory potential recorded from the human scalp, *Proc. Natl. Acad. Sci. USA* 78:2643-2647.

Ghose, G.M., and Freeman, R.D., 1992, Oscillatory discharge in the visual system: does it have a functional role? *J. Neurophysiol.* 68:1558-1574.

Gochin, P.M., Miller, E.K., Gross, C.G., and Gerstein, G.L., 1991, Functional interactions among neurons in inferior temporal cortex of the awake macaque, *Exp. Brain Res.* 84:505-516.

Gray, C.M., and Singer, W., 1987, Stimulus-specific neuronal oscillations in the cat visual cortex: a cortical functional unit, *Soc. Neurosci. Abstr.* 13:1449.

Gray, C.M., and Singer, W., 1989, Stimulus-specific neuronal oscillations in orientation columns of cat visual cortex, *Proc. Natl. Acad. Sci. USA* 86:1698-1702.

Gray, C.M., König, P., Engel, A.K., and Singer, W., 1989a, Oscillatory responses in cat visual cortex exhibit inter-columnar synchronization which reflects global stimulus properties, *Nature* 338:334-337.

Gray, C.M., Raether, A., and Singer, W., 1989b, Stimulus-specific intercolumnar interactions of oscillatory neuronal responses in the visual cortex of alert cats, *Soc. Neurosci. Abstr.* 15:798.

Gray, C.M., Engel, A.K., König, P., and Singer, W., 1990, Stimulus-dependent neuronal oscillations in cat visual cortex: receptive field properties and feature dependence, *Eur. J. Neurosci.* 2:607-619.

Gray, C.M., and Viana Di Prisco, G., 1993, Properties of stimulus-dependent rhythmic activity of visual cortical neurons in the alert cat, *Soc. Neurosci. Abstr.* 19:868.

Hebb, D.O., 1949, "The Organization of Behavior", Wiley, New York.

Hubel, D.H., and Wiesel, T.N., 1965, Binocular interaction in striate cortex of kittens reared with artificial squint. *J. Neurophysiol.* 28:1041-1059.

Hubel, D.H., and Wiesel, T.N., 1970, The period of susceptibility to the physiological effects of unilateral eye closure in kittens. *J. Physiol.* 206:419-436.

Innocenti, G.M., 1980, The primary visual pathway through the corpus callosum: morphological and functional aspects in the cat, *Arch. Ital. Biol.* 118:124-188.

Köhler, W., 1930, "Gestalt Psychology", Bell and Sons, London.

König, P., 1994, A method for the quantification of synchrony and oscillatory properties of neuronal activity, *submitted*.

König, P., and Schillen, T.B., 1991, Stimulus-dependent assembly formation of oscillatory responses: I. synchronization, *Neural Comput.* 3:155-166.

König, P., Engel, A.K., Löwel, S., and Singer, W., 1993, Squint affects synchronization of oscillatory responses in cat visual cortex, *Eur. J. Neurosci.* 5:501-508.

König, P., Engel, A.K., and Singer, W., 1994, The relation between oscillatory activity and long-range synchronization in cat visual cortex, *submitted*.

Kreiter, A.K., and Singer, W., 1992, Oscillatory neuronal responses in the visual cortex of the awake macaque monkey, *Eur. J. Neurosci.* 4:369-375.

Kreiter, A.K., Engel, A.K., and Singer, W., 1992, Stimulus dependent synchronization in the caudal superior temporal sulcus of macaque monkeys, *Soc. Neurosci. Abstr.* 18:12.

Krüger, J., and Aiple, F., 1988, Multimicroelectrode investigation of monkey striate cortex: spike train correlations in the infragranular layers, *J. Neurophysiol.* 60:798-828.

Levi, D.M., and Klein, S.A., 1985, Vernier acuity, crowding and amblyopia, *Vision Res.* 25:979-991.

Livingstone, M.S., 1991, Visually-evoked oscillations in monkey striate cortex, *Soc. Neurosci. Abstr.* 17:176.

Livingstone, M., and Hubel, D., 1988, Segregation of form, color, movement, and depth: anatomy, physiology, and perception, *Science* 240:740-749.

Löwel, S., and Singer, W., 1992, Selection of intrinsic horizontal connections in the visual cortex by correlated neuronal activity. *Science* 255:209-212.

Maunsell, J.H.R., and Gibson, J.R., 1992, Visual response latencies in striate cortex of the macaque monkey, *J. Neurophysiol.* 68:1332-1344.

Michalski, A., Gerstein, G.L., Czarkowska, J., and Tarnecki, R., 1983, Interactions between cat striate cortex neurons, *Exp. Brain Res.* 51:97-107.

Milner, M.P., 1974, A model for visual shape recognition, *Psychol. Rev.* 81:521-535.

Munk, M.H.J., Nowak, L.G. and Bullier, J., 1993, Spatio-temporal response properties and interactions of neurons in areas V1 and V2 of the monkey, *Soc. Neurosci. Abstr.* 19:424.

Murthy, V.N., and Fetz, E.E., 1992, Coherent 25- to 35-Hz oscillations in the sensorimotor cortex of awake behaving monkeys, *Proc. Natl. Acad. Sci. USA* 89:5670-5674.

Nakamura, K., Mikami, A., and Kubota, K., 1992, Oscillatory neuronal activity related to visual short-term memory in the monkey temporal pole, *Neuroreport* 3:117-120.

Nelson, J.I., Salin, P.A., Munk, M.H.J., Arzi, M., and Bullier, J., 1992, Spatial and temporal coherence in cortico-cortical connections: a cross-correlation study in areas 17 and 18 in the cat, *Vis. Neurosci.* 9:21-37.

Pantev, C., Makeig, S., Hoke, M., Galambos, R., Hampson, S., and Gallen, C., 1991, Human auditory evoked gamma-band magnetic fields, *Proc. Natl. Acad. Sci. USA* 88:8996-9000.

Ribary, U., Ioannides, A.A., Singh, K.D., Hasson, R., Bolton, J.P.R., Lado, F., Mogilner, A., and Llinás, R., 1991, Magnetic field tomography of coherent thalamocortical 40-Hz oscillations in humans, *Proc. Natl. Acad. Sci. USA* 88:11037-11041.

Roelfsema, P.R., Engel, A.K., König, P., Sireteanu, R., and Singer, W., 1993, Squint-induced amblyopia is associated with reduced synchronization of cortical responses, *Soc. Neurosci. Abstr.* 19:867.

Roelfsema, P.R., König, P., Engel, A.K., Sireteanu, R., and Singer, W., 1994, Reduced neuronal synchrony: a physiological correlate of strabismic amblyopia in cat visual cortex, *submitted*.

Rougeul, A., Bouyer, J.J., Dedet, L., and Debray, O., 1979, Fast somato-parietal rhythms during combined focal attention and immobility in baboon and squirrel monkey, *Electroenceph. Clin. Neurophysiol.* 46:310-319.

Sanes, J.N., and Donoghue, J.P., 1993, Oscillations in local field potentials of the primary motor cortex during voluntary movement, *Proc. Natl. Acad. Sci. USA* 90:4470-4474.

Schillen, T.B., and König, P., 1991, Stimulus-dependent assembly formation of oscillatory responses: II. desynchronization, *Neural Comput.* 3:167-178.

Sheer, D.E., and Grandstaff, N., 1970, Computer analysis of electrical activity in the brain and its relation to behavior, *in*: "Current Research in Neurosciences", H.T. Wycis, ed., Karger, Basel.

Singer, W., 1993, Synchronization of cortical activity and its putative role in information processing and learning, *Annu. Rev. Physiol.* 55:349-374.

Tovée, M.J., and Rolls, E.T., 1992, The functional nature of neuronal oscillations, *Trends Neurosci.* 15:387.

Ts'o, D.Y., and Gilbert, C.D., 1988, The organization of chromatic and spatial interactions in the primate striate cortex, *J. Neurosci.* 8:1712-1727.

Ts'o, D.Y., Gilbert, C.D., and Wiesel, T.N., 1986, Relationships between horizontal interactions and functional architecture in cat striate cortex as revealed by cross-correlation analysis, *J. Neurosci.* 6:1160-1170.

von der Malsburg, C., 1981, "The Correlation Theory of Brain Function", Internal Report 81-2, Max-Planck-Institute for Biophysical Chemistry, Göttingen.

von der Malsburg, C., 1986, Am I thinking assemblies? *in*: "Brain Theory", G. Palm, and A. Aertsen, eds., Springer, Berlin.

von der Malsburg, C., and Schneider, W., 1986, A neural cocktail-party processor, *Biol. Cybern.* 54: 29-40.

von Noorden, G.K., 1990,"Binocular Vision and Ocular Motility. Theory and Management of Strabismus", Mosby, St. Louis.

Young, M.P., Tanaka, K., and Yamane, S., 1992, On oscillating neuronal responses in the visual system of the monkey, *J. Neurophysiol.* 67:1464-1474.

Zeki, S., 1993, "A Vision of the Brain", Blackwell, Oxford.

OSCILLATORY AND NON-OSCILLATORY SYNCHRONIZATIONS IN THE VISUAL CORTEX OF CAT AND MONKEY

Reinhard Eckhorn

Department of Biophysics
Philipps-University
Renthof 7, 35032-Marburg
Germany

INTRODUCTION

Our visual system can associate parts of a figure into a perceptual whole if a minimal feature contrast of the figure against the background is present, and if certain rules of Gestalt properties are fulfilled. It is, for example, easy for us to recognize the triangles in figure 1 even though they are presented only as parts. Such type of feature association does not require previous knowledge of the figure. The association from its parts is a pre-attentively ("bottom-up") acting process. In contrast, the Dalmatine dog in figure 2 can only be discovered quickly and easily if one is already familiar with the picture. Here, the difficulty of preattentive feature association is due to the complexity of the figure, and its discovery is made even more difficult by the similarity of the features of dog and background. Feature association and recognition, in this case, requires visual memory or other "top-down" support.

Psychophysicists have become increasingly interested in visual feature associations during the last years, paricularly in preattentive mechanisms. Interest was stimulated by neuroscientists' discovery of oscillations in the visual cortex and by the hypothesis of "feature association by synchronization" (Eckhorn et al. 1988; Gray and Singer 1989). A good example of perceptual feature association related to neurophysiological findings is the work of Field and coworkers (Field et al., 1993), who found evidence for "local association fields" for contour formation among strings of Gabor elements (fig. 3). Associations among local line (Gabor-) elements into contours were induced within elongated "association fields" around the Gabor elements only if neighboring elements had similar orientations and were aligned on first order curvatures. These findings parallel observations in cat and monkey visual cortex that

Oscillatory Event-Related Brain Dynamics, Edited
by C. Pantev, Plenum Press, New York, 1994

led to the definition of "association fields of local populations of visual neurons" (Eckhorn et al. 1990b; discussed below).

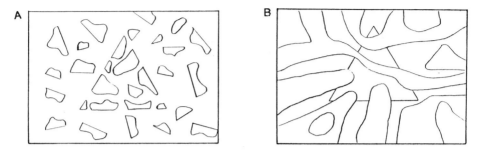

Figure 1. Association of a figure of aligned elements by the human visual system. **A:** The triangle is easily recognized even though the "background" consists of similar elements. **B:** The visible elements of the partly occluded triangle are easily associated into a figure.

Figure 2. Association and detection of a figure by the human visual system with help of "top-down" mechanisms. Figure and ground are composed of similar elements, and the figure is not a simple visual object (dog). Its recognition requires help, e.g. from visual memory or by advice (figure: Dalmatine dog with its head to the left and its tail to the right side; modified from James, 1966).

Are feature associations supported by neural synchronization ?

The neural mechanisms of associations are still unknown. For the visual system the main question is how the local cortical feature detectors, characterized by their receptive field (RF) properties, are combined such that their various properties are associated into coherent perceptual events. The "synchronization hypothesis" seems a promising approach: it states that those neurons participate in the representation of a visual object whose activities engage in a common synchronized state in response to stimulation by that object. The synchronization hypothesis attracted attention when stimulus-specific synchronized oscillations of 35-90 Hz were found in the visual cortex of anesthetized cats (Eckhorn et al. 1988, Gray et al. 1989, 1990) and awake monkeys (Freeman and van Dijk 1987; Kreiter and Singer 1992; Eckhorn et al. 1993b).

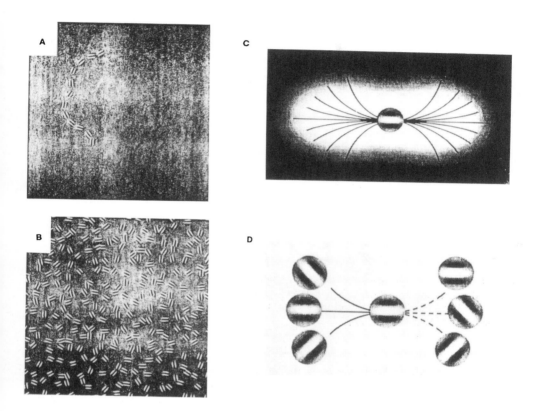

Figure 3. Contour integration by the human visual system: evidence for a local "association field". **A:** Path of elements the subject must detect when (**B**) embedded in an array of randomly oriented elements. **C:** The "association field". It represents the rules by which the elements in the path are associated and segregated from the background (the brighter the shading of the association field the stronger the associative forces). **D:** The curves represent the specific rules of alignment. Grouping occurrs only when the orientation of elements conforms to first-order curves (left side); elements like those at the right will not be associated. (Fig. modified from Field et al., 1993).

Recording synchronized activities simultaneously from several visual cortical areas

Locally, coding in the visual cortex can be characterized by receptive field (RF) properties of single neurons. Globally, features of a visual object are represented in a distributed fashion in several cortical areas. If we want to investigate the neural mechanisms associating the distributed local feature representations into a global coherent percept we have to record neural activities in parallel from several cortical areas. This was made in our Marburg group in the typical experimental situation shown in figure 4. Fiber-microelectrodes were inserted into the visual cortex areas 17, 18 and 19 of the cat and into V1 and V2 of the monkey. The signal correlations between recording positions with overlapping and neighboring receptive fields in the same and in different cortical areas can, in this setup, systematically be varied and related to various receptive field properties.

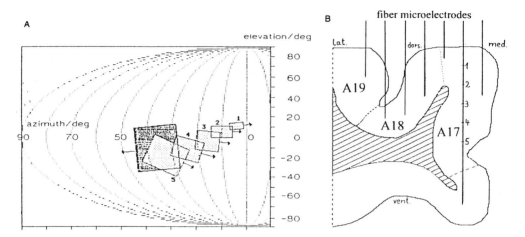

Figure 4. Schema of experimental situation in multiple electrode recordings from several cortical areas. **A:** Receptive fields in the left visual hemisphere. **B:** Frontal section of cat visual cortex, area 17, 18 and 19. The numbers at the tips of the schematically drawn fiber electrodes correspond to the respective receptive field numbers in A.

Figure 5 shows the three different signal types that were generally extracted from the extracellular broad band recording (Eckhorn 1992). The synchronized signal components of local cortical cell populations were of particular interest with respect to the "feature-linking" hypothesis. Such a signal is the local slow wave field potential (LFP; 12-120 Hz band pass). LFP provide an estimate of the average activities of postsynaptic signals on dendrites and somata near the electrode tip (Mitzdorf 1987). It is an averaged signal that mainly contains the synchronized components at the inputs of the local neural population near the electrode tip (range < 1 mm) because the synchronized components superimpose to relatively high amplitudes while statistically independent signals average out.

The synchronized components at the outputs of a local population were obtained by multiple unit recordings (MUA) from the same microelectrodes that recorded LFP. MUA is spatially more confined (range < 150 μm) and gives an estimate of the average spike activity near the electrode tip. High MUA amplitudes occur preferentially if the outputs of the local population are synchronized. Single unit activity (SUA) was also recorded by the same microelectrodes using conventional spike amplitude window discrimination. The capability of recording synchronized components of input (LFP) and output (MUA) signals of local populations enabled us to study their interactions at higher signal-to-noise ratios than is possible using single-unit spike trains alone.

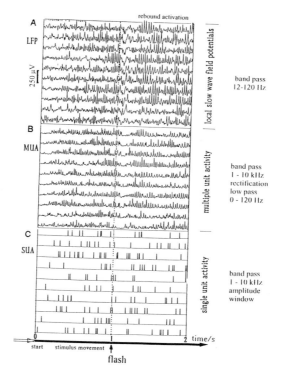

Figure 5. Three types of neural signals from each micro-electrode. **A:** Local slow wave (12-120 Hz) field potentials (LFP). **B:** Multiple unit spike activities (MUA). **C:** Single unit spike trains. All signals were recorded by a single electrode in response to identical stimulus repetitions. A whole-field grating stimulus (1 cycl/deg) started moving at t = 0 and continued at constant velocity (2°/s) throughout the shown sweep to t = 2 s. At about t = 1 s a diffuse short photoflash was applied at the stimulation screen. This caused inhibition of oscillatory events (visible in LFP and MUA), and induced after a delay of 100-200 ms oscillations at higher amplitudes as before the flash. (Fig. from Eckhorn et al. 1992).

RESULTS OF STIMULUS-SPECIFIC SYNCHRONIZATIONS IN CAT AND MONKEY VISUAL CORTEX

In the following paragraphs results obtained from the visual cortex of anesthetized cat and awake monkey are explained. The data are not presented and discussed separately, as the effects observed in both species do not differ essentially.

High-frequency oscillations in the visual cortex of the awake monkey

An example of the simultaneous recording of LFP, MUA and single unit spikes with a single microelectrode from the primary visual cortex V1 of an awake monkey is shown in figure 6. LFP and MUA had mainly low frequency components at moderate amplitudes while the moving light stimulus was outside the receptive fields (left sides of Figs. 6A,B,C). When the stimulus entered the receptive fields, high frequency LFP and MUA oscillations appeared (right side panels), indicating that many neurons are active synchronously. The single unit spike train in this example was barely involved in the rhythmic activity of the neighboring neurons. This is evident from the lack of a pronounced peak in the spectrum at the oscillation frequency, as is present for LFP and MUA (details of single neuron participation in oscillatory states are presented below). This example is typical for the visual cortex of cat and monkey as it shows that oscillations occur nearly exclusively during direct activation of the neurons within their receptive fields (Eckhorn et al. 1988; Gray and Singer 1989; Kreiter and Singer 1992; Eckhorn et al. 1993b).

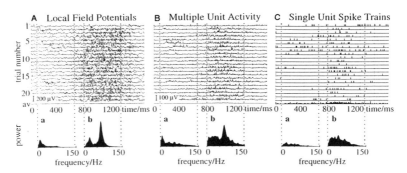

Figure 6. High-frequency (60-90Hz) oscillations in primary visual cortex of awake monkey. Oscillatory activity during stimulation. Stimulus: light bar (2.47 x 0.12°) appearing at t = 0 and sweeping at 1°/s for 2 s in the direction preferred by the neurones at the recording position. **A:** Local field potentials (10-100 Hz). **B:** Multiple unit activity (1-10 kHz, rectified, low-passed at 150 Hz). **C:** Single cell spike trains (spectra were calculated after folding of the spikes by a Gaussian). Upper panels: 22 signal courses in response to identical stimulus repetitions, recorded by a single electrode. Lowermost traces (**av**) show the averages of the 22 responses (same amplitude scale). Lower panels: average power spectra calculated from the 22 single response epochs marked by dotted lines (arbitrary power scales). **a:** 512 ms data epochs before stimulation, **b:** 512 ms epochs during stimulation of the RFs. Note the strong high frequency LFP and MUA oscillations and its weak appearance in SUA. Note also, that in the average traces (av) the oscillations are not present because they are not phase-locked to the stimulus. (Fig. modified from Eckhorn et al. 1993).

Orientation specific oscillations in monkey visual cortex

The validity of the "linking-by-synchronization" hypothesis requires that oscillations are occurring when the neurons are activated by their preferred stimuli. Such stimulus specificity was extensively shown for orientation and movement direction of a contrast border in anesthetized cats (e.g., Eckhorn et al. 1988; Gray and Singer 1989; Gray et al. 1989, 1990; Engel et al. 1990, 1991a, b; Eckhorn and Brosch 1993). An example of a characteristic for the direction of stimulus movement measured for the oscillatory components in monkey primary visual cortex is seen in figure 7. The oscillatory components of LFP and MUA were generally more sharply tuned (black characteristics) than the respective integrated responses (dotted characteristics; which show the conventional measure). This finding of narrower directional characteristics for oscillatory signals is supportive for the "synchronization hypothesis" because it shows that the synchronized population signals code the directional properties of the stimulus better than the respective conventionally measured population responses.

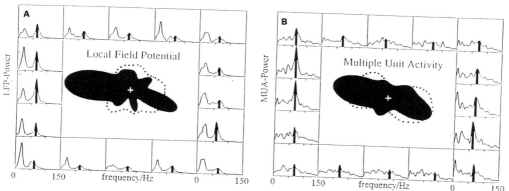

Figure 7. High frequency oscillations in V1 of awake monkey are orientation/direction specific. Directional tuning of oscillatory components in two different types of population activities. Simultaneous recordings of MUA and LFP from the same electrode. Stimulus: as for fig.6; movement in 16 different directions at 1°/s, each direction 10 times presented in random order. Outer panels: Average power spectra of 512 ms response epochs during which the stimulus crossed the RFs (arbitrary power scales, identical for all panels). Black bars denote the range ± 4 Hz around the actual spectral maximum between 50 and 150 Hz for each stimulus direction; (note that these dominant frequencies were very similar at different stimulus directions in cases where the oscillatory power was high). Centre curve: direction tuning characteristic of oscillatory components in response to different stimulus movement directions. Amplitudes are proportional to the area of the black bars in the outer spectra. Center shading: direction tuning characteristic of the entire response power. The direction characteristics were smoothed by a sinc-interpolation. **A:** Local field potentials, **B:** Multiple unit activity. Note the similarity of the MUA and LFP tuning curves of the high frequency oscillatory components and MUA overall power and their similarity with the appearance of 20 Hz LFP components in contrast to the broad tuning of LFP overall power. (Fig. modified from Eckhorn et al. 1993).

Spatial profile of cortical coherence in cat visual cortex

The "synchronization hypothesis" also requires that neighboring parts of the same visual object cause synchronized signals in the neurons representing these parts. This could be tested experimentally with a spatially extensive stimulus and parallel recordings distributed across the "visual map" of a single cortical area (fig. 8). Brosch and coworkers found in cat area 17 and 18 that the average spatial coherence declined with cortical distance with a space constant of about 4 mm, corresponding to about 12° visual angle (Brosch et al. 1991). At a distance of 6 mm coherence was down to 20% of its initial value. The decline of coherence between oscillatory events was generally due to the increasing phase jitter with distance. It was not due to a spatial reduction in the oscillation amplitudes because these were, on average, similar in different cortical positions due to the large field stimulation. Closer inspection of the data in fig. 8 revealed that the large variation of coherence at a single recording seperation was mainly due to differences in receptive field properties. Paired recording positions with similar receptive field preferences had, on average, considerably higher values of coherence than recording pairs with differing properties.

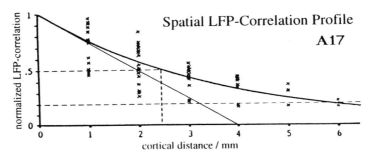

Figure 8. Spatial correlation profile of oscillatory LFP activities in cat visual cortex area 17. (Modified from Brosch et al. 1991)

The spatial decline of coherence indicates that oscillations do not occur in an ideally synchronized way within the range of representation of a larger visual object. According to the "feature association hypothesis", the coding of spatial continuity of a visual object by local groups of neurons in area 17 of the cat is restricted to a range of about 12-15° visual angle. However, such spatial confinement in the definition of continuity by a local group does not imply that it is impossible to define a larger figure on the basis of correlated activities of neurons within area 17. Coding of continuity would already be possible if the different neural subgroups, actually representing an object, would have some of their signal components synchronized with other subgroups of the same assembly. This spatial confinement of synchronized oscillations parallels, for example, the restricted area of the association fields induced by Gabor line elements in the psychophysical experiments described above (Field et al. 1993). However, in those experiments the perception of contours of local elements is not confined to the "local association field" of a single stimulus element.

Spatial relations of synchronized oscillations among different visual cortex areas

The presence of stimulus specific synchronization between different visual cortical areas is another prediction of the "feature association hypothesis". Inter-areal synchronization is expected because different features from the same position of a visual object and spatially dispersed features of the same extensive object have to be associated by synchronization. Inter-areal synchronization of oscillations was indeed observed (including positions in which the receptive fields do not overlap; Eckhorn et al. 1988; Engel et al. 1991b; Eckhorn and Brosch 1993). An example of synchronization between recording pairs from area 17 and 18 of the cat at various distances is given in figure 9. Highest degrees of inter-areal synchronization were present at positions with overlap of the receptive fields, and up to separations of about 10° visual angle. Synchronization in these LFP recordings was present up to 15-20° visual angle (on average with a moderate level of confidence). This means in terms of the "synchronization hypothesis", that feature associations are supported over a broad visual range by signals occurring synchronized in different cortical areas (here area 17 and 18).

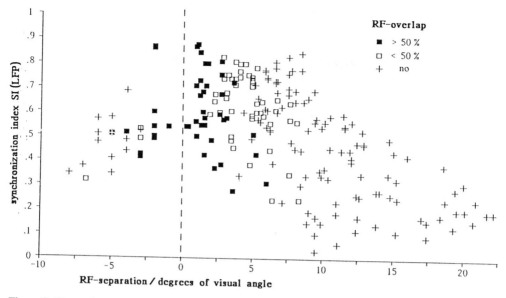

Figure 9. Dependence of area 17-18 synchronization index of local field potentials on the separarion of the receptive field centers. Different symbols indicate the 3 classes of RF overlap and separation, respectively (see insert). RF separation was measured from the area 18 reference RF positions. Positive values indicate area 17 RF positions in peripheral retinal direction relative to the area 18 RF, while negative values indicate RF positions nearer to the central retinal position. (Modified from Eckhorn and Brosch, 1993).

Synchronization of oscillations in different cortical areas depends on stimulus orientation and movement direction

The broad range of synchronization indices at a given distance of receptive fields in figure 9 can again be explained by the differences in preferred stimulus properties, because pair recordings were made between neurons of any type of receptive field properties. The more similarities in receptive field properties, the more probable were high degrees of synchronization among given neurons. An example of stimulus-specific synchronization among two cortical areas (A17 and 18 of the cat) is shown in figure 10 : movement of a large grating in one direction synchronized neurons in the recording pair that preferred this direction, whereas the others were hardly activated or not at all (fig. 10A). Change of the stimulus movement to the orthogonal direction synchronized another A17-18 pair of neurons, while the previously synchronized pair became inactive.

Visual area 18 can dominate the synchronization with area 17

The stimulus-specific synchronizations observed between cat areas 17 and 18 (Eckhorn and Brosch 1993) had another interesting specificity. They occurred with highest amplitudes and probability if the stimulus movement direction matched that of the area 18 neurons (e.g. fig. 10, 11). This might be interpreted as a domination of area 18 neurons over those in area 17 (fig. 11). As area 18 neurons in the cat have larger receptive fields, and prefer broader ranges of stimulus movement velocities in comparison with neurons in area 17, one might argue that area 18 of the cat strongly supports associations of the finer area 17 features by virtue of its coarser ones. Further speculations lead to parallels in psychophysical observations, in which dominance of coarse over fine spatial features is often found in feature association tasks.

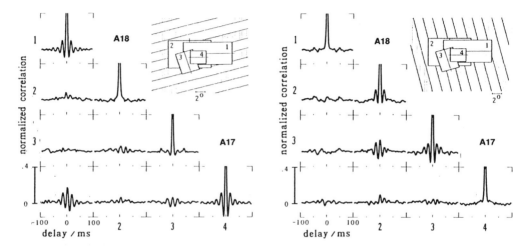

Figure 10. Area 17-18 synchronization is direction specific. Example for the synchronization between area 17-18 spike activities SI(MUA) and its dependence on the orientation preferences of the neurons. Simultaneous recordings from two electrodes in area 17 (3,4) and two in area 18 (1,2). Autocorrelograms (clipped at 0.4) are plotted on the diagonals (upper left to lower right), cross-correlograms in the respective off-diagonal positions. The inserts show the RFs and (moving) grating stimuli drawn to scale. **A:** Correlograms obtained with moving grating oriented near horizontally, **B:** near vertically. Electrode 1: RF type not determined; el.2 and 3: simple-like RFs; el. 4: complex-like RF. Note the oscillatory modulations (autocor.) and synchronizations (crosscor.) of those recording pairs where the neurons are stimulated near their preferred orientations (A: 1 and 4; B: 2 and 3). Same stimulation as for fig. 4. (Fig. modified from Eckhorn and Brosch, 1993).

Oscillatory events in different cortical areas occur at zero phase difference

Average phase differences between various positions of the same cortical area, and among different cortical areas (cat area 17, 18 and 19, and monkey V1 and V2) were approximately zero (< 1 ms; Eckhorn and Brosch, 1993; Eckhorn et al. 1993a). Such in-phase oscillations were also demonstrated by us between visual areas V1 and V2 of an awake monkey (Eckhorn et al. 1993a). Figure 12 shows an example and average values of V1-V2 phase differences. They were, in the data recorded so far, narrowly distributed around zero (fig. 12A). Phase differences of individual recording pairs, however, had slightly broader distributions (SD=3.2 ms).

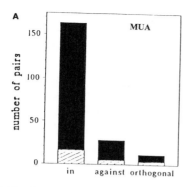
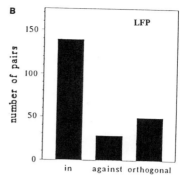

Figure 11. Numbers of area 17-18 recording pairs in which synchronization was maximal in/against/orthogonal to/ the preferred stimulus direction of the area 18 site. **A:** Data for MUA. The hatched areas indicate the number of cases where only those recording positions from area 17 were selected in which the neurons had an orthogonal orientation preference (to the area 18 neurons) and were direction and orientation selective. **B:** With LFP. (Fig. modified from Eckhorn and Brosch 1993).

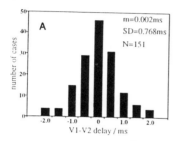
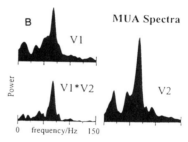

Figure 12. Phase-Locked High Frequency oscillations between visual cortical areas V1 and V2 of an awake monkey have zero phase difference. **A:** Distribution of average phase differences between oscillatory events in V1 and V2. **B:** Example of power spectra and cross-power spectra of V1 and V2 multiple unit activities. Experimental parameters as for fig. 6. (From Eckhorn et al. 1993).

The zero phase difference among V1 and V2 is of particular interest, because these visual areas are serially arranged with respect to the afferent visual stream, and because V1-V2 conduction delays are several milliseconds. A possible explanation of zero phase difference would be a common oscillatory input source to V1 and V2. At present it is more probable, however, that zero phase synchronization is mainly due to temporally symmetric feedback interactions among the cortical areas (Engel et al. 1991a; Eckhorn 1991, 1992; Gerstner and van Hemmen, 1993). In conclusion, associations of features, the representations of which are dispersed among different cortical areas, could be efficiently supported by oscillations at zero phase difference because primarily the synchronized signals support one another at the spike encoder of a neuron while uncorrelated activities are transmitted with much less efficiency.

Single neurons are differently involved in stimulus-specific oscillations in cat visual cortex

The neural mechanisms and structures that generate and synchronize fast cortical oscillations have not yet been identified. We were therefore trying to clarify how and when single cortical neurons participate in oscillatory activities of remote as well as nearby neural populations (Eckhorn and Obermueller, 1993). We asked, in addition, if different types of participation can occur, and if they depend on visual stimulation and receptive field properties.

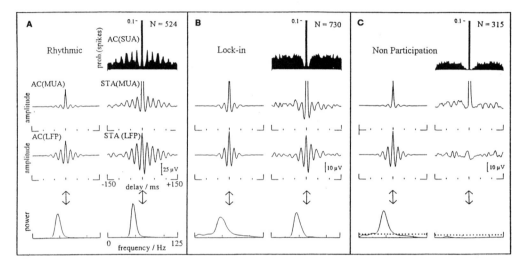

Figure 13. Three different states of single cell couplings with oscillatory population activities. **A:** rhythmic, **B:** lock-in, and **C:** non-participation states of three different neurons. Each recording triple is from a single electrode. AC: auto-correlation histograms of single cell spikes (SUA), of multiple unit activity (MUA), and of local field potential (LFP). STA denotes spike-triggered averages of multiple unit activity or local field potentials. N: number of single cell spikes, equal to the number of averages in the STAs. According to the classification, STAs have oscillatory modulations in the rhythmic and lock-in states, and lack such modulation in the non-participation state. Note that in the rhythmic state (A) the single cell correlogram (top) is clearly modulated at 44 Hz, while in the lock-in (B) and the non-participation states (C) rhythmic modulations in the range 35-80 Hz are not visible (by definition). Lowest row of panels: power spectra for the above row of correlograms, calculated by Fast-Fourier transformation. The dotted lines in the power spectra of panel C indicate the respective confidence levels (4 times background variance). (Modified from Eckhorn and Obermueller, 1993).

Three states of single cell participation in oscillations can be distinguished in spike-triggered averages of LFP or MUA from the same electrode (Figure 13): (1) rhythmical states are characterized by the occurrence of relatively regular spike intervals at frequencies of 35-80 Hz, and these rhythms are correlated with LFP and MUA oscillations; (2) lock-in states lacking rhythmic components in single-cell spike patterns, although many spikes are phase-coupled with LFP or MUA oscillations; (3) during non-participation states, LFP or MUA oscillations are present but single-cell spike trains are neither "rhythmic" nor phase coupled to these oscillations. Stimulus manipulations (ranging from "optimal" to "sub-optimal" for the generation of oscillations) often led to systematic transitions between these states (i.e., from rhythmic, to lock-in, to non-participation). Single-cell spike coupling was generally associated with negative peaks in LFP oscillations, irrespective of the cortical separation of single cell and population signals (0 - 6 mm). These results suggest that oscillatory cortical population activities are not only supported by local and distant neurons which display rhythmic spike patterns, but also by those showing irregular patterns in which some of the action potentials are phase-locked to oscillatory field events.

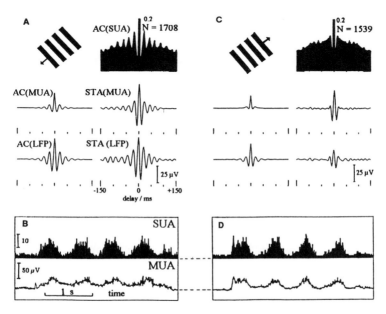

Figure 14. Transition from rhythmic to lock-in state with reversing stimulus movement direction. Simple cell from area 17, sensitive for stimulus orientation but not for movement direction (defined by the number of spikes per stimulus). Stimulus: Square wave grating of 0.7 cycles/°, moving at 2.4 °/s for 3.5 s across the receptive fields at preferred orientation. **A** and **C** are correlograms as in fig. 13, both from the same recording position at two reverse stimulus movement directions (indicated by arrows at the grating schema). **B** and **D** show average single unit and multiple unit activities in response to 14 identical stimulus repetitions. Note the similar spike activation levels in A and B versus C and D while the simple cell changes its state from rhythmic (A) to lock-in (C). It is also noteworthy that the oscillatory modulation of the multiple unit activities and local field potentials are lower (hence significant) in the lock-in than in the rhythmic state in this example. (Modified from Eckhorn and Obermueller, 1993).

State transitions of single cell coupling with oscillatory population activities depend on stimulation and receptive field properties. Figure 14 is an example of changing a strong rhythmic state (14A) into a lock-in state (14B) by reversing the movement direction of a grating stimulus. Such reversals of stimulus movement direction cause often transitions from rhythmic to lock-in states and from lock-in to non-participation states.

Summarizing the influences of RF-properties on state transitions, we have found that state transitions in the direction from rhythmic to lock-in to non-participation states often occurred by changing a stimulus from "optimal" to "sub-optimal"

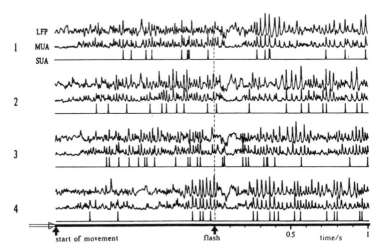

Figure 15. Rhythmic signals in cat visual cortex induced by a moving stimulus (and inhibited by a flash). Responses to 4 identical stimulus repetitions of three types of neuronal signals recorded in parallel with the same microelectrode. Stimulus: grating of 0.25 cycl/deg moving at v = 4.5 deg/s. Arrows below mark beginning of stimulus movement (left) and instance of a light flash (middle) that was given while the grating continued to move. Note the 200 ms post flash depressions of LFP high frequency components and of MUA and single unit spike discharges. (From Eckhorn and Schanze, 1991).

for the generation of oscillatory population activity. State changes in the reverse direction rarely occurred under these conditions. Here the term "optimal" versus "sub-optimal" is used with respect to the stimulus preferences of the neurons (tested were orientation, direction and velocity of a moving stimulus, binocular disparity, and temporal modulation frequency).

Two components of cortical synchronization: rhythmic and nonrhythmic

If internally generated synchronized rhithms play a role in perceptual feature-linking relations, one may assume that other types of stimulus related synchronizations probably also play a role. Most important in this respect are synchronizations that are due to sudden retinal image shifts. Such stimuli evoke stimulus-locked synchronizations at short response delays in the visual cortex (Eckhorn et al.1990a). Synchronized oscillations, on the other hand, often occur at longer delays and last longer. It would be of advantage in temporally critical situations for such stimulus-locked responses to be able to "overwrite" presently ongoing rhythmic synchronizations, in order to obtain an updated view of the visual scene. Such behavior has indeed been seen in the cat's visual cortex (fig. 15): stimulus-locked cortical synchronizations could inhibit immediately oscillatory activities immediately if the stimulus was strong and fast enough (Eckhorn and Schanze 1991; Kruse et al. 1991). The occurrence of stimulus-locked response components in the cortex can be explained by the simultaneous firing, due to sudden image changes, of those neurons that have similar receptive fields and are stimulated simultaneously by similar features of the visual object (for simulations of stimulus-locked synchronizations in laterally coupled neural networks see Pabst et al., 1989).

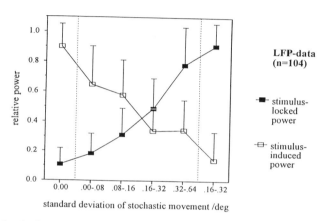

Figure 16. Stimulus-locked response components gradually inhibit stimulus-induced oscillations. Local field potential recordings from cat visual cortex. Stimulus: grating, moving with a random component superimposed on a constant slow velocity. Open symbols: average normalized response power of the high-frequency oscillatory components; filled symbols: normalized power of the stimulus-locked components. Amplitudes (standard deviation) of the random stimulus movement are indicated at the abscissa. Each value is the average of N=104 single curves (modified from Kruse and Eckhorn 1991).

In order to study interactions between externally driven (stimulus-locked) and internally induced (oscillatory) signals in some detail, we applied stimuli that elicited both types of activities (Kruse et al. 1991). Strong oscillations (40-90 Hz) were preferentially induced in our experiments by sustained slowly moving grating or bar stimuli that did not evoke brisk transient responses in cortical neurons. Externally driven (stimulus-locked) responses occurred in the range 1-40 Hz, and were evoked in our experiments by (1) sudden stimulus movements (jerks), (2) random movements with variable amplitudes, and (3) light flashes (fig. 15). Gradual increases in stimulus-locked response amplitudes were paralleled by gradual decreases in the amplitudes and occurrence probability of oscillatory events (fig. 16): the stronger the stimulus-locked response components, the stronger the inhibition of oscillatory events. After periods of inhibition, oscillatory events usually occurred with enhanced amplitudes and probabilities, compared with situations where no stimulus-locked responses interfered with their generation. It can therefore be assumed that strong stimulus-locked responses dominate cortical synchronization.

After a certain "relaxation" time, however, when the evoked stimulus response has partly worn off, a process of self-organization might lead to oscillatory synchronization. Similar behavior was observed in neural network models (Eckhorn et al. 1990b; Arndt 1993).

Periods of strong stimulus-locked and oscillatory stimulus-induced synchronizations might characterize extreme processing states of the visual cortex. In natural vision, short stimulus shifts are often followed by phases with more stationary retinal images, e.g., in saccade-fixation sequences or when a visual object suddenly moves and stops again. In both visual situations, the primary stimulus-locked responses that occur simultaneously in many cortical neurons might help to signal relatively crudely, but quickly, the "when?", "where?" and "what?" of an ongoing visual event. After sudden shifts of an object, or after saccades, post-inhibitory rebound activations might be useful for the sensitive generation of synchronized oscillations in just those parts of the cortical network in which the postsynaptic influence of the linking connections is still present.

We therefore assume stimulus-locked and oscillatory synchronizations occurring in alternation during natural vision if the stimulus-locked responses are strong. However, both components are simultaneously present whenever sustained and moderately transient stimulation drives the neurons. Both types of synchronized activities might, therefore, be capable of contributing to the perceptual integration in proportion to their actual amplitudes.

Feature associations might also be supported by coupled oscillations among different frequencies

Fast oscillations in cat and monkey visual cortex were often found to occur together with multiple spectral peaks at different frequencies. It is interesting for the hypothesis of "feature association by synchronization" whether the features to be linked have to do so at the same frequency or whether phase coupling at different frequencies might also be able to support it. Coupling of relatively low-frequency oscillations with those in the high-frequency range are of particular interest because the former have been monitored for many decades in human and animal EEGs, and because low-frequency waves have been attributed to a broad variety of higher brain functions (for example the theta activity of the hippocampus for memory formation: for an overview, see Basar 1980). Thomas Schanze in my group has followed this question by asking whether the frequency components recorded intracranially in the visual cortex of cat and monkey are phase-coupled, in order to reveal processes which were previously assumed to be independent (e.g., coupling between rhythms in the alpha and gamma frequency ranges of EEG; Schanze and Eckhorn, 1993). Such phase-coupling among different frequencies cannot be detected in the conventional power spectrum, but the bi-spectrum is able to extract them if the correlations are due to quadratic phase couplings (Nikias and Raghuveer 1987). Bi- and power spectra were calculated of multiple-unit activities (MUA) and local field potentials (LFP) for the visual cortex of anesthethized cats (A17, 18 and 19), and an awake monkey (V1; fig. 17). For the cat and monkey cortical areas analyzed so far, the bi-spectra often were stimulus dependent and, in about 60% of the cases, revealed significant quadratic phase couplings of oscillations between two or several pairs of frequencies: (1) within the gamma-frequency band (30-90Hz); (2) between gamma and low frequency (1-30 Hz) rhythms; and (3) among rhythms at lower frequencies (1-30Hz).

We conclude that phase-coupling of oscillatory events may not be restricted to high frequency (35-90 Hz) events at a common frequency. This suggests that rhythmic processes at different time scales can be coupled in the visual system and, according to the "association hypothesis", are thus candidate signals for the support of multi-modal feature associations in sensory and association cortices.

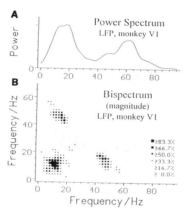

Figure 17. Phase coupling between oscillations at different (incommensurate) frequencies. **A:** Typical power spectrum of a local field potential recording from the visual cortex (V1) of an awake monkey during stimulation of the receptive fields at the recording position. Note the peaks at low and high frequencies. **B:** Normalized amplitudes of the respective bispectrum. The black dots indicate frequency pairs that are phase coupled. In this example, this is the case within the low frequency band (7-15 Hz) and between low and high frequency components (strong coupling, e.g., around 11 and 46 Hz (= 57 Hz). (Modified from Schanze and Eckhorn, 1993).

Visual association fields of local cortical populations

In order to bring mechanisms of synchronization among neurons representing a visual object into correspondence with perceptual capabilities of feature associations, the concept of the "linking or association field" of a local neural assembly was introduced (Eckhorn et al. 1990b). The linking field extended the concept of single-cell receptive fields (local coding) to neural ensemble coding. The association field of a local assembly of visual neurons is defined as that area in visual space where appropriate local stimulus features induce synchronization of activities within that assembly. Association fields are constituted both by the RF-properties of the respective assemblies and by the properties of their linking interconnections (feed-forward, lateral, feedback). This implies that the association field of a given group of visual neurons is generally much broader than the super-imposed receptive fields of the neurons that constitute the "association field assembly".

CONCLUSIONS

It has been postulated that the perceived association of visual features is based on the synchronization of neural signals activated by a coherent visual object. Two types of synchronized cortical signals were recorded in the visual cortex of both cat and monkey, and were proposed as candidates for feature association: (1) stimulus-locked signals, evoked by transient retinal stimulation, and typically non-rhythmic; (2) oscillatory signals, induced by sustained stimuli, and typically not locked in their oscillation phases to stimulus events. Both types of signals can occur synchronously in neurons activated by a common stimulus. Synchronized activities were found in paired recordings within vertical cortex columns, in separate columns of the same cortical area, and even between different cortical areas or hemispheres. The average phase difference between such common oscillatory events was typically close to zero. Oscillatory events in any two assemblies were more closely correlated the more similar their receptive field properties were, and the better a common stimulus activated the assemblies simultaneously. Phase coupling of oscillatory events occurring simultaneously at different frequencies was also observed. It might also play a role in associative mechanisms. Our results can explain some neural mechanisms of perceptual feature-linking, including mutual enhancement among similar, spatially and temporally dispersed features, definitions of spatial and temporal continuity, scene segmentation, and figure-ground discrimination. We further propose that mutual enhancement and synchronization of cell activities are general principles of temporal coding by assemblies. We assume that these mechanisms are also used within and among other sensory modalities as well as between sensory and motor systems. However, these assumtions have to be tested in future experiments.

Acknowledgements

I am thankful to Prof. H. J. Reitboeck, the head of our Marburg Biophysics Department, for his support in developing both concepts and the multiple-electrode instrumentation. This article could not have been written without the numerous discussions with my colleagues and without their extensive help in experiments, data acquisition and processing, as credited in the text and figure captions. Expert help in care and preparation of our cats and monkeys and help in experimental techniques came from U. Thomas, J.H. Wagner and W. Lenz. The financial support by the Deutsche Forschungsgemeinschaft is also greatly acknowledged (Ec 53/4, Ec 53/6 and Ec 53/7).

REFERENCES

Arndt, M. (1993) Representation of spatial and temporal continuity by synchronization of neural signals (in German). Doctoral Thesis, Philipps-University, Marburg, Germany.

Arndt, M., Dicke, P., Erb, M. and Reitboeck, H.J. (1991) A multilayered, physiology-oriented neural network model that combines feature-linking with associative memory. In: Elsner, N. and Penzlin, H. (eds), Synapse, Transmission, Modulation. Thieme Verlag, Stuttgart, New York, 587.

Basar, E. (1980) EEG-Brain Dynamics. Amsterdam, New York, Oxford, Elsevier, North-Holland Biomedical Press.

Brosch, M., Bauer, R., Schanze, T. and Eckhorn, R. (1991) Stimulus-induced oscillatory events and their spatial correlation profiles in cat visual cortex. (Suppl) Europ. J. Neurosci., Suppl. 4: 54,1235.

Eckhorn, R., Bauer, R., Jordan, W., Brosch, M., Kruse, W., Munk, M. and Reitboeck, H.J. (1988) Coherent oscillations: A mechanism of feature linking in the visual cortex? Multiple electrode and correlation analysis in the cat. Biol. Cybernetics 60: 121-130.

Eckhorn, R., Reitboeck, H.J., Arndt, M. and Dicke, P. (1989) A neural network for feature linking via synchronous activity: Results from cat visual cortex and from simulations. In: Models of Brain Function. Cotterill, R.M.J. (ed.) Cambridge University Press, pp 255-272.

Eckhorn, R., Brosch, M., Salem, W. and Bauer, R. (1990a) Cooperativity between cat area 17 and 18 revealed with signal correlations and HRP. In: Brain and Perception, Elsner, N. and Roth, G. (eds.), Thieme, Stuttgart New York, p 237.

Eckhorn, R., Reitboeck, H.J., Arndt, M. and Dicke, P. (1990b) Feature linking among distributed assemblies: Simulations and results from cat visual cortex. Neural Computation 2:293-306.

Eckhorn, R. (1991) Stimulus-evoked synchronizations in the visual cortex: Linking of local features into global figures? In: Neural Cooperativity, Krüger, J. (ed.). Springer Series in Synergetics. Springer-Verlag, Berlin, Heidelberg, New York, pp 184-224.

Eckhorn, R. and Schanze, T. (1991) Possible neural mechanisms of feature linking in the visual system: Stimulus-locked and stimulus-induced synchronizations. In: Babloyantz, A. (ed.), Self-Organization, Emerging Properties and Learning, Plenum Press, New York, pp 63-80.

Eckhorn, R. (1992) Principles of global visual processing of local features can be investigated with parallel single-cell- and group-recordings from the visual cortex. In: Information Processing in the Cortex. Aertsen, A. and Braitenberg, V. (eds.). Springer-Verlag, Berlin, Heidelberg, New York, pp 385-420.

Eckhorn, R., Dicke, P., Arndt, M. and Reitboeck, H.J. (1992) Feature linking of visual features by stimulus-related synchronizations of model neurons. In: Basar, E. and Bullock, T.H. (eds.). Induced Rhythms in the Brain. Brain Dynamics Series, Birkhäuser, Boston, Basel, Berlin, pp 397-416.

Eckhorn, R., Schanze, T., Brosch, M., Salem, W. and Bauer, R. (1992) Stimulus-specific synchronizations in cat visual cortex: Multiple microelectrode and correlation studies from several cortical areas. In: Basar, E. and Bullock, T.H. (eds.), Induced Rhythms in the Brain. Brain Dynamics Series, Birkhäuser, Boston, Basel, Berlin, pp 47-82.

Eckhorn, R. and Brosch, M. (1993) Synchronization between oscillatory activities in areas 17 and 18 of the cat's visual cortex. Exper. Brain Res. (in press).

Eckhorn, R., Frien, A., Bauer, R., Kehr, H. and Woelbern, T. (1993a) Phase-locked high frequency oscillations between visual cortical areas V1 and V2 of an awake monkey. In: Elsner, N. and Heisenberg, M. (eds.), Gene, Brain, Behaviour. Thieme Verlag, Stuttgart, New York, p 444.

Eckhorn, R., Frien, A., Bauer, R., Woelbern, T. and Kehr, H. (1993b) High frequency (60-90 Hz) oscillations in primary visual cortex of awake monkey. NeuroReport 4: 243-246.

Eckhorn, R. and Obermueller, A. (1993) Single neurons are differently involved in stimulus-specific oscillations in cat visual cortex. Exp. Brain Res. 95:177-182.

Engel, A.K., König, P., Gray, C.M. and Singer, W. (1990) Stimulus-dependent neuronal oscillations in cat visual cortex: Inter-columnar interaction as determined by cross-correlation analysis., Europ. J. Neurosci. 2: 588-606.

Engel, A.K., König, P., Kreiter, A.K. and Singer, W. (1991a) Interhemispheric synchronization of oscillatory neuronal responses in cat visual cortex. Science 252:1177-1179.

Engel, A.K., Kreiter, A.K., König, P. and Singer, W. (1991b) Synchronization of oscillatory neural responses between striate and extrastriate visual cortical areas of the cat. Proc. Natl. Acad. Sci. USA 88:6048-6052.

Field, D.J., Hayes, A. and Hess, R.F. (1993) Contour integration by the human visual system: evidence for a local "association field". Vision Res. 33:173-193.

Freeman, W. and van Dijk, B. W. (1987) Spatial patterns of visual cortical fast EEG during conditioned reflex in a rhesus monkey. Brain Res. 422:267-276.

Gray, C.M. and Singer, W. (1989) Stimulus-specific neuronal oscillations in orientation columns of cat visual cortex. Proc. Natl. Acad. Sci. USA, 86:1698-1702.

Gray, C.M., König, P., Engel, A.K. and Singer, W. (1989). Oscillatory responses in cat visual cortex exhibit inter-columnar synchronization which reflects global stimulus properties. Nature 338:334-337.

Gray, C.M., Engel, A.K., König, P. and Singer, W. (1990) Stimulus-dependent neuronal oscillations in cat visual cortex: receptive field properties and feature dependence. Europ. J. Neurosci. 2:607-619.

Kreiter, A.K. and Singer, W. (1992) Europ. J. Neurosci. 4: 369-375.

Kruse, W., Eckhorn, R. and Schanze, T. (1991) Two modes of processing and their interactions in cat visual cortex: Stimulus-induced oscillatory and stimulus-locked synchronizations. In: Elsner, N. and Penzlin, H. (eds.), Synapse, Transmission, Modulation. Thieme Verlag, Stuttgart, New York, p. 217.

Mitzdorf, U. (1987) Properties of the evoked potential generators: current source-density analysis of visually evoked potentials in the cat cortex. Intern. J. Neurosci. 33:33-59.

Nikias, C.L. and Raghuveer, M.R. (1987) Bispectrum estimation: a digital signal processing framework. Proc. IEEE 75:869-891.

Pabst M, Reitboeck HJ, Eckhorn R (1989) A model of pre-attentive region definition in visual patterns. In: Models of Brain Function. Cotterill, R.M.J. (ed.), Cambridge University Press, pp 137-150.

Schanze, T. and Eckhorn, R. (1993) Phase coupling between oscillations at different frequencies in the visual cortex of cat and monkey. In: Elsner, N. and Heisenberg, M. H. (eds.), Gene, Brain, Behaviour. Thieme Verlag, Stuttgart, New York, p 442.

Stoecker, M., Eckhorn, R. and Reitboeck, H.J. (1991) Oscillatory synchronizations in neural networks: Responses to stimuli that induce the perception of subjective contours in humans. In: Elsner, N. and Penzlin, H. (eds.). Synapse, Transmission, Modulation. Thieme Verlag, Stuttgart, New York, p 216.

EVENT-RELATED CHANGES IN THE 40 HZ ELECTROENCEPHALO-GRAM IN AUDITORY AND VISUAL REACTION TIME TASKS

Hennric Jokeit[1], Ralf Goertz[2], Erika Küchler[2], and Scott Makeig[3]

[1]Institute for Medical Psychology, University of Munich, Germany
[2]Department of Psychology, Humboldt-University, Berlin, Germany
[3]Naval Health Research Center, San Diego, USA

INTRODUCTION

Mean motor response times to expected stimuli can differ considerably between subjects even when experimental conditions are stable. It is well known that central rather than peripheral processes are mainly responsible for these differences (Glickstein, 1972; Luce, 1986). However, questions remain open whether and which measures of neuroelectrical brain activity can reveal significant aspects of the underlying sensorimotor processes. A number of studies using event-related brain potentials have yielded only small or insignificant correlations between electrophysiological measures and motor reaction times (Kutas et al., 1977; Goodin et al., 1990; Ortiz et al., 1993).

However, averaged evoked responses reveal only one aspect of event-related brain dynamics (Makeig, 1993). An important distinction can be made between evoked EEG activity, consisting of potential shifts which are both time- and phase-locked to experimental events and which are captured in ERPs, and induced EEG activity, also including oscillations time-locked but not necessarily phase-locked to experimental events (Eckhorn et al., 1988). Induced activity which is not phase-locked to experimental events does not appear in averaged evoked responses, but its mean tendencies can be measured by averaging event-related spectra.

It has been suggested that oscillatory brain activity in the gamma band (roughly 25-90 Hz) is intimately involved in elementary sensorimotor and cognitive processing (Singer, 1993). Electroencephalographic (EEG) studies have reported spontaneous cortical activity in the 40 Hz range during periods of increased alertness and vigilance (Boyer et al., 1987; Steriade et al., 1991), during accurate performance of a conditioned response (Freeman and Dijk, 1987), and during periods of focused arousal during cognitive task performance (Sheer, 1989) including reaction time

Oscillatory Event-Related Brain Dynamics, Edited
by C. Pantev, Plenum Press, New York, 1994

tasks (Krieger and Dillbeck, 1987). Suppression of 40 Hz activity has been noted during central anaesthesia (Madler and P ppel, 1987) and delta sleep (Llinas and Ribary, 1993). While stable gamma band features have been identified in averaged event-related potentials (ERPs) time- and phase-locked to auditory stimuli (Basar et al., 1976; Galambos et al., 1981; Makeig, 1990; Pantev et al., 1991), visual and olfactory studies in animals have found stimulus-induced cortical gamma band activity that typically takes the form of brief, irregular bursts (Freeman and Skarda, 1985; Basar-Eroglu and Basar, 1991; Singer, 1993) which follow, but are not reliably phase-locked to stimulus onsets. Therefore, it is reasonable to wonder whether the temporal dynamics of 40 Hz activity in the human EEG are related to particular sensorimotor tasks and individual different performances.

The experiments reported here were conducted to examine event-related changes in spectral power in the 40 Hz range induced by a simple auditory and a visual priming reaction time task. More detailed analyses of both experiments are being documented (Jokeit and Makeig, in press; Goertz, Jokeit, and Küchler, in preparation).

A SIMPLE AUDITORY REACTION TIME TASK

Methods

In the first experiment, 23 right-handed adults (ages 20-53) were tested in two conditions. First, 110 clicks (2 ms width, 75 dB SPL) were presented binaurally at random inter-stimulus intervals (ISIs) of 3 to 7 s. Subjects were then asked to react as quickly as possible to a second set of 275 clicks (with the same intensity and ISI distribution) by pressing a response button with the index finger of their right hand. EEG was collected from a scalp electrode located at the vertex (Cz) referred to linked mastoids.

EEG epochs of 640 ms beginning 128 ms prior to each click were recorded with a sampling rate of 2000 Hz using a 12 bit A/D converter with an analog high pass filter cutoff of 0.67 Hz and a 50 Hz line frequency notch filter. To exclude large eye movements and muscle activity, epochs in which potential anywhere exceeded ±70 µV were eliminated from the analysis. Reaction time (RT) was recorded separately with a temporal resolution of 1 ms. Trials with RTs shorter than 100 ms or longer than 600 ms were also eliminated. On average, 80% of the trials were retained for analysis. Averaged evoked gamma band responses (GBRs) were computed by applying a 20 to 60 Hz bandpass filter to the epochs before averaging. All filtering used symmetric Butterworth filters with 24 dB/octave slopes.

To compute 40 Hz band event-related spectra (ERS), each response epoch was first band pass filtered from 20-60 Hz. The epochs were then divided into 68 overlapping 24-point (96 ms) time windows with a shift interval of 8 ms. After tapering using a Gaussian window function, and zero-padding to 64 points, each window was converted to spectral power using an FFT. Since gamma band power was interrupted by the 50 Hz notch filter used in the recording, the range 35 to 43 Hz was chosen for analysis. Power in this range was integrated for each time point using a Hamming window function. The 40 Hz ERS transform of each epoch thus consisted of 68 power estimates at 8 ms intervals.

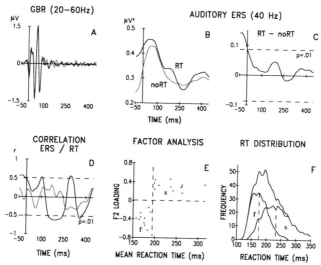

Figure 1. Responses to clicks in passive listening (noRT) and speeded motor response (RT) task conditions. Grand means of responses from 23 adult subjects (site Cz referred to linked mastoids). (A) Grand mean gamma band responses (GBRs), bandpass filtered between 20 and 60 Hz. Abscissa is time re the onset of the click. The ordinate is potential in μV. Note the above baseline activity after stimulus onset. (B) Grand mean event-related spectra (ERS) for two task conditions: speeded motor responses (RT) and passive listening (noRT). Each trace plots mean time-varying power near 40 Hz. Abscissa: time re stimulus onset. Ordinate: power in μV^2. (C) ERS difference curve between both task conditions which differs significantly only prior to the click onset. (D) Pearson correlation coefficients between ERS power at each time point in the RT condition and mean response times. The correlation for the noRT condition (dotted) is also shown. Abscissa: time re stimulus onset. Ordinate: Pearson correlation coefficient r. Dotted line shows limits of significant correlations ($p < .01$). (E) Factor analysis on subjects (Q-Factor Analysis) performed on a Pearson correlation matrix on task differences between individual event-related spectra (ERS) at 40 Hz. Output submitted to Varimax rotation. Scatter plot of the second factor F2 against individual mean reaction time (RT). Abscissa: mean RT in ms. Ordinate: the subject's loading on F2. Note that F2 separates subjects by mean RT: no subject in the 'slow' group had a lower mean RT than any of the 'fast' group. (F) Reaction time distribution of 23 subjects and sub-distributions of the 10 fast ('f') and 12 slow ('s') responders respectively.

Statistical significance of response differences in the measures was tested by repeated measures analysis of variance (RM-ANOVA), by Bonferroni-corrected t-tests for dependent samples, and by t-tests for independent samples. Threshold of significance was considered to be $p < 0.01$.

Results

Grand mean GBRs for the speeded response (RT) and the passive listening (noRT) condition (Fig. 1A) each contain above-baseline activity during approximately the first 100 ms after click presentation. As shown in Fig. 1A the GBR did not vary as a function of task. Although the GBR was unaffected by task, there were task differences in the 40 Hz ERS (Fig. 1B,C). In the RT condition the subjects had significantly more 40 Hz power just prior to the stimulus than in the noRT condition.

Following the stimulus, both ERS curves reach their maximum during the first 100 ms after stimulus onset, paralleling the activity visible in the GBR (Fig. 1A).

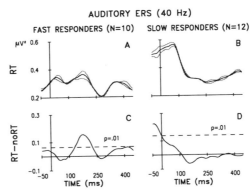

Figure 2. Grand mean 40 Hz event-related spectra (ERS) for two subject subgroups: 10 fast responders and 12 slow responders from the original 23 subjects. Task conditions: speeded motor responses (RT) and passive listening (noRT). Each trace plots mean time-varying power at 40 Hz at site Cz (referred to linked mastoids). See text for details of group selection and spectral transform parameters. Abscissa: time re stimulus onset. Ordinate: power in μV^2. (A) Mean ERS of 10 fast responders and (B) Mean ERS of 12 slow responders, both in the RT task condition. The dotted traces show the ERS averages for above- and below-RT median epochs from the fast and slow responders, respectively. The solid vertical lines represent grand mean RTs, and the dotted vertical lines, the means of above- and below-median RTs, respectively. (C) Difference wave between the RT and noRT task responses for the fast responders. (D) Difference wave for the slow responders. Dotted lines in C and D show limits of significant difference (p<.01).

To reveal possible dependencies between the temporal dynamics of 40 Hz activity and mean motor response time, Pearson correlation coefficients between each data point of each subject's mean ERS curve and their mean reaction time were computed. Fig. 1D (bold trace) shows that the mean 40 Hz ERS power is significantly correlated with the mean response time at several latencies. Moreover, the existence of a temporal chain of significant positive and negative correlation coefficients indicates high temporal dependencies within the mean ERS curve. In contrast, the correlation function of the ERS curves of the noRT condition re reaction times in the RT condition (light trace) did not reveal any significant correlations.

To attempt to better characterize the correlations between the ERS curve of the RT condition and the mean reaction times, the individual mean ERS difference curves (RT-noRT) were submitted to a Q-factor analysis on subjects with Varimax rotation. With one exception, all subjects loaded positively on the first factor (F1), which explained 46% of the total variance. The second factor (F2), explaining 25% of variance, clearly split the other 22 subjects into two groups by its (+/-) sign of loading. Plotting the 22 individual F2 loadings against individual mean RTs (Fig. 1E), two subject subgroups emerge clearly. The ten subjects whose mean RTs were less than 195 ms load in the opposite direction to the twelve remaining subjects, whose mean RTs were longer than 195 ms. In fact, the factor analysis separated subjects by mean RT: no subject in the slow responder group had a lower mean RT than any of the fast responder group, and accordingly the mean RT of the 10 fast responders (176+/-44 ms) was significantly faster than that of the 12 slow responders (244+/- 78 ms). The resulting decomposition of the sample RT distribution into two significantly different sub-distributions is shown in Fig. 1F.

Grand mean GBRs, and ERS responses were then calculated for the two subject groups. There were no significant group differences in the GBRs. Figures 2A and 2B show mean 40 Hz ERS responses for the fast and slow responders in the RT condition. While mean ERS power did not differ significantly between the two groups, there was a significant group difference in its temporal dynamics, and the interaction of group and dynamics was also significant.

Group comparison of difference waves between ERS responses in RT and noRT conditions (Fig. 2C,D) shows the effects of task and group affiliation on 40 Hz ERS dynamics. The dotted lines in Figs. 2C and 2D give limits of significant difference (p=0.01). The most prominent component in the difference wave for the fast responders is a relative peak at 200 ms, followed by a second peak at about 400 ms which also reaches significance. In contrast, the (RT-noRT) ERS difference for the slow responders (Fig. 2D) contains no peaks following stimulus presentation. Instead, in the RT condition these subjects had significantly more 40 Hz power just prior to the stimulus than in the noRT condition.

To test the consistency of the group differences, and to assess effects of varying RTs within groups, response epochs in the RT condition were separated into relatively fast-RT and slow-RT subsets, depending on whether RT was longer or shorter than the subject median. The dotted lines in Figs. 2A and 2B show the ERS averages for these fast- and slow-RT epochs from the two groups. The 40 Hz ERS differences between the two groups are clearly maintained, even between the relatively slow-RT epochs of the fast responders (RT=201+/-48 ms), and the relatively fast-RT epochs of slow responders (RT=202+/-36 ms), data subsets in which mean RT did not differ significantly. Statistical comparison confirmed that there were no significant within-group effects of RT subset on ERS dynamics. Finally, the two ERS data subsets were again correlated with the mean reaction time. Neither the ERS curves of the fast nor of the slow responders elicited further reaction time dependencies, revealing that the observed correlation of the whole sample was based on the different ERS patterns of the slow and fast responders.

A VISUAL PRIMING REACTION TIME TASK

Methods

In the second experiment, involving a visual priming paradigm, 16 right-handed adults (ages 19-35) were asked to react as quickly as possible to three different sets of visual stimuli: digits, vowels, and consonants. Each set was restricted to five stimuli and was assigned to a separate response button and response finger. The stimuli were displayed in the center of a computer screen. One second prior to the target, a priming stimulus was presented at the same location. The information content of the priming stimulus, and the assignment of the response button was varied between blocks. In the Information (INFO) condition, the priming stimulus was either a "b", indicating a letter, or a "z", indicating that a digit was to follow. Digits were used as distractor items and were removed from analysis. In the cueing (CUE) condition, an asterisk "*" was presented in place of the priming stimulus. In two of the four presented blocks priming information was given. A block consisted of 400 items, so each subject had to respond to 1,600 items. Interstimulus intervals varied between 3 and 7 seconds.

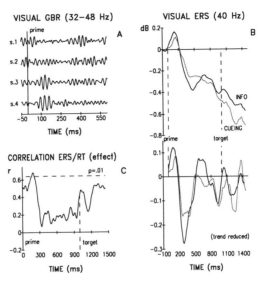

Figure 3. Responses to visual stimuli in a priming reaction time task. Grand means of responses from 15 adult subjects (site Cz referred to linked mastoids). (A) Individual mean averaged gamma band responses (GBRs), bandpass filtered between 32 and 48 Hz. Abscissa is time re the onset of the priming stimulus. The ordinate is z-transformed potential (0+/-1). Note the above baseline activity after stimulus onset in subjects s3 and s4. (B) Grand mean event-related spectra (ERS) for two priming conditions: in which the priming stimulus gives information about the target (INFO condition), or simply cues the appearance of a target (CUE condition). Each trace plots mean time- varying power at 40 Hz. Abscissa: time re stimulus onset. Ordinate: amplitude in dB re prestimulus interval. The lower traces show the dynamics of the upper two traces after minimizing the linear trend by differentiation. Note the relative increases of 40 Hz power after prime and target onset, followed by secondary peaks. (C) Pearson correlation function between ERS difference between INFO and CUE conditions and reaction time effect (weighted difference in mean RT between the same conditions). The dotted line shows limits of significant correlation (p<.01). The uniformly positive correlations suggest that the presence of high 40 Hz amplitudes is associated with faster reaction times.

The EEG was collected from Cz referred to linked mastoids. The recording epoch began 250 ms prior to the onset of the priming stimulus and lasted 1750 ms, i.e. to 500 ms after the onset of the target stimulus. A 50 Hz notch filter and an analogue band pass filter with cut-offs at 3 and 70 Hz were used during data collection. The data were sampled with a 12 bit A/D converter at a rate of 1000 Hz. EOG was also collected, and was used to reject epochs containing artifacts as in the first experiment.

Each recorded epoch was digitally band pass filtered using a symmetric Butterworth filter (12 dB/octave) with cut-offs at 20 and 60 Hz. An analyzing window of 100 ms was shifted across the whole epoch in steps of 10 ms. Data within the overlapping windows were Fourier transformed to obtain 40 Hz amplitude. Thus an ERS epoch consisted of 166 data points. The estimates for each window were separately averaged across trials for both conditions. To obtain baseline-independent dynamics, the amplitudes were divided by the mean amplitude of the prestimulus interval, and the decimal logarithm of this quotient was multiplied by 20. The difference between the two conditions was also computed. This difference can be regarded as the 40 Hz activity due to processing of information given by the priming stimulus. A repeated-measures ANOVA for both conditions was calculated using information as a factor. A separate repeated-measures ANOVA was computed for the difference between conditions.

Averaged evoked gamma band responses (GBR) were separately computed for both conditions using a 32 to 48 Hz bandpass filter with 24 dB/octave slopes. Response times were recorded with a temporal resolution of 1 ms. The data of the correct responses were statistically screened by applying the two-tailed Thompson rule to remove extreme values. To estimate how well a subject used the given information, the effect size was computed as the difference between the mean response times under both conditions, divided by the weighted and pooled standard deviation of the response times.

To investigate the influence of the 40 Hz event-related spectrum on effect size, we computed a running (Pearson) correlation between the 40 Hz ERS difference curves (INFO minus CUE conditions) beginning 10 ms after priming stimulus onset and the reaction time effect size. Threshold of significance was considered to be $p < 0.01$.

Results

Due to data loss, data from only 15 persons could be analyzed. For these subjects, on average 80 percent of all trials were judged free of artifacts. As expected, subjects responded significantly more quickly under the information condition: Mean RT in the INFO condition was 674 ms and in the CUE condition, 723 ms. All but three subjects seemed to take advantage of the given information, i.e. their responses were quicker in the INFO condition.

A recognizable averaged evoked gamma band response (GBR) was found for 8 of the 15 subjects. Figure 3A illustrates the high inter-individual variability of the GBR in four subjects. Subjects s3 and s4 show clear above baseline activity after prime onset. The onset latency of visual GBR activity was longer than gamma band responses to auditory stimuli, beginning about 100 ms after priming stimulus onset, whereas its amplitude was smaller than corresponding auditory response. The average responses of the other two subjects (s1 and s2) appear to contain no above-baseline phase-coherent gamma band response activity.

The 40 Hz event-related spectrum showed a greater variability between subjects than in the auditory task, but for most subjects, an increase in 40 Hz amplitude immediately after priming stimulus onset was observed, followed by a suppression after about 250 ms (Fig. 3B). This was followed by a second peak at about 600 ms. The trend-reduced ERS grand average in Fig. 3B discloses a similar biphasic structure after the onset of the target stimulus. The first increase seems to be related to the occurrence of the GBR. Repeated measures ANOVAs for the both conditions, however, revealed significant temporal changes in the 40 Hz ERS, but neither the information factor, nor the interaction between time and information were significant. A repeated measures ANOVA for individual ERS difference between INFO and CUE conditions, however, showed a significant effect of time. Figure 3C gives the correlation between the ERS difference between the two information conditions, and the information effect on mean response time. All correlations were positive, but only the values in the interval 170 to 250 ms exceeded significance threshold. In other words, subjects whose mean reaction time became shorter when priming information was presented also had increased EEG amplitude at 40 Hz in the interval 170-250 ms following the priming stimulus.

DISCUSSION

The task and response mode-dependent temporal patterns of gamma band EEG activity reported here appear to capture significant dynamics of cerebral activity in the 40 Hz range involved in task-related cognitive and sensorimotor processing, and demonstrate that these dynamic patterns can depend on subject, task, modality and level of performance. The results of both experiments indicate that dynamics of 40 Hz activity are intimately related to early stages of sensorimotor and cognitive processing.

The experiment on simple auditory reaction times reveals stable individual and group differences in stimulus/response brain processing associated with distinctive temporal patterns of stimulus-induced changes in the 40 Hz EEG. Although there were no task-related differences in time- and phase-locked stimulus-evoked gamma band activity between passive listening and speeded response tasks, reliable time-locked and task-dependent patterns of changes in ERS power near 40 Hz were induced by auditory stimuli. Moreover, two stable patterns of event-related dynamics of 40 Hz EEG activity occur in separate subsets of subjects, one with relatively fast mean response times, and another consisting of subjects who, on average, respond more slowly.

Figs. 2A-D show convincingly that similar mean spectral EEG power may result from quite different event-related patterns of spectral activity, and that at least two different 40 Hz activity patterns (Fig. 2A,B) accompany speeded reactions in behaviorally distinct subject subgroups (Fig. 1E). Although there was no overall difference in 40 Hz power between relatively fast and slow responders, the dynamics of 40 Hz activity preceding and following the stimuli were qualitatively different for the two groups.

In differences between speeded response and passive listening conditions, slow responders produced relatively greater power near 40 Hz only prior to the auditory stimulus; following it, 40 Hz power did not differ between the two task conditions (Fig. 2D). In contrast, the fast responders' ERS patterns contained equal amounts of 40 Hz power prior to the stimuli in both tasks, but in the speeded response condition they displayed phasic relative increases in 40 Hz power peaking near 200 and 400 ms after stimulus presentation (Fig. 2C). Since above-baseline 40 Hz activity in the evoked response lasts less than 200 ms (Figs. 1A), later ERS activity must reflect spectral activity regularly induced following stimulus presentations but not reliably phase locked to them. The independence of the two groups' response patterns from reaction time per se is shown by the high consistency of the mean ERS responses for the fast versus slow-RT comparisons for each subject group (Fig. 2A,B dotted lines). The apparent dissociation of individual RTs and ERS dynamics, as revealed by the fast- versus slow-RT comparisons, implies that this differentiating late 40 Hz activity is also not time locked to the execution of the response. Therefore, it is reasonable to conclude that stable subject differences in neurocognitive processing can occur even in a simple speeded response task resulting in equivalent levels of performance.

The significantly larger pre-stimulus 40 Hz activity in the slow responders might reflect a kind of heightened anticipation or focused preparation to facilitate controlled sensorimotor processing (Sheer, 1989; Crick and Koch, 1990). One might speculate that an increased state of anticipation of the stimulus in slow responders (Fig. 2C), might in turn require more fully elaborated and time-consuming stimulus processing, accompanied by a more distinct perception of the stimulus, prior to initiation of a motor response. The existence of qualitatively different response modes, as opposed to graded individual differences in response speed, might also relate to the theoretical dichotomy between automatic and controlled processing which has been proposed to categorize qualitative differences in performance in many task and training conditions (Schneider and Shiffrin, 1977; Norman and Shallice, 1986).

The second experiment, employing a visual priming task, showed that a time- and phase- locked gamma band response to visual stimuli is also observable in some subjects. However, the latency of the GBR, as well as that of the first peak in the ERS curve, is about 100 ms later than for auditory stimuli. Possibly, longer transduction and processing times in visual compared with auditory pathways may lead to a much higher inter-trial and inter-individual phase variability in the GBR, resulting in noisy and small GBR amplitudes. This view is supported by the fact that subjects without a noteworthy GBR in some cases had a prominent increase in event-related 40 Hz activity during the same time range. The ERS transforms show also phasic increases following the target stimuli (Fig. 3B) although a grand mean GBR cannot be detected.

The most prominent feature of the highly correlated ERSs in both conditions is the decrease of the 40 Hz amplitude at 250 ms following the first peak after prime presentation. A slow negative-going shift was observed in the averaged event-related slow potential. However, the onset of this contingent negative variation (CNV) occurred about 600 ms after prime onset. In the auditory experiment, the average ERS of the slow responders of the RT condition (Fig. 2B) also showed an apparent post-stimulus suppression of 40 Hz activity, though on a shorter time scale. However, this apparent suppression was actually due to an increase in 40 Hz prestimulus power, as shown by comparison with the reference noRT condition. Therefore, the significant prestimulus difference between RT and noRT conditions and the absence of any difference following the stimulus in the grand and slow responders average 40 Hz ERS may be distinguished from event- related desynchronisations of ongoing brain activity in alpha and beta bands of the EEG (Pfurtscheller, 1977; Kaufman et al., 1990). It is reasonable to speculate that in the visual priming task, expectation of a priming stimulus is associated with an above-baseline increase in 40 Hz activity prior to the priming stimulus.

The 40 Hz ERS in both task conditions are highly correlated ($r=0.95$). Therefore, we assume that the observed temporal pattern of 40 Hz activity is primarily related to a task-defined processing mode. This could also explain why a repeated measures ANOVA on ERS data failed to reveal a significant effect of the priming information, or an interaction of temporal dynamics and information. In addition, the large inter-individual variability in 40 Hz dynamics, paralleling similarly large between-subject differences in performance, has to be taken into account. However, individual differences between the INFO and CUE conditions show still a significant effect of time, indicating the existence of changes in these patterns due to the processing of response relevant priming or alerting information or with preparation for response selection and processing.

The fact that correlations between the ERS difference wave and reaction time effect sizes were positive means that subjects whose mean reaction time became shorter when priming information was presented also tended to have increased EEG amplitude at 40 Hz during the epoch. In other words, the presence of high 40 Hz amplitudes is associated with good performance. Interestingly, the interval from 170-250 ms after priming stimulus onset appears to be the most important time for utilizing priming information as indicated by significant correlation coefficients. It can be assumed that by 170 ms encoding processes have already been terminated, and processes of decision making as well as response preparation would have started. In the auditory reaction time experiment, we found that the observed correlations between 40 Hz activity and reaction times were based primarily on stable group differences between slow and fast responders. The absence of significant correlations between 40 Hz ERS and response times within the fast and slow subgroups in the auditory experiment, however, does not contradict the occurrence of significant correlations in the visual priming task, since in the visual condition, the response time effect size (the difference between the two conditions) and not the raw response times, as in the auditory experiment, were found to yield significant correlations.

In summary, it appears that time-frequency averaging methods applied to 40 Hz EEG data recorded during auditory and visual experiments of different degrees of complexity may be useful for studying brain processes involved in performance of sensorimotor and cognitive tasks. From the results of both experiments we suggest that the dynamics of 40 Hz activity are related to stimulus processing and response preparation. In the auditory reaction time task, only relatively slow responders had increased 40 Hz activity prior to the stimulus onset, which we interpret as meaning that these subjects required more fully elaborated and time-consuming stimulus processing, accompanied by a more distinct perception of the stimulus, before initiation of a motor response. Such a response mode, however, seems to be more appropriate for more complex tasks like that presented in the visual priming paradigm. In that experiment our results suggest that an exhaustive stimulus processing and focused response preparation, accompanied by increase in circa 40 Hz activity leads to faster responses.

Acknowledgements

The authors would like to thank Peter Bartsch and his colleagues at the Department of Neurophysiology (Charit , Berlin) for the opportunity to perform the experiments on simple auditory reaction times and Rainer Kniesche for programming visual priming task. This research was supported by DFG (Jo 242/1-1, Po 121/17-1), by 'Friedrich-Baur-Stiftung'. Dr. Makeig's collaboration was supported by NMRDC grant 62233N.

REFERENCES

Basar, E., G nder, A., and Ungan, P., 1976, Important relation between EEG and brain evoked potentials. I. Resonance phenomena in subdural structures of the cat brain, Biol. Cybern. 25:27.

Basar-Eroglu, C., and Basar, E., 1991, A compound P300-40 Hz response of the cat hippocampus, Intern. J. Neuroscience 60: 227.

Boyer, J.J., Montaron, M.F., Vahn e, J.M., Albert, M.P., and Rougeul, A., 1987, Anatomical localization of cortical beta rhythms in cat, Neuroscience 22:863.

Crick, F., and Koch, C., 1990, Towards a neurobiological theory of consciousness, Seminars in the Neurosciences 2:263.

Eckhorn, R., Bauer, R., Jordan, W., Brosch, M., Kruse, W., Munk, M., and Reitboeck, H.J., 1988, Coherent oscillations: A mechanism of feature linking in the visual cortex? Biol. Cybern. 60:121.

Freeman, W., and Skarda, C., 1985, Spatial EEG patterns, non linear dynamics and perception: the neo-sherringtonian view, Brain Res. Rev. 10:147.

Freeman, W.J., and Dijk, B.W. v., 1987, Spatial patterns of visual cortical fast EEG during conditioned reflex in a rhesus monkey, Brain Res. 422: 267.

Galambos, R., Makeig, S., and Talmachoff, P., 1981, A 40 Hz auditory potential recorded from the human scalp, Proc. Natl. Acad. Sci. USA. 78:2643.

Goertz, R., Jokeit, H., Küchler, E., 1994, Dynamics of 40 Hz power in a visual reaction time task, submitted for publication.

Glickstein, M., 1972, Brain mechanisms in reaction time, Brain Res. 40:33.

Goodin, D.S., Aminoff, M.J., and Shefrin, S.L., 1990, Organization of sensory discrimination and response selection in choice and nonchoice conditions: A study using cerebral evoked potentials in normal humans, J. Neurophysiol. 64:1270.

Jokeit, H., and Makeig, S., Differing event-related patterns of gamma band power in brain waves of fast and slow reacting subjects, Proc. Natl. Acad. Sci. USA (in press).

Kaufman, L., Schwartz, B., Salustri, C., and Williamson, S.J., 1990, Modulation of spontaneous brain activity during mental Imagery, J. Cog. Neuroscience 2:124.

Krieger, D., and Dillbeck, M., 1987, High frequency scalp potentials evoked by a reaction time task, Electroenceph. Clin. Neurpohysiol. 67:222.

Kutas, M., McCarthy, G., and Donchin, E., 1977, Augmenting mental chronometry: The P300 as a measure of stimulus evaluation time, Science 197:792.

Llinas, R., and Ribary, U., 1993, Coherent 40-Hz oscillation characterizes dream state in humans, Proc. Natl. Acad. Sci. USA. 90: 2078.

Luce, R.D., 1986, Response Times, Oxford University Press, New York.

Madler, C., and Pöppel, E., 1987, Auditory evoked potentials indicate the loss of neuronal oscillations during general anaesthesia, Naturwissenschaften 74:42.

Makeig, S., 1990, A dramatic increase in the auditory middle latency response at very slow rates, in: Psychophysiological Brain Research, C.H.M. Brunia, A.W.K. Gaillard, and A. Kok, ed., University Press, Tilburg.

Makeig, S., 1993, Auditory event-related dynamics of the EEG spectrum and effects of exposure to tones. Electroencephal. Clin. Neurophysiol. 86: 283.

Norman, D.A., and Shallice, T., 1986, Attention to action, in: Consciousness and Self-Regulation, R.J. Davidson, G.E. Schwartz, and D. Shapiro, ed., Plenum Press, New York.

Ortiz, T.A., Goodin, D.S., and Aminoff, M.J., 1993, Neural processing in a three-choice reaction-time task: A study using cerebral evoked-potentials and single-trial analysis in normal humans, J. Neurophysiol. 69: 1499.

Pantev, C., Makeig, S., Hoke, S., Galambos, R., Hampson, S., and Gallen, C., 1991, Human auditory evoked gamma band magnetic fields, Proc. Natl. Acad. Sci. USA. 88: 8996.

Pfurtscheller, G., 1977, Graphical display and statistical evaluation of event-related desynchronization (ERD), Electroencephal. Clin. Neurophysiol. 43:757.

Schneider, W., and Shiffrin, R.M., 1977, Controlled and automatic human information processing I: Detection, search, attention, Psychol. Rev. 84:1.

Sheer, D.E., 1989, Sensory and cognitive 40-Hz event-related potentials: Behavioral correlates, brain function and clinical application, in: Brain Dynamics, Springer Series in Brain Dynamics 2, E Basar and T.H. Bullock, ed., Springer-Verlag, Berlin.

Singer, W., 1993, Synchronization of cortical activity and its putative role in information processing and learning, Annu. Rev. Physiol. 55:349.

Steriade, M., Dossi, R.C., Par , D., and Oakson, G., 1991, Fast oscillations (20-40 Hz) in thalamocortical systems and their potentiation by mesopontine cholinergic nuclei in the cat, Proc. Natl. Acad. Sci. USA. 88: 4396.

ATTENTIONAL EFFECTS ON IMAGE-MOTION RELATED POTENTIALS AND SPECTRAL PERTURBATIONS

Mitchell Valdes-Sosa, Maria A. Bobes, Carlos Sierra, Maribel Echevarria, Leonel Perez, Jorge Bosch, Pedro Valdes-Sosa

Cuban Center for Neuroscience, Havana, Cuba

ATTENTION AND IMAGE MOTION

Attention operates on the products of early stages of visual perception. Electrical recordings from the scalp in man offer a window into the brain mechanisms of attentional processes, with the possibility of simultaneous psychophysical measurements. This chapter will be concerned with electrophysiological indexes of attention to image motion, with emphasis on the interplay of bottom-up and top-down processes.

The interaction of low-level perceptual processes and attention allocation (which is cognitively driven) has been extensively studied for visual texture discrimination (Treisman and Gelade, 1980; Beck, 1983; Nothdurft 1991; Julesz, 1981, 1986, 1990). A basic conclusion is that certain gradients of features (textons) can be detected pre-attentively, causing borders to effortlessly "pop-out". On the other hand, gradients based on more complex conjunctions of features are only detected after careful scrutiny. This distinction between early, bottom-up, parallel processes and later, top-down, serial processes has important theoretical implications.

Although image motion is considered a source of pre-attentive information (a texton, Julesz, 1986, 1990), its attentional processing has not been studied as extensively as that of two-dimensional static textures. As recognized by the Gestalt psychologists in the principle of "Common Fate", elements of the same object move together at an identical (or a similar) speed (Bruce and Green, 1990). This coherence in motion is considered to be an important clue in scene segmentation and figure-ground segregation, which are critical processes in early vision (Nakayama 1985; Crick and Koch, 1990). Thus the influence of attentional processes on the detection of motion coherency is of great interest.

Our studies were carried out with displays of randomly placed bars that were set into motion for several seconds. The critical variable manipulated was the percent-

Oscillatory Event-Related Brain Dynamics, Edited by C. Pantev, Plenum Press, New York, 1994

age of coherent motion, that is the proportion of the moving bars that traveled in the same direction and at the same speed. This kind of display, like that of random-dot-kinematograms, can yield vivid subjective experiences in spite of being devoid of familiarity cues, and thus free from the influence of previous "knowledge" (Julesz 1990).

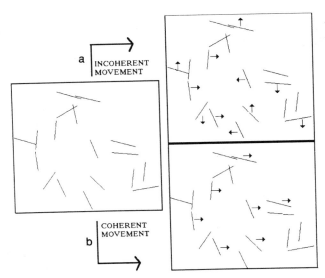

Figure 1. Two extremes of the motion stimuli used in this chapter. After the presentation of a static pattern of randomly oriented lines, motion-onset could lead to: a) incoherently moving lines (in this case with 0% coherency) or b) coherently moving lines. Actual stimuli were white lines on black background. The other condition used, 50% coherency, is not shown.

When a completely coherent display (100% coherent motion) was used (fig. 1), the subjects reported perceiving a "moving surface", or "one object in motion". When a completely uncorrelated display (0% coherent motion) was used (fig. 1), the subjects could not perceive the illusion of a moving object and instead reported seeing a random pattern (eg "a bunch of moving ants").

Intermediate amounts of coherent motion produced the illusion that the correlated lines comprised a surface that slid over (or under) the randomly moving lines. However this was easy to detect only if the proportion of coherent bars was high (over 70%). The difficulty in segregating the subset of coherent bars will be demonstrated in the next section.

AMOUNT OF COHERENT MOTION AND TASK DIFFICULTY

Before moving on to the electrophysiological studies we will describe our psychophysical methods. The general design of the experiments reported in this chapter is as follows (fig. 2). The subjects looked at a fixation spot on a 20 inch CRT monitor. The beginning of each trial was initiated by drawing thirty-three white lines on the screen. After this, the subject could initiate the motion of the lines by pressing a mouse-button. A delay of 500 ms ensued, and the lines would begin to move slowly (all at about 2 degrees/sec). The motion continued for 2000 ms, after which the screen was blanked. The direction of motion for each line could be determined independently.

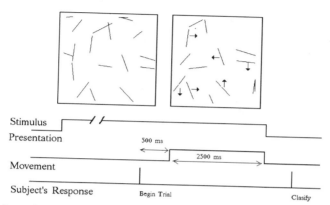

Figure 2. Time chart of the trials comprising the experiments in this chapter. After observation of a static pattern, the subjects could initiate movement onset. Responses were delayed (as in the figure) when electrical recordings from the scalp were taken. Actual stimuli were white lines on black background.

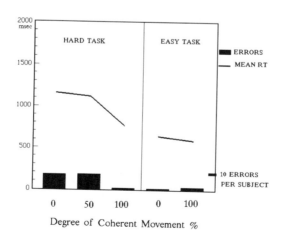

Figure 3. Data from the psychophysical experiment. The mean values of five subjects. In this experiment the difficulty of the task was a within subject factor and responses were required to be as fast as possible. The electrophysiological data was obtained in separate subjects, and task difficulty was a between subject factor.

In order to manipulate the need for focal attention, the discrimination between displays with different amounts of coherent motion was performed in two different tasks. In the first task, designed as "easy", completely uncorrelated motion of the bars had to be distinguished from completely correlated motion (0% vs. 100% coherence). This discrimination was performed rapidly, with mean reaction times under 700 ms in a sample of 5 subjects. Few mistakes were made and no differences were apparent in error rates between the two conditions (fig. 3).

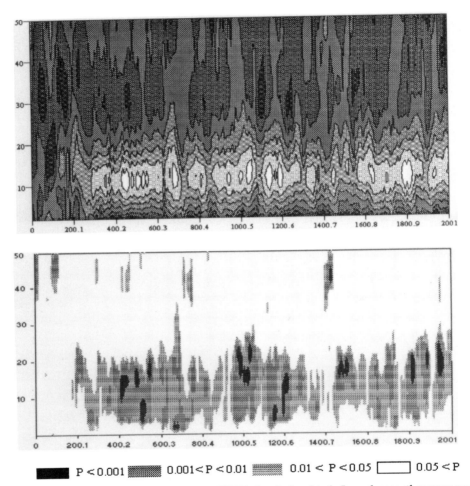

Figure 4A. Contour plot of the grand average ERSPs for the hard task. In each case, the upper panel shows the logarithmic transform of the time-varying spectrum as a function of frequency (2-50 Hz) and time (0- 2000 ms). Remember that this is the plot after subtracting pre-stimulus baseline levels. The bottom panel of each plot shows the significance levels of these perturbations, calculated from the sample t tests. *The coherency of motion is 0%.*

In the second task, designed as "hard", the identification of the uncorrelated motion was made more difficult by introducing trials with 50% correlated motion. In spite of being equidistant from completely correlated and completely uncorrelated motion, 50% coherence is subjectively very similar to the 0% correlation condition. Thus trials with 0%, 50%, and 100% coherence in motion were interspersed.

As seen in fig. 3, in the hard task, subjects took a long time (mean RT over 1000 ms) to respond to 0% and 50% coherence, and a much shorter time to respond to the 100% coherence condition. Mistakes were also more frequent for the 0% and 50% states. On debriefing, subjects referred that discriminating 0% and 50% coherence was difficult, while 100% coherency "popped out".

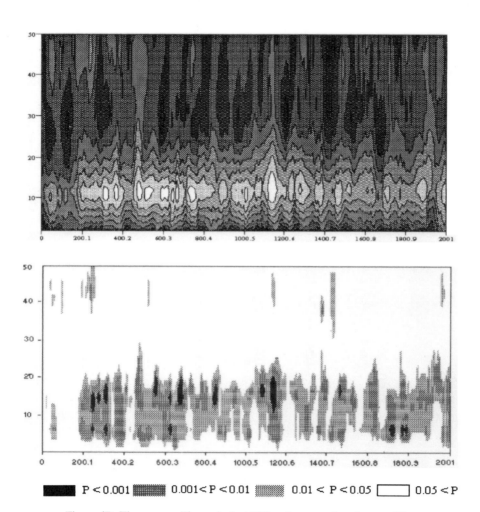

P < 0.001   0.001 < P < 0.01   0.01 < P < 0.05   0.05 < P

Figure 4B. The same as Figure 4a, but *50% coherency of motion condition.*

The reaction times for identifying the same stimulus, displays with 0% correlated motion, differed in the easy and hard tasks. An almost 500 ms increment in reaction time is observed for the hard task. We explain the increase in reaction time as reflecting the greater scrutiny necessary in discriminating the 0% and 50% displays.

Thus the segregation of the two subsets of bars (correlated and uncorrelated) seems to be an effortful endeavour. It does not depend on bottom up mechanisms, that produce "pop-out" effects. However the kind of visual scrutiny involved can not depend on focal space-based attention as that posited for the detection of feature conjuntions (Posner, 1980; Treisman and Gelade, 1980; Julesz, 1990). Instead it seems to correspond to what Duncan (1984) calls object-based attention. Object-based attention can encompass wide regions of the visual field, and seems to be directed by goal-directed processes. This type of attention can be directed to a collection of randomly moving points, immersed in another subset of irrelevant elements, as shown by Pylyshyn and Storm (1988) and Yantis (1992).

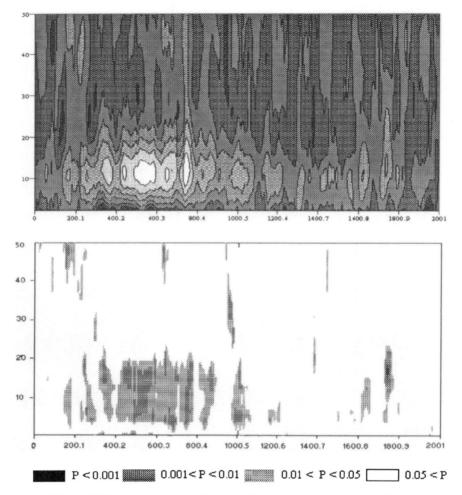

Figure 4C. The same as Figure 4a, but *100% coherency of motion condition.*

Several distinct mechanisms can be considered as examples of object-based attentional processes, including the motion-filter proposed by McLeod et al (1991), the attention based motion system proposed by Cavanagh (1992, 1993) and the non-rigid grouping model advanced by Yantis (1992).

ELECTROPHYSIOLOGICAL STUDIES OF MOTION PROCESSING

In this paper we look at clues about the brain mechanisms of visual attention that can be obtained from the scalp recorded electroencephalogram (EEG). We do this by examining brain electrical responses in subjects performing the tasks described above, while manipulating the amount of correlated motion and the allocation of attentional resources. However in the experiments described below responses were delayed for the 2000 ms. of bar motion, after which the screen was blanked and the subjects indicated the amount of coherence they had perceived. This was done to eliminate contamination of the recordings with movement related signals. Task difficulty (easy vs. hard) was tested in a between subjects design.

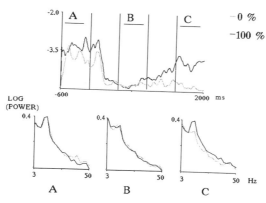

Figure 5. Different slices of the ERSPs in two conditions of the hard task in one subject. The alpha power against time (which corresponds to maintaining frequency constant while varying time) is shown above. Observe the high prestimulus alpha in window A, the desynchronization of alpha which reaches it's maximum in window B, and the faster recovery of alpha for the 100% coherency condition in window C. The power spectrum (varying the frequency while maintaining time constant) is shown below for the same time as before.

The EEG was recorded on magnetic disk with a MEDICID III/E system. The full 10/20 system of electrode placement was used. A bandwidth from 0.05 to 70 Hz (-3dB points) was used. The subjects observing responses, and discriminatory responses, as well as the stimulus onset and offset were marked on the digitized recordings. Two types of electrical responses are examined: Event Related Potentials and Event Related Spectral Perturbations.

By aligning EEG segments with a certain event and taking the ensemble average, we attempt to discard all activity which is not phase locked with the event, activity usually considered to consist entirely of "noise". The resultant average contains a signal (often considered "the" signal) which consists of components which are more or less strictly phase-locked with the triggering event. This is the average Event Related Potential (ERP; Valdés, 1984).

It is important to recognize that significant electrical activity may be time-locked to an event, even if it is not phase-locked. This is the case for amplitude changes in an on-going oscillation (brain rhythm). This type of activity will not be extracted by averaging, but should not be eschewed as noise. An appropriate label for this kind of phenomenon could be "Event Related Spectral Perturbation" (ERSP), as coined by Makeig (1993). Following Pfurtscheller (1992) special cases of ERSPs would be Event Related Desynchronization (ERD) and Event Related Synchronization (ERS), respectively referring to power decrements and enhancements of specific brain rhythms.

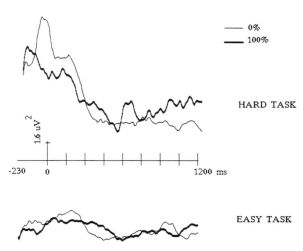

Figure 6. Grand average of alpha power against time for the two tasks. Above the plots for the hard task. Below the plots for the easy task. Large values of power point up.

Special techniques are necessary to visualize ERSPs such as those proposed by Pfurtscheller and cols, (Pfurtscheller, 1992; Pfurtscheller and Klimesh, 1991), or Makeig (1993) and Makeig and Galambos (1989). In this paper we estimate a time varying spectrum from the ensemble time-varying autoregressive coefficients, using a maximum likelihood method. The estimation eliminates the contribution of the ensemble mean (the ERP). The results are plotted as the difference from a pre-stimulus baseline (see appendix 1). This representation has the advantage over plotting power changes in a limited bandwidth, as in the plots by Pfurtscheller (1992), in that no assumptions are made on the EEG frequencies in which the event related changes are to be found. Once areas of interest are localized on the time-frequency plane, cross-sections of the graphs can be made to observe more details.

As befits the nature of this volume, special emphasis will be placed on effects of coherency of motion and attention allocation on ERSPs. However since we believe that scalp recordings of rhythmic activity and averaged ERPs are telling different aspects of the same story, stress will be laid on the comparison of both.

MOTION RELATED ERSPS AND ATTENTION

The ERSPs related to motion displayed robust effects. A long lasting diminution of alpha and beta power (from about 7 to 20 Hz) was observed in the hard task. This effect was largest at Oz, and lasted for several seconds. It was clearly visible and robust in the grand average over subjects in both the power and the probability plots (fig. 4). It was also visible in most subjects (7 out of 10 tested). However few significant effects were observed in the gamma band (20-50 Hz).

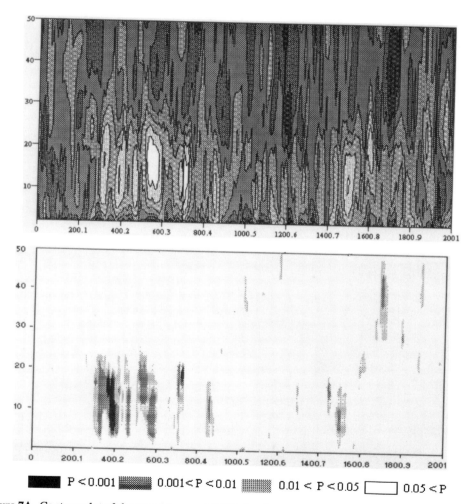

Figure 7A. Contour plot of the grand average ERSPs for the easy task: *0% coherency of motion condition*. Conventions as in figure 4. Remember that in the graph above pre-stimulus baseline levels have been subtracted.

155

Moreover the duration of the alpha/beta ERD in the hard task was very short for the 100% condition, whereas it was longer for the 0% and 50% conditions (compare fig. 4a and 4b with fig. 4c). The detailed time profile of the alpha/beta ERD can be better observed in plots of the average power in the alpha (8-13 Hz) band in slices that cut across time. The plots of alpha power vs. time from one subject, for two levels of coherent motion during the hard task, are shown in fig. 5. The amount of alpha desynchronization is similar for 0% and 100% coherent motion. However the return to baseline levels is faster in the 100% coherence trials. The same effect can be seen in the grand average plot of alpha power vs. time (fig. 6a).

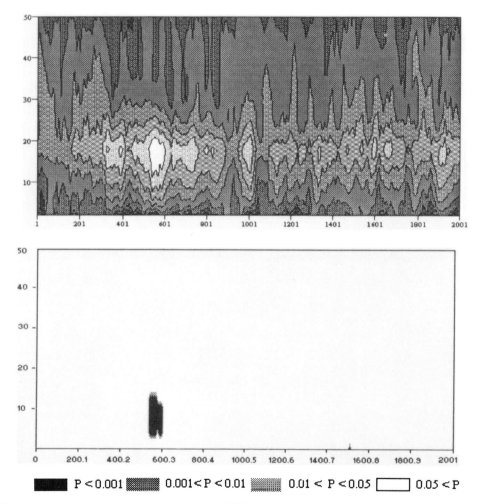

Figure 7B. Contour plot of the grand average ERSPs for the easy task: *100% coherency of motion condition*. Conventions as in figure 4. Remember that in the graph above pre-stimulus baseline levels have been subtracted.

This phenomenon was related to changes in EEG power spectra, as shown in the lower part of figure 5. Whereas no differences exist between 100% and 0% spectra in the pre-stimulus and early post-stimulus periods, in the late post-stimulus period a significant difference can be found in the alpha and beta bands. This corresponds to relatively more energy in these bands for the 100% trials which can be seen to originate from a slower disappearance of the alpha blocking for trials with uncorrelated motion (0%).

However in the easy task the alpha/beta ERD was difficult to observe in the individual cases (only visible in 1 out of 10 examined), and very slight in the grand average results (fig. 6b and 7). Also the plots of trials with different amounts of coherence did not differ. This means that the differential duration of alpha blocking is strongly affected by the task structure, and only present when sufficient attentional demands are made. The data suggests that the pre-stimulus alpha power was lower in the easy task, a finding that must be replicated in a within subject design, and that is not easy to explain. It is important to note that no effects of coherency were found in the gamma band for this task.

ERD was maximal at occipital sites and restricted in distribution. In the group average map of scalp topography (fig. 8) this component was larger over the right side of the head. This effect was also present in most subjects.

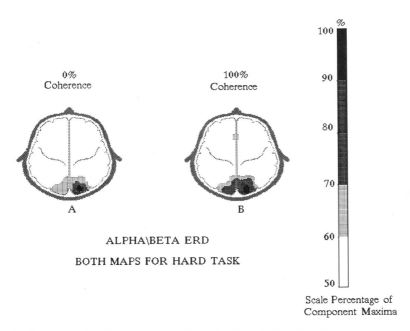

Figure 8. Scalp topography of grand average (n = 10 subjects) alpha/beta ERD components, for the hard task. The average amplitude in 50 ms around the peak from all sites of the 10/20 system was interpolated and expressed as a contour plot of the magnitude of the ERD percentages of the maximum value. Maps for both 0% and 100% coherency of motion condition are displayed.

These results confirm the reports of Pfurtscheller and colleagues (Pfurtscheller, 1992) and Kaufman and colleagues (1989) who describe event-related alpha/beta desynchronization with maxima at occipital sites after the onset of patterned visual stimulation. Our results show the same type of time-locked changes in EEG power for the onset of image-motion. However it contradicts Vijn et al (1991), who report a reduction of EEG power for all frequencies up to 70 Hz as the result of a the rotation of a "dartboard" pattern (compared to when the pattern is stationary). The ERD is small in the easy task, that does not require much focal attention. The ERD is large in the hard task which does require attention. ERD duration in the hard task (but not its amplitude) is related to processing time: it is shorter for 100% coherency which "pops-out" and is longer for the difficult to discriminate 0% and 50% correlation trials.

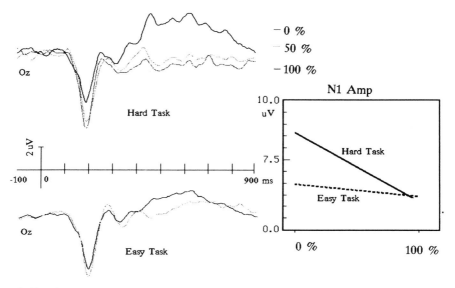

Figure 9. Grand average ERPs from two groups of subjects, one engaged in the hard task and another in the easy task. Recordings are from Oz, and positive deflections point up. ERPs associated with different conditions are overlaid for each group. In the inset the grand average N150 amplitude (measured with respect to mean pre-stimulus values) is plotted.

These results have led us to speculate that the ERD reflects the time-envelope corresponding to the use of attentional resources, whereas it would relatively insensitive to the amount of resources mobilized. This corresponds to the idea expressed by Pollen and Trachtenberg (1972), who considered alpha blocking to be related to the degree of attentiveness to detail and to mental effort. Several authors (Klimesch et al, 1990; Boeiten et al, 1992; Van Dijk et al, 1992) have found that greater computational demands lead to larger ERD magnitudes. However Van Dijk et al (1992) also describe that beyond a certain level of task complexity ERD magnitude does not increase. This saturation effect could explain the insensitivity of ERD amplitude to trial difficulty.

The idea that the duration of alpha ERD reflects processing time has also been posited by Kaufman and collaborators (Schwartz et al, 1989; and Kaufman et al, 1989). According to Pfurtscheller (1992) the alpha ERD is specific to visual processing. If so, mental effort (like arithmetic) will lead to alpha ERD if it involves imagery. Some support for this idea was found by Kaufman et al (1989). They showed that after an initial ERD due to word presentation, a prolonged alpha suppression was present when images were generated to words, and not when rhyme production was required.

MOTION RELATED ERPS AND ATTENTION

Following motion-onset the transient ERPs contained a P100, N150 and P200 sequence for all conditions in all subjects (fig. 9). The N150 was very prominent and largest over the occipital sites. A late positivity, with latency near 500 ms, was also observed in some conditions (see below). The structure of this ERP is very similar to that described by other authors who have used as stimuli the onset of slowly moving borders and long inter-stimulus intervals (Göpfert et al, 1990; Kuba and Kubova, 1992a, 1992b). The use of higher speeds reduces the N150 and enhances the P100 (Kuba and Kubova, 1992a,b).

The scalp topography of N150 and the alpha ERD are very similar. These responses are larger at occipital sites (see figs. 8 and 10). It is interesting that a tendency for larger responses on the right side of the scalp was also found for N150. This suggests that these average ERP components and the changes in alpha/beta ERD originate in the same cortical areas. An interesting finding was that in all cases where a clear onset to the alpha ERD was measurable, this latency was larger than that of the corresponding P100, or the N150 (see fig. 11).

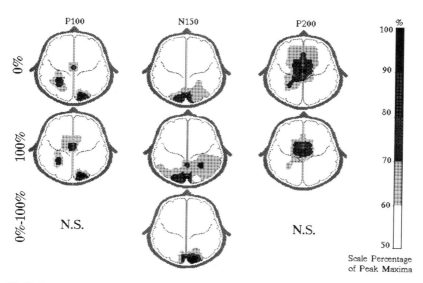

Figure 10. Scalp topography of grand average (n = 10 subjects) ERP components, for the hard task. The average amplitude in 50 ms around the peak from all sites of the 10/20 system was interpolated and expressed as a contour plot of percentages of the maximum value. In the first row the maps for the 0% coherency of motion condition are displayed. In the second row the maps for the 100% coherency of motion. In the last row the difference map of the amplitude of N150 obtained by subtracting these two conditions. The difference maps for P1 and P2 are not shown since the subtraction did not produce components significantly different from noise.

159

The results described above suggest that N150 and the alpha ERD index different aspects of the activity of the same cortical areas. The ERP components would reflect earlier, transient, events, related to the arrival of information about motion-onset, while the alpha ERD would reflect a later and longer lasting process.

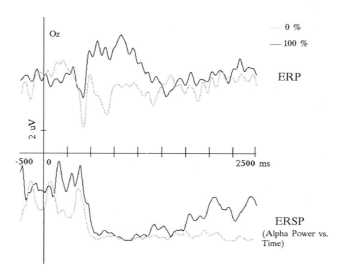

Figure 11. Comparison of ERPs and ERSPs from the hard task in one subject. The waveforms from the 0% and 100% coherency of motion conditions are overlaid (the 50% condition is omitted). Positive deflections in the ERPs, as well as higher power in the ERSPs point up. The average power in the alpha band (7-13 Hz) is plotted against time as a representation of the complete ERSP (see figure 4).

Large modulations of N150 amplitude were observed related to attention. In the hard task N150 amplitude was small after coherent stimuli (100% coherence) and larger after the more incoherent stimulus (0% and 50% coherence). However this difference is reduced (but not eliminated) in the easy task: the N150 was small in both conditions. Again these effects can not be due to obligatory, bottom-up, mechanisms. They are probably related to the changes in attentional demands imposed by the two tasks.

An enhancement of N150 is associated to the greater processing effort elicited by the completely incoherent (0%) and partially incoherent stimuli (50%) in what we have called the hard task. It is reasonable to assume that this enhancement of N150 is a consequence of deploying a larger amount of some processing resource, not necessary in identifying the completely coherent stimuli which pop-out through pre-attentive mechanisms. It is tempting to speculate that the processing resource allocation reflected by the N150 enhancement is associated with object-directed attention (Duncan, 1984; Yantis, 1992). In this case the "object" is the perceptual group consisting of the subset of coherently moving bars in the 50% coherency condition. An important point is that since the enhancement is obtained for both the 0% and 50% coherent stimuli, N150 amplitude can not reflect the output of the search for the coherent subset. This search is successful only for the latter stimuli, although the enhancement occurs for both. Thus N150 amplitude can only reflect the initiation of the attentive search for the coherent subset.

Hillyard and collaborators (1990,1993) have found that small flashed patterns at attended locations in visual space elicit enlarged P1-N1 components over the occipital scalp. This is a space-directed focal type of attention. Topographic mapping of the non-attended P1 and N1 shows the same distribution as the difference waveforms corresponding to these enlargements. This indicates that the P1-N1 enhancement is really the modulation of a pre-existing peak, and not the effect of an independent but overlapping component (a conclusion also supported by the enhancement of both a positive and negative wave in close contiguity).

The attentional modulation of the motion-onset elicited N150, found in our study, is similar to the N1 modulation described by Hillyard and collaborators in several respects. It also has an relatively early onset, and the difference waveform corresponding to the enhancement associated with attention has the same scalp topography as the pre-existing component. However it is not associated with a significant effect on the preceding positive peak (P100). The modulation of N150 extends to object-based attentional processes, the type of result previously described for space-based attention.

Moreover a prominent late positivity is differentially enhanced by the 100% condition only in the hard task (fig. 9). This component is very small for all other conditions in the two tasks. This enhanced late positivity would be consistent with an increased salience (due to the pop-out effect) of the completely coherent motion in the hard task. The salience (and perhaps the relative subjective frequency) are consistent with the antecedent conditions described in the literature for an enhanced P300 (Donchin and Coles, 1988).

DISCUSSION

Coherency of motion by itself apparently did not produce any effects independent from changes in attentional demands as manipulated by task difficulty. The critical test of this was in the easy task in which attentional differences were not found between conditions. This suggests that bottom-up processes have a limited expression on the EEG phenomenon that we studied. We were particularly disappointed because evidence for increased synchronization of visual neurons, as the consequence of more coherent motion, was not obtained. In particular no variation of the EEG gamma-band as a function of the degree of coherency was found.

Recent work (reviewed in Singer, 1993 and Koch, 1993) suggests that increased coherency of motion in kinematograms could lead to increased synchronization of gamma band neuronal oscillatory activity. This follows from the report that neurons of several visual cortical areas (including V1 and V5) fire rhythmically (ca 40 Hz) and in synchrony if the respective receptive fields are simultaneously stimulated by bars that move in the same direction and at the same velocity (Singer, 1993; Engel et al, 1991a, 1991b, 1992; Gray et al, 1992; Kreiter and Singer, 1992; Eckhorn et al, 1988). These synchronized oscillations have attracted attention as the basis for a solution to the so called "binding-problem" (Koch, 1993)

Thus if adequate stimuli are used (full-field presentation of a large number of moving bars, as used in this paper), large assemblies of neurons oscillating in phase could be expected. This should leak out to the scalp as enhanced power in the EEG gamma band. Our data contradict this prediction.

While our negative findings could be construed as evidence against the existence of gamma band synchronization in response to increased motion coherency in man, several alternative explanations must be examined first. The negative results could arise from several causes including the geometry of the involved neurons, inadequate stimuli and the temporal macro-structure of the coherent neural activity. However these results strike a cautionary note which should be considered together with the difficulty reported by some in finding gamma band oscillations in visual neurons of the monkey (Young et al, 1992; Tovee and Rolls, 1992; Bair, 1993; Koch, 1993).

On the other hand, task difficulty did have clear-cut effects on the EEG, which were expressed both in changes of the ERSPs and of the average ERPs. The N150, was larger for those trials were more focal attention was required (0% and 50% coherence in the hard task). In the same trials the alpha/beta ERD was more prolonged. We have speculated that N150 amplitude reflects the initial deployment of attentional resources for motion analysis, and the ERD the time that these resources are used. The scalp topography of these components is similar suggesting that the generators are in the same cortical areas. Establishing the identity of these areas is a difficult task, but a review of the literature suggests some hypothesis.

Mangun et al (1993) found evidence that P1 to flashed patterns (latency about 70-80 ms) could originate from lateral extrastriate areas (areas 18 and 19), while N1 (latency about 130-190 ms) from other extrastriate areas. An earlier component which apparently originates in primary visual cortex (area 17) is not affected by focal visual attention. These results converge with findings from single cell studies in monkeys (Desimone et al, 1990), and metabolic mapping in man (Corbeta et al, 1991; Grady et al, 1992), in suggesting that attentional selection on the basis of spatial location modulates the information flow from V1 to other visual areas. Attentional selection on the basis of other features, such as color, orientation, brightness or shape (instead of location) is manifested by later components (latencies about 150-300 ms; Harter and Ajne, 1984; Wijers et al, 1989; Kenemans et al, 1993).

As originally postulated by Desimone and Ungerleider (see Desimone et al., 1990), two processing streams may originate in V1. One, related with object recognition (form and color processing), would proceed ventrally to V2, V3, V4 and the inferior temporal areas. The pattern elicited P1-N1 would reflect the activity of these areas. The other, related with spatial and motion processing would travel dorsally to V5 (MT), MST and the inferior parietal lobe. The motion-onset related N150 could reflect activity of the second pathway. Results consistent with this idea have been obtained by Gilhuijs, Spekreijse and cols. (personal communication) who have studied dipole fits for N150 associated with motion-onset, using information from SPECT images from subjects viewing moving bars as a priori constraints. The SPECT images show enhanced blood flow in lateral-superior extrastriate areas, regions in which dipoles associated with N150 are placed. Both N150 and SPECT activity was larger over the right hemisphere.

In this case the modulation of the motion-onset N150 by task difficulty could reflect processes similar to those reflected by the pattern P1-N1, but in a different visual processing stream. In both cases the amplitude of the ERP components serves as an index of the information flow out from V1 into the respective processing streams for object recognition and motion processing. This proposal depends critically on the purported localization of the generators for P1-N1 and N150. On the other hand the alpha/beta ERD could reflect other aspects of the activity of the same areas. In any case the larger ERP and ERSP components over the right hemisphere may be indications of hemispheric specialization of the cortical areas that are involved.

Thus alpha/beta ERD duration and N150 amplitude offer supplementary windows into the mechanisms of visual processing. This is an interesting finding. After all the mean (ERP) and variance (ERD) of the same variable (EEG) do not have to necessarily reflect different processes. The ERD is initiated after the occurrence of P100 and N150. This would suggest that the changes in cortical rhythms associated with processing are a consequence of the attention modulated events that are indexed by N150. The tasks carried out by our human subjects are well within the capabilities of monkeys. Laminar recordings in monkey cortex would permit analysis of Current Source Density and of Multiunit Activity associated with N150 and alpha/beta ERD. This would permit the elucidation of the respective cellular origin, the interaction, and the functional significance of these phenomena.

CONCLUSIONS

No sign of differential response to bottom-up manipulation (i.e. changing the amount of coherent motion), beyond those related to attentional factors, were found. In particular no evidence of gamma band EEG power increase was found as the bars moved more coherently. Top-down influences, such as attention allocation to detect the degree of coherence in motion, have powerful influences on two image-motion related electrical events. These events are the N150 ERP component (seen in the EEG average), and the duration of alpha-beta desynchronization (seen as the EEG variance). The first may reflect the amount of processing resources allocated at the beginning of a trial, and the second the time that such resources are necessary. Thus measures of the first moment (ERPs) and the second moment (ERSPs) of the statistical distribution of the EEG offer distinct windows into brain function that supplement each other.

Appendix: Statistical estimation of event related spectral perturbations

Let $x^m(t)$ denote the EEG/MEG signal, from which the mean (classical ERP) has been subtracted, and corresponding to the m-th trial at any given time instant t ($m = 1,2,...M$, and $t=1,2,...T$). The following model will be assumed to hold for the data:

$$X^m(t) = \sum_{k=1}^{p} \alpha_k(t) X^m(t-k) + \varepsilon^m(t)$$

Thus each single trial data will be considered as the addition of a strictly phase locked signal , which is eliminated with the mean ERP, plus a non stationary background stochastic activity.. The non-stationary background activity is itself modeled as **time varying autoregressive process (TVAR)**. This approach views the mean-corrected data as the output of a time varying IIR filter driven by gaussian white noise (t) with variance ²(t) that is also time varying. The filter is characterized by it's coefficients $_k$(t). This model was initially introduced by Kitagawa and Gersch (1985), who also proposed a Kalman filtering technique for the estimation of the parameters $_k$(t) and ²(t) . Our group has modified the procedure in order to take into account multiple trials m. This modification allows the direct application of the maximum likelihood estimators for the parameters. From the TVAR parameters, the instantaneous spectrum (Priestley, 1988) is estimated by «INCRUSTAR Equation»

$$f_1(\omega) = \frac{\sigma^2(t)}{\left|1 - \sum_{k=1}^{p} \alpha_k(t) e^{j\omega k}\right|^2}$$

A particular case of the model occurs when the TVAR parameters do not vary in time. In this case the background activity is stationary and not time locked at all.

In calculating the ERSPs in this chapter a value of 10 was set for p. In addition the 400 ms preceding the stimulus were used as a baseline and the spectra from these time points were averaged and subtracted from the post-stimulus spectra.

REFERENCES

Bair, W., Koch, C., Newsome, W., Britten, K., 1993, Power spectrum analysis of MT neurons in the awake monkey, in: "Computation and Neural Systems 92", F. Kluwer, ed., Academic Publishers, in Press.

Beck, J., 1983, Textural segmentation, second-order statistics, and textural elements, Biol. Cybern. 48:125-130.

Boiten, F., Sergeant, J., Geuze, R., 1992, Event-related desynchronization: the effects of energetic and computational demands, Electroencephal. & Clin. Neurophysiol. 82:302-309.

Bruce, V. and Green, P., 1990, Visual Perception: Physiology, Psychology and Ecology, Lawrence Erlbaum Associates, Hove and London (UK).

Cavanagh, P., 1992, Attention-based motion perception, Science. 257:1563-1565.

Cavanagh, P., 1993, The perception of form and motion, Current Opinion in Neurobiol. 3:177-182.

Crick, F., Koch, C., 1990, Some reflections on visual awareness, Cold Spring Harb Symp Quant. Biol., 55:953- 962.

Corbetta, M., Miezin, F.M., Dobmeyer, S., Shulman, G.L., Petersen, S.E., 1991, Selective and divided attention during visual discriminations of shape, color and speed: functional anatomy by positron emission tomography, J. Neurosci. 11:2383-2402.

Desimone, R., Wessinger, M., Thomas, L. and Schneider, W., 1990, Attentional control of visual perception: cortical and subcortical mechanisms, Cold Spring Harb Symp Quant. Biol. 55:963-971.

Donchin, E. and Coles, M., 1988, Is the P300 component a manifestation of context updating?, Behav. & Brain Sci. 11:357-374.

Duncan, J., 1984, Selective attention and the organization of visual information, J. Exp. Psychol.: Gen. 113:501-517.

Eckhorn, R., Bauer, R., Jordan, W., Brosch, M., Kruse, W., Munk, M.,and Reitboeck, H.J., 1988, Coherent oscillations: a mechanism of feature linking in the visual cortex?, Biol. Cybern. 60:121-130.

Engel, A.K., Kreiter, A.K,. König, and Singer, W., 1991, Synchronization of oscillatory neuronal responses between striate and extrastriate visual cortical areas of the cat, Proc. Natl. Acad. Sci. 88:6048-6052.

Engel, A.K., König, P., and Singer W., 1991, Direct physiological evidence for scene segmentation by temporal coding, Proc. Natl. Acad. Sci. 88:9136-9140.

Engel, A.K., König, P., Kreiter, A.K., Schillen, T.B., and Singer, W., 1992, Temporal coding in the visual cortex: new vistas on integration in the nervous system, Trends Neurosci. 15(6):218-226.

Göpfert, E., Müller, R., and Simon, E.M., 1990, The human motion onset VEP as a function of stimulation area for foveal and peripheral vision, Doc. Ophthalmol. 75:165-173.

Grady, C.L., Harby, J.V., Horwitz, B., Schapiro, M.B., Rapoport, S.I., 1992, Dissociation of object and spatial vision in human extrastriate cortex: age-related changes in activation of regional blood flow measured with (150) water and positron emission tomography, J. Cogn. Neurosci. 4:23-34.

Gray, C.M., Engels, A.K., König, P., and Singer, W., 1992, Synchronization of oscillatory neuronal responses in cat striate cortex: temporal properties, Vis. Neurosci. 8:337-347.

Harter, M.R., Ajne, C.J., 1984, Brain mechanisms of selective attention, in: "Varieties of Attention", R. Parasuraman & D.R. Davies, eds., Academic Press, London.

Hillyard, S., Mangun, G., Luck, S.J. and Hans-Jochen, H., 1990, Electrophysiology of visual attention, in: "Machinery of the Mind", E. Roy John, ed., Birkhauser, Boston

Hillyard, S., 1993, Electrical and magnetic brain recordings: contributions to cognitive neuroscience, Current opinion in Neurobiol. 3:217-224.

Julesz, B., 1981, A theory of preattentive texture discrimination based on first-order statistics of textons, Biol. Cybern. 41:131-138.

Julesz, B., 1986, Texton gradients: the texton theory revisited, Biol. Cybern. 54:245-251.

Julesz, B., 1990, Early vision is bottom-up, except for focal attention, Cold Spring Harb Symp Quant. Biol. 55:973-978.

Kaufman, L., Glanzer, M., Cycowiz, Y.M., Williamson, S., 1989, Visualizing and rhyming cause differences in alpha suppression, in: "Advances in Biomagnetism", S. Williamson, M. Hoke, G. Stroink, and M. Kotani, eds. , Plenum Press, New York.

Kenemans, J.L., Kok, A., Smulders, F.T.Y., 1993, Event related potentials to conjuctions of spatial frequency and orientation as a function of stimulus parameters and response requirements, Electoencephal. & Clin. Neurophysiol. 88:51-63.

Kitagawa G., Gersch W., 1985, A smoothness priors time varying AR coefficient modeling of non stationary time series, IEEE Trans. Automatic Control, AC-30:48-56

Klimesch, W., Pfurtscheller, G., Mohl, W., and Schimke, H., 1990, Event-related desynchronization, ERD-mapping and hemispheric differences for words and numbers, Int. J. Psychophysiol. 8:297-308.

Kreiter, A.K., and Singer, W., 1992, Oscillatory neuronal responses in the visual cortex of the awake macaque monkey, Eur. J. Neurosci. 4:369-375.

Koch, C., 1993, Computational approaches to cognition: the bottom-up view, Current Opinion in Neurobiol. 3:203-208.

Kuba, M., & Kubová, Z., 1992a, Visual evoked potentials specific for motion onset, Doc. Ophthalmol. 80:83-89.

Kuba, M., & Kubová, Z., 1992b, Clinical application of motion onset visual evoked potentials, Doc. Ophthalmol. 81:209-218.

Mangun, G.R., Hillyard, S.A., Luck, S.J., 1993, Electrocortical substrates of visual selective attention, in: "Attention and Performance", MIT Press, Cambridge, Mass.

Makeig, S., Galambos, R., 1989, The CERP: event-related perturbations in steady-state responses, in: "Brain Dynamics:Progress & Perspectives", E. Basar & T.H. Bullocks ,eds., Springer-Verlag, Berlin.

Makeig, S., 1993, Auditory event-related dynamics of the EEG spectrum and effects of exposure to tones, Electroencephal. & Clin. Neurophysiol. 86:283-293.

McLeod, P., Driver, J., Dienes, Z., Crisp, J., 1991, Filtering by movement in visual search, J. Exp. Psychol: Hum. Percept. Perform. 17:55-64.

Nakayama, K., 1985, Biological image motion processing a review, Vision Res. 25(5):625-660.

Nothdurft, H.C., 1991, Texture segmentation and pop-out from orientation contrast, Vision Res. 31(6):1073-1078.

Pfurtscheller, G. and Klimesch, W.,1991, Event-related desynchronization during motor behavior and visual information processing, in: "Event-Related Brain Research (EEG Suppl. 42)", C.H.M. Brunia, G. Mulder & M.N. Verbaten, eds., Elsevier Sci. Publishers, B.V.

Pfurtscheller, G., 1992, Event-related synchronization (ERS): an electrophysiological correlate of cortical areas at rest, Electroencephal. & Clin. Neurophysiol. 83:62-69.

Pollen, D.A., Trachtenberg, M.C., 1972, Some problems of occipital alpha block in man, Brain Res. 41:303-314.

Posner, M.I., 1980, Orienting of attention, Quart. J. Exp. Psychol. 32:3-25.

Priestley, MB., 1988, Non-linear and Non-stationary Time Series Analysis. Academic Press , New York.

Pylyshyn, Z.W. and Storm, R.W., 1988, Tracking multiple independent targets: evidence for a parallel tracking mechanism , Spatial Vision. 3:151-224.

Schwartz, B.J., Salustri, C., Kaufman, LL., Williamson, S.J., 1989, Alpha suppresion related to a cognitive task, in:" Advances in Biomagnetism", S.J. Williamson M. Hoke, G. Stroink, and M Kotani, eds., Plenum Press, New York.

Singer, W., 1993, Synchronization of cortical activity and its putative role in information processing and learning, Annu. Rev. Physiol. 55:349-74.

Treisman, A., Gelade, G., 1980, A feature-integration theory of attention, Cog. Psychol. 12:97-136.

Tovée, M.J., Rolls, E.T., 1992, Oscillatory activity is not evident in the primate temporal visual cortex with static activity, Neuro Report. 3:369-372.

Valdés P., 1984, Neurometric assesment of brain dysfunction in neurological patients, in: "Functional Neuroscience", T. Harmony, ed., Lawrence Erlbaum Associates, Hilsdale, New Jersey.

Van Dijk, J.G., Caekebeke, J.F.V., Jennekens-Schinkel, A., and Zwinderman, A.H.,1992, Background EEG reactivity in auditory event-related potentials, Electroencephal. & Clin. Neurophysiol. 83:44-51.

Vijn, P.C.M., van Dijk, B.W., and Spekreijse, H., 1991, Visual stimulation reduces EEG activity in man, Brain Research. 550:49-53.

Yantis, S., 1992, Multi-element visual tracking: attention and perceptual organization, Cogn. Psychol. 24:295-340.

Young, M.P., Tanaka, K., Yamane, S., 1992, On oscillating neuronal responses in the visual cortex of the monkey, J. Neurophysiol. 67:1464-1474.

Wijers, A.A., Mulder, G., Okita, T., Mulder, L.J.M., 1989, Event-related potentials during memory search and selective attention to letter size and conjunctions of letter size and color, Psychophysiol. 26:529-547.

RETINAL AND CORTICAL OSCILLATORY RESPONSES TO PATTERNED AND UNPATTERNED VISUAL STIMULATION IN MAN*

Walter G. Sannita

Center for Neuropsychoactive Drugs, Department of Motor Sciences-Neurophysiopathology, University of Genova
and Center for Cerebral Neurophysiology, National Council of Research, Genova, Italy
Department of Psychiatry, State University of New York
Stony Brook, NY, USA

INTRODUCTION

Visual information is processed throughout the nervous system via functional subsystems serving as detectors of luminance change, contrast, color, motion, edges and shapes. Stimulus-related responses are generated at uni- and multicellular level in concomitance with functional activation at discrete stations of the visual system and can be recorded under proper experimental conditions in *in vitro* models as well as in vertebrates and in man. Noninvasive recordings in humans are restricted to retinal and scalp levels; sequential and parallel functions can nevertheless be investigated by properly manipulating the physical properties of the stimulus, analysing the scalp distribution of evoked responses, or interfering with neurotransmitter-receptor systems to identify component-specific drug effects; putative generators and driving mechanisms of several distinct, though partially overlapping components of retinal and cortical evoked responses have been identified (e.g.: Jeffreys and Axford, 1972; Parker and Saltzen, 1977; Zemon et al., 1980; Celesia et al., 1980; Bodis-Wollner et al., 1986; Regan, 1982,1983,1989, Maier et al., 1987; Ossenblok and Spekrejise, 1991; Sannita, 1991; Arakawa et al. 1993; Sannita et al., 1988a,1993b). Oscillatory responses time-locked to the stimulus can be recorded at retinal and scalp level in animal and man after stimulation with unpatterned stimuli (flash). Evidence will also

---

* Biomagnetic measurements of flash-evoked potentials (e.g. Fig. 4) were performed at the Institute of Solid State Electronics, CNR, Rome, in collaboration with L. Lopez, M.D., V. Pizzella, Ph.D. and G.L. Romani, Ph.D (see Lopez et al, 1993 for technical detail).

Oscillatory Event-Related Brain Dynamics, Edited
by C. Pantev, Plenum Press, New York, 1994

be given in this paper that cortical oscillatory responses reflecting visual functions can be evoked in man by patterned stimulation.

OSCILLATORY RESPONSES TO UNPATTERNED STIMULATION

Flash stimulation evokes responses in the retina, optic tract, lateral geniculate nucleus (LGN), and visual cortices. Components in the retinal response (electroretinogram; flash-ERG) reflect activation of receptors (a wave), Müller glial cells (b wave) and pigment epithelium cells (c wave) respectively, and have functional and clinical relevance when recorded in standardized conditions (Armington, 1974) (Fig. 1). None of the waves identified in the flash-ERG is postsynaptic in origin, while both a and b waves depend on K currents and changes in membrane conductance (Dowling, 1987; Tomita and Yanagida 1981; Newman and Frishman, 1991). The flash-ERG reflects the retinal area subjected to luminance shifts as well as the stimulus intensity. Latencies decrease with increasing stimulus intensity; the amplitude of a-wave increases linearly over a wide range of stimulus intensities, while a saturation effect on b-wave amplitude is evident at high intensities (Armington, 1974; Fulton, 1991). Rod and cone responses can be evoked separately depending on adaptation and stimulus characteristics. Both a and b waves vary in relation to age, sex, pO_2, pCO_2, acid-basic balance, spontaneous changes in metabolic factors such as glucose availability, and action of drugs interfering with membrane and/or synaptic mechanisms (e.g. Peterson et al., 1968; Weleber, 1981; Francois and De Rouck, 1978; Niemeyer and Steinberg, 1984; Niemeyer et al., 1987; Sannita, 1991; Sannita et al., 1989,1993a).

Short (20-30 msec) bursts of fast frequency (100-140 Hz) rhythmic oscillations, the retinal oscillatory potentials (OPs), are superimposed to the b wave of the electroretinogram evoked by high intensity flash stimulation, especially after mesopic or scotopic adaptation (Ogden, 1973; Wachtmeister and Dowling, 1978; Wachtmeister, 1987) (Fig. 1). Retinal OPs are thought to be postsynaptic potentials originating from neurons in deeper retinal layers independent of b-wave, but generators have not been identified unambiguously and the available experimental evidence points at amacrine and interplexiform retinal cells (Marchiafava and Torre, 1978; Karwoski and Kawasaki, 1991) or bipolar cells (Heynen et al., 1985) as candidates sources. A role of inhibitory feedback mechanisms in amacrine cells has been suggested (Wachtmeister and Dowling, 1978).

Retinal OPs reflect the stimulus intensity, with latencies decreasing and amplitudes increasing at increasing stimulus intensity, with saturation phenomena comparable to ERG (Peachey et al., 1989), and change during life and as the result of variations of glucose availability (Sannita et al., 1989,1993a). Retinal OPs are routinarily applied in clinical neurophysiology and have applications in human neuropharmacology. Although individual OPs wavelets have been suggested to express distinct phenomena and selective drug effects on different wavelets have been reported (e.g. Harnois et al., 1988), experimental evidence exists that early waves are peculiarly sensitive to manipulations of the GABA system, while later wavelets reflect glycine transmission (Wachtmeister, 1987).

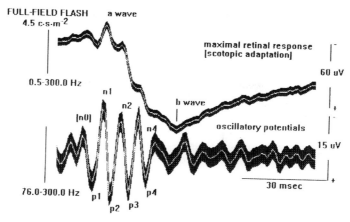

FULL-FIELD FLASH a wave
4.5 c·s·m⁻²

maximal retinal response
[scotopic adaptation]

0.5-300.0 Hz

n1

n2 b wave

[n0]

n4 oscillatory potentials

76.0-300.0 Hz p1

p2 p3

p4

60 uV

15 uV

30 msec

Figure 1. Human electroretinogram (maximal retinal response; top) and retinal oscillatory potentials (OPs; bottom) to full-field stimulation with high intensity white flashes at 0.12 Hz, as derived from the same signal after broad- (0.5-300.0 Hz) and narrow-band (76.0-300.0 Hz) filtering. Average on 15 epochs and 95% confidence interval. ERG a and b waves are indicated according to the conventional nomenclature, while OPs wavelets are labelled by polarity [(p)ositive or (n)egative] and order of appearance (Sannita et al., 1988a). Note that ERG and OPs confidence interval is narrower compared to background. Right eye; skin electrode at the lower orbital margin versus linked mastoids. Dark adaptation.

Human data also indicate differential responses to drug action of early and late waves (Sannita, 1991, for bibliography), in agreement with the dichotomy described in regard to the effect of stimulus intensity, light/dark adaptation, ON/OFF mechanisms, etc., and suggesting that two main components exist in this otherwise oscillatory phenomenon. Early and late OPs waves also are differently affected by retinal pathologies, while statistical approaches and factor analysis indicate that consecutive individual waves of OPs are not independent from each other and cluster into two homogeneous and distinct components, with a cutoff consistent over time and across subjects and a differential sensitivity to varying stimulus intensity and to spontaneous fluctuations of metabolic factors such as glucose availability and ammonia plasma concentration (Sannita et al., 1993a). Different processes in the generation of these components are conceivable.

Oscillatory potentials comparable in morphology to retinal OPs have been recorded in response to flash stimuli from LGN in alert monkeys, in concomitance with a negative wave at approximately 25 msec from the stimulus (Schroeder et al., 1992). Available experimental evidence suggests these oscillations to originate in the LGN lamina 6, with current sinks and sources reflecting transmembrane current flow, and to grade off rapidly few millimeters ouside this structure. Unlike other potentials originating in the LGN (e.g. the n25 wave), the contribution of these oscillations to scalp-recorded evoked potentials (VEP) is thought to be minimal (Kraut et al., 1985; Schroeder et al., 1992). A sequence of consecutive oscillations can be recorded in cat from area 17 as late components or after-discharge of the classically defined primary cortical response. This response is reportedly independent of flash-evoked potentials in LGN and optic tract and is suggested to originate in primary visual cortex based on evidence that increasing cortical excitability concurrently increases the late components and after-discharge (Steriade and Demetrescu, 1966; Steriade, 1968).

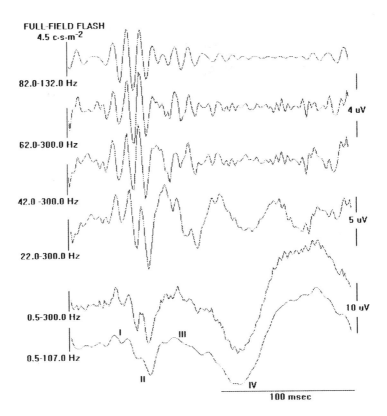

Figure 2. Effects of high-pass filters on scalp-recorded responses to full-field flash stimulation. Same subject and experimental conditions as in Figure 1. The oscillatory response is already detectable in the conventional, broad-band VEP (e.g 0.5-300.0 Hz) and is enhanced by filtering at any high-pass cutoff compatible with the response frequency content. Right eye stimulation; electrode derivation: O2-Fpz; ground at the vertex. Averages on sixty epochs. VEP waves are labeled according to Ciganek (1961).

Oscillatory potentials at 100-110 Hz are superimposed in latency to the earlier waves of flash-VEP recorded from the scalp in man, with latency of the first wave of approximately 50-60 msec, and broad scalp distribution with preponderance on posterior areas (Cobb and Dawson, 1960; Cracco and Cracco, 1978; Siegfried and Lukas, 1981; Whittaker and Siegfried, 1983). This response is enphasized when the signal is high-pass filtered at approximately 70-90 Hz, but does not depend on filter distortion (Fig. 2). Scalp distribution and the existing differences from retinal OPs (e.g. in frequency content, sensitivity to light wavelength, and peak latencies from stimulus) concur in ruling out volume conducted contamination from precortical sources. The existing difference in man of the latencies of the first oscillation recordable at retinal and scalp level (16-20 and 50-60 msec respectively, depending on stimulus intensity), as opposed to the small difference in latency between retinal and LGN oscillatory responses, apparently exceeds axonal transport through the optic radiations and suggests complex cellular events and interactions.

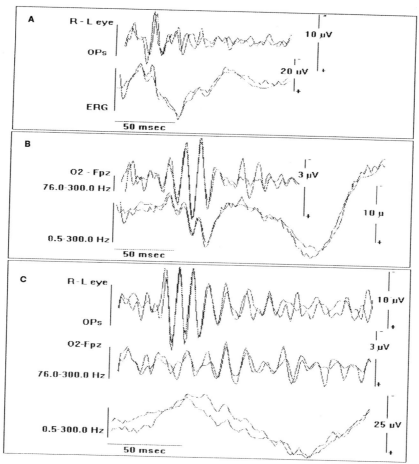

Figure 3. A: Retinal ERG and OPs to flash stimulation in photopic conditions (4.5 c.s/m2 at 1.9/sec); note the short-latency of photopic b-wave. **B**: Broad- band VEP and oscillatory response recorded at scalp location simultaneously to the ERG and OPS shown in A; the scalp-recorded oscillatory response is more regular and includes a higher number of wavelets compared to retinal OPs. **C**: Retinal OPs (top) and scalp-recorded broad-band VEP (bottom) and oscillatory response (middle) to flash stimulation (1.33 c.s/m2 at 0.12 Hz) after scotopic adaptation. Note the high amplitude retinal OPs in the absence of any sizeable scalp-recorded oscillatory response or short-latency VEP wave. The longer latency from stimulus of retinal OPs as compared to A is due to lower stimulus intensity. Retinal potentials were recorded using the non-stimulated eye (closed) as reference; other stimulation and recording specifications as in Figure 1.

Oscillatory potentials time-locked to flash stimuli and showing a dipolar structure of the main wavelets have been recorded in man with reference-free biomagnetic methods (Lopez et al., 1993). This observation further supports a cortical origin, consistent with the theoretical (Williamson and Kaufman, 1990) and experimental (Okada et al., 1987) evidence that only primary currents related to intracellular activity are recorded magnetically. This oscillatory response consistenly occurred at 100-110 Hz (i.e. approximately 10 Hz slower than retinal OPs) and the early wavelets were in phase with retinal OPs (though with an average delay of about 25-35 msec depending on stimulus intensity), while later waves were not (Lopez et al., 1993).

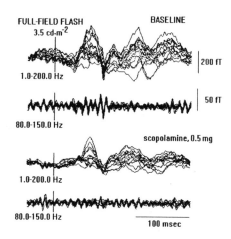

Figure 4. Reference-free broad-band conventional flash-VEP and oscillatory response recorded magnetically in baseline conditions and 30 min after im administration of scopolamine, 0.5 mg. Right eye stimulation; right occipital recording at 0.16-250.0 Hz (high-pass filter: 6 dB/octave; low-pass filter: 48 dB/octave). Oscillatory responses are shown after filtering at 80.0-150.0 Hz. Note the postdrug decrement of VEP late waves and increase of early wave (III or n2, after Ciganek, 1961) as previously reported using conventional electrophysiological measurements (Sannita et al., 1993b). The oscillatory response is virtually abolished by scopolamine.

The hypothesis that the so-called cortical response originates from striate visual cortex (Steriade and Demetrescu, 1966; Steriade, 1968) appears applicable to scalp-recorded oscillatory responses as well, in agreement with the results of intracerebral recordings in awake monkeys (Schroeder et al., 1992) and in man (Ducati et al., 1988). In man, the scalp-recorded oscillatory response is larger and includes more components at high stimulus intensity and fast stimulation rate, unlike retinal OPs that are more evident in scotopic conditions at low stimulation rate (Fig. 3). Retinal OPs depend on stimulus intensity in both photopic and scotopic conditions, and a differential role of rod and cone receptors in triggering mechanisms producing scalp-recorded oscillations is conceivable. The higher amplitude of cortical oscillatory response when stimuli are frequent and of high intensity also implies a role of the state of cortical excitability (Steriade, 1968).

The cortical origin of the scalp-recorded oscillatory response to flash is further indicated by the effects of manipulating cholinergic transmission, as reported in animals (Steriade and Demetrescu, 1966) and replicated in man after scopolamine-induced blockade of muscarinic receptors at acute intramuscular doses that do not affect accomodation and focussing significantly. In these experimental conditions, the cortical oscillatory response recorded magnetically is virtually abolished (Fig. 4), in concomitance with the reported amplitude reduction of the late waves of broad band VEP and in the absence of sizeable effects on flash-evoked ERG and OPs (Sannita et al., 1988a,1993b). Acetylcholine action in the cat visual cortex is tentatively located on layers IV and V of striate visual cortex (Sato et al., 1987), and layer IV is proposed as possible source of the cortical oscillatory response based on intracortical recordings in monkeys (Kraut et al., 1985) and in man (Ducati et al., 1988).

PATTERNED STIMULATION

Subsystems of the direct retinocortical pathway are known to exist and to be stimulus specific (Bodis-Wollner et al., 1986; Regan, 1989 for general bibliography). When patterned stimuli (e.g. pattern-reversal) are applied at constant luminance, the influence of stimulus factors such as contrast, spatial frequency, orientation, etc. reflects the functional geometry of retina and visual cortex and the characteristics of visual processing through mechanisms of center-sorround interaction, spatial tuning, receptive field organization, etc. Owing to antagonism between populations of neurons for the center and sorround parts of the receptive field, bipolar and ganglion cells of the retina are more sensitive to spatial contrast than to abrupt luminance changes over large retinal areas (i.e. flash-ERG) (Regan, 1982,1983,1989; Bodis-Wollner et al., 1986; Dowling, 1987). The electrophysiological response that is recorded at retinal level after pattern-reversal stimulation therefore reflects mainly field potentials generated at foveal level in proximal retina, at the level of ganglion and amacrine cells (Riggs et al., 1964; Maffei and Fiorentini, 1981,1982; Sieving and Steinberg, 1987; Baker et al., 1988). Due to the dipole location and limited number of neurons involved, the electroretinogram to patterned stimulation (pattern-ERG) recorded by macroelectrodes in humans appears simplified compared to flash-ERG, and oscillatory components have not been described.

Oscillatory activity at 30-60 Hz, with intercolumnar synchronization, was observed in the cat primary visual cortex (area 17) when stimulating with preferentially oriented moving light bars, and is attributed to oscillations of membrane potential produced by rhythmic activity in synaptic inputs (Gray et al., 1989; Jagadeesh et al., 1992). These observations and the low frequency (15-40 Hz) firing rate in response to luminance changes of amacrine and interplexiform cells presumably generating retinal OPs (Sakai and Naka, 1988; Karkowski and Kawasaki, 1991) concur in suggesting that cortical oscillatory responses at lower frequency than those induced by flash stimuli may exist. The reported origin of these oscillations from striate cortex (Jagadeesh et al., 1992), where at least some components of VEP evoked by pattern stimulation are believed to originate (Jeffreys and Axford, 1972; Phelps et al., 1981; Celesia et al., 1982; Kraut et al., 1985; Maier et al., 1987; Kushner et al., 1988; Ducati et al., 1988; Regan, 1989), also suggests that cortical oscillatory phenomena may be evoked by patterned stimulation.

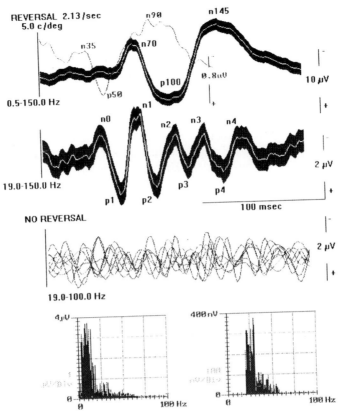

Figure 5. Broad-band VEP and high-pass (19.0-150.0 Hz) filtered oscillatory response to vertical gratings with sinusoidal luminance profile (9 degree; 80% contrast; spatial frequency: 5.0 c/deg) counterphase reversed at 2.13 Hz. Average on 120 epochs and 95% confidence interval. Wavelets of the oscillatory response are indicated by polarity and order of appearence as for retinal OPs in Fig. 1. The concomitant pattern-ERG is superimposed to the VEP for comparison. Averaged waveforms recorded while the same stimulus was not reversing on the screen are also shown superimposed. Power spectra corresponding to the broad-band VEP (left) and narrow- band oscillatory response (right) are shown (bottom). Scalp active electrode 5 cm lateral to inion, with Fpz reference and ground at the vertex. The pattern-ERG was recorded by skin electrodes from the stimulated eye, using the contralateral (closed) eye as reference. Amplifiers bandwidth is indicated; A/D conversion at 512 Hz.

Following this rationale, oscillatory potentials time-locked to the stimulus and reflecting contrast and spatial frequency effects were identified after monocular foveal stimulation with vertically oriented gratings at sinusoidal luminance profile that were produced on a CRT screen and square-wave counterphase reversed (VENUS Systems; Neuroscientific). With this stimulus modality, conventional broad-band pattern-reversal VEP were electrically recorded from the scalp in healthy volunteers, with a predominant spectral distribution of power on the lower frequency interval (below 20 Hz) and peak frequency at 4 to 8 Hz (Fig. 5). A smaller power peak at 20-35 Hz was also evident in most broad band VEP recordings, with peak frequency at approximately 25 Hz. A sequence of rhythmic oscillations time-locked to the stimulus was identified after high-pass digital filtering, with a 55.7±8.6 mean latency of the first replicable wavelet (Fig. 5; Table I). Nine wavelets were identified, hereafter referred to as n(egative)0, p(ositive)1, n1, p2, n2, p3, n3, p4, and n4 upon polarity and order of appearance after the stimulus. This oscillatory response is enhanced by high-pass filtering, but is not generated by filter distortion.

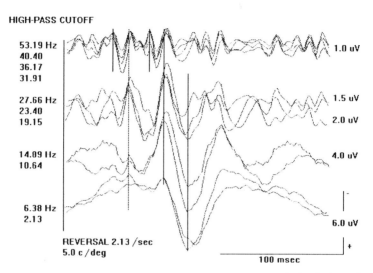

Figure 6. Effects of filter on scalp-recorded responses to pattern-reversal stimulation. The oscillatory response is detectable after high-pass filtering at any cutoff value between approximately 19.0 and 27.7 Hz, whereas the n70 waveis enhanced or a sequence of higher- frequency (approximately 40-50 Hz), poorly replicable wavelets with short latency after stimulus become evident with lower or higher high-pass cutoffs respectively. Right eye stimulation; electrode: 5 cm lateral to inion, with reference at Fpz; ground at the vertex.

Table1. Scalp-recorded oscillatory responses superimposed to the conventional, broad-band pattern-reversal VEP. Latency and peak-to-peak amplitude; normative values across the ranges of spatial frequency and contrast used in the study. Average across subjects (n=16) and standard deviation [in parentheses].

LATENCY (msec)			AMPLITUDE (microV)		
n0	55.67	[8.61]	n0-p1	1.10	[0.63]
p1	70.17	[8.33]	p1-n1	1.77	[1.07]
n1	85.95	[9.19]	n1-p2	1.97	[1.37]
p2	103.96	[9.48]	p2-n2	1.80	[1.31]
n2	122.18	[11.77]	n2-p3	1.44	[0.83]
p3	140.55	[14.34]	p3-n3	1.08	[0.59]
n3	158.35	[18.34]	n3-p4	0.94	[0.54]
p4	173.86	[18.72]	p4-n4	0.88	[0.58]
n4	191.29	[18.57]			

Table 2. Effects of stimulus contrast and spatial frequency on the amplitude and latency values of scalp-recorded oscillatory responses superimposed to VEP. Nonlinear regression analysis; slope, t values and statistical significance.

| | SPATIAL FREQUENCY | | | CONTRAST | | |
	slope	t value	p<	slope	t value	p<
LATENCY						
n0	0.013	0.719	ns	-0.112	-7.171	0.000001
p1	0.005	0.383	ns	-0.084	-7.625	0.000001
n1	-0.006	-0.484	ns	-0.083	-9.410	0.000001
p2	-0.035	-3.230	0.005	-0.061	-7.015	0.000001
n2	-0.044	-4.105	0.0001	-0.051	-5.147	0.000005
p3	-0.033	-2.712	0.01	-0.049	-4.429	0.0005
n3	-0.029	-1.938	ns	-0.064	-5.057	0.000005
p4	-0.030	-1.974	0.05	-0.052	-4.101	0.0001
n4	-0-033	-1.893	n.s.	-0.038	-3.257	0.005
AMPLITUDE						
n0-p1	0.223	2.707	0.01	0.248	3.332	0.005
p1-n1	0.450	5.998	0.000001	0.345	5.047	0.000005
n1-p2	0.504	6.857	0.000001	0.294	3.640	0.0005
p2-n2	0.393	5.612	0.000001	0.207	3.586	0.0005
n2-p3	0.336	4.671	0.00001	0.218	2.892	0.005
p3-n3	0.383	4.017	0.0001	0.004	1.760	ns
n3-p4	0.435	4.473	0.00005	0.145	1.457	ns
p4-n4	0.453	4.559	0.00005	0.068	0.655	ns

The sequence of wavelets was detectable at high-pass cutoffs between approximately 19 and 35 Hz; low-voltage wavelets at approximately 40-50 Hz were also observed occasionally when high-pass filters with cutoff higher than 35-40 Hz were applied (Fig. 6) and might conceivably reflect attentive phenomena related to vision (Bouyer et al., 1981,1987). Scalp-recorded oscillatory potentials proved replicable within- and between-subject, although with a higher variability of late compared to earlier wavelets. Two wavelets, notably n0 and p1, had shorter mean latencies than wave n70 of broad-band VEP, with significant (p<0.0001) latency differences between waves. The latency decreased non-monotonically as a function of both contrast and spatial frequency, while amplitude increased. Higher amplitude and shorter latency values were observed at 2.5-5.0 c/deg and 30-60% contrast values (Table II). The amplitude of the shortest latency waves (e.g. from n0 to p1) was an exception to this general trend in that higher amplitude values were observed on the entire 2.5-10.0 c/deg range. Both amplitude and latency measures of the oscillatory responses were linearly correlated with those of broad band VEP p100, while an inverse correlation was observed with the n70 amplitude and latency. The effect of contrast on the oscillatory response differed from that on conventional VEP (Bodis-Wollner et al., 1986; Regan, 1989), while the the tuning of amplitude at 5.0 c/deg was comparable to, though sharper than broad-band VEP.

DISCUSSION

The observations reported in this paper, though unsystematic, provide evidence that oscillatory responses time-locked to the stimulus and consistent with experimental data can be recorded in man and tentatively attributed to distinct functions and/or structures of the visual system. Patterned and unpatterned stimuli are equivalent in this respect in evoking oscillatory responses of suggested cortical origin, that are recorded from the scalp with similar morphology and comparable latencies though with frequency contents at circa 30 and 110 Hz respectively for pattern and unpatterned stimulation. These discrete frequency intervals allow the classification of these oscillatory potentials as mid- and high-gamma response respectively (Galambos, 1992) and should be tentatively attributed to functional dissimilarities between visual subsystems and processes sensitive to luminance variations or to contrast. Several mechanisms, e.g. involving membrane potential, rhythmic activity in synaptic afferent sources, and recurrent excitatory/inhibitory interacting phenomena in pools of neurons, have been considered to have a role in the generation of the oscillatory activities that are recorded in the visual cortex at single- and multi-unit level or as local field potential (Llinas and Sugimori, 1980; Gray et al., 1989; Bressler, 1990; Connors and Guntick, 1990; Jagadeesh et al., 1992; Basar and Bullock, 1992). In this regard, synchronization across relatively large populations of neurons is necessary for any neural activitiy to be driven into patterns of oscillation recordable via macroelectrodes, while also being a common functional characteristic of visual cortex. The oscillatory activity induced in the area 17 of cat by a bright bar moving along a preferential direction synchronizes across spatially segregated colums (Gray et al., 1989) and evidence was provided in the monkey that oscillations in the 20-40 Hz range occur in coherent spatial patterns over an area of the visual cortex larger than 10 cm[2] (Freeman and van Dijk, 1987). This mass effect suggests patterns of oscillation modulated over time and space (Bressler, 1990) and a prominent cortical origin of the oscillatory response recorded from the scalp in man, consistent with the results of intracerebral recordings in monkeys (Schroeder et al., 1992), the latency of scalp-recorded oscillatory responses to flash and pattern as opposed to the minimal difference in latency between retinal and LGN oscillatory events (Schroeder et al., 1992), and the biomagnetic recording in man of oscillatory responses to flash stimulation that are dramatically reduced in amplitude after scopolamine administration (Lopez et al., 1993).

An effect of stimulus contrast and spatial frequency on the oscillatory response to patterned stimulation was observed, as expected based on available electrophysiological and psychophysical evidence (Bodis-Wollner et al., 1986; Regan, 1982,1983,1989) and consistent with the inter-columnar synchronization of the oscillatory responses in cat visual cortex reportedly reflecting the stimulus properties (Gray et al., 1989). The existing differences between broad-band VEP and oscillatory response further add to the current hypothesis that parallel and independent, though partly overlapping in time, phenomena follow in the visual system upon stimulation (e.g.: Jeffrey and Axford, 1972; Parker and Saltzen, 1977; Celesia et al.,

1980; Regan, 1982,1983,1989; Maier et al., 1987; Ossenblok and Spekrejise, 1991). Additional evidence in this regard, and inference on the possible cortical generators can be extrapolated when comparing the effects of scopolamine on the oscillatory response with those on late waves of broad-band flash-VEP (Sannita et al., 1993b), in the framework of the hypothesis that drug effects on VEP can be component-specific (Zemon et al., 1980; Arakawa et al. 1993; Sannita et al., 1993b).

Retinal oscillatory potentials to flash stimulation are thought to be postsynaptic (Karwoski and Kawasaki, 1991), and this hypothesis can be tentatively set for the oscillatory responses superimposed to VEP as well. It also remains matter of speculation whether the latencies of oscillatory responses reflect travelling of information through the visual system, or these responses originate upon sensory input at each station in the visual system and latencies reflect the modalities of information processing at each location. The phase-locking described in the visual cortex between unit activity and local field potentials in the gamma band (Eckhorn et al., 1988; Gray et al., 1989) provide some clue as to the possible roles and interactions of these two electrophysiological phenomena in transferring visual information whithin and between structures of the CNS. The physiological basis of the different frequency ranges at which the oscillatory responses to patterned and unpatterned stimuli occur in spite of their comparable latencies is also to be actively investigated, and the existence of interferences between processes generating mid-gamma and high-gamma band responses to visual stimulation (Galambos, 1992) should be considered.

REFERENCES

Arakawa, K., Peachey, N.S., Celesia, G.G. and Rubboli, G., 1993, Component-specific effects of physostigmine on the cat visual evoked potential. Exp. Brain Res. 5:271

Armington, J.C., 1974, "The Electroretinogram", Academic Press, New York.

Baker, C.L., jr., Hess, R.R., Olsen, B.T. and Zrenner, E., 1988, Current source density analysis of linear and non-linear components of the primate electroretinogram. J. Physiol. 407:155

Basar, E. and Bullock, T, eds, 1992, "Induced rhythms in the brain", Birkheuser, Boston

Bodis-Wollner, I., Ghilardi, M.F. and Mylin, L.H., 1986, The importance of stimulus selection in VEP practice: the clinical relevance of visual physiology, in: "Evoked Potentials", R.Q. Cracco and I. Bodis-Wollner, eds., Alan R. Liss Inc., New York

Bouyer, J.J., Montaron, M.F. and Rougeul, A., 1981, Fast fronto- parietal rhythms during combined focused attentive behavior and immobility in cat: cortical and thalamic localizations, Electroenceph. Clin. Neurophysiol. 51:244

Bouyer, J.J., Montaron, M.F., Vahnee, J.M., Albert, M.P. and Rougeul, A., 1987, Anatomical location of cortical beta rhythms in cat. Neuroscience 22:863

Bressler, S.L., 1990, The gamma wave:a cortical information carrier, Trends Neurosci. 13:161

Celesia, G.G., Archer, C.R., Kuroiwa, Y. and Goldfader, P.R., 1980, Visual function of the extra-geniculo-calcarine system in man. Arch. Neurol. 37: 704

Celesia, G.G., Polcyn, R.E., Holden, J.E., Nickles, R.J., Gatley, J.S., and Koeppe, R.A., 1982, Visual evoked potentials and positron emission tomographic mapping of regional cerebral blood flow and cerebral metabolism: can the neuronal potential generators be visualized? Electroenceph. Clin. Neurophysiol. 54:243

Ciganek, L. 1961, The EEG response (evoked potentials) to light stimulus in man. Electroenceph. Clin. Neurophysiol. 13:165

Cobb, W.A. and Dawson, G.D., 1960, The latency and form in man of the occipital potentials evoked by bright flashes. J. Physiol. 152:108

Connors, B.W. and Gutnick, M.J., 1990, Intrinsic firing patterns of diverse neocortical neurons. Trends Neurosci. 13:99

Cracco, R.Q. and Cracco, J.B., 1978, Visual evoked potentials in man: early oscillatory potentials. Electroenceph. Clin. Neurophysiol. 45:731

Dowling, J.E., 1987, The retina, Harvard University, Cambridge

Ducati, A., Fava, E. and Motti, E.D.F., 1988, Neural generators of the visual evoked potentials: intracerebral recording in awake humans. Electroenceph. clin. Neurophysiol. 71:89

Eckhorn, R., Bauer, R., Jordan, W., Brosch, M., Kruse, W., Munk, M. and Reitboeok, H.J., 1988, Coherent oscillations: a mechanisms of feature linking in the visual system?, Biol. Cybern. 60:121

Francois, J. and DeRouck, A., eds., 1978, "Electrodiagnosis, Toxic Agents and Vision". Doc. Ophthalmol. Proc. Series 15, Dr W. Junk, The Hague

Freeman, W.J. and van Dijk, B.W., 1987, Spatial patterns of visual cortical fast EEG during conditioned reflex in a rhesus monkey, Brain Res. 422:267

Fulton, A. B., 1991, Intensity relations and their significance, in: "Principles and Practice of Clinical Electrophysiology of Vision, J.R. Heckenlively and G.B. Arden, eds., Mosby-Year, St. Louis, 260

Galambos, B. 1992, A comparison of certain gamma band (40-Hz) brain rhythms in cat and man, in: E. Basar and T.H. Bullock, eds., "Induced rhythms in the brain", Birkhauser, Boston, 201

Gray, C.M., Konig, P., Engel, A.K. and Singer, W., 1989, Oscillatory responses in cat visual cortex exibit inter-columnar synchronization which reflects global stimulus properties. Nature 338:334

Harnois, C., Marcotte, G., and Bedard, P.J., 1988, Alteration of monkey retinal oscillatory potentials after MPTO injection. Doc. Ophthalmol. 67:363

Heynen, H,, Wachtmeister, L. and van Norren D., 1985, Origin of oscillatory potentials of the retina. Vision Res. 10:1365

Jagadeesh, B., Gray, C.M. and Ferster, D., Visually evoked oscillations of membrane potentials in cells of cat visual cortex. Science 257:552

Jeffreys, D.A. and Axford, J.C., 1972, Source locations of pattern-specific components of human visual evoked potentials. I. Components of striate cortical origin. II. Components of extrastriate cortical origin. Exp. Brain Res. 16:1

Karwoski, C. and Kawasaki, K., 1991, Oscillatory potentials, in. J.R. Heckenlively and G.B. Arden, eds., "Principles and Practice of Clinical Electrophysiology of Vision". Mosby-Year Book, St. Louis, 125

Kraut. M.A., Arezzo, J.C., and Vaughan jr, H.G., 1985, Intracortical generators of the flash VEP in monkeys. Electroenceph. clin. Neurophysiol. 62:300

Kushner, M.J., Rosenquist, A., Alavi, A., Rosen, M., Dann, R., Fazekas, F., Bosley, T., Greenberg, J. and Reivich, M., 1988, Cerebral metabolism and patterned visual stimulation: A positron emission tomography study of the human visual cortex. Neurology 38:89

Llinas, R. and Sugimory, M., 1980, Electrophysiological properties of in vitro Purkinje cell dendrites in mammalian cerebellar slices. J. Physiol. (Lond.), 1305:197

Lopez, L., Pasquarelli, A., Romani, G.L., Torrioli, G. and Sannita, W.G., 1993, Magnetic recording of oscillatory potentials in response to flash stimulation in man, presented at the IXth International Conference on Biomagnetism, Vienna (Austria), August 15-21, 1993

Maffei, L. and Fiorentini, A., 1981, Electroretinographic responses to alternating gratings before and after section of the optic nerve. Science 21:953

Maffei, L. and Fiorentini, A., 1982, Electroretinographic responses to alternating gratings in the cat. Exp. Brain Res. 48:327

Maier, J., Dagnelie, G., Spekreijse, H. and Van Dijk, B.W., 1987, Principal component analysis for source location of VEPs in man, Vision Res. 27:165

Marchiafava, P.L. and Torre, V., 1978, The response of amacrine cells to light and intracellularly applied currents, Doc. Physiol. 276:83

Newman, E.A. and Frishman, L.J., 1991, The b-wave, in: "Principles and Practice of Clinical Electrophysiology of Vision, J.R. Heckenlively and J.B. Arden, eds., Mosby-Year Book, St. Louis, 101

Niemeyer, G. and Steinberg, R.H., 1984, Differential effects of pCO_2 and pH on the ERG and light peak of the perfused cat eye, Vision Res. 24:275

Niemeyer, G., Cottier, D. and Gerber, U., 1987, Effects of beta- agonists on b- and c-waves implicit for adrenergic mechanisms in cat retina, Doc. Ophthalmol. 66:373

Ogden, T.E., 1973, The oscillatory waves of the primate electroretinogram, Vision Res. 13:1059

Okada, Y., Lauritzen, M. and Nicholson C., 1987, MEG source models and physiology, Phys. Med. Biol. 32:43

Ossenblok, P. and Spekreijse H., 1991, The extrastriate generators of the EP to checkerboard onset. A source localization approach. Electroenceph. clin. Neurophysiol. 80:181

Parker, D.M. and Salzen, E.A., 1977, The spatial selectivity of early and late waves within the human visual evoked response. Perception, 6:85

Peachey, N.S., Alexander, K.R. and Fishman, G. A., 1989, The luminance-response function of the dark-adapted human electroretinogram. Vision Res 29:263

Peterson, H., 1968, The normal b-potential in the single flash electroretinogram. Acta Ophthalmol. 99(suppl.):5

Phelps, M.E., Kuhl, D.E., and Mazziotta, J.C., 1981, Metabolic mapping of the brain response to visual stimulation: studies in humans. Science 211:1445

Regan, D., 1982, Visual information channeling in normal and disordered vision, Psychol. Rev. 89:407

Regan, D., 1983, Spatial frequency mechanisms in human vision: VEP evidence. Vision Res. 23:1401

Regan, D., 1989, "Human Brain Electrophysiology", Elsevier, Amsterdam

Riggs, L.A., Johnson, E.P. and Schick, A.M.L., 1964, Electrical responses of the human eye to moving stimulus patterns. Science 144:567

Sakai, H. and Naka, K.I., 1988, Neuron network in catfish retina:1968-1987, Prog. Ret. Re. 7:149

Sannita, W.G. Neuropsychiatric drug effects on the visual nervous system, in: J.R. Heckenlively and G. Arden, eds., "Principles and Practice of Clinical Electrophysiology of Vision", Mosley-Year, St. Louis, 167

Sannita, W.G., Fioretto, M., Maggi, L. and Rosadini, G., 1988a, Effects of scopolamine parenteral administration on the electroretinogram, visual evoked potentials and quantitative electroencephalogram of healthy volunteers, Doc. Ophthalmol. 67:379

Sannita, W.G., Maggi, L., Fioretto, M., 1988b, Retinal oscillatory potentials recorded by dermal electrodes. Doc. Ophthalmol. 76:371

Sannita, W.G., Maggi, L., Germini, P.L., and Fioretto, M., 1989, Correlation with age of flash-evoked electroretinogram and oscillatory potentials, Doc. Ophthalmol. 71:413

Sannita, W.G., Balestra, V., Di Bon, G., Gambaro, M., Malfatto, L. and Rosadini, G., 1993a, Spontaneous variations of flash- electroretinogram and retinal oscillatory potentials in healthy volunteers are correlated to serum glucose. Clin. Vision Sci. 8:147

Sannita, W.G., Balestra, V., Di Bon, G., Marotta, V. and Rosadini, G., 1993b, Human flash-VEP and quantitative EEG are independently affected by acute scopolamine, Electroenceph. Clin. Neurophysiol. 86:275

Sato, H., Hata, J., Masui, H. and Tsumoto, T. 1987, A functional role of cholinergic innervation to neurons in the cat visual cortex. J. Neurophysiol. 58:765

Schroeder, C.E., Tenke, C.E. and Givre, S.J., 1992, Subcortical contributions to the surface recorded flash-VEP in the awake macaque, Electroencephal. Clin. Neurophysiol. 84:219

Siegfried, R.H. and Lukas, J., 1981, Early wavelets in the VECP. Invest. Ophthal. Visual Sci. 20:125

Sieving, P.A. and Steinberg, R.G., 1987, Proximal retinal contribution to the intraretinal 8 Hz pattern ERG of cat. J. Neurophysiol. 57:104

Spekreijse, H., Estevez, O. and Reits, D., 1977, Visual evoked potentials and the physiological analysis of visual processes in man, in: J.E. Desmet, ed., "Visual Evoked Potentials in Man: New Developments", Clarendon, Oxford

Steriade, M., 1968, The flash-evoked afterdischarge, Brain Res. 9:169

Steriade, M., and Demetrescu, M., 1966, Postprimary cortical responses to flashes and their specific potentiation by steady light, Electroenceph. Clin. Neurophysiol. 20:576

Tomita, T. and Yanagida T., 1981, Origin of the ERG waves, Vision Res. 21:1703

Wachtmeister, L., 1987, Basic research and clinical aspects of the oscillatory potentials of the electroretinogram. Doc. Ophthalmol. 66:187

Wachtmeister, L. and Dowling, J.D., 1978, The oscillatory potentials of the mudpuppy retina, Invest. Ophthalmol. Vis. Sci. 17:1176

Weleber, R.G., 1981, The effect of age on human cone and rod ganzfeld electroretinogram, Invest. Ophthalmol. 20:392

Whittaker, S.G. and Siegfried, J.B., 1983, Origin of wavelets in the visual evoked potential, Electroenceph. Clin. Neurophysiol. 55:91

Williamson, S.J. and Kaufman, L., 1990, Theory of neuroelectric and neuromagnetic fields, in: F. Grandori, M. Hoke and G.L. Romani, eds., "Auditory Evoked Magnetic Fields and Electric Potentials", Advances in Audiology Series 6, Karger, Basel

Zemon, V., Kaplan, F. and Ratliff, F., 1980, Bicuculline enhances a negative component and diminishes a positive component of the visual evoked cortical potential in the cat. Proc. Natl. Acad. Sci. USA, 77: 7476

LOW TEMPORAL FREQUENCY DESYNCHRONIZATION AND HIGH TEMPORAL FREQUENCY SYNCHRONIZATION ACCOMPANY PROCESSING OF VISUAL STIMULI IN ANAESTHETIZED CAT VISUAL CORTEX

Bob W. van Dijk,[1, 2] Peter C.M. Vijn,[1, 2] and Henk Spekreijse[1]

Graduate School Neurosciences Amsterdam
[1]Netherlands Ophthalmic Research Institute, Department of
Visual System Analysis and
[2]University of Amsterdam, Faculty of Medicine; the Laboratory of
Medical Physics and Informatics

INTRODUCTION

When our brains process a sensory event, a spatio-temporal pattern of active neurons results. It is likely that none of the essential features of the event is carried by one single neuron, rather it is carried by the activity of a subpopulation. As a consequence, processing of sensory information by the brain should be studied with techniques that reveal both the temporal pattern and the spatial pattern of the activity induced by the stimulus ("the neural image").

The common understanding of visual processing by the cortex, in extenso the processing that occurs at peripheral levels, is that of parallel streams through feed-forward connections of static local cortical circuits, detecting or signalling specific local features or feature gradients (e.g.: Livingstone and Hubel, 1988; Bolz et al., 1989). Global vision then occurs by somehow comparing the information in the different streams through "linking features" in areas that lie more central or "higher" in the cortical hierarchy.

In disagreement with these concepts, there is growing evidence, in different species and in different sensory cortical regions, that, even at the "lowest" cortical levels, global features and behavioral conditioning play an important role and modulate the responses to the local features, indicating that strictly hierarchical models can not be correct (e.g.: Freeman and Skarda, 1986; Freeman and van Dijk, 1987; Allman et al., 1985; Vaadia et al., 1989; Knierim and van Essen, 1990; Fox et al., 1990; Lamme et al., 1993ab). These findings may reflect the strong cortico-cortical

Oscillatory Event-Related Brain Dynamics, Edited
by C. Pantev, Plenum Press, New York, 1994

183

feedback that is present for every "forward" connection. In cat, for instance, all the visual cortical areas that lie on the outer surface of the cortex (areas 17, 18, 19 and 21) are connected reciprocally (Rosenquist, 1985; Mumford, 1992).

Recently, many groups have proposed that coherent spiking behaviour of neurons plays an important role in global sensory processing. Coherency in neural populations can be analysed using time-time correlation functions, which reflect the synchrony of the activity recorded at different locations. Indeed numerous papers have demonstrated that strong correlation functions can be found during long time intervals and over rather large distances in the cortex. These correlation functions are strongly dependent on the state of the cortex and can be modulated by presenting a stimulus to the sensory system (e.g.: Eckhorn et al., 1988; Gray et al., 1989; Nelson et al., 1992).

It is not clear which mechanisms underlie the observed coherence and its modulation. A number of different mechanisms has been proposed (a.o.: von der Malsburg and Singer, 1988; Tononi et al., 1992). The phenomenology of the observed phenomena (their topology, stimulus dependence, spatial character, temporal character, etc.) has been described too unsystematically to allow critical evaluation of such models. The importance of synchrony in global sensory processing therefore remains unsure. A systematic description of coherent phenomena with its consequences for the underlying mechanisms is the primary goal of the research described in this paper.

In our study we used (local) mass responses recorded simultaneously at different locations from the surface of the cortex or intra-cortically to study synchronous and coherent activity that occurs in the visual cortex of anaesthetized cat. In particular we studied how synchrony that is present in the ongoing activity changes upon the presentation of visual stimuli. This is done at two different spatial scales: at a relatively larger scale using the ElectroCorticoGram (ECoG), and at a more local scale using Current Source Density (CSD) techniques and Multi-Unit Activity (MUA) recordings from within the cortex.

METHODS

Experiments were done on adult female cats (weight approx. 3 kg). In total 13 cats were used for these experiments; 11 cats were used in the low temporal frequency experiments, 2 cats were used in the high temporal frequency experiments. The experiments were acute; the animals were killed by high doses of Barbiturate (i.v.) immediately after the experimental sessions. Some of the consequences of our results were tested in awake human volunteers (scalp recordings). Four subjects cooperated (1 female, 3 male). These subjects all had (corrected to) normal vision, and no neurological history.

Preparation

Cats were initially anaesthetized with Ketamine hydrochloride (30 mg/kg; i.m.). During surgery anaesthesia was supplemented with Thiopental on signs of awakening or pain (5 mg doses i.v.). During recording the cats were kept at light anaesthesia levels by constant infusion with Urethane (4 mg/kg per hr). Wound margins were infiltrated with Xylocaine (0.8%). The corneas were protected from drying by covering them by contact lenses. Oxybuprocaine (0.2%) eye drops were used for local anaesthesia of the corneas.

Visual stimulation

Atropine (1%) and Phenylephrine (5%) eye drops were administered to dilate the pupils, block accommodation and retract the nictitating membranes. Neutral contact lenses covered the corneas. Using spectacle lenses the conjugacy of both retinas and a screen placed 70 cm in front of the cat was arranged. Projections of the optic disk, the major vessels of the fundus and the visual streak were marked on this screen. All stimuli were presented to one of the eyes while the other eye was covered.

Stimuli were generated by a special purpose video generator (Neuroscientific, VENUS model 1020) that drove an electrostatic B/W CRT display (Hewlett Packard, model 1321A) at 192.9 Hz frame rate. Mean screen luminance was 120 Cd/m^2.

In the experiments on cats we used three modes of stimulation: i. a nonmodulated equiluminous screen ("non-stimulated"); ii. flashes of 3 frames (11.6 ms) maximum intensity on 190 frames of minimum intensity ("stroboscopic flashes"); iii. a sinusoidally moving oriented white bar (30' wide, 8 degrees long) on a black background. Contrast was 100%. The bar was oriented optimally. The amplitude of the bar's motion was 7.5 degrees, the period 1.8 s. As the centre of the stimulus was aligned on the visual field projection of the recorded region, the maximum velocity of 3.5 deg/s occurred when the bar moved through the projection area ("moving bar stimulus").

In the experiments on humans we used two different stimuli. In the first experiment the left half field of a dartboard pattern rotated clockwise while the right half field was stationary or vice versa. Equal periods of 9.6 s of rotation of the left and right half fields were alternated. Thus the period of this stimulus was 19.2 s. Contrast was 90%; rotation velocity was 215.35 deg/s. In the second experiment the dartboard pattern within one quadrant of the circular display rotated, while the pattern in the other three quadrants was stationary. The velocity of the pattern within the rotating quadrant was 192.9 deg/s. The position of the quadrant, to which rotation was restricted, itself rotated slowly; one full turn (360 degrees) in 18.7 s.

Data acquisition

Human scalp recordings. Ag-AgCl coated cup electrodes were attached to the posterior half of the head. In total 31 electrodes were used in 6 rows perpendicular to the mid-line spaced 3 cm apart. Each odd row had 6, each even row 5 electrodes, except for the most frontal row which had only 3 electrodes. The spacing between the electrodes in each row was also 3 cm. The most occipital row was 4.5 cm beneath the inion. All data were re-referenced to the mid-line electrode of the fourth row.

The EEG was amplified and filtered (first-order high-pass at 0.5 Hz, fourth-order low-pass at 70 Hz). The signals were digitized (CED 1401) at a rate of 192.9 Hz, time locked to the frame refresh of the CRT. Digital data and timing signals describing the different phases of stimulation were stored on the hard disk of a PC.

Electro Cortico Gram. For the local recordings small Ag-AgCl balls (diameter 500 μm) fixed in a teflon frame were pressed on the dura. Slits in the frame allowed passage of a micro-electrode. In the experiments where a larger area of the brain was sampled, including both hemispheres, stainless steel screws (diameter 1500 μm) were used that penetrated the skull. Care was taken that these screws touched the dura. Amplification, filtering and storage of signals was done as with the human scalp EEG data.

Intra-cortical mass responses. A commercially available linear electrode array (Otto Sensors, C1 probe) was inserted radially to the cortical surface. The array consisted of 16 50 μm x 50 μm Ag-AgCl islands (electrodes) spaced 150 μm apart. The laminar position of the electrode was verified prior to the experiments. Verification was based on the electrical characteristics of the signals while very slowly moving the electrode downward. Stimulus locked responses to stroboscopic flash stimuli yielded a typical pattern of synaptic currents that was used to reproduce the electrode position (for a more detailed description see Lamme et al., 1993b).

Signals were amplified (1.0 Hz - 10 kHz band-pass) by special purpose low current noise FET amplifiers built in our laboratory. Analog hardware performed temporal and spatial filtering to yield 12 MUA and 12 CSD signals, that were digitized and stored just as the EEG or ECoG data.

For the CSD data the 2nd-order spatial derivative of low-pass filtered (at 70 Hz) signals was obtained using a five-point approximation (Freeman and Nicholson, 1975). The resulting signals show the relative strength of current sinks and sources at 12 equidistant laminar positions in one cortical column. EPSPs and to a lesser extent IPSPs contribute to these sources and sinks, while action potentials do not (Mitzdorf, 1985; Kraut et al., 1985). Thus we interpret these signals as local weighted averages of synaptic activity[1].

For the MUA data the signals were high-pass filtered (1 kHz). Then the first-order spatial derivative was calculated using a three-point formula. The resulting data were full-wave rectified and low-pass filtered (70 Hz). In the resulting 12 signals prior to the integrative stage, spikes from individual cells could be identified. We interpret these signals as local weighted averages of spiking activity. CSD and MUA data were digitized and stored on hard disk.

Single unit recordings. A glass isolated tungsten electrode was inserted through a small slit in the dura. Signals were 10,000x amplified and band-pass filtered (300 Hz - 3 kHz) and were fed to a loudspeaker and oscilloscope. Using a window discriminator spikes from a single unit could be isolated. The shaped spikes were stored on hard disk on one of the event channels.

---

[1]It would have been more appropriate to calculate the full Laplacian instead of the 2nd-order radial derivative. Yet the tangential components of the Laplacian are one order of magnitude smaller than the radial component. Also it should be noted that our signals are proportional to the current source and sink density, assuming that the radial conductivity of the cortex is homogenous.

186

Data analysis

All raw signals were stored on hard disk. They were off-line analyzed. Routinely the records were scanned by eye for artifacts from poor electrical contacts, motion, etc., or dominant rhythms suggesting too deep anaesthesia levels, deep sleep state or cortical damage. Upon doubt data were discarded.

Human scalp data. Discrete Fourier Transform periodograms were calculated for overlapping consecutive epochs. Each epoch was 1.67 s for the half field rotation stimulus, and 1.87 s for the rotating quadrant stimulus. Stimulus and phase locked activity was removed from the data by setting the amplitudes of all the harmonics of the stimulus to zero. The periodograms of epochs that had corresponding positions in the stimulus cycle were averaged resulting in an estimate of the dynamics of the spectral density during the stimulus cycle. These densities were normalized with respect to the average response for each channel and expressed in dB. Data are presented either in time-frequency representation, or as the distribution over the scalp of the relative amplitude within a narrow frequency band. Both types of 3-D plots were obtained after cubic spline interpolation.

Low temporal frequency synchrony; non-stimulated. Synchrony was estimated by means of the cross-correlation functions of single unit activity and ECoG data. This is equivalent to a spike triggered average of the ECoG. Whenever spikes occurred quickly in succession, successive ECoG epochs overlapped. The spike-triggered average of the single unit activity itself yielded an estimate of the auto-correlation function of the shaped spike-train. To obtain an estimate of signal quality a "plus-minus" average (Schimmel, 1967) was also estimated. Since many ECoG epochs overlapped, which would result in an underestimate of the noise, we adapted this technique: we added and subtracted in alternation a fixed number (e.g. 7) of consecutive epochs. The spike-triggered averages (A) of each of the ECoG channels were converted into "synchrony signals" (s) by:

$$s(t) = \frac{A^{+}(t)}{\sqrt{m}\sqrt{\frac{1}{n-1}\sum_{t'=0}^{n-1}\left[A^{+}(t') - A^{+/-}(t')\right]^2}}$$

where m is the number of averaged epochs, n is the number of samples per epoch, the + or +/- indices indicate normal averages and plus-minus averages respectively, and the overscore indicates the ensemble average. Since plus-minus averaging results in the removal of time-locked components, and since the variance of the plus-minus average decreases with the square root of the number of averaged epochs, the denominator is a first order estimate of the rms value of the ongoing (non-time locked) signal. The synchrony values thus indicate the strength of the time locking of the ECoG with respect to the ongoing ECoG. Synchrony values can vary between $-\infty$ and $+\infty$, but for estimates based on a large number of epochs the values will converge to the range from -1 to +1. Values around zero indicate absence of synchrony.

Low temporal frequency synchrony; stimulated. When a stimulus drives both the single unit activity and the ECoG, partialization of the cross-correlation functions with respect to the stimulus is necessary to remove spurious (non-endogenous) correlations. For this reason we calculated the stimulus triggered averages of the single unit activity (this is the peri-stimulus time histogram), and the ECoG data (this is the epi-durally recorded visual evoked potential). The time reversed convolution of these two signals is a linear estimate of the correlation that occurs through time- and phase-locking to the stimulus of the two processes (Priestley, 1981). These estimates were subtracted from the spike-triggered ECoG to yield the partial cross-correlation functions. The procedure corresponds to correction of the data with the average of all possible shift predictors (for a mathematical prove of this statement see: Vijn, 1992).

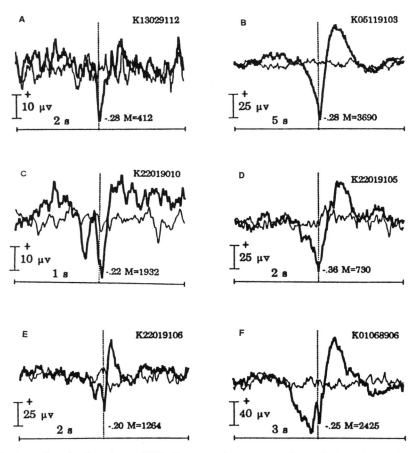

Figure 1. Examples of spike-triggered ECoG temporal patterns. In A the single unit and ECoG were from area 18, in B the unit was in area 19, while ECoG was from area 17, in C, D and E the single units and ECoG were in area 18, in F the single unit was in area 18, the ECoG from area 17. The numbers in each of the traces depict the maximum synchrony value obtained. M is the number of averaged epochs in each pattern. Note that time scales differ between the plots.

High temporal frequency synchrony. For each of the 12 MUA and 12 CSD channels we calculated the auto-spectral density functions using the Bartlett estimate (Priestley, 1981). Likewise we calculated cross-spectral density functions for each of the 144 pairs of one CSD and one MUA signal. From the spectral density functions we calculated coherency and cross-coherency functions, cross-covariance functions and cross-correlation functions. Only the cross-correlation functions are shown in this paper. Each of the 144 cross-correlation functions contains 200 samples from -193 ms to +193 ms, with a resolution of 1.93 ms.

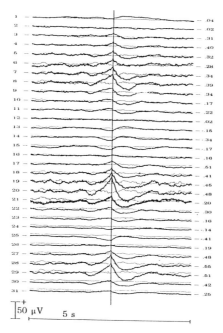

Figure 2. Separability of the spatio-temporal pattern of synchrony in temporal and spatial correlation functions. The figure shows 31 spike-triggered ECoG traces. The single unit was in area 18; the ECoG was recorded at 32 sites within a 1 cm diameter circular recording area. The tip closest to the micro-electrode was used as reference to accentuate differences between traces.

RESULTS I. LOW TEMPORAL FREQUENCIES

Non-stimulated

Temporal patterns. Significant correlations were found in all cats and all units tested in the absence of stimulation. The maximum synchrony values ranged from 0.15 to 0.50.

Figure 1 shows 6 examples of spike-triggered ECoG. As in the following figures thick lines show normal averages, while thin lines show plus-minus averages. These examples are representative in many ways. We did not observe any narrow-band synchrony phenomena in the ECoG recordings. The patterns contained all frequencies in a normal EEG. The spectrum showed 1/f behaviour. The major excursion of the spike-triggered ECoG was always negative with respect to a mid-frontal reference electrode, and was simultaneous (within ±10 ms) with the single unit activity.

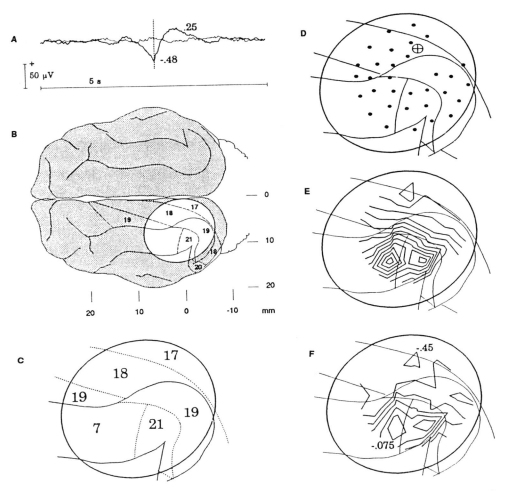

Figure 3. Spatial pattern of low frequency synchrony. A the correlation function found; B schematic top view of the cat cortex, the position and size of the trephan hole; C schematic map of the exposed cortical areas within the trephan hole; D the positions of the epi-dural electrodes (small filled circles) and the single unit-electrode (crossed open circle); E a contour plot of spike triggered ECoG distance between contours is 4 µV; F a contour plot of synchrony values, distance between the contours is 0.075.

The patterns were never symmetric with respect to this time instant. The observed patterns were very stable. Repeated recordings after (0.5 hr) intervals of stimulation always yielded identical patterns. The patterns, however, varied appreciably between animals and within animals. They could be mono-phasic, bi-phasic, tri-phasic or

190

more complex. They could precede or succeed the single unit activity. The examples D and E show that even the correlation functions with two single units that were recorded in succession (without moving the micro-electrode) could vary.

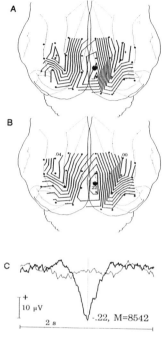

Figure 4. Spatial distribution of the spike-triggered ECoG over both hemispheres. A contour plot of spike-triggered ECoG, distance between contours 2.5 µV; B contour plot of synchrony values, distance between contours .025; C temporal pattern of synchrony.

Separability. Figure 2 shows an example of spike-triggered ECoG recorded at 31 channels simultaneously. It shows that the time functions recorded at different electrodes are all of the same shape. The spatio-temporal correlation functions can therefore be written as the product of a temporal and a spatial function. This held true for all correlation patterns observed.

Spatial patterns. Figures 3 and 4 show examples of the spatial distribution of the correlation functions and synchrony values in the non- stimulated condition.
Figure 3 shows results for a local recording. The single unit of this example was in area 18. Figure 3A shows the temporal pattern of synchrony; figures 3B, C and D show the recording positions, and the layout of the cortical areas that were exposed by our trephan hole. Areas 17, 18, 19 and 21 are visual cortical areas, while area 7 is not (strictly) a visual area. The small filled circles in fig. 3D show the positions of the epi-dural electrodes, while the large crossed circle shows the tangential position of the single unit electrode. Figure 3E shows the distribution of the spike-triggered ECoG over the exposed cortex. It shows a gradual fall of the amplitude over all vi-

sual areas and a much steeper decline over area 7, the only non-visual area covered by our electrodes. Figure 3F shows the distribution of synchrony values. It shows a behaviour that is quite similar to fig. 3E. Maximum synchrony values were -.45 for the electrodes on the border between areas 17 and 18, the lowest synchrony value was -.075 for the electrodes above area 7.

Figure 4 shows an example of the spatial distributions observed when a larger area of the cortex was sampled at smaller resolution. In this example the single unit recording was from area 17 in the right hemisphere. Figure 4A shows the distribution of the spike-triggered ECoG, fig. 4B the distribution of synchrony. Both of these show symmetrical patterns extending over all the visual cortical areas from both hemispheres. Both synchrony and spike-triggered ECoG show slightly larger values over the right hemisphere. Figure 4C shows the temporal pattern observed.

The two examples show that significant correlations of ECoG with single unit activity can be observed over large distances. The patterns show a gradual decrease over visual cortices, including both hemispheres, and a much more rapid decrease over non-visual areas. The spatial patterns were very stable and showed very little variation both within and between animals.

Stimulated

Cat. Figure 5 shows the auto-correlation function of the spike train of a single unit from area 18 in   stimulated (moving bar) and non-stimulated conditions. The figure shows that stimulus driven spikes are carried on a constant background of on-going spikes. The maintained spiking activity therefore is additive to stimulus driven (time and phase locked) activity. Similar observations have been reported earlier (Tolhurst et al., 1983).   The spurious correlation that results from a common driving force for single unit and ECoG activity can therefore be removed by a linear partial-ization procedure. The additive nature of ongoing spike activity was observed in all our recordings. Presumably this observation is closely tied to the observation that none of our stimuli drove the units to near maximal firing rates, while the spontaneous firing rates are low (below 5 spikes/s).

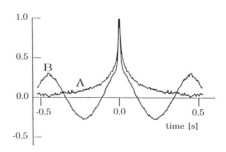

Figure 5. Auto-correlation functions obtained during non-stimulated (A) and stimulated (B) epochs. Same unit as in fig.6 right column (from layer VI of area 18).

Figure 6 shows two examples of the influence of a moving bar on temporal synchrony patterns. The left column shows results for ECoG recorded from area 18 and single unit activity in area 18; the right column shows trans-cortical potentials (derived from the linear electrode array) from area 18 and again single unit activity from area 18. In both these experiments 5 s epochs during which a stimulus was presented were alternated by 5 s epochs of stationary stimulation. Rows 1 to 5 in the figure show the data obtained during stimulation, while row 6 shows the spike-triggered averages during the no-stimulation epochs. Row 1 shows the stimulus-triggered averages of the ECoG and trans-cortical potential (the VEP); row 2 shows the PSTHs obtained, row 3 shows the spike-triggered mass potentials, row 4 the linear prediction of the stimulus induced spurious correlation of spike activity and mass potentials (the time-reversed convolution of rows 1 and 2), row 5 shows the partial correlation functions of spike and mass activity (the difference of rows 3 and 4). In both examples we see that the strength of the correlation function between ongoing spikes and ongoing ECoG is much reduced during stimulation.

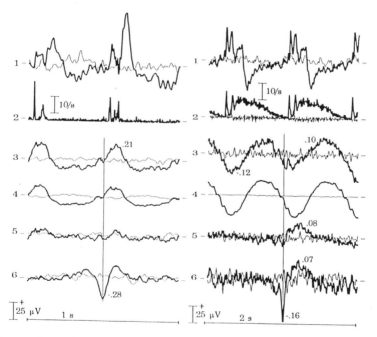

Figure 6. Temporal correlation patterns during stimulated (row 1-5) and non-stimulated (row 6) epochs that were alternated in one recording. Rows 5 and 6 should be compared as they show the spike-triggered correlation of the ECoG that is not phase locked to the stimulus. Both examples show strong desynchronization. See text for details.

In the example shown in the left column no significant correlations remained, while in the example of the right column only a small component remained. In all records studied we observed the same effect: the correlation of spike activity and mass potentials became much smaller or vanished during stimulation.

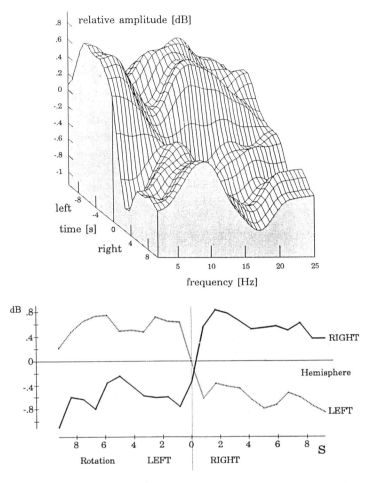

Figure 7. (top) Relative spectral density changes in human scalp EEG recorded with electrodes on the left half of the head for the alternating left half / right half field rotation stimulus. Data are averaged over 3 bipolar derivations each over the left hemisphere. X-axis temporal frequency, Y-axis time with respect to the alternation from left to right half field, Z-axis relative amplitude with respect to the mean spectrum over all epochs. (bottom) Relative EEG amplitude in the 15 to 20 Hz band as a function of time during the stimulus cycle for electrodes over the right (heavy line) and left (thin line) side of the head. The analysis window was 1.67 s, subsequent epochs overlapped one half window. (Adopted from Vijn et al., 1992)

Human scalp data. The EEG reflects the dynamic behaviour of neuronal populations within the "field" of the recording electrodes. We may assume that when a considerable number of neurons within this population is involved in processing a stimulus this will be reflected in the ongoing EEG, not just in the time and phase locked components that form the evoked potential. The experiments described in this paper are designed to show changes in ongoing EEG dynamics upon visual stimulation. Stimuli were designed such that they drive a large fraction of the neurons in the primary visual cortex, while at the same time, by driving subpopulations of neurons out of phase, these stimuli do not yield a considerable VEP.

The observation that in cat visual cortex the ongoing EEG desynchronizes with single unit activity upon visual stimulation, leads to the prediction for human scalp EEG that the amplitude of the ongoing (not phase locked) EEG decreases upon stimulation. If such a decrease is observed we can also measure how local this phenomenon occurs by stimulating a small region of the visual field and recording the topology of the desynchronization.

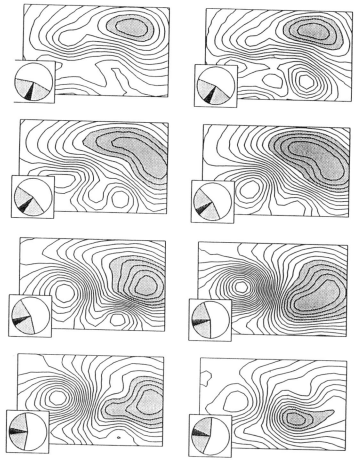

Figure 8. Contour plots of the distribution of relative EEG amplitude in the 15-30 Hz frequency band in eight successive overlapping epochs during part of the stimulus cycle in the rotating quadrant experiment. Distance between neighbouring contour lines is 0.1 dB. The shaded areas indicate those regions where the relative EEG amplitude is - 1.2 dB or smaller. Contour plots are made using an angle preserving map of the best fitting sphere through the electrodes to a plane. The X-axis runs parallel to the line connecting mastoids; the Y-axis parallel to the mid-line. (Adopted from Vijn et al., 1992)

Figure 7 (top) shows the dynamics of the relative spectral density recorded above the left hemisphere of a human subject during different epochs within the stimulus cycle. The figure clearly shows a broadband reduction of amplitude while the right half of the dartboard pattern was rotating. The difference in the ongoing EEG amplitude between rotation of the right half and rotation of the left half fields is most outspoken for the low γ-band, and least outspoken for the α-band (presumably due to a prominent α-rhythm in the recording). Figure 7 (bottom) shows the changes in EEG amplitude in the 15-20 Hz band recorded from electrodes above the right (heavy line) and left (dashed line) hemispheres. Considering the length of the analysis window (1.67 s) this plot shows that the EEG amplitude instantly decreases upon local stimulation; in this case by as much as 1.5 dB. The EEGs recorded above the two hemispheres are modulated in counterphase.

Figure 8 shows contour plots of the relative EEG amplitude in the 15-30 Hz band recorded in subsequent overlapping epochs of the rotating quadrant experiment. The shaded areas indicate those regions on the scalp where the EEG amplitude within this band was more than 1.2 dB smaller than the average EEG recorded for that particular region. The small insets of each of the figures show the beginning

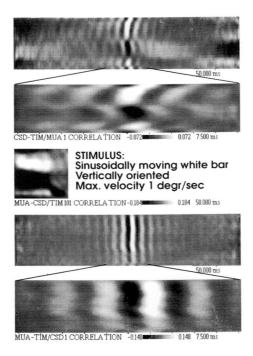

Figure 9. Five slices through the correlation image of a moving white bar. The top two panels show correlation functions of the 12 different CSD signals with a single MUA reference signal at two different time scales. The middle panel shows the correlation strength of the 144 pairs of one CSD and one MUA signal at zero time shift. The bottom two panels show the correlation functions of the 12 different MUA signals with a single CSD reference. The time scales are indicated beneath the panels. Time scales refer to the distance between the small tick marks. The panels are 10 tick intervals wide, i.e. either 500 ms or 75 ms. See text for further details.

and end positions of the slowly rotating quadrant in the epoch. The figure shows that the largest reduction of EEG in the low γ-band follows the projection of the visual stimulus (left half of the visual field projects to right hemisphere and vice versa; bottom half of the visual field projects to ventral, top half of the visual field to caudal in primary visual cortex). The data of figs. 7 and 8 correspond to the prediction based on anaesthetized cat data. It suggests that the desynchronization of the ongoing EEG occurs locally and instantaneously (in less than 0.8 s).

DISCUSSION I. SLOW DESYNCHRONIZATION

The data presented show that in the non-stimulated state the ECoG of cat is strongly correlated over wide spatial regions and during long time intervals. This synchrony stretches over both hemispheres even though most of the connections between the hemispheres in cat are restricted to a small strip around the vertical meridian (Segraves and Rosenquist, 1982). Our data are in accordance with the quite common finding of significant correlations of the activity of pairs of single units in different visual cortical areas or hemispheres of the cat (Nelson et al., 1992).

Upon visual stimulation the synchrony becomes much weaker or even disappears completely. This desynchronization occurs within our data segment length (<0.8 s). The data recorded at the human scalp suggest that desynchronization occurs locally.

One could argue that the synchrony that we observed is due to the general anaesthesia. We can not exclude this; our data, however, suggest a functional role. The modulation of synchrony is one argument, the findings of EEG amplitude reduction in non-anaesthetized subjects a second.

Reductions of the amplitude of (rhythmic) EEG activity have been described earlier by many authors (e.g. Morrell 1960, see many of the contributions elsewhere in this volume). These reductions occur with many different tasks involving a number of different cortical areas. In general they occur at well circumscribed scalp regions and can be very stimulus specific (e.g. Pfurtscheller and Klimesch, 1992; Kaufman et al., 1992). Our data support the notion that desynchronization leads to an amplitude reduction of the ongoing EEG, as suggested in most of these papers.

Which mechanisms underlie the strong synchrony and its modulation? This remains a question. The reduction is not due to a decrease of ongoing neuronal activity or an increase of the burstiness of the activity upon strong sensory input, as can be inferred by comparing the auto-correlation functions of spike activity with and without stimulation. Two possible mechanisms remain:

i. synchrony is due to the strong inter-connectivity of cortical neurons, and visual stimulation changes the effective wiring of the cortex, e.g. by activation of a cell class that is silent in absence of stimulation, in such a way that the neurons tend to desynchronize over large distances;

ii. synchrony is due to a common input, that can be modulated locally and instantly upon stimulation.

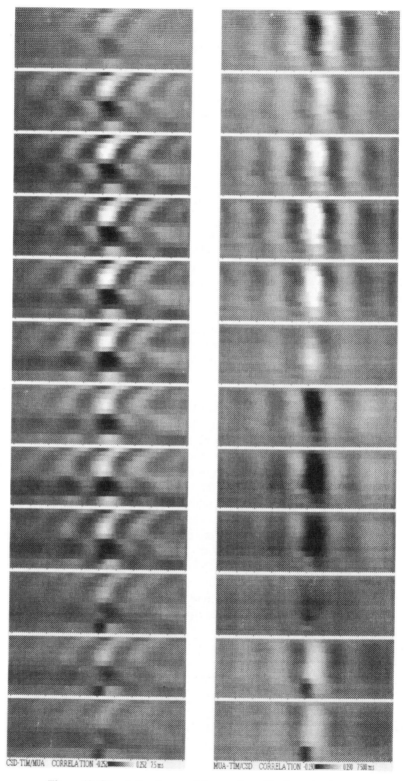

CSD-TIM/MUA CORRELATION -0.252▮▮▮▮▮ 0.252 7.5 ms

MUA-TIM/CSD CORRELATION -0.190▮▮▮▮▮ 0.190 7500 ms

Figure 10. Correlation image of the moving white bar stimulus.

RESULTS II. HIGH TEMPORAL FREQUENCIES

In a second set of experiments we recorded the ongoing activity from the cortex at a much more local scale. In these experiments we recorded the laminar profile of CSD signals, which reflect the local synaptic activity, and the laminar profile of MUA signals, which reflect the local spiking activity. We compared such profiles recorded in area 18 for three different stimuli: no stimulus, stroboscopic flashes and sinusoidally moving bars.

All 12 CSD signals were correlated with all 12 MUA signals, yielding the correlation strength of MUA and CSD signals at the different levels, for different time shifts.

Moving white bar stimuli

Figure 9 shows an example obtained for the moving bar state. The two panels in the top of this figure show slices parallel to the CSD and time shift axes. Such a slice shows the time dependence of the synaptic activity at the different layers in the column that are correlated with one and the same (in this example the most superficial) local spiking activity signal. The two panels have different time scales. From top to bottom of the figure correlation functions of CSD signals recorded from superficial to deep layers of the cortex are depicted. The horizontal axis is the time shift in the cross-correlation, positive time shifts correspond to CSD signals lagging behind MUA signals. Correlation strength is depicted in a pseudo brightness scale. White corresponds to the maximum positive correlation values, black to the maximum negative correlation values and gray to insignificant correlations. Each panel shows oscillatory cross-correlation functions which are most outspoken in the superficial and deep layers. The period of the oscillations was 15.6 ms. The correlation functions at the different depths all have the same oscillation frequency, but the phases are different between layers. The latency difference between middle and superficial layers in this example was approximately 9 ms, latency differences of up to 20 ms were found.

The middle panel shows the instantaneous correlation pattern (at zero time shift) between all pairs of one CSD signal (vertical) and one MUA signal (horizontal). Going from superficial to deep layers in the cortex corresponds to going from top to bottom or from left to right in this figure. The most striking feature in slices such as these are horizontally oriented "blobs". This indicates that spiking activity in the different layers are similarly correlated with the synaptic activity in any layer.

The two bottom panels show this even better. In these panels the time dependence of the correlation function between one CSD signal (the most superficial) and the 12 MUA signals is depicted. The format of these panels is much the same as that of the top panels; again positive time shifts indicate CSD signals lagging behind MUA signals; a different pseudo black and white scale is used. The most salient feature is the presence of almost entirely vertical alternating black and white stripes.

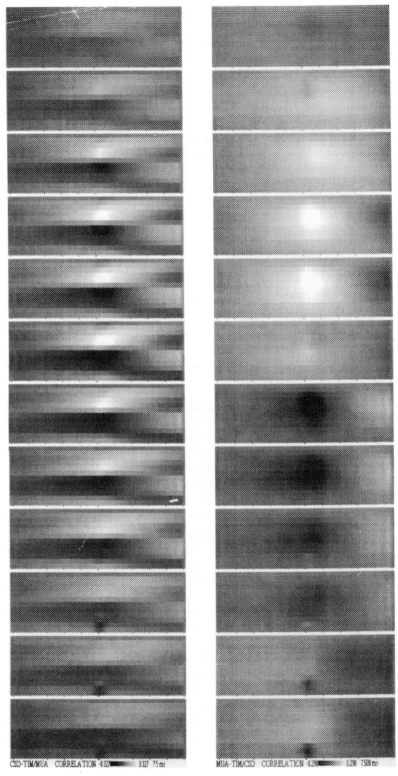

Figure 11. Correlation image when no stimulus is presented.

In our interpretation this shows that the local spiking activities in the different layers of the cortex, that are correlated with synaptic activity in the most superficial layer, are simultaneous. Time shifts between the correlation functions of different MUA signals with one and the same CSD signal were smaller than one sample period (~2 ms).

Figure 10 shows the complete correlation image of the sinusoidally moving white bar stimulus. The two columns in the figure contain the same data organized in two different ways. Each of the panels of the left column is comparable to the top panels of fig. 9, each of the panels in the right column is comparable to the bottom panels of fig. 9. The left column shows the correlation functions of the 12 CSD signals with one particular MUA signal (the reference) in each panel, while the right column shows the correlation functions of the 12 MUA signals with one particular CSD reference in each panel. The 12 panels per column correspond to the 12 possible reference signals.

Again going from top to bottom in the figure corresponds to going from superficial to deep layers in the cortical column. The figure shows that oscillations can be found with about every reference signal; the oscillations are very weak only for the MUA signals from the input layer (6th panel of the right column) and for the deepest CSD signal (bottom panel of the left column).

Non-stimulated and stroboscopic flash data

Figures 11 and 12 show the complete correlation images for the ongoing activity and the stroboscopic flash stimuli respectively. The most obvious conclusion from these images is that they are almost exactly equal (apart from the strength of the correlations which was in the range from -.4 to +.4 for the flashes and from -.3 to +.2 for the ongoing activity).

As with the moving bar stimulus, correlation functions of different CSD signals with the same MUA reference could show phase reversals and considerable time shifts (up to 7 ms), while the correlation functions of different MUA signals with the same CSD reference were simultaneous and did not show phase reversals. The entire correlation pattern can be explained by a single CSD "dipole" around the 6th electrode resulting in simultaneous spikes throughout all cortical layers.

DISCUSSION II. FAST SYNCHRONIZATION

The data presented here on fast synchrony are from a limited data set; yet they seem to allow some preliminary conclusions:

At a local scale visual stimuli may evoke synchronous activity in an endogenous rhythm (i.e. time locked but not phase locked activity). We made sure by numerous tests that the oscillation frequency was not related to any of the harmonics of the moving bar stimulus.

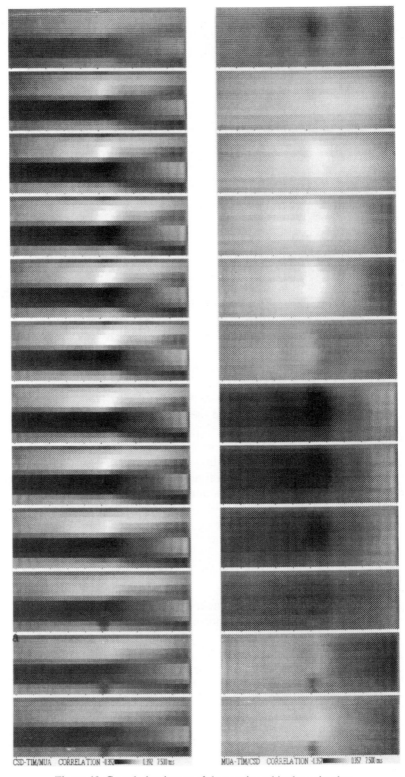

CSD-TIM/MUA CORRELATION -0.392 ▬▬▬ 0.392 7.500 ms   MUA-TIM/CSD CORRELATION -0.357 ▬▬▬ 0.357 7.500 ms

Figure 12. Correlation image of the moving white bar stimulus.

Our data suggest that, owing to the phase differences between the different CSD signals correlated to one and the same MUA channel, it will be difficult to find this endogenous oscillatory activity from a large distance, for example from the scalp or when using epi-dural electrodes.

We observed that synaptic activity in each of the cortical layers results in simultaneous patterns of spiking activity throughout all the layers of the column. This suggests that the column sends the same message over all of its different efferent paths, i.e. to other columns within the same cortical area, to other cortical areas and to sub-cortical structures. This would imply that we should consider a column of the visual cortex as a (non-linear) system with many inputs but a single output.

The finding that stroboscopic flashes yielded the same correlation image as obtained during ongoing activity, indicates that in these two states the column receives a similarly organized input. In other words that most of the ongoing activity of the column is due to input arriving (from the LGN?) at the same layers as the input evoked by light flashes. Combined with our data on low-temporal frequency desynchronization, this suggests the presence of a common input to the cortex from a sub-cortical structure, that can be modulated locally by visual stimulation.

The difference between the correlation images of the moving white bar compared to the two other stimuli indicates that the effective connections within the column can be changed instantaneously by appropriate visual stimuli or alternatively that the dynamics of the population activity within the column can be pushed to a different state by such stimuli (see the paper by Aertsen within this volume). The narrow widths of the cross-correlation functions and the distribution over the different layers make it unlikely that the observed oscillations are the result of mutual interactions between oscillatory systems.

REFERENCES

Allman, J., Miezin, F., McGuiness, E., 1985, Stimulus specific responses from beyond the classical receptive field: Neurophysiological mechanisms for local-global comparisons in visual neurons, Ann. Rev. Neurosc., 8:407-430.

Bolz, J., Gilbert, C.D. and Wiesel, T.N., 1989, Pharmacological analysis of cortical circuitry, TINS, 12:292-296.

Eckhorn, R., Bauer, R., Jordan, W. et al., 1988, Coherent oscillations: A mechanism of feature linking in the visual cortex? Biol. Cybern., 60:121-130.

Fox, J.M., Delbruck, T., Gallant, J.L. et al., 1990, Modulation of classical receptive field responses by moving texture backgrounds in monkey striate cortex: spatial and temporal interactions, Soc. Neurosc. Abstr., 16:#523.5

Freeman, J.A. and Nicholson, C., 1975, Experimental optimization of current-source-density technique for anuran cerebellum, J. Neurophysiol., 38:369-382.

Freeman, W.J. and Skarda, C.A., 1986, Spatial EEG patterns, non-linear dynamics and perception: the neo-Sherringtonian view, Brain Res. Rev., 10:147-175.

Freeman, W.J. and van Dijk, B.W., 1987, Spatial patterns of visual cortical fast EEG during conditioned reflex in a rhesus monkey, Brain Res., 422:267-276.

Gray, C.M., Konig, P., Engel, A.K. and Singer, W., 1989, Oscillatory responses in cat visual cortex exhibit inter-columnar synchronization which reflects global stimulus properties, Nature, 338:334-337.

Kaufman, L., Curtis, S., Wang, J.-Z. and Williamson, S.J., 1992, Changes in cortical activity when subjects scan memory for tones, Electroenceph. clin. Neurophysiol., 82:266-289.

Knierim, J.J. and Van Essen, D.C., 1990, Spatial organization of suppressive surround effects in neurons of area V1 in alert monkeys, Soc. Neurosc. Abstr., 16:#523.6

Kraut, M.A., Arezzo, J.C. and Vaughan, H.G., jr., 1985, Intracortical generators of the flash VEP in monkeys. Electroenceph. clin. Neurophysiol., 62:300-312.

Lamme, V.A.F., van Dijk, B.W. and Spekreijse, H., 1993a, Contour from motion processing occurs in primary visual cortex, Nature, 363:541-543.

Lamme, V.A.F., van Dijk, B.W. and Spekreijse, H., 1993b, Organization of texture segregation processing in primate visual cortex, Visual Neurosc., 10:781-790.

Livingstone, M.S. and Hubel, D.H., 1988, Segregation of form, color, movement, and depth: anatomy, physiology and perception, Science, 240:740-749.

Mitzdorf, U., 1985, Current source density method and application in cat cerebral cortex: Investigation of evoked potentials and EEG phenomena, Physiol. Reviews, 65: 37-100.

Morrell, F., 1960, Electrical signs of sensory coding, in: the Neurosciences: a study program", Rockefeller University Press, New York.

Mumford, D., 1992, On the computational architecture of the neocortex II: the role of the corticocortical loops, Biol. Cybern., 66:241-251.

Nelson, J.I., Salin, P.A., Munk, M.H.J. et al., 1992, Spatial and temporal coherence in cortico-cortical connections: a cross-correlation study in areas 17 and 18 in the cat, Visual Neurosc., 9:21-37.

Pfurtscheller, G. and Klimesch, W., 1992, Functional topography during a visuo-verbal judgement task studied with event related desynchronization mapping, J. Clin. Neurophysiol., 9:120-131.

Priestley, M.B. "Spectral analysis and time series, Vol. I Univariate series", Academic Press, 1981.

Rosenquist, A.C., 1985, Connections of visual cortical areas in the cat, in: "Cerebral cortex, vol. 3: Visual cortex", A. Peters and E.G. Jones, eds., Plenum Press, New York, London.

Schimmel, H., 1967, The (+/-) reference: Accuracy of estimated mean components in average response studies. Science, 157:92-94.

Segraves, M.A. and Rosenquist, A.C., 1982, The afferent and efferent callosal connections of retinotopically defined areas in cat visual cortex, J. Neurosc., 2:1090-1107.

Tolhurst, D.J., Movshon, J.A. and Dean, A.F., 1983, The statistical reliability of signals in single neurons in cat and monkey visual cortex, Vision Res., 23:775-785.

Tononi, G., Sporns, O. and Edelman, G.M., 1992, Reentry and the problem of integrating multiple cortical areas: simulation of dynamic integration in the visual system, Cerebral cortex, 2:310-335.

Vaadia, E., Bergman, H. and Abeles, M., 1989, Neuronal activities related to higher brain functionstheoretical and experimental implications, Trans. Biom. Eng., 36:25-35.

Von der Malsburg, C. and Singer, W., 1988, Principles of cortical network organization, in: "Neurobiology of neocortex", P. Rakic and W. Singer, eds., John Wiley and sons Ltd., London.

Vijn, P.C.M. "Coherent neuronal activity underlying the EEG", Ph-D. thesis University of Amsterdam, 1992.

Vijn, P.C.M., van Dijk, B.W. and Spekreijse, H., 1992, Topography of occipital EEG reduction upon visual stimulation, Brain Topogr., 5:177-181.

HIGH-FREQUENCY ACTIVITY (600 HZ) EVOKED IN THE HUMAN PRIMARY SOMATOSENSORY CORTEX: A SURVEY OF ELECTRIC AND MAGNETIC RECORDINGS

Gabriel Curio[1], Bruno-Marcel Mackert[1],
Klaus Abraham-Fuchs[2] and Wolfgang Härer[2]

[1]Dept. of Neurology, Klinikum Steglitz, Berlin
[2]Siemens, AG Medical Engineering, Erlangen

INTRODUCTION

The purpose of this paper is twofold:

Firstly, it shall draw attention to the phenomenon of evoked high-frequency responses in the cerebral somatosensory system; in this regard the frequency range of interest is between 300 and 1000 Hz. These brain activities can be measured non-invasively in man using either *electrical* somatosensory evoked potentials (SEP) or recently introduced *magneto*-encephalographic somatosensory evoked field technology (SEF); the respective studies will be surveyed here with regard to their particular merits and shortcomings.

Secondly, it will be demonstrated that SEF-recordings (due to the physics of MEG) provide some advantages in comparison to electric SEP measurements: MEG allows an easy and quick identification of those signal sources which reside in or near fissural parts of the neocortex (such as area 3b) and contribute to the multi-component electric burst.

Basically, electric shock stimulation of the median nerve at the wrist evokes early activities in the first somatosensory cortex. In man this primary cortical response has an onset latency of 16 ms and peaks at about 20 ms; it is called N20 and can be detected non-invasively using magnetic as well as electric recordings (Wood et al., 1985). The parietal N20 component of the human SEP is generated by excitatory postsynaptic potentials in area 3b which is located in the anterior wall of the postcentral gyrus (Allison et al., 1991). Scalp SEP reflect *extra*cellular potential distributions; the observed combination of parietal SEP negativity with frontal positivity at 20 ms latency is, therefore, consistent with concomitant *intra*cellular parietofrontally directed postsynaptic currents. These currents flow in the main

Oscillatory Event-Related Brain Dynamics, Edited
by C. Pantev, Plenum Press, New York, 1994

apical dendrites of pyramidal cells being tangentially oriented to the skull and parallel to each other. This population of pyramids activated in parallel can be approximated by a tangential equivalent current dipole (ECD) when being viewed from a distance great as compared to its extension; this ECD is pointing parietofrontally with an orientation orthogonal to the postcentral gyrus.

If the pick-up coil of a magnetic sensor is placed tangentially to the scalp, it will detect the field component normal to the surface of the volume conductor (B_r); it is a central feature of MEG physics that *only tangential but not radial* current sources in the brain contribute to B_r (Williamson and Kaufman, 1987). Hence, in contrast to scalar electric potential measurements, the vectorial character of neuromagnetic fields permits selective recordings from fissural neocortex without superposition from radially oriented sources such as pyramidal neurons located at the crown of a gyrus.

When several current generators in the brain are simultaneously active their particular contributions to electric potential or magnetic field distributions as measured at the scalp surface may be separated by means of (i) spatiotemporal dipole modeling, (ii) physiological manipulations (e.g. twin pulse stimulations; state changes such as sleep or anesthesia), or (iii) specially adapted bandpass filtering of the signals in the time domain, in case that at least some of the current sources generate activities in a clearly discriminable frequency range.

With regard to the source structure of cerebral SEP it turned out that, indeed, simple highpass filtering in the time domain could isolate a distinct high-frequency component (Eisen et al., 1984); later studies showed that its susceptibility against physiological manipulations was different from the behavior of the concomitant low-frequency component (Emerson et al., 1988; Yamada et al., 1988; Emori et al., 1991). In the following only those SEP studies will be reviewed which have relied on digital filtering (thus avoiding problems of peak latency distortions known from analog filter procedures); then recent results from a magnetic study will be reported which permits focussing on a neocortical component of this high-frequency activity.

HIGH-FREQUENCY ELECTRIC SEP-RECORDINGS

Five SEP-studies and two SEF-studies (see below) were summarized in Table 1. All studies used conventional repetitive transcutaneous single shock stimulation of the median nerve at wrist with the intensity set above motor threshold. Only one study (Eisen et al., 1984) also examined peroneal nerve SEP and (in accordance with the longer path of conduction) found the onset of the high-frequency activity delayed in comparison to median nerve stimulation.

An illustrative example of a recent **combined SEP and SEF measurement** is displayed in Fig.1: After data acquisition with a wide bandpass (0.5-1500 Hz), off-line averaging of 8000 responses and digital highpass filtering with a binomial filter (corner frequency 423 Hz; Link and Trahms, 1992) an oscillatory low amplitude burst could be isolated from the underlying N20 activity for both the electric and the magnetic recording. Notably all peaks of the bursts are contained already in the wide bandpass versions as small ripples with identical latencies indicating that this filter procedure has neither generated spurious peaks nor introduced phase distortions.

Table 1. Non-invasive EEG and MEG studies on cerebral somatosensory evoked high-frequency activities

study	location of recordings	bandpass and averages	oscillation features	specific physiology	suggested generators	critical points
EEG:						
Eisen 1984	C3-Fpz (median), Cz-C3 (peroneal)	300-2500 Hz n = 1024	770 Hz (for median nerve)	peron. < median frequency	"recurrent intrathalamic oscillations"	montage mixes fronto-parietal sources
Green 1986	C3 vs. Cz, M1 or clavicle	200-3000 Hz n = 2000	"inflections"	raising corner frequency gives additional inflect.	not discussed	subjects "frequently sleeping"
Emerson 1988	C3 vs. C4	30-3000 Hz n = 6000	sharp inflections on N20 upstroke	inflections vanished in sleep (but N20 did not)	restricted to the contralateral centro-parietal region	no highpass filtering; no source analysis
Yamada 1988	F3 or C'3 vs. linked ears (or: knee reference)	250-2500 Hz n = 1500	fast frequency potentials (FFP) >500 Hz	in sleep frontal FFP decreased more than parietal FFP	"regionally specific thalamo-cortical projection system"	montage mixes cortical with subcortical contributions
Emori 1991	C'3 vs. linked ears	200-2500 Hz n = 1000	FFP>500 Hz	paired stimuli: "progressively longer ISIs for full recovery of the later FFP peaks"	"cortically generated via a closely situated polysynaptic network", or: short latency differences by parallel inputs	montage mixes cortical with subcortical contributions
MEG:						
Tiihonen 1989	N20m field extrema (over upper and lower part of the contralateral central sulcus)	0.05-2000 Hz n =1000 noise estimate: 8-10 fT	between N20m and P27m only one significant peak ("P22m")	"all deflections (N20m, P22m, P27m) reverse polarity"	P22m: magnetic correlate of the mainly radial P22 SEP dipole at sensorimotor primary cortex (*not* at 3b)	"P22m source could not be determined"; but: no highpass filtering; low number of averages
Curio 1993	inferior N20m field extremum (for 1 subj.: mapping over central sulcus); SEP: C'3-C4	400-1500 Hz n = 8000 noise estimate: 2-4 fT	4-7 peaks; max. power near 600 Hz; amplitude: 5-30% of N20m	magnetic burst cortically colocalized with N20m at 3b; its time course can differ from SEP-burst	action potential burst in thalamocortical afferences; ultra-fast non-NMDA EPSPs; post-synaptic neocortical bursting	burst in 7/9 Ss; source analysis (single case): dipole fitting only at oscillation peaks

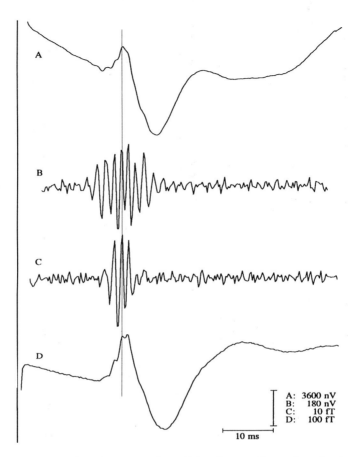

A:	3600 nV
B:	180 nV
C:	10 fT
D:	100 fT

10 ms

Figure 1. Simultaneous non-invasive magnetic and electric recordings of cerebral somatosensory responses (n = 8000 averages) after shock stimulation of the left median nerve at wrist (recordings performed in cooperation with the PTB Biomagnetism Laboratory, Berlin). Electric SEP (traces A/B) were obtained using a C'4-C'3 derivation (negativity at C'4 upward); magnetic SEF (traces C/D) were recorded at the N20 field maximum over the lower part of the right central sulcus (upward deflections indicating magnetic field lines entering the head). Traces A/D are original records using a wide bandpass for data acquisition (0.5-1500 Hz); traces B/C are the respective digitally high-pass filtered versions (binomial filter with a 423 Hz corner frequency). Solid line to the left indicates time of median nerve shock stimulation (shock artifacts during the first milliseconds have been clipped); thin line at 21 ms demonstrates the peak latency preservation (before vs. after digital filtering) for both electric (A/B) and magnetic (C/D) recordings. Moreover, note that the magnetic (C) high-frequency burst in this subject is substantially shorter in duration than the simultaneously recorded electric burst (B) which starts earlier than its magnetic counterpart and outlasts it. Hence the *electric* burst must be considered to be of a *hetero*geneous nature, and (due to the physics of the MEG) *magnetic* recordings can easily isolate activity from a *subset* of generators which must be superficially located and tangentially oriented (i.e. in or near fissural neocortex such as area 3b in the anterior wall of the postcentral gyrus).

It is evident from Table 1 that the filter settings as well as the number of averages varied widely between the studies; this introduced substantial differences in the signal-to-noise ratio. Given this limitation each study contributes to the understanding of where and how these wavelets may be generated, but is also afflicted with some methodological shortcomings:

Eisen et al. (1984) found a 770 Hz burst at 20 ms for median nerve stimulation with up to 4 consistent negative and corresponding positive peaks. The authors postulate that a "recurrent intrathalamic oscillatory mechanism" is causally involved in the generation of the high-frequency burst which itself was assumed to reflect a synchronous cortical activation. Their C3-Fpz scalp electrode montage, indeed, made direct detection of deep thalamic activity unlikely since at the scalp this should show up as a widespread "far-field" which in this kind of bipolar derivation will be largely cancelled. A drawback of this montage, however, is the mixing of parietal activity with possible frontal contributions (cf. below Yamada et al., 1988). Furthermore, it is remarkable that the average interpeak latency difference within the burst was distinctly lower for peroneal nerve stimulation (peroneal: 1.8 ms; median nerve: 1.3 ms). In this regard it is noteworthy that the compound action potential (CAP) is highly synchronized only initially after being triggered by the transcutaneous electric nerve shock. One might speculate that the peripheral and spinal conduction pathways up to the dorsal column nuclei, which are longer for the peroneal than for the median nerve response, could have led to a substantially different dispersion of the propagating CAP. In addition, the thalamocortical conduction distance can differ slightly for the hand area of the postcentral gyrus at the lateral hemispherical convexity and for the foot area beneath the vertex in the interhemispheric fissure. Notwithstanding, this study was the first to firmly establish the presence of electric high-frequency activity superimposed on the N20 component of median nerve SEP by means of digital bandpass filtering; however, a strong hypothesis on how and where this burst was generated could not be derived from these early results.

Green et al. (1986) reported that raising the corner frequency of the highpass filter above 200 Hz gives additional inflections riding on the N20 but did not quantify this effect. Furthermore, they used a vertex, mastoid or clavicle reference; the latter two montages do not differentiate between cortical and subcortical contributions, the former represents a rather closely spaced bipolar recording possibly reducing the amplitude of some smaller signal components below the noise level. In addition, their results may be not directly comparable to other studies since their subjects were "frequently sleeping", and even shallow sleep was reported to decrease the burst amplitude (Yamada et al., 1988).

Emerson et al. (1988) used a C3 vs. C4 montage; this minimized contributions from subcortical as well as frontal sources while largely preserving the peak amplitude of signals from the 'active' electrode placed over the receiving hemisphere due to the wide (biparietal) electrode spacing. They found "sharp inflections on the initial N20 upstroke" which were "topographically restricted to the contralateral centroparietal region"; these inflections vanished in sleep but N20 did not. There are two limitations of this study: (i) the lack of appropriate highpass filtering which would facilitate studying details in the scalp topography of the low amplitude inflections at a higher amplification isolated from the much larger N20; and (ii) no spatial source analysis could be performed beyond a broad regional description due to the low number of recording channels. Taken together, however, this SEP-study provides strong evidence firstly for a strictly cortical component of the high-frequency burst originating near the central sulcus, and secondly for functional independence of the neural systems generating the N20 proper and the superimposed burst.

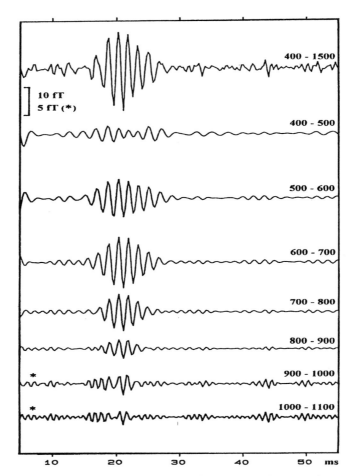

Figure 2. Time-frequency analysis of evoked somatosensory magnetic fields (average of n = 8000 epochs) recorded over the hemisphere contralateral to a median nerve stimulated transcutaneously using strong electric single shocks. Top trace displays the highpass version (400-1500 Hz) of an original wideband record (0.5-1500 Hz; see Figure 1). The sequence of lower traces show narrow bandpass versions (100 Hz nominal bandwidth) with ascending center frequencies as indicated. Note that (i) the main energy is contributed from bands between 500 and 800 Hz, and (ii) that the high-frequency activity is strictly confined in time to an interval around 20 ms corresponding to the arrival time of thalamocortical action potential volleys at the primary somatosensory cortex and the subsequent cortical primary responses.

Yamada et al. (1988) described fast frequency potentials (FFP) above 500 Hz which decreased already in sleep stage II and did even more so in deep slow wave sleep (stage IV); interestingly, FFP bursts reappeared in REM-sleep but the amplitude was smaller than in the waking state. The sleep dependent FFP variation was said to be pronounced frontally (at F3/4) more than parietally (at C3/4); thus the involvement of "*regionally specific* thalamocortical projection systems" was proposed (italics added). Notably, however, the 'active' electrodes were referenced against linked ears (or knee); thus cortical as well as subcortical sources could have contributed to these FFP.

Emori et al. (1991) applied paired shock stimuli with variable delay and found progressively longer interstimulus intervals to be necessary for full recovery of the later FFP peaks. Based on this result they suggested a "cortical generation via a closely situated polysynaptic network". They also mentioned, however, the possibility of short latency differences between axonal volleys arriving via parallel input pathways with slightly different conduction velocities. Since a linked ear reference was again used, cortical and subcortical contributions could not be differentiated.

HIGH-FREQUENCY MAGNETIC SEF-RECORDINGS

At present, two studies have been reported using a bandpass for data acquisition wide enough for reliable magnetic measurement of somatosensory high-frequency responses. This is a reflection of the fact that today only a few systems provide a noise level low enough to achieve an acceptable signal-to-noise ratio using a reasonable number of averages.

Tiihonen et al. (1989) used a wide bandpass open up to 2000 Hz; despite the relatively low number of averages (n = 1000) their system still warranted a final noise estimate of only 8-10 fT. Within this noise background they could, however, identify only one significant peak between N20 and P27 which was found in 6/6 subjects, showed a wide intersubject amplitude variability and was tentatively named P22 according to its peak latency. They recorded over the upper and lower part of the central sulcus contralateral to the stimulated median nerve where the B_r field extrema for a tangential dipolar current source are to be expected when this ECD is oriented orthogonally to the postcentral gyrus in the hand area of area 3b. Actually, a field polarity reversal was found for *all* deflections (N20, P22, P27), but (in contrast to the N20 and P27) the "P22 source could not be determined"; notably, however, no attempt at additional off-line digital highpass filtering for the isolation of high-frequency components was reported. It was suggested that P22 may represent the magnetic correlate of a mainly (but sometimes not completely) radial SEP dipole which is assumed to reside within the sensorimotor primary cortex but outside the tangentially oriented pyramid population of area 3b which itself generates the N20.

Curio et al. (1994a) could use a newly developed dc-SQUID magnetometer which introduces only 2.2 fT/√Hz system noise in the upper frequency region (Drung and Koch, 1993). Given a hardware bandpass of 0.5-1500 Hz (analog-digital conversion at 4 kHz) and an 8/sec stimulation rate the recording time is about 15 minutes for the acquisition of 8000 responses (which can be tolerated not only by motivated research subjects but also by patients). The residual experimental noise for the averaged trace amounts to about 4 fT (see Fig.1); this is consistent with a rough estimate based on the relevant set-up parameters:

[crest factor] times [system noise figure] times [sqrt (bandwidth)] divided by [sqrt (no. of averages)] $\approx$ 4 x 2.2 fT x $\sqrt{(1500-423)}$ / $\sqrt{8000}$.

Interestingly, this implies that at these high frequencies it was mainly the system which contributed noise, and any endogenous background activity of the brain was still below this instrumental noise level; hence, in case of future system improvements, e.g. in SQUID or dewar technology, a substantial (directly proportional) reduction of the residual noise may be expected in the averaged responses.

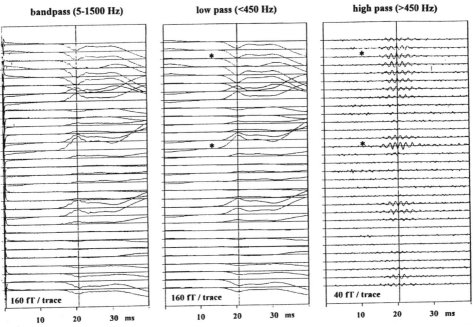

Figure 3. Pilot multichannel recordings of evoked somatosensory magnetic fields. From left to right: original wideband record, lowpass version for isolation of the N20, and highpass version for the burst activity. A vertical line is provided at 21 ms for cross-referencing of the responses between channels as well as between the different bandpass versions. Note the reversal of field polarity in identical channels for both the low- and the highpass traces (asterisks) indicating a close resemblance of the amplitude distribution across the sensor array (26 mm center-to-center intersensor distances; 196 mm array diameter) as it is expected for a colocalisation of the respective current sources (recordings performed at the Siemens Laboratories, Erlangen).

The magnetic burst activity is a brief transient riding on the slower and larger N20 wave. For an analysis of the spectral location of the main energy a Fourier transform would, therefore, be less informative than a frequency analysis of the time domain signals by means of a series of narrow bandpass filters (100 Hz nominal bandwidth) with successively ascending center frequencies (Fig.2): the major contribution to the burst at 20 ms comes from frequency bands between 500 and 700 Hz; thus, it was called a 600 Hz activity. For most subjects (7/9), several peaks were found clearly above noise level. The number of SEF peaks could differ by one between the two hemispheres of a particular subject; interindividually it varied between three and seven peaks in the train, but did not covary with the response strength of the concomitant N20 wave. The longest burst was a 7-peak train of about 10 ms, the largest response measured about 34 fT peak-to-peak. Simultaneous recordings of electric SEP (C'3 vs. C'4) and of magnetic SEF in 2 of 4 subjects showed slight but definite differences either in burst morphology or even a lower number of peaks within the magnetic burst (see fig.1 B/C). This indicates that the magnetic recordings may sample from a *subset* of the electric SEP generators. The burst amplitude expressed as a fraction of the N20 varied from 5 to 30 % between subjects, but *intra*individually it was relatively constant across the sensor array. Hence, a similar spatial field distribution may be assumed for both the N20 proper and the high-frequency oscillations.

This agrees with recent pilot measurements (Curio et al., 1993) using the experimental prototype of a new *multi*channel sensor (Schneider et al., 1993) which provided a comparably low noise level at high frequencies. In these multichannel records small notches riding on the N20 peak can be recognized even when using the widely opened bandpass (Fig.3, left). The lowpass (middle) isolates the N20 proper from the high-frequency burst which itself is extracted by the highpass filter (right). Based on an amplitude mapping across the sensor array for these filtered responses, equivalent current dipoles were fitted to the field distributions around the peak of the N20 and at the burst peaks obtained after highpass filtering, and, indeed, a close cortical colocalisation was found for the N20 and the burst sources (Curio et al., in prep.).

Taken together, magnetic recordings of evoked high-frequency response bursts in the cerebral somatosensory system have become technically feasible by means of two different new low-noise magnetometer systems. Simultaneous SEP and SEF recordings indicate that the electric SEP burst is not a homogeneous entity; rather it appears to contain a component which can be easily isolated by MEG measurements. This subset is cortically colocalized with the generator of the N20 wave which itself is situated in the primary somatosensory hand area in the anterior wall of the postcentral gyrus.

MODES OF GENERATION FOR HIGH-FREQUENCY CEREBRAL SEP/SEF

Several generators have been suggested, each based on the results of a particular study (see Table 1); these proposals have been rearranged with respect to the locus of generation and are presented in Table 2; some further arguments will be discussed in the following, but it should be stressed that, at present, it does not appear possible to arrive at a definitive conclusion.

In and near the **thalamus,** activity can be detected in the appropriate high-frequency range and at the required latency interval as is well-known from both animal (Arezzo et al., 1979) and human depth recordings (Katayama and Tsubokawa, 1987; Morioka et al., 1989). This activity is assumed to reflect the initial bursts of impulses in the **thalamocortical radiations.** However, as the recording site moved rostrally, these components were rapidly reduced in amplitude. Nevertheless, the small notches on the ascending slope of the first major parietal scalp SEP negativity were tentatively related to this deep activity (Katayama and Tsubokawa, 1987). Postsynaptic membrane potentials at the soma of thalamic neurons are less likely to directly contribute to the scalp activity since the more spherical neuronal geometry would generate closed or only partially open field patterns which contrasts with the elongated structure of both thalamocortical axons and cortical pyramidal cells.

Table 2. Possible causes for cerebral evoked somatosensory high-frequency activities

---

A) Thalamus:

- **postsynaptic membrane potential** oscillations at thalamic neurons due to
  - high-frequency action potential discharges in lemniscal afferences
  - recurrent intrathalamic oscillations
  - intrinsic membrane potential oscillations
- high-frequency **action potential** discharges of thalamocortical relay neurons conducted in the initial near-soma axon segment: early depolarisation dipole detectable as far-field; subsequent repolarisation causes a *quadru*polar source aspect with an immediate *drop* in far-field amplitude

B) Thalamocortical radiation:

- **deep:** virtual dipoles near bendings in corticopetal, regionally specific fiber tracts
- **superficial:** virtual dipoles where axons are curving from radial path direction (within white matter) into tangential course <u>before</u> penetrating into the gray matter of area 3b in the anterior wall of the postcentral gyrus

C) Primary somatosensory cortex:

- high-frequency **action potential** burst conducted within **terminal segments** of thalamocortical afferences: quadrupolar (depol/repol) source aspect changes into an isolated repolarisation dipole after cessation of the leading depolarisation phase
- ultrafast **EPSPs** mediated by non-NMDA receptors
- high-frequency discharges of **action potentials in the first postsynaptic cortical cells** (which would need to be precisely phase-locked to the shock stimulus as well as highly synchronized within the responsive cell population)
- **sequential (trisynaptic) activation** of neurons within the cortical column receiving the thalamic afference (progressing mainly from infra- to supragranular layers)
- intracortical **local circuit** reverberatory activity

However, even if thalamic membrane oscillations do not represent the actual generator locus for the bursts observed at the scalp, they might be causally involved by triggering them. In any case, this deep activity would substantially contribute to electric scalp recordings only when a mastoid or extracephalic reference is used; for bipolar SEP derivations (e.g C3/C4), it should be largely cancelled due to its widespread far-field distribution, and, according to theoretical analysis, radially oriented generators such as the deep segments of the thalamocortical radiation cannot be detected by magnetic recordings at all.

Several **superficial sources** may contribute to both electric and magnetic recordings; they can be ordered according to their sequential occurrence:

presynaptic action potentials, cortical EPSPs, burst discharges of the first-order postsynaptic cortical cell population, rapid sequential activation of second- and higher-order neurons or even reverberatory activity within the local cortical microcircuitry.

High-frequency bursts of action potentials conducted in **thalamocortical afferences** may be detected if the usual quadrupolar (i.e. depol./repol.) source configuration with its low-amplitude far-field characteristic is broken up either by (i) bending of the axon before entering the grey matter causing a stronger dipolar field aspect even for minimal deviations from axial alignment (Burghoff and Curio, 1994), or by (ii) cessation of the leading depolarisation phase when running into the terminal intracortical axon segments.

Ultrafast evoked non-NMDA (AMPA/KA-) **EPSCs** with a 10-90% rise-time of 0.4 ms and a decay time constant of $\tau = 1.3$ ms have been described recently for the mossy fibre to cerebellar granule cell synapse (Silver et al., 1992); in non-pyramidal layer IV neurons of rat primary visual cortex (probably stellate cells) a distribution of rise-times (for 20-80% peak amplitude) was found ranging from 0.1 to 0.8 ms (peak at 0.2 ms) along with a mean decay time constant $\tau = 2.4$ ms (Stern et al., 1992). Assuming a maximally synchronized input as provided by the electric shock stimulation of the peripheral nerve, at least the leading activity, which is conducted in the fastest glutamatergic thalamocortical transmission pathway, could eventually trigger a comparably synchronized ultrafast subtype of EPSCs in recipient cells of area 3b; this could be superimposed on currents gated by other non-NMDA receptors (with a slower kinetic) as well as on coactivated still slower NMDA-EPSCs (rise time 5.4-6.6 ms) both of which may be related to the N20 underlying the high-frequency burst.

High-frequency discharges of **action potentials in postsynaptic cortical cells** after electrical peripheral nerve stimulation have been described in cat and monkey primary somatosensory cortex (Amassian and Thomas, 1952; Amassian, 1953); for strong stimuli, up to seven spikes could be evoked in 13 ms. Additional experiments using microstimulation of the sensory thalamic relay nucleus (ventralis posterior) found cortical units responding with eight spikes at a repetition rate of 600/sec (Li et al., 1955). Since a majority of axons ascend side-by-side to targets in supragranular layers within the respective cortical column, their activity might be recordable at the scalp as is the case for any tangential current dipole in area 3b. More recently, 600 Hz action potential bursts were described for visual cortex cells which are active with a 40 Hz interburst periodicity in vivo as well as in vitro (with synaptic transmission being blocked) indicating the possibility that also intrinsic membrane oscillations might contribute to the oscillatory behavior of these "chattering cells" (McCormick et al., 1993).

Trisynaptic progression of excitation towards the supragranular layers within a neocortical column has been described for in vivo as well as for in vitro recordings from primary visual cortex (Mitzdorf, 1985; Bode-Greuel et al., 1987). Inputs from thalamic afferences to layer IV were relayed (among other routes) via strong local connections to layer III and from there to layer II. The layer-specific sinks in a slice current source density analysis were sequentially delayed with respect to each other; interlayer onset-of-activation differences ranged between 0.8 and 1.6 ms. Thus, this sequential passage of activity could contribute, at least in principle, to the high-frequency bursts recorded non-invasively from the scalp.

Intracortical **local circuit reverberatory activity** in a high-frequency range may also be considered as a possible burst component: white matter stimulation in primary visual cortex slices evoked repetitive population spike bursts which are tightly phase-locked to the stimulation (Langdon and Sur, 1990). The firing of layer III pyramidal cells is proposed to be "driven and synchronized by short-latency, excitatory neurotransmission mediated by recurrent collaterals". Even electrical stimulation of the optic tract in vivo was reported to evoke brief bursts of spike-like potentials that repeat at 300-400 Hz in supragranular visual cortex suggesting that "phase-locked firing bursts in layer III may not require direct stimulation of the cortex" and that it "may reveal features of neocortical circuitry that play a role in the normal function of cortex".

SYNTHESIS AND PERSPECTIVES

Using new low-noise recording devices, a high number of averages (n = 8000) and a precisely defined digital highpass filter technique, recent MEG measurements contributed significantly to the understanding of where and how the high-frequency (600 Hz) burst is generated, which is superimposed on the primary cortical SEP response to median nerve electric shock stimulation:

Concerning the **'where?'**, MEG recordings showed the source of the magnetic burst to be cortically colocalized with the magnetic counterpart of the electric N20 which is assumed to be generated by EPSPs in area 3b. Concerning the **'how?'**, it became evident that, at least for some subjects (2/4 in a pilot study), the electric and magnetic burst can differ either in gross (envelope) morphology or even in the number of peaks within the burst (magnetic < electric); this is the first indication that the *electric* high-frequency burst should be considered as a potentially *hetero*geneous event.

In this context two features of MEG physics are relevant: (i) The attentuation of fields from relatively remote sources if the magnetic sensors are operated in a gradiometer mode, and (ii) the intrinsic insensitivity to radially oriented current sources. Accordingly, deep sources such as thalamic projection neurons or initial segments of thalamocortical radiation fibers are unlikely to contribute substantially to MEG records; they may be incorporated, however, in a 'parietal' (C'3 or C'4) SEP recording in case a mastoid or extracephalic reference is used. Furthermore, superficial generators with a major radial component such as events in area 1 or 2 will not contribute to magnetic SEFs but to electric SEPs (and due to the steepness of their scalp potential gradients they will do so even if a 'bipolar' electrode montage, e.g. C'3 vs. C'4, is used).

It may appear a trivial corollary to emphasize the possibility that for electric burst recordings deep, superficial tangential and superficial radial current sources need not contribute single peaks to the burst in a strictly successive manner allowing an independent peak-by-peak analysis; rather a repetitive discharge in thalamo-cortical projection neurons may still continue when the cortical responses first in area 3b, then in areas 1/2, will occur. Hence, activity from all these sources might superimpose in the SEP record. This partial overlap may relate to the crescendo/decrescendo burst envelope which is found in some subjects; certainly, it would complicate the interpretation of SEP records in a way which is basically avoided by magnetic measurements.

But even the magnetic burst itself might not represent an unitary phenomenon since pre- vs. postsynaptic contributions at the cortical level (as discussed above and listed in Table 2) could not be differentiated at present. The next analytic steps will include to replicate earlier studies now with simultaneous EEG and MEG recordings using double pulse stimulation, to look for burst modulations by vigilance changes or even by focussed attention and to intervene pharmacologically, e.g. by applying receptor antagonists; much might also be learned by the examination of patients presenting with well-defined brain diseases such as circumscribed cortical lesions (or more diffuse system degenerations like cortical dementias) and basal ganglia disorders.

From a more general point of view, the presynaptic as well as the diverse post-synaptic modes of burst generation appear equally interesting: (i) If the 600 Hz activity represents a volley of high-frequency discharges conducted in axons of thalamo-cortical projection neurons, then it would become possible to record non-invasively in man telencephalic action potentials at the multiunit level as well as to monitor the membrane status of human thalamic neurons with respect to their mode of impulse generation. (ii) On the other hand, if the burst represents, at least in part, neocortical EPSPs triggered by the ultrafast non-NMDA subtype of the glutamate receptor family, then the low- and high-pass filtering (with a corner frequency around 400 Hz) of the original wide-band recordings (0.5-1500 Hz) may be considered as a kind of pharmacological dissection tool which permits monitoring of the response status of receptor subgroups with slow or fast kinetics, respectively. (iii) And finally, if an in-tracortical spreading of excitation (be it serial or reverberatory) is contributing to the burst generation, then a new level of non-invasive electromagnetic studies on human neocortical brain function in health and disease would become accessible.

REFERENCES

Allison, T., McCarthy, G., Wood, C.C., and Jones, S.J., 1991, Potentials evoked in human and monkey cerebral cortex by stimulation of the median nerve : a review of scalp and intracranial recordings. *Brain* 114: 2465-2503.

Amassian, V.E., 1953, Evoked single cortical unit activity in the somatic sensory areas. *Electroenceph. clin. Neurophysiol.* 5: 415-438.

Amassian, V.E., and Thomas, L.B., 1952, Evoked single cortical unit activity in the primary cortical receiving areas. *Nature* 169: 970-971.

Arezzo, J.C., Legatt, A.D., and Vaughan, H.G. Jr., 1979, Topography and intracranial sources of somatosensory evoked potentials in the monkey: I. Early components. *Electroenceph. clin. Neurophysiol.* 46: 155-172.

Bode-Greuel, K.M., Singer, W., and Aldenhoff, J.B., 1987, A current source density analysis of field potentials evoked in slices of visual cortex. *Exp. Brain Res.* 69: 213-219.

Burghoff, M., and Curio, G., 1994, The high variability of nerve injury field patterns can be simulated by minimal alterations of source parameters in a realistic double dipole configuration. Proc. 9th Int.Conf. on Biomagnetism, Vienna, August 1993 (in press.).

Curio, G., Mackert, B.-M., Burghoff, M., Koetitz, R., Drung, D., and Marx, P., 1994, Neuromagnetic detection of cerebral high-frequency activity (600 Hz) superimposed on the N20-activity from SI in man. Proc. 9th Int.Conf. on Biomagnetism, Vienna, August 1993 (in press.).

Curio, G., Mackert, B.-M., Burghoff, M., Koetitz, R., Drung, D., Abraham-Fuchs, K., Haerer, W., and Marx, P., 1993, Neuromagnetic detection of high-frequency (600 Hz) oscillations evoked in the human primary somatosensory cortex. *Europ. J. Neuroscience*, suppl.6: 228.

Drung, D., and Koch, H., 1993, An electronic second order gradiometer for biomagnetic applications in clinical shielded rooms. *IEEE Trans.Appl.Superconductivity* 3: 2594-2597.

Eisen, A., Roberts, K., Low, M., Hoirch, M., and Lawrence, P., 1984, Questions concerning the sequential neural generator theory of the somatosensory evoked potential raised by digital filtering. *Electroenceph.clin.Neurophysiol.* 59: 388-395.

Emerson, R.G., Sgro, J.A., Pedley, T.A., and Hauser, W.A., 1988, State-dependent changes in the N20 component of the median nerve somatosensory evoked potential. *Neurology* 38: 64-68.

Emori, T., Yamada, T., Seki, Y., Yasuhara, A., Ando, K., Honda, Y., Leis, A.A., and Vachatimanont, P., 1991, Recovery functions of fast frequency potentials in the initial negative wave of median SEP. *Electroenceph.clin.Neurophysiol.* 78: 116-123.

Green, J.B., Nelson, A.V., and Michael, D., 1986, Digital zero-phase-shift filtering of short-latency somatosensory evoked potentials. *Electroenceph.clin.Neurophysiol.* 63: 384-388.

Katayama, Y., and Tsubokawa, T., 1987, Somatosensory evoked potentials from the thalamic sensory relay nucleus (VPL) in humans: correlations with short latency somatosensory evoked potentials recorded at the scalp. *Electroenceph.clin.Neurophysiol.* 68: 187-201.

Langdon, R.B., and Sur, M., 1990, Components of field potentials evoked by white matter stimulation in isolated slices of primary visual cortex: spatial distribution and synaptic order. *J. Neurophysiol.* 64: 1484-1501.

Li, C.-L., Cullen, C., and Jasper, H.H., 1955, Laminar microelectrode studies of specific somatosensory cortical potentials. *J. Neurophysiol.* 19: 111- 130.

Link, A., and Trahms, L., 1992, Highpass filters for detecting late potentials, *in*: "Computers in Cardiology", R. Werner, ed., IEEE Computer Society Press, Washington, 159-162.

McCormick, D.A., Gray, C., and Wang, Z., 1993, Chattering cells: a new physiological subtype which may contribute to 20-80 Hz oscillations in cat visual cortex. *Soc. Neurosci. Abstr.* 19, in press.

Mitzdorf, U., 1985, Current source density method and application in cat cerebral cortex: investigations of evoked potentials and EEG phenomena. *Physiol. Rev.* 65: 37-100.

Morioka, T. Shima, F., Kato, M., and Fukui, M., 1989, Origin and distribution of thalamic somatosensory evoked potentials in humans. *Electroenceph.clin.Neurophysiol.* 74: 186-193.

Schneider, S., Abraham-Fuchs, K., Seifert, H., and Hoenig, H.E., 1993, Current trends in biomagnetic instrumentation. *Applied Superconductivity* 1: 1791-1812.

Silver, R.A., Traynelis, S.F., and Cull-Candy, S.G., 1992, Rapid-time-course miniature and evoked excitatory currents at cerebellar synapses in situ. *Nature* 355: 163-165.

Stern, P., Edwards, F.A., and Sakmann, B., 1992, Fast and slow components of unitary EPSCs on stellate cells elicited by focal stimulation in slices of rat visual cortex. *J. Physiol.* 449: 247-278.

Tiihonen, J., Hari, R., and Hämäläinen, M., 1989, Early deflections of cerebral magnetic responses to median nerve stimulation. *Electroenceph.clin.Neurophysiol.* 74: 290-296.

Williamson, S.J., and Kaufman, L., 1987, Analysis of neuromagnetic signals, chapter 14 *in*: "Methods of Analysis of Brain Electrical and Magnetic Signals. EEG Handbook (Revised Series, Vol.1)", A.S.Gevins and R.Rémond, eds., Elsevier, Amsterdam, 405-448.

Wood, C.C., Cohen, D., Cuffin, B.N., Yarita, M., and Allison, T., 1985, Electrical sources in human somatosensory cortex: identification by combined magnetic and potential recordings. *Science* 227: 1051-1053.

Yamada, T., Kameyama, S., Fuchigami, Y., Nakazumi, Y., Dickins, Q.S., and Kimura, J., 1988, Changes of short latency somatosensory evoked potential in sleep. *Electroenceph. clin. Neurophysiol.* 70: 126-136.

THE TRANSIENT AUDITORY EVOKED GAMMA-BAND FIELD

Christo Pantev and Thomas Elbert

Institute of Experimental Audiology, University of Münster, Germany

INTRODUCTION

A stimulus elicits an afferent volley of action potentials, which is transmitted to the brain where it generally evokes cooperative neural mass activity. This means that transient sensory stimulation will, locally in time and space, enhance average neural depolarization which results in the slow event-related potentials (Elbert, 1993). Furthermore, coordinated patterns of neural mass activity will result in fluctuations in the gamma band, i.e. in the frequency range above 20-30 Hz. The overall sum of depolarizations manifests itself in EEG and MEG, and the corresponding fluctuations are referred to as induced rhythms, a term which was introduced to call attention to a large variety of oscillatory responses that follow either clearly timed stimuli or less sharply timed state changes such as attention, sleep, expectations and seizures. Induced rhythms may be tightly or loosely time-locked to the triggering event and, hence, may or may not appear in averaged responses or only in single trials. Bullock, in his chapter of this book, suggested considering evoked rhythms as a subset of induced activity. Evoked rhythms follow clearly timed stimuli or events. They are more or less tightly time-locked and thus may be extracted by conventional averaging across many responses, i.e. by averaging in the time domain. The term evoked is generally not used for responses to less sharply timed state changes or events such as those just named. There is, however, no sharp border between the two terms, and therefore it may be useful to add descriptors, such as "phase locked" or "nonphase locked" etc.

The evoked gamma-band activity is stimulus- and phase-locked and thus it can be extracted from the noise in the time-domain using the averaging procedure. The present chapter reports results from studying the evoked, i.e. the time-locked transient oscillatory gamma-band activity, arising from the auditory cortex in humans.

The first oscillatory activity in the 40-Hz range arising from the auditory cortex and the hippocampus of the cat as response to auditory stimulation was recorded by Basar (1972; Basar and Özesmi (1972); Basar and Ungan, (1973) and investigated in the following years by Basar and his group (Basar et al., 1987; Basar, 1988; Basar-

Oscillatory Event-Related Brain Dynamics, Edited
by C. Pantev, Plenum Press, New York, 1994

Eroglu and Basar, 1991). Human auditory evoked steady state responses in the gamma-band were intensively studied by Galambos et al. (1981); Mäkelä and Hari (1987); Galambos and Makeig (1988); Hari et al. (1989); Ribari et al. (1989) and Romani et al. (1982). Both cortical and subcortical sources for the steady-state responses have been proposed (Mäkelä and Hari, 1987). Recently transient electric and magnetic gamma-band responses time locked to the onset of auditory stimuli that increases in amplitude as stimulus rate is decreased below one per second and their sources located in the auditory cortex have been shown (Makeig and Galambos, 1989; Pantev et al., 1991; Pantev et al., 1993).

The goal of the present study was to increase our understanding of the transient auditory gamma-band responses (GBR) as well as the function that GBR may serve for sensory processing, particularly in the human auditory cortical areas. Therefore, we carried out several parametric and attentional studies. The functional behavior of the GBR in the time domain and the spatial locations of its generators were compared with those of the best investigated evoked responses of the auditory system – the slow auditory evoked responses (sAER). We will report here results from three different sets of experiments: the study of *stimulus repetition rate*, of *tonotopy* (dependence upon stimulus frequency) and of *attentional variations*.

METHODS

Subjects

A total of 12 subjects with no history of otological or neurological disorders and normal audiological status, aged between 23 and 42, participated in the experiments. All subjects were right-handed. Informed consent was obtained from the subjects after the nature of the particular experiment was fully explained to them.

Stimulation

Stimuli were delivered to the right ear. They were presented through a non-magnetic and echo-free stimulus delivery system with an almost linear frequency characteristic (deviations less than ±4 dB in the range between 200 and 4000 Hz). During stimulus presentation the subjects were asked to stay awake and to keep their eyes open.

The **recovery function** of the GBR and of the sAER was studied in four subjects at stimulus rates of 1/8, 1/4, 1/2 and 1 Hz, interstimulus intervals of 8, 4, 2 and 1 s, respectively. Stimuli were Gaussian tone-pulses with half-value time of 6 ms, a carrier-frequency of 1000 Hz, and an intensity of 60 dB nHL (normative hearing level). They were presented in two blocks of 256 stimuli each.

The dependence upon the stimulus frequency of the GBR and the sAER, characterizing one of the most important features of the auditory system - its **tonotopic organization** – was studied in two subjects to stimulus frequencies of 250, 1000 and 4000 Hz. Stimuli were tone burst of 500 ms duration, 15 ms rise and fall times, 4 s interstimulus interval, 60 dB nHL intensity and the above mentioned carrier frequencies. They were presented in three blocks of 128 stimuli each.

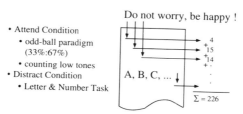

Figure 1. Stimulus design of the odd-ball paradigm: probability of appearance of the rare stimuli (low pitch) has 33%, of the frequent stimuli (high pitch) 67%. During the attend condition the subject concentrated on the appearance of the rare stimuli and had to count them. During the distract condition the subject performed a letter-to-number test (example shown above), involving mental arithmetic and considerable demand on memory.

Effects of attention and processing on the GBR and the sAER was studied in six subjects. In a classical odd-ball paradigm rare stimuli (low pitch of 667 Hz, 33% of all stimuli) were randomly interspersed in a sequence of frequent stimuli (high pitch of 1500 Hz; 67%). Stimuli were Gaussian tone-pulses of the above mentioned frequencies with half-value time of 6 ms and an intensity of 60 dB nHL. The inter-stimulus interval varied randomly from 1.5 to 2.5 s. The stimuli were presented in twelve blocks, with one block having a random length between 150 and 250 stimuli. During one block, the subjects task was either to attend to the oddball stimuli or to process a completely different task in order to distract from the oddball stimuli. During the attend condition, the subject was instructed to concentrate on the appearance of rare stimuli and to count them. The counting results were controlled. During the distract condition, the subject had to perform a numbers-to-letter task involving mental arithmetic with considerable demand on memory (c.f. Fig. 1). With the begin of the stimulus block (distract condition), the subjects started to calculate the sum of the numbers corresponding to the rank of the letters of certain well known sentences. The results of their calculations were also controlled. Task-sequence was randomized across blocks.

Neuromagnetic recordings

Recordings were carried out in a magnetically shielded room using a 37 channel SQUID (Super Conducting Quantum Interference Device) biomagnetometer (Biomagnetic Technologies). Detection coils of the biomagnetometer are arranged in a circular concave array with a diameter of 144 mm, and a spherical radius of 122 mm. The axes of the detection coils are normal to the surface of the sensor array. The distance between the centers of two adjacent coils is 22 mm; each coil itself measures 20 mm in diameter. The sensors are configured as first-order axial gradiometers with a baseline of 50 mm. The spectral density of the intrinsic noise of each channel is between 5-7 fT/√Hz in the frequency range above 1 Hz. The subjects lay in a right lateral position with their head, neck and upper part of the body supported by a specially fabricated vacuum cushion. A sensor position indicator system determined the spatial locations of the sensors relative to the head and indicated if head movements occurred during the recordings. The sensor array was centered over a point about 1.5 cm superior to the position T3 of the 10-20 system for electrode placement and was positioned as near as possible to the subject's head. Using a bandwidth from 1 to 100 Hz and a sampling rate of 520.8 Hz (16 bit ADC), stimulus-related epochs between 1000 and 1200 ms for the different experiments (including 200 ms prestimulus baselines) were recorded and stored for further analysis. The slope was 6 dB/oct for the analog high-pass filter and 60 dB/oct for the on-line digital low-pass filter.

Data Analysis

Wide band responses to each block were first averaged after rejection of arti-fact-contaminated epochs due to eye blinks or muscle activity. Since the waveforms of the averaged responses recorded in the different blocks were highly reproducible, their grand average was used for further evaluation. The wide-band responses were digitally filtered with passbands 0.1-20 Hz, and 24-48 Hz (first order Bessel filter, 6 dB/oct) in order to obtain the slow auditory evoked field (sAEF) and the gamma-band field (GBF), respectively. To avoid phase shifts, the data were filtered twice (forward and backward); thus, the effective slope of the filter characteristic was 12 dB/octave. Root-mean-square (RMS) field values over the 37 recording channels were calculated for every sampling point.

Source analyses based on a single moving Equivalent Current Dipole (ECD) model in a spherical volume conductor (Sarvas, 1987) was applied to both magnetic field distributions studied: bandwidths 0.1-20 Hz and 24-48 Hz. The size of the sphere was determined by a fit to the scalp in the area of measurement. It was in this area that the shape of the scalp was digitized with the sensor position indicator. The center and the radius of the best fitting sphere were determined using a least squares fit algorithm. The relative angulation of the pickup coils and the influence of the volume currents were taken into account for the source analysis. Source locations, their confidence volumes and the dipole moment were estimated for each sampling point in a head-based coordinate system. The origin of this coordinate system was set at the midpoint of the medial-lateral axis (y-axis) which joined the center points of the entrance to the acoustic meatuses of the left and the right ears (positive to-wards the left ear). The posterior-anterior axis (x-axis) was oriented from the origin to the nasion (positive towards the nasion) and the inferior-superior axis (z-axis) was perpendicular to the x-y plane (positive towards the vertex).

The estimated source locations (x, y, z coordinates) of the major wave of the sAEF – N1m and of the three largest peaks of the transient GBF were determined by means of the following manipulations. For each wave-peak the time interval (hereafter termed "evaluation interval") containing the field maximum and mini-mum, the RMS field maximum, and the goodness of fit between the theoretical field generated by the model and the measured field were determined. The spatial ECD coordinates (x, y, z) for the N1m wave of the sAEF were assigned to the correspond-ing averaged values of seven adjacent sampling points around the center of the ob-tained evaluation interval. For the GBF, the "evaluation interval" was determined for each of the three largest peaks (number of adjacent sampling point usually 3-4) and the corresponding averaged values were averaged across the three peaks. However, only those x, y and z values were included in the average which further fulfilled the following conditions based on statistical and anatomical considerations:

- goodness of fit above 95% for the N1m wave of the sAEF and above 90% for the GBF
- distance of the ECD to the midsagittal plane more than 2 cm
- inferior-superior value greater than 1 cm
- confidence volume of the source location smaller than 0.3 cm³

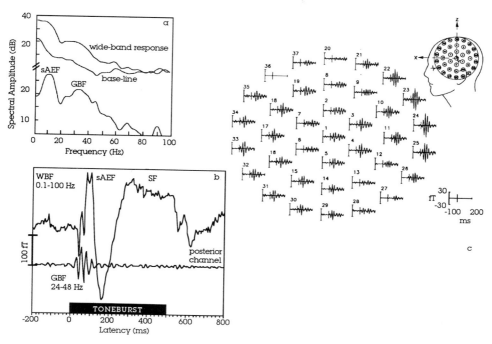

Figure 2. (a) Spectrum of the wide-band responses averaged over three consecutive data blocks from each of seven subjects. The top two traces were constructed by averaging the Fourier amplitude spectra of 200 ms Hanning-windowed epochs and smoothing the result with a rectangular 3 Hz window. Epochs selected were (-35 to 165 ms) for the wide-band response (top trace), and (-200 to 0 ms and 300 to 500 ms) for the baseline estimate (middle trace). The bottom trace is the log difference between the response and baseline spectra, clearly showing separate spectral peaks for the slow auditory evoked field (sAEF) and the gamma-band field (GBF) components. (b) Auditory evoked gamma-band field as compared to the corresponding wide-band field (WBF). The coincidence of GBF peaks in the unfiltered and filtered data is visible. (c) Auditory evoked gamma-band field distribution over the left hemisphere (N=384).

RESULTS

Fig. 2a shows a spectrum grand-averaged across 384 stimulus-related epochs for each of the 37 magnetic channels over the supratemporal cortex for 7 different subjects in response to 500 ms tone-burst stimulation. The wide-band response contains spectral energy in two distinct frequency ranges. One peak around 10 Hz representing the sAEF, and another one between 30-40 Hz reflecting the GBF. A smaller peak at around 70 Hz also seems evident but will not be examined here any further.

To extract the time function of the transient oscillatory gamma-band response from the wide-band activity, responses were digitally filtered with a 24-48 Hz pass-band. An example of the wide-band (01.-100 Hz) and the gamma-band response (24-48 Hz) is shown in Fig. 2b. In the wide-band response not only the large slow wave components but also noticeable gamma-band activity during the first 100 ms after the stimulus onset can be clearly discerned. The sustained field and the "off" response can also be well detected. Finally, this figure represents the excellent coincidence between the GBF peaks in the wide-band and in the narrow-band version of the data. Fig. 2c displays the gamma-band field distribution after the tone-burst stimulation, as obtained through the 37 recording magnetic channels. The field distribution shows field minimum and maximum at similar locations as the sAEF, suggesting a source of the GBF in the supra-temporal cortex as well. The gamma-band field has clear dipolar pattern, thus a single moving ECD model can be used in order to estimate the underlying sources. It was shown by (Pantev et al., 1991), that the GBF equivalent source is located in cortical regions on the floor of the Sylvian fissure (Brodman's areas 41, 42 and 22), more medial and anterior than the source location of the major N1m-component of the sAEF.

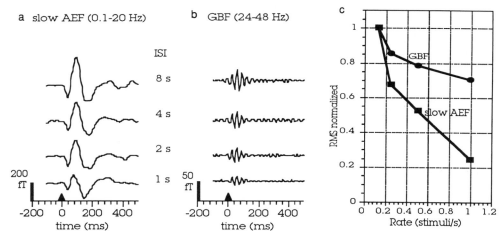

Figure 3. Experimental set of data for four different interstimulus intervals of 8, 4, 2, and 1 s. (a) Slow auditory evoked fields (b) Gamma-band fields. (c) This panel indicates the rate-amplitude-characteristics of the slow AEF and GBF based on the mean RMS-values averaged across subjects as relative presentation referred to the maximum value obtained at ISI of 8 s.

Figures 3a, b illustrate the sAEF and the GBF at the different 4 rates with interstimulus intervals of 8, 4, 2 and 1 s. The appearance of the stimulus is marked by a filled triangle on the abscissa. The amplitude of N1m of the sAEF and this one of the GBF appear to decrease as the stimulus rate increases; however, the N1m is decreasing faster than the GBF. In order to estimate the recovery function of the GBF and the N1m, the dependence of the RMS field amplitude calculated over the 37 magnetic channels and averaged across subjects, upon the stimulus rate (termed "rate-amplitude characteristic") was determined. Because of the very different magnitudes of N1m and the GBF, the comparison of their characteristics is proper after they have been normalized. The normalized "rate-amplitude characteristics", which are shown in Fig. 3c, have different slopes, indicating different recovery patterns as well different generation mechanisms of the sAEF and the GBF.

Using the source localization procedure described in the Methods section, the equivalent source locations for the N1m and the GBF were determined for each stimulus rate. The consistency of the localization results across subjects allowed calculation of the grand means of the estimated source locations for each stimulus rate, as shown in Fig. 4.

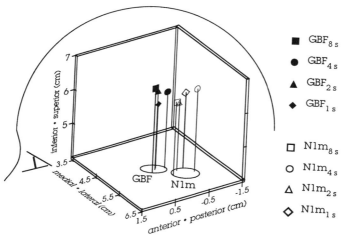

Figure 4. Mean values of the estimated source locations of the slow AEF and GBF for the different interstimulus intervals. Anterior-posterior, medial-lateral and inferior-superior axes of the 3D-plot represent the axes of the head-based coordinate system as described in the Methods section.

Two separate groups of source locations can be clearly recognized. The first one (open symbols) includes the locations of N1m; this group is most lateral and posterior. The second group (filled symbols) represents the mean locations of the GBF, also suggesting a source in the supra temporal cortex somewhat medial and anterior to N1m. The data suggest that the GBF and the sAEF have different origins and may arise from different processes in the auditory pathway.

The tonotopy is one of the general principles of functional organization of the auditory system (Romani et al., 1982; Pantev et al., 1988; Pantev et al., 1989; Bertrand et al., 1991). The depth of the N1m source indicates this tonotopic organization: the higher the stimulus frequency, the deeper the equivalent source. Moreover, in most subjects the depth of N1m increases linearly with the logarithm of the stimulus frequency (Fig. 5, upper part). Opposite to this finding for the N1m component of the sAEF was that the depths of the equivalent GBF source (Fig. 5, lower part) obtained from the same data sets of the two investigated subjects do not change with an increase in the stimulus frequency.

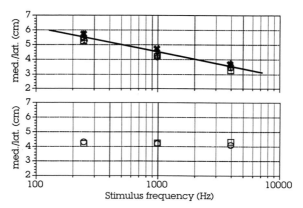

Figure 5. Dependence of the y-coordinate of the source location (corresponding to the depth of the equivalent source) of the N1m of the slow AEF (top) and the GBF (bottom) upon the stimulus frequency (250, 1000 and 4000 Hz) as obtained from two different subjects.

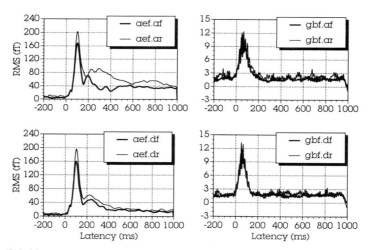

Figure 6. RMS field values over the 37 recording channels and grand averaged across subjects for the rare (thin line) and the frequent (thick line) stimuli of the attend (top) and distract (bottom) conditions for the slow AEF (left) and the gamma-band field (right). af – attend condition, frequent stimulation; ar – attend condition, rare stimulation; df – distract condition, frequent stimulation; dr – distract condition, rare stimulation

The results of the attention experiment for the "odd-ball" stimulus paradigm are illustrated in the next two figures. Fig. 6 shows the RMS field values over the 37 recording channels, which were averaged across the six subjects investigated and which are displayed separately for the rare stimuli (thin line) and frequent stimuli (thick line) of the attend condition (top panels) and distract condition (bottom panels) for the sAEF (left side) and the GBF (right side). A clear difference was obtained within the time-range of P300 for the attend and distract conditions, indicating that the subjects complied well with the experimental task. Higher amplitudes of N1m were obtained for the rare stimuli in both attend and distract conditions. No substantial difference was obtained for the GBF, neither between the rare and frequent stimuli, nor between the attend and distract conditions.

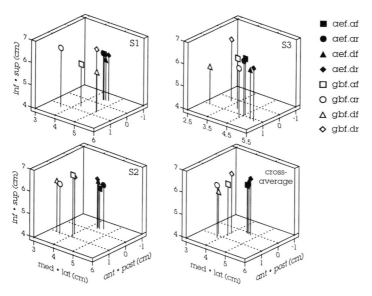

Figure 7. 3D-source locations of the N1m of the slow AEF and the GBF for the different conditions: af, ar, df, and dr (c.f. Fig. 6) as obtained from three different subjects (S1, S2, S3). The right low panel shows the cross-average source locations obtained from all subjects investigated. Anterior-posterior, medial-lateral and inferior-superior axes represent the axes of the head-based coordinate system as described in the Methods section.

The results of the corresponding source analysis of N1m and the GBF for rare and frequent stimuli as well as attend and distract conditions are presented in Fig. 7 for the 3 best subjects (S1, S2, and S3 were the subjects with the best signal-to-noise ratio). The grand mean of the corresponding source locations of the six subjects are shown at the bottom right corner. As already demonstrated by Fig. 4 the GBF sources are more anterior and more medial than those of N1m. Neither the N1m nor GBF differed in source location between conditions (rare/frequent or attend/distract comparisons by means of paired t-test)

DISCUSSION

The auditory evoked gamma-band response seems to be an event related rhythmic response which persists within the first 100 to 120 ms after the stimulus onset. The individual waves of the GBF are visible in the subject's wide-band unfiltered data and show a clear spectral peak between 30 and 40 Hz. After separation of the wide-band activity in sAEF and GBF, the moving single equivalent current dipole model accounts for each data set (sAEF and GBF) almost completely. The obtained source locations for the sAEF and the GBF differ significantly, but both responses originate within the supratemporal cortex. The GBF sources for the different stimulus rates were found to be medial and anterior to the sources of the N1m of the sAEF.

If the N1m and GBF, two components of the complex human auditory cortical response, represent responses to information of the same aspects of the auditory pathway, then their rate-amplitude characteristics should also be similar. The pronounced difference in the slopes of their rate-amplitude characteristics, however, indicates different recovery patterns and, hence, different generator mechanisms for the N1m and GBF. The finding of separate habituation patterns for the auditory N1m and GBF is paralleled by the results of an investigation of slow and gamma-band evoked responses in the somatosensory system (Kaukoranta and Reinikeinen, 1985).

Tiitinen et al. (1993) investigated the effect of *selective* attention on the transient auditory gamma-band potentials by having subjects listen to tone pips presented in one ear while ignoring concurrent sequence of tone pips in the other ear. They found that the gamma-band response was significantly larger when subjects paid attention to stimuli rather than when they ignored them and that this attention effect was most pronounced over the frontal and the central scalp areas. This attention effect was found only for the frequency band with central frequency near 40 Hz. The authors thus concluded that this transient auditory gamma-band response demonstrates a physiological correlate of selective attention in humans. In our manipulation of global attention, we were not able to show a variation of the transient auditory GBF upon attentional variables neither in the GBF-amplitude in time domain, nor in the GBF-source location. A possible explanation for this difference seems to be the technical details: Tiitinen et al. (1993) used EEG-recordings and a much smaller frequency band (8 Hz around the central frequency of 40 Hz) as compared to the 24 Hz band (24-48 Hz) of MEG used in our studies. More likely, however, the difference in experimental conditions may have caused the different outcome: Tiitinen et al. (1993) varied *selective* attention, which means that a specific subset of stimuli must be processed while another subset should be ignored, whereas the present experiment included a variation in *global* attention, meaning that all of the incoming stimuli must be attended in one case or were ignored in the other condition.

The experimental results of the auditory evoked, time and phase locked gamma band activity presented in this chapter allow us to make the following speculations: The different recovery pattern of the sAEF and the GBF as well as their different origins point out that these two different kinds of cortical activity may have different generator mechanisms and may arise from different steps in the course of processing of auditory information. This view is supported by the present results of different short-term habituation for the sAEF and the GBF and receives additional confirmation by the difference in the dependence on the stimulus repetition rate of both phenomena. Whereas the slow auditory evoked activity is probably representing cortical analysis of auditory stimulus features, the missing tonotopic representation in the GBF source locations lets us assume that the gamma-band activity may represent more general perception mechanisms in the central part of the auditory system. Due to the fact that gamma-band activity appears relatively early and does not demonstrate cortical tonotopic representation, it could be concluded that the gamma-band activity reflects a cortical process synchronizing different auditory areas. According to the suggestions outlined by Roelfsema et al. and Eckhorn (see respective chapters in this volume) this more global cortical synchronization could serve as a dynamic link of separate subregions of the auditory cortex in order to combine complex auditory features into unitary auditory percepts, which can also be regarded as a kind of recognition of auditory objects. Synchronizing activity is also predicted by models of brain functioning (e.g. Pulvermüller et al., this volume). Grossberg (1980, 1987) for example suggests that a dynamic state which he refers to as "adaptive resonance" occurs between successive stages of processing. Such a short-term memory resonance is needed to drive changes in long-term memory and might express itself as gamma-band activity. Therefore, the relationship of the GBF to memory functions should be explored in future experiments.

Acknowledgement

Research was supported by the Deutsche Forschungsgemeinschaft. We appreciate the assistance of and critical discussions with many colleagues and friends, in particular of Olivier Bertrand, Carsten Eulitz, Scott Hampson, Bernd Lütkenhöner, Scott Makeig, Brigitte Rockstroh, and Bernhard Roß.

REFERENCES

Basar, E. A study of the time and frequency characteristics of the potentials evoked in the acoustical cortex. *Kybernetik*, 1972, 10: 61-64.

Basar, E. EEG-Dynamics and evoked potentials in sensory and cognitive processing by the brain. Berlin-Heidelberg-New York: Springer-Verlag, 1988, 1: 311-318.

Basar, E. and Özesmi, C. The hyppocampal EEG activity and a systems analytical interpretation of averaged evoked potentials of the brain. *Kybernetik*, 1972, 12: 45-54.

Basar, E., Rosen, B., Basar-Eroglu, C. and Greitschus, F. The associations between 40 Hz-EEG and the middle latency response of the auditory evoked potential. *Int. J. Neurosci.*, 1987, 33: 103-117.

Basar, E. and Ungan, P. A component analysis and principles derived for the understanding of evoked potentials of the brain: Studies of the hyppocampus. *Kybernetik*, 1973, 12: 133-140.

Basar-Eroglu, C. and Basar, E. A compound P300-40 Hz response of the cat hippocampus. *Int. J. Neuroscience*, 1991, 60: 227-237.

Bertrand, O., Perrin, P. and Pernier, J. Evidence for a tonotopic organization of the auditory cortex observed with auditory evoked potentials. *Acta Otolaryngol* (Stockh), 1991, Suppl. 491: 116-123.

Crossberg, S. How does a brain build a cognitive code? *Psychol. Rev.*, 1980, 87: 1-51.

Crossberg, S. The adaptive brain. Amsterdam: Elsevier, 1987, I. and II.

Eckhorn, R. Oscillatory and non-oscillatory synchronizations in the visual cortex of the cat and monkey. (*this volume*).

Elbert, T. Slow cortical potentials reflect the regulation of cortical excitability. New York, London: Plenum Publishing Corp., 1993, 235-252.

Galambos, R. and Makeig, S. Dynamic changes in steady-state responses. Berlin: Springer, 1988, 103-122.

Galambos, R., Makeig, S. and Talmachoff, P.J. A 40-Hz auditory potential recorded from the human scalp. *Proc. Natl. Acad. Sci.*, 1981, 78: 2643-2647.

Hari, R., Hämäläinen, M. and Joutsiniemi, S.L. Neuromagnetic steady-state responses to auditory stimuli. *J Acoust Soc Am*, 1989, 86: 1033-9.

Kaukoranta, E. and Reinikeinen, K. Somatosensory evoked magnetic fields from SI: An interpretation of the spatiotemporal field pattern and effects of stimulus repetition rate. Helsinki University of Technology, Low Temperature Laboratory, 1985

Makeig, S. and Galambos, R. The 40-Hz band evoked responses lasts 150 msec and increases in size at slow rates. Phoenix 9.30.89: 1989, 15: 113.

Mäkelä, J.P. and Hari, R. Evidence for cortical origin of the 40 Hz auditory evoked response in man. *Electroencephalogr Clin Neurophysiol*, 1987, 66: 539-46.

Pantev, C., Elbert, T., Makeig, S., Hampson, S., Eulitz, C. and Hoke, M. Relationship of transient and steady-state auditory evoked fields. *Electroenceph. clin. Neurophysiol.*, 1993, 88: 389-396.

Pantev, C., Hoke, M., Lehnertz, K., Lütkenhöner, B., Anogianakis, G. and Wittkowski, W. Tonotopic organization of the human auditory cortex revealed by transient auditory evoked magnetic fields. *Electroenceph. clin. Neurophysiol.*, 1988, 69: 160-170.

Pantev, C., Hoke, M., Lütkenhöner, B. and Lehnertz, K. Tonotopic organization of the auditory cortex: pitch versus frequency representation. *Science*, 1989, 246: 486-8.

Pantev, C., Makeig, S., Hoke, M., Galambos, R., Hampson, S. and Gallen, C. Human auditory evoked gamma-band magnetic fields. *Proc Natl Acad Sci*, 1991, 88: 8996-9000.

Ribari, U., Llinas, R., Kluger, A., Suk, J. and Ferris, S.H. Neuropathological dynamics of magnetic, auditory steady-state responses in Alzheimer's disease. New York: Plenum Press, 1989, 311-314.

Roelfsema, P.R., Engel. A.K., König, P., Singer, W. Oscillations and synchrony in the visual cortex: evidence for their functional relevance. (*this volume*).

Romani, G.L., Williamson, S.J., Kaufman, L. and Brenner, D. Characterization of the human auditory cortex by the neuromagnetic method. *Exp Brain Res*, 1982, 47: 381-93.

Sarvas, J. Basic mathematical and electromagnetic concepts of the biomagnetic inverse problem. *Phys. Med. Biol.*, 1987, 32: 11-22.

Tiitinen, H., Sinkkonen, J., Reinikeinen, K., Alho, K., Lavikeinen, J. and Näätänen, R. Selective attention enhances the auditory 40-Hz transient response in humans. *Nature*, 1993, 364: 59-60.

STIMULUS FREQUENCY DEPENDENCE OF THE TRANSIENT OSCILLATORY AUDITORY EVOKED RESPONSES (40 HZ) STUDIED BY ELECTRIC AND MAGNETIC RECORDINGS IN HUMAN

Olivier Bertrand[1] and Christo Pantev[2]

[1]Brain Signals and Processes Laboratory, INSERM U280, Lyon, France
[2]Institute of Experimental Audiology, University of Münster, Germany

INTRODUCTION

In the last years, several studies aimed to propose a specific functional role to electrical brain activities characterized by oscillations around 40 Hz, the so-called gamma-band activity. Eckhorn et al. (1988) and Gray et al. (1989) have shown that, in the visual cortex, a spatially disparate subset of cells fires in coherent bursts (in the range 40- 60 Hz) after appropriate visual stimuli. It has thus been suggested that these synchronous gamma-band activities participate to fundamental mechanism underlying feature binding in object recognition. The observation of oscillatory response in the gamma-band has also been reported in human auditory modality, elicited by either steady-state (Galambos et al. 1981) or transient stimulation (Pantev et al. 1991) recorded over the scalp. How these gamma-band responses, recorded on the head surface and certainly due to large populations of neurons, relate to the synchronous rhythmic activities recorded intracranially and representing few neurons in very focal cortical area, is still a question of debate.

Pantev et al. (1991) revealed from multichannel magnetic recordings that the auditory transient 40 Hz averaged response is generated at least partially in the auditory cortex. He suggested that "a conceivable function of a global gamma-band response (GBR) excitation could be to synchronize and dynamically link separate subregions of auditory cortex in order to combine separately represented auditory features into unitary percepts". However, the relation between the generators of the different auditory evoked components needs detailed investigations. Is the GBR a restricted version of a complex wide-band response including the middle-latency and slow auditory components ? In this case, it should share the same functional properties. Or, is it an additional activity which involves distinct generators and thus may have its proper functional role ?

In more recent studies, Pantev et al. (1993) have found that the neural source of the magnetic GBR had a distinct location than those of the slow auditory response and of the magnetic steady-state response. Furthermore, the amplitude characteristics of the magnetic GBR and N100 with stimulus rate seem to be different, thus suggesting the contribution of neural generators having distinct habituation patterns. Similar results were obtained by Forss et al. (1993) with 40 Hz click train stimuli. More recently, a specific attentional effect has been reported (Tiitinen et al. 1993) in the gamma-band range during the early part (before 50 ms) of the auditory response. At last, it has been suggested (Ribary et al. 1991) that auditory 40 Hz magnetic responses may be reset and enhanced by sensory stimuli, and seem to sweep over the cortex. This interpretation was proposed because a fronto-occipital phase shift of the oscillating waves was observed over the head. However, this same data may be explained as well by several active areas in or near the auditory cortex slightly delayed in time.

In the present study, we aimed to investigate whether the GBR has specific properties related to the stimulus frequency, or at least whether these properties are distinct to those reported for other auditory components, i.e., the middle latency response (MLR) and the slow auditory evoked responses (sAER). Previous studies (Pantev et al. 1988, Bertrand et al. 1991, Pantev et al. 1994) have described the frequency dependence of these components both from multichannel electric and magnetic recordings. Distinct tonotopic organizations of the underlying generators of the Pa/Pam and N100/N100m have been demonstrated from scalp potential and scalp current density mapping, and more accurately by means of dipole source analysis of the magnetic data.

In the past literature, the gamma-band transient auditory response has been investigated by means of magnetoencephalographic recordings, but studies based on multichannel electric recordings are still lacking. Multichannel magnetic and multichannel electric measurements were combined in the present study to investigate both the topographical characteristics of the response and the source locations of the underlying generators. The same set of EEG/MEG data were used to understand the distinct tonotopic organizations of the sources of the middle latency and the slow waves. All the following analysis has been performed separately on the major peaks of the gamma-band response. Finally, the data of all GBR peaks were gathered to facilitate a global comparison with the sources of the MLR and sAER.

METHODS

Subjects

Seven female and five male right handed subjects between the ages of 20 and 33 years with no history of otological or neurological disorders and normal audiological and neurological status participated in this study. Informed consent was obtained from each subject after the nature of the study was fully explained. The subjects were paid for their participation.

Stimulation

Short tone-bursts of 50 ms duration (3 ms rise and decay time, cosine function) and 60 dB nHL (normative hearing level) were presented to the subject's right ear (contralateral to the MEG-investigated hemisphere) with an interstimulus interval randomized between 600 and 800 ms. The carrier frequencies were 500, 1000 and 4000 Hz. Blocks of 500 stimuli for each frequency were presented three times in random order; thus, each frequency appeared a total of 1500 times. The stimuli were presented through a non-magnetic and echo-free stimulus delivery system with an almost linear frequency characteristics (deviations less than ±4 dB in the range between 200 and 4000 Hz). During stimulus presentation the subjects were asked to keep their eyes open, and to stay awake and fixate their gaze onto small complex pictures.

MEG Recordings

Recordings were carried out in a magnetically shielded room using a 37 channel biomagnetometer (Biomagnetic Technologies). The detection coils of the biomagnetometer are arranged in a circular concave array with a diameter of 144 mm, and a spherical radius of 122 mm. The axes of the detection coils are normal to the surface of the sensor array. The distance between the centers of two adjacent coils is 22 mm; each coil itself measures 20 mm in diameter. The sensors are configured as first-order axial gradiometers with a baseline of 50 mm. The spectral density of the intrinsic noise of each channel is between 5-7 fT/√Hz in the frequency range above 1 Hz. The subjects lay in a right lateral position with their head, neck and upper part of the body supported by a specially fabricated vacuum cushion. A sensor position indicator system determined the spatial locations of the sensors relative to the head and indicated if head movements occurred during the recordings. The sensor array was centered over a point about 1.5 cm superior to the position T3 of the 10-20 system for electrode placements and was positioned as near as possible to the subject's head. Using a bandwidth from 1 to 100 Hz and a sampling rate of 520.8 Hz, stimulus-related epochs of 300 ms (including 100 ms prestimulus baselines) were recorded and stored for further analysis. The slope was 6 dB/oct for the high-pass filter and 60 dB/oct for the low-pass filter.

EEG Recordings

Simultaneous with the magnetic recordings, electric potentials were recorded from 26 Ag-AgCl electrodes placed over the whole head according to the 10-20 International System and at additional intermediate locations. The reference electrode was at the nose. Electrode impedance for all electrodes was below 5 kOhms. After preamplification, the electric signal underwent the same type of processing by amplifiers, filters and A/D converters, as described for the magnetic signals.

Data Analysis

For each stimulus frequency three blocks of 500 stimulus-related epochs were available for averaging. Epochs contaminated by muscle or eye blink artifacts (signal variations of more than 3 pT in the MEG, 100 μV in the EEG) were automatically rejected from the averaging procedure. Since the waveforms of the averaged responses of the three blocks reproduced excellently, the grand average of the responses across blocks was used for further evaluation. The baseline was corrected for each channel according to the mean value of the signal during the 100 ms prior to the stimulus. The responses were processed with a 100 Hz low-pass filter for the middle-latency components and with a 24 Hz or 48 Hz low-pass filter for the slow components. Gamma-band responses were obtained with a 24-48 Hz band-pass filter (second order zero-phase shift Butterworth filter, 12 dB/oct). After baseline correction, root mean squared (RMS) values over the 37 magnetic or 26 electric signals were computed for every sampling point.

Mapping. Scalp potentials (SP) maps were computed with a spherical spline interpolation algorithm (Perrin et al. 1989). The electrodes were considered to be located on a spherical surface following the 10-20 System strategy, with T3, Fpz, T4 and Oz being equispaced on the equatorial circle and Cz located at the upper pole. The spline interpolation was performed on the sphere, and 2D maps were constructed by computing the radial projection from a viewpoint (T3 for left lateral views) that respects the length of the meridian arcs (Perrin et al. 1990). In addition, scalp current density (SCD) maps were estimated by computing the second derivative (Laplacian) of the interpolated potential distribution. SCDs have the property of being reference-free, and of having sharper peaks and valleys than scalp potential distributions. This facilitates interpretation in cases in which there are multiple overlapping sources.

Source Analysis. The data recorded as mentioned above have been analyzed in a previous study (Pantev et al. 1994) to investigate the tonotopic organization of the neural generators underlying the Pa/Pam middle-latency components compared to that of the sources of the N100/N100m components. In the present study, source analysis was additionally performed on magnetic gamma-band filtered field data, and compared to the results obtained from the low-pass filtered data. These analyses were based on a single moving Equivalent Current Dipole (ECD) model in a spherical volume conductor (Sarvas 1987). The size of the sphere was determined by a fit to the scalp in the area of measurement. It was in this area that the shape of the scalp was digitized with the sensor position indicator. The center and the radius of the best fitting sphere were determined using a least square fit algorithm. The relative angulation of the pickup coils and the influence of the volume currents were taken into account for the source analysis. Source locations and dipole moments were estimated for each sampling point in a head-based coordinate system, as well as relative residual error (RRE) computed as the quadratic difference between experimental and model data normalized by the quadratic energy of the signal. The origin of this coordinate system was set at the midpoint of the medial-lateral axis (y-axis) which joined the center points of the entrance of the acoustic meatus of the left and the right ears (positive towards the left ear). The posterior-anterior axis (x-axis) was positively oriented from the origin to the nasion and the inferior-superior axis (z-axis) was perpendicular to the x-y plane (positive towards the vertex).

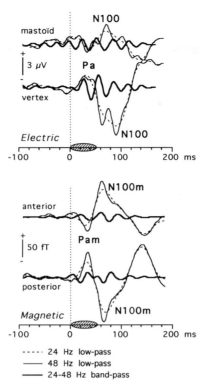

Figure 1. Example of electric and magnetic signals simultaneously recorded from the same subject. 48 Hz low-pass, 24 Hz low-pass and 24-48 Hz band-pass filtered curves are shown from two particular couples of electrodes and SQUIDs. The duration of the tone-burst is indicated on the time scale.

For each subject and each test frequency, a dipolar source was considered only at those latencies where the following criteria, based on statistical and anatomical considerations, were fulfilled : maximum of the RMS value and minimum of the RRE value, RRE below 10%, distance of the ECD to the midsagittal plane greater than 2 cm, and inferior-superior position greater than 1 cm.

RESULTS

Time Analysis

Figure 1 shows an example of auditory electric and magnetic responses obtained in the same subject after 1000 Hz stimuli. Two electric and two magnetic channels were chosen to show clear polarity reversals (mastoid/vertex and anterior/posterior respectively) of the wide-band response. Several components can be clearly identified on the low-pass filtered curves : Pa (mean extremum latency 30.7 ms, standard error of the mean, s.e.m. 0.9 ms) and N100 (89.0 ms, s.e.m. 1.6 ms) having respectively a positive and a negative maximum on fronto-central electrodes, and a polarity reversal on postero-lateral electrodes; as well as Pam (28.4, s.e.m. 0.5 ms) and N100m (87.8 ms, s.e.m. 1.8 ms) having respectively ingoing and outgoing fields at the anterior channel, and posteriorly reversed in polarity.

The gamma-band filtered curves revealed oscillations, both on electric and magnetic signals, mostly occurring during the first 100 ms following stimulus onset. These oscillations may account for some deflections observed on the 48 Hz low-pass filtered responses that are generally extensively smoothed by classically used low-pass filtering under 30 Hz. Although these oscillating components have a certain inter-subject variability, both in terms of latency and amplitude, they differ from the prestimulus noise level. RMS values computed at each time instant over space have been used to characterize these oscillations in further statistical analysis.

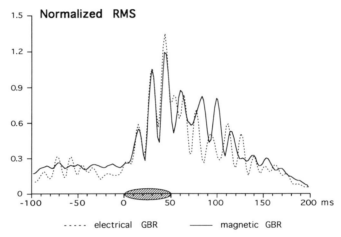

Figure 2. RMS values computed from electrical and magnetic multichannel recordings, averaged over subjects and normalized by the first major peak observed around 30 ms.

Figure 2 presents the electric and magnetic RMS curves, grand averaged over the 12 subjects, for 1000 Hz stimuli. To facilitate comparisons, they were normalized such that the first major peak observed on both RMS curves around 30 ms was set to 1. Each successive peak of the RMS may roughly be related to the peaks of alternating polarity in the original time signals. It appears that the early part of the RMS curves are quite identical for electric and magnetic data, while after 50 ms some differences occur in latency and amplitude of the peaks. This figure should be interpreted with care since averaging all the RMS curves of individual subjects would mainly enhance the peaks which are strongly phase-locked to the stimulus. It seems that the RMS peaks occurring after 50 ms have a more pronounced interindividual latency jitter.

The first three peaks were found in 10 subjects over 12 at the mean latencies of 15.9 ms (s.e.m. 0.94 ms), 30.1 ms (s.e.m. 0.83 ms), and 46.5 ms (s.e.m. 1.16 ms) in the electric RMS signal, and at 16.2 ms (s.e.m. 0.49 ms), 31.0 ms (s.e.m. 0.81 ms), and 45.5 ms (s.e.m. 1.28 ms) in the magnetic RMS signal. No statistical difference was found between electric and magnetic latencies (paired t-test, p>0.23). The amplitude of those peaks were found to be significantly higher than the mean RMS value of the prestimulus period (paired t-test, p<0.03 for electric data and p<0.003 for magnetic data). In the electric RMS curve, the third peak was followed by several peaks of smaller amplitude. Only one peaking at 65.6 ms (s.e.m. 2.4 ms) was present in 9 subjects over 12, and was significantly different from the noise level, while the two other peaks were either not identifiable in enough subjects, or they had a rather great intersubject amplitude and latency variability. Similarly, in the magnetic RMS curve, several peaks were discernible after 50 ms. Only two were present in more than 9 subjects out of 12, and significantly different from noise, with a rather stable latency : 62.8 ms (s.e.m. 2.0 ms) and 78.9 ms (s.e.m. 1.6 ms).

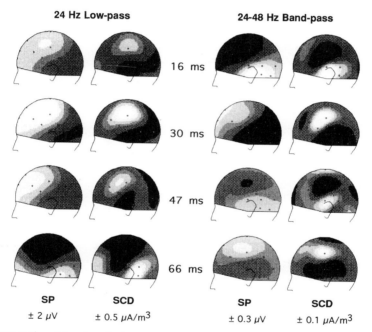

Figure 3. Topography of the electrical data computed from the grand average response. Scalp potential (SP) and scalp current density (SCD) maps are drawn for 24 Hz low-pass and 24-48 Hz band-pass filtered signals after 1000 Hz stimuli. The four selected latencies correspond to SCD maps having clear dipolar-like pattern with an alternating polarity. Amplitude scale ranges from positive values (white) to negative values (black).

Similar results were found for 500 Hz and 4000 Hz stimulus frequencies for both electric and magnetic data. By means of a two-way ANOVA with frequency and peak as factors, no significant frequency effect (p>0.3), nor frequency x peak interaction (p>0.9) were found on these RMS peak latencies.

Topographical Analysis

For the mapping analysis of electrical data, the first four reliable RMS peak will be considered, ranging from 16 to 66 ms. Figure 3 shows scalp potential (SP) and scalp current density (SCD) maps over the left hemisphere, contralateral to the stimulated ear, after 1000 Hz stimuli with both 24 Hz low-pass and 24-48 Hz band-pass filtered electric signals. To exhibit the most characteristic topographical features, those maps have been computed from the grand-average electrical response over subjects, at those four selected latencies. One may note that the spherical spline interpolation technique, combined with a high enough density of electrodes, especially over the infero-lateral area, allow to produce reliable SCD estimates. The gray scale ranges from white positive values to black negative values, centered around zero. The amplitude scale is the same across latencies.

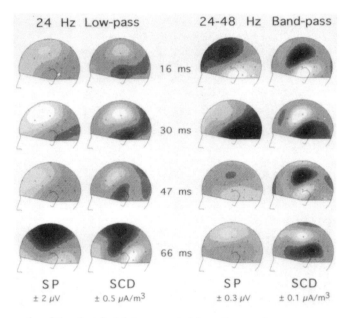

Figure 4. Topography of the electrical data computed from the grand average response. Scalp potential are drawn for 24 Hz low-pass and 24-48 Hz band-pass filtered signals after 500, 1000 and 4000 Hz stimuli. The four selected latencies are the same as for figure 3. Amplitude scale ranges from positive values (white) to negative values (black)..

On the left hand side of the figure, maps drawn for low-pass signals show the classical SP and SCD pattern of components Pa (peaking around 30 ms) and N100 (peaking around 90 ms, but having an onset around 60 ms) components. The Pa topography has a large front-central positivity with a polarity reversal across an area roughly delineating the Sylvian fissure. At 66 ms, the N100 topography, although not maximum, is already present, and characterized by a large central negativity accompanied by a lateral positivity. The SCD topographies at 30 and 66 ms are also clearly interpretable. Although Pa and N100 distributions seem to be just reversed in polarity, it results from a closer visual inspection that they are not identical : the zero iso-contour line is slightly tilted counterclockwise for N100.

On the right hand side of the figure, the SP and SCD maps of the gamma-band response confirm the existence of topographies of alternating polarity with time. The SCD maps clearly exhibit a rather dipolar-like topography characterized by a sink/source pattern across the Sylvian fissure, which seems to keep a stable position with time. This suggests underlying neural generators of alternating orientation in a rather focal area in or near the auditory cortex. Similar topographies were found over the right hemiscalp at the same latencies.

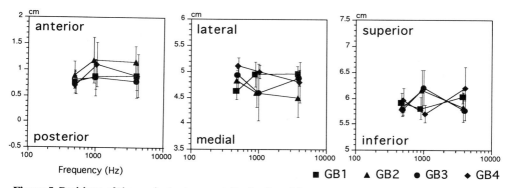

Figure 5. Positions of the equivalent current dipoles found from magnetic signals for the 4 first major peaks of the gamma-band response versus the logarithm of the stimulus frequency. Average values over subjects and standard errors of the mean are drawn for the 3 directions of the coordinate system.

To evaluate the influence of the stimulus frequency on the potential topographies, SP maps are presented at the same 4 selected latencies for both low-pass and band-pass filtered grand-averaged signals (figure 4). Again, the low-pass data shows the classical topography of the Pa and the onset of the N100, on which the influence of the stimulus frequency is characterized by an increase of the postero-lateral potential extremum as described in previous studies (Bertrand et al. 1991, Pantev et al. 1994). On the gamma-band SP maps, no clear effect is discernible. At 66 ms, the postero-lateral positivity increase clearly seen for the low-pass data is not at all visible on the band-pass data. At 30 ms, the topography does not vary in the same way as for the low-pass data.

Source Analysis

To better evaluate the influence of the stimulus frequency on the neural generators underlying those oscillation waves, a dipole source analysis has been performed for the 4 RMS peaks found previously, starting from the first prominent one around 30 ms. This was performed on magnetic data for it is known that MEG provides an accurate way of localizing generators in the auditory cortex which are oriented approximately parallel to the scalp surface. For each subject and stimulus condition, the dipole locations have been considered only in those cases where they fulfilled the requirements specified in the method section. In particular, the latencies of interest have been chosen where the RRE was minimal near each RMS peak. In the following, these components will be called GB1 (mean latency 29.9 ms, s.e.m. 0.54), GB2 45.6 ms, s.e.m. 1.63 ms), GB3 (64.0 ms, s.e.m. 2.2 ms), and GB4 (80.0 ms, s.e.m. 1.58 ms). Under those conditions, and for certain stimulus frequencies, the dipolar fit was considered acceptable in 10 subjects for peaks GB1 and GB2, and in 8 subjects for peaks GB3 and GB4. The rejection of cases was very often due to a RRE greater than 15-20%. In the remaining cases, the RRE was rather good (mean value 4%, s.e.m. 0.2 %).

Figure 5 shows the mean dipole parameters in the x (anterior-posterior), y (medial-lateral) and z (inferior-superior) axis with their standard error of the mean. They are drawn for the 3 test frequencies with a logarithmic frequency scale. By means of two-way ANOVAs with peak and frequency as factors, no significant frequency effect (p>0.34), nor frequency-peak interaction (p>0.31) was found for the x, y, and z parameters. Furthermore, one-way ANOVAs with peak as factor, performed for each test frequency, did not allow to find any significant difference between the dipole locations of the different peaks (p>0.89 for x, p>0.26 for y, p>0.63 for z).

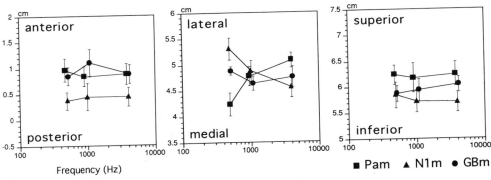

Figure 6. Positions of the equivalent current dipoles found from magnetic signals for Pam, N1m components and average positions of the dipoles corresponding to the 4 first major gamma-band peaks. Average values over subjects and standard errors of the mean are drawn for the 3 directions of the coordinate system.

Since no significant difference could be found between GB1, GB2, GB3, and GB4 sources, and to globally compare the frequency dependence of the gamma-band response to that previously reported by Pantev et al. (1994) for the Pam and N100m components, dipole locations have been averaged across peaks for each frequency. The final result is presented in Figure 6 where the mean gamma-band source positions are superimposed to those of the Pam and the N100m. From that previous study, it has been found that the Pam source was located more anteriorly (around 5 mm) than the N100m source, and, overall, that they had distinct frequency dependence in the medial-lateral direction. N100m source is located deeper in the supra-temporal plane for high frequency stimuli while Pam source has an opposite be-haviour. Compared to this, the gamma-band mean sources (GBm) do not seem to show any frequency dependence. Nevertheless, the GBm source seem to be located as anteriorly as the Pam source, but is between Pam and N100m in the vertical di-rection. In the medial-lateral direction where the frequency effect was prominent for Pam and N100m sources, the GBm source seems again to be located somewhere in between, without any clear frequency dependence.

DISCUSSION

A transient 40 Hz oscillatory response, elicited by acoustic stimuli, appears simi-larly on electric and magnetic recordings after time averaging. In the present data set, the gamma-band oscillation is characterized by a maximum amplitude before 100 ms. Electric and magnetic RMS peaks are nicely synchronized in the early part of the response while the waveforms become different after 50 ms.

The topography of the electric gamma-band distributions is quite stable at each peak from 16 to 66 ms, especially on SCD maps. Furthermore, the sink/source pat-terns observed here are consistent with a major underlying neural activity in or near the auditory cortex similarly to what is classically observed for low-pass Pa and N100 components, although they are not strictly identical to them. This probably suggests distinct locations for the gamma-band sources compared to the Pa and N100 sources.

Whereas the topographical modifications of the low-pass Pa and N100 compo-nents are rather clear when the stimulus frequency varies, the effect is not so obvious on gamma-band maps for the 4 peaks under study. After 100 ms, it has been ob-served that the gamma-band topographies were no longer dipolar-like.

MEG, having a greater sensitivity to tangentially oriented generators, is an ap-propriate neuronal information to perform dipole source analyses with a better ac-curacy than with EEG. In terms of equivalent current dipoles found in the vicinity of the primary auditory cortex, the results are twofold : no significant difference could be found between the source locations of the 4 first gamma-band peaks, and no sig-nificant frequency effect could be found on these source locations. This double neg-ative result should be interpreted with care. No statistically significant effect does not mean necessarily that there is no effect at all. It might also be interpreted as a lack of sensitivity of the approach, or, most likely, as due to a great intersubject vari-ability of the data. The latter can be related to the rather poor signal-to-noise ratio of the gamma-band response. It may also be interpreted as due to a greater intersub-ject anatomical variability of the part of the auditory cortex involved in the genera-tion of those waves.

Anyhow, it may also involve groups of neurons, still in the auditory cortex, but having no tonotopic organization. It has been previously demonstrated (Pantev et al. 1994) that middle-latency (Pam) and slow (N100m) components have clear but dif-ferent frequency dependence, with their source varying regularly in depth with the logarithm of the stimulus frequency. If we assume that the gamma-band oscillations are only a narrow-band version of the 48 Hz low-pass auditory response, it should share the same frequency property as Pam and N100m. At least the source of the early peaks of the gamma-band response (GB1, GB2) should behave as the Pam source, and the source of the later peak (GB4) as the N100m source. We have rather found that the source of all the 4 peaks cannot be distinguished in terms of frequency dependence. This finding suggests that the generators of the gamma-band response might be functionally distinct to those of the wide-band response. A peak by peak analysis was necessary to address this issue. The difference between magnetic and electric RMS curves after 50-100 ms can be due to the presence of additional generators which may not be identically seen by electric and magnetic recordings.

Acknowledgment

This study was supported by Deutsche Forschungsgemeinschaft and in part by grants of the European Science Foundation (European Neuroscience Programme). We appreciate the assistance and the critical discussions with our colleagues: Thomas Elbert, Chantal Verkindt, Carsten Eulitz, Scott Hampson and Bernhard Ross.

REFERENCES

Bertrand, O., Perrin, F., and Pernier, J. Evidence for a tonotopic organization of the auditory cortex observed with electrical evoked potentials, 1991, *Acta Otolaryngol. Suppl.*, 491: 116-123.

Eckhorn, R., Bauer, R., Jordan, W., Brosch, M., Kruse, W., Munk, M., and Reitboeck, H.J. Coherent oscillations : a mechanism of feature linking in the visual cortex ?, 1988, *Biol. Cybern.*, 60: 121-130.

Forss, N., Mäkelä, J.P., Mc Evoy, L., and Hari, R. Temporal integration and oscillatory responses of the human auditory cortex revealed by evoked magnetic fields to click trains, 1993, *Hearing Research*, 68: 89-96.

Galambos, R., Makeig, S., and Talmachoff, P., 1981, A 40 Hz auditory potential recorded from the human scalp, *Proc. Natl. Acad. Sci. USA* 78: 2643-2647.

Gray, C.M., König, P., Engel, A.K., and Singer, W. Oscillatory responses in cat visual cortex exhibit inter-columnar synchronization which reflects global stimulus properties, 1989, *Nature*, 338: 334-337.

Pantev, C., Hoke, M., Lehnertz, K., Lütkenhöner, B., Anogianakis, G., and Wittkowski, W. Tonotopis organization of the human auditory cortex revealed by transient auditory evoked magnetic fields. *Electroenceph. Clin. Neurophysiol.*, 1988, 69: 160-170.

Pantev, C., Makeig, S., Hoke, M., Galambos, R., Hampson, S., and Gallen, C., 1991, Human auditory evoked gamma-band magnetic fields, *Proc. Natl. Acad. Sci. USA* 88: 8996-9000.

Pantev, C., Elbert, T., Makeig, S., Hampson, S., Eulitz, C., and Hoke, M. Relationship of transient and steady-state auditory evoked fields, 1993, *Electroenceph. Clin. Neurophysiol.*, 88: 389-396.

Pantev. C. Bertrand, O., Eulitz, C. Verkindt, C., Hampson, S., Schuirer, G., and Elbert, T. Mirror-image tonotopy of different areas of human auditory cortex revealed by simultaneous magnetic and electric recordings, 1994, *Electroenceph. Clin. Neurophysiol.*, *(in press)*.

Perrin, F., Pernier, J., Bertrand, O., and Echallier, J.F. Spherical splines for scalp potential and current density mapping. *Electroenceph. Clin. Neurophysiol.*, 1989, 72: 184-187.

Perrin, F., Bertrand, O., Giard, M.H., and Pernier, J. Precautions in topographic mapping and in evoked potential map-reading, 1990, *J. Clin. Neurophysiol.*, 7 (4): 498-506.

Sarvas, J. Basic mathematical and electromagnetic concepts of the biomagnetic inverse problem, 1987, *Phys. Med. Biol.*, 32: 11-22.

Tiitinen, H., Sinkkonen, J., Reinikainen, K., Alho, K., Lavikainen, J., and Näätänen, R., Selective attention enhances the auditory 40 Hz transient response in humans, *Nature*, 364: 59-60.

GAMMA-BAND RESPONSES REFLECT WORD/PSEUDOWORD PROCESSING

Friedemann Pulvermüller[1], Hubert Preißl[1], Carsten Eulitz[2], Christo Pantev[2], Werner Lutzenberger[1], Bernd Feige[2], Thomas Elbert[2] and Niels Birbaumer[1]

[1]Institute of Medical Psychology and Behavioral Neurobiology, University of Tübingen, Germany
[2]Institut of Experimental Audiology, University of Münster, Germany

CELL ASSEMBLIES: POSSIBLE BUILDING BLOCKS OF COGNITION

According to one brain-theoretical view, higher brain functions are based on processing units called cell assemblies. Cell assemblies are large groups of cortical pyramidal neurons with strong and reciprocal internal connections. Cell assemblies develop in a randomly connected neuronal network when sets of neurons are frequently active simultaneously so that their connections strengthen (Hebb's law, Hebb, 1949, Gustafsson et al., 1987, Bonhöffer et al., 1989, Ahissar et al., 1992). Neurons making up one assembly do not need to be located in a small cortical area. They may be distributed widely over various cortical regions. Such *transcortical assemblies* are likely to be held together through long axons of pyramidal cells which are well-known to connect distant cortical areas (Pandya and Yeterian, 1985, Braitenberg and Schüz, 1991, Deacon, 1992a, Deacon, 1992b). Due to strong intra-assembly connections, excitation of some neurons of an assembly leads to spreading activation in the network and, finally, to an "ignition" of the whole assembly. Overshooting activity will occur in the cortex when too many assemblies ignite at the same time. Therefore, a regulation mechanism is required which keeps the level of cortical excitation close to a target value (Braitenberg, 1978). This regulation mechanism guarantees that only one or a limited number of assemblies will ignite at a time.

What is the purpose of transcortical assemblies? One obvious answer is the following: transcortical assemblies make it possible to "bind" information represented in different parts of the cortex. Such binding may occur between neurons in the visual system stimulated by features of a perceived object. If assemblies include various local clusters of neurons, they may well represent entities composed of numerous features, such as a complex object. Binding in widely distributed cell assemblies may also be relevant for associations between sensory modalities. Such associations are necessary for the representation of objects that can be perceived through various sensory channels. Transcortical assemblies may also be the neurobiological counterpart of sensori-motor associations, a prerequisite of language acquisition. For example, in order to repeat syllables and words the child must associate the auditory pattern perceived with a motor (articulatory) pattern. Binding of information represented in distant parts of the cortex is necessary for solving complex tasks, such as object recognition, gestalt perception, using and understanding language, and reasoning. It appears that these higher or cognitive brain capacities require devices that bridge distances in the cortex and allow for fast inter-area exchange of information. Candidate machines that may serve this function are transcortical assemblies.

If Hebb's rule holds true, formation of cell assemblies takes place during ontogenesis. This can be illustrated using language representation as a paradigm case (Braitenberg, 1980, Braitenberg and Pulvermüller, 1992). During early language development, the infant frequently perceives sounds that lead to an "imprinting" of phonetic perception (Kuhl et al., 1992). Presentation of language sounds causes coactivation of neurons responding to features of these sounds, and neurons co-activated by one phoneme develop into an assembly, the neurobiological correlate of the phoneme. In a later stage of development, the infant repeatedly articulates sound sequences and word forms. These articulations are caused by activity in the motor cortex. Articulations cause acoustic signals which are fed back to the auditory cortex where they lead to additional neuronal activation. Thus, during early articulations specific patterns of activity are present almost simultaneously in three cortical regions, the motor, the somatosensory, and the auditory system. Because there are strong cortico-cortical projections connecting these systems, simultaneously activated neurons will strengthen their connections and develop into an assembly. This argues that early articulations trigger the formation of transcortical assemblies corresponding to specific syllables or word forms. Note that such cell assemblies must be assumed to comprise neurons of *distant* cortical regions located in both hemispheres. However, it is well-known that language is normally lateralized to the left hemisphere in most right-handers, suggesting that the majority of neurons comprised in language assemblies are located in the left hemisphere. If transcortical assemblies are the neurobiological correlates of individual words, it may be assumed that connections between such assemblies are the physical realization of rules determining word sequences. In this case, the hierarchy of linguistic structures (phoneme, morpheme, word, sentence) has its biological equivalent in a hierarchy of cell assemblies corresponding to these cognitive entities (Braitenberg and Pulvermüller, 1992, Pulvermüller, 1993, 1994).

At this point, it must be emphasized that the assumption of transcortical assemblies is still speculative to some degree. However, there is a large body of evidence from various fields which can be accounted for on the basis of the cell assembly approach. For example, the cell assembly theory of language processing accounts for data from aphasia research (Pulvermüller and Preißl, 1991, Pulvermüller, 1992, Pulvermüller and Schönle, 1993, Pulvermüller and Preißl, 1994), from psycholinguistic experiments (Pulvermüller et al., 1993, Mohr et al., 1994), and from electrophysiological investigations of language processes (Lutzenberger et al., 1994, Pulvermüller et al., 1994a, Pulvermüller et al., 1994b). Therefore, the concept of cell assemblies is speculative but useful.

The major advantage of a theory bridging neurobiology to cognition is the following: it allows for predictions which cannot be formulated on the basis of biological or cognitive theories alone. In this article, we discuss an implication of the cell assembly theory and some recent data obtained to test this prediction.

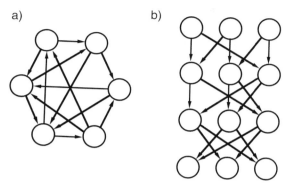

Figure 1. Simple models of strongly connected neuron ensembles. A reciprocally connected cell assembly (a) and a synfire chain (b) are sketched. Circles represent neurons, arrows represent connections between neurons. All neurons have a threshold of two, i.e. they need two simultaneous inputs to become active. For explanation, see text.

SYNCHRONIZATION, SPATIO-TEMPORAL ACTIVITY PATTERNS AND PERIODICITY

If a cell assembly has strong internal connections, its ignition will take place instantaneously so that all or at least many of its neurons become active almost simultaneously. The question of how activity spreads through the assembly has at least two possible answers. First, after activation of a subset of the assembly neurons, almost all remaining assembly neurons will be activated synchronously a few time-steps later. Figure 1a presents a sketch of such an "assembly". If all "neurons" of this network have a threshold of 2 (i.e. they need 2 excitatory inputs to become active), simultaneous activation of four out of the six neurons will lead to synchronous activity of all assembly neurons a few time-steps later. However, as a second possible answer, the internal connections of a neural network may also determine a "stepwise" mode of activation. In the network schematized in Figure 1b, activity spreads from one row to the next after all neurons of the uppermost row have been activated. In this case, the activation process does not lead to synchronous activation of all assembly neurons, but to a well-ordered spatio-temporal pattern of activity within the network. Therefore, only a few assembly neurons become active at exactly the same time. The more likely case is that two neurons become active one after the other with a fixed delay. Although these simple examples are far from providing an adequate picture of cortical mechanisms, they can illustrate an important point. Activation of strongly coupled sets of neurons may take place synchronously or stepwise, according to a well-ordered spatio-temporal pattern of activity. The architecture of the network determines the activation process.

If cell assemblies have strong and reciprocal internal connections, these connections warrant that activity is retained for some time within an activated assembly. Using a terminology proposed by Braitenberg (1978), the assembly "holds" after its activation. The processes taking place while activity is retained may vary as a function of assembly architecture. For example, the network depicted in Figure 1a stays active; that is, all of its neurons remain excited after their simultaneous activation. The network illustrated in Figure 2a (which is taken from Palm 1982) shows a simple periodic activity pattern. If only two of the four neurons are active, the other two will become excited at the next time step. (Again, neurons are assumed to have a threshold of 2.) One time-step later, the original pair of neurons become excited and so on. Each neuron of this network repeatedly becomes active and inactive, i.e. it oscillates. A similar process takes place in the assembly illustrated in Figure 2b. Since these neurons form a "circle", activity circulates within the network after one row of neurons has been activated. This results in repeated occurrences of the same spatio-temporal pattern. Again, each assembly neuron will show repetitive or periodic activity.

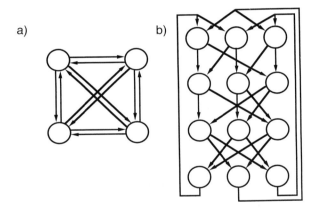

Figure 2. Models of cell assemblies generating periodic neuronal activity. All neurons have a threshold of two. If two neurons of the assembly in (a) are active, the other two neurons will be active at the next time step etc. If three neurons of the reverberating synfire chain in (b) are active simultaneously, a wave of activity will circulate in the network.

Another mechanism generating periodic activity requires inhibitory connections (Wilson and Cowan, 1973). Consider the simple networks depicted in Figure 3a, where each neuron has an activation threshold of 1. The uppermost neuron receives constant excitatory input, so that it will be active initially. This leads to activation of the lower, inhibitory neuron, which, in turn, will lead to an inhibition of the upper neuron. When the upper neuron is switched off by the inhibitory input, the inhibitor itself becomes inactive one time-step later. At this point, the constant input can again activate the upper neuron which again inhibits itself through the inhibitory loop, thus resulting in oscillatory or periodic activity. If the excitatory neuron is interpreted not as a single neuron but as an assembly of neurons, such as the one depicted in Figure 1a, coherent and repetitive activation and deactivation of these neurons can be assumed. Coherent oscillations will occur even if each excitatory neuron of such a network has its own inhibitor (and additional conditions are met; Schuster and Wagner, 1990). If one of the excitatory neurons of Figure 3b is oscillating, it will force its sister neurons into the same oscillations after a short time lag.

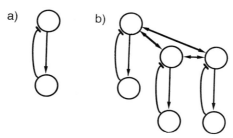

Figure 3. Simple oscillating circuits including inhibitory elements. All neurons have a threshold of 1. If the upper neuron in (a) receives continuous input, the neurons will become active and inactive periodically. If one of the three coupled oscillators in (b) receives constant input, oscillations of all three neuron pairs will occur with a short phase lag.

Activation of strongly connected sets of neurons may lead to synchronous, patterned and/or periodic activity. There is strong electrophysiological evidence indicating that all three phenomena play a role in cortical processing (Abeles, 1982, Gerstein et al., 1989, Abeles, 1991, Engel et al., 1992). If cell assemblies are defined as neuron sets with strong *reciprocal* connections (Braitenberg, 1978), a feed forward net such as the one depicted in Figure 1b is excluded by this definition. Ignition of an assembly with reciprocal connections will, most likely, include periodic activity. If the architecture of the assembly is similar to the networks in Figure 2a and b, periodic activity will occur. If the assembly architecture itself does not determine periodic activity (see Figure 1a), it is plausible that inhibitory regulation processes will generate periodic activity, as illustrated for the simple networks in Figure 3a and b[1]. Thus, it can be assumed that ignition of an assembly includes periodic activity of a large group of neurons. This periodicity may have the form of repetitive spatio-temporal activity patterns in a reverberating assembly and/or of coherent oscillations

---

[1] Note that periodicity may be caused by cortical or subcortical mechanisms. The oscillator in Figure 3a could have its analogon in excitatory and inhibitory neurons in the cortex. However, inhibition could also be provided by complex loops involving neurons in the basal ganglia and/or the thalamus.

due to inhibitory regulation processes. In the case of a transcortical assembly, periodic activity should occur in various cortical areas.

A PREDICTION OF THE CELL ASSEMBLY THEORY

If cell assemblies are the basic units of cognition, neuronal populations must become active when cognitive processes take place. This implies correlated activity of numerous neurons. Correlated postsynaptic activity in apical dendrites of numerous pyramidal cells will also affect the local field potential and, if the number of neurons participating in the pattern is large enough, the weighted sum of postsynaptic currents will lead to a surface potential which is visible in the EEG (Speckmann and Elger, 1982, Mitzdorf, 1985, Birbaumer et al., 1990). Magnetic fields caused by the intracellular currents flowing from the dendritic tree towards the soma can also be picked up in the MEG primarily if pyramidal cells are located in Sulci, thus causing currents to flow tangentially to the surface of the head (Cuffin and Cohen, 1979, Hari and Lounasmaa, 1989). Ignition of a cell assembly implies fast activation of a large neuron population. This predicts a sharp rise in EEG and MEG. After activation, an assembly with reciprocal connections will retain its activity and correlated periodic firing of assembly neurons takes place. In this case, EEG and MEG peaks may occur repeatedly, so that the EEG and MEG signal will include enhanced spectral power in at least one frequency band. If, for example, the ignition process implies synchronous 40 Hz oscillations of assembly neurons, the spectral power in the 40 Hz range should specifically be affected. If ignition consists of repetitive spatio-temporal activity patterns, a more complex change results which may affect several frequency bands. Nevertheless, such a more complex change will also be visible in specific frequencies. Equal numbers of neurons may be active at each time-step and their activity may contribute equally to the recorded potential or field. Nevertheless, if numerous neurons are active at each time step, but contribute differently to the surface potential, it is likely that periodic activity will be seen at the recording electrode. On the basis of theoretical considerations it is hardly possible to determine the exact frequency range which will be affected by assembly ignition. However, the spreading of activity and reverberation must take place very quickly, possibly in the range of few to several milliseconds. This suggests that changes in spectral power will occur in the gamma-band (20 Hz and up). In addition, sinchronous periodic activity of large neuron ensembles will possibly cause slow changes of the evoked potential and magnetic field. In this case, a slow wave shift (in addition to changes in the gamma-band) can be expected (Elbert, 1993).

The cell assembly theory predicts ignition when cognitive processing takes place. In this case, large numbers of neurons become active simultaneously and possibly join a periodic pattern thereafter. This may be visible in the gamma-band response of the cortex. If cognitive processing does not take place, cell assemblies do not become active. More precisely, cell assemblies may be stimulated and, therefore, become slightly active, but no full activation or ignition takes place. Therefore, gamma-band responses should be reduced. In summary, gamma-band spectral power should be strong when cognitive processing takes place, but reduced when such processing does not occur.

IS GAMMA-BAND ACTIVITY AN INDICATOR OF COGNITIVE PROCESSING?

Gamma-band responses have been observed in various mammals (including humans) using different methods, such as single and multiple unit recordings, local field potentials, and EEG and MEG recordings. Gamma-band activity can be elicited by visually perceived moving bars (Eckhorn et al., 1988, Engel et al., 1992), by simple auditory stimuli such as tones (Pantev et al., 1991), by somatosensory stimuli (Ahissar and Vaadia, 1990) and by odors (Bressler and Freeman, 1980, Freeman, 1991). They also accompany manipulative movements (Murthy and Fetz, 1992). However, this does not imply a specifically cognitive function of gamma-band responses. Very simple stimuli, such as bars or tones, are unlikely to elicit cognitive processing. Such stimulations trigger perception processes, but it is unclear whether additional processes follow.

To decide whether synchronized and/or repetitive activity of neurons serves a specifically cognitive function, it is necessary to compare brain responses between two paradigms only one of which invokes cognitive processing of a certain kind. In recent experiments it was found, that two bars moving in the same direction lead to synchronous and fast oscillatory activity of neurons activated by the stimuli (Eckhorn et al., 1988, Gray and Singer, 1989). These neurons may be located in distant cortical areas. In contrast, if two bars move in different directions, two neurons responding to one of the stimuli, respectively, do not respond synchronously (Gray et al., 1989, Engel et al., 1990). While this indicates that cortical spatio-temporal responses change with *Gestalt features* (Engel et al., 1992, Singer, 1994), such as continuity, it is still unclear how these responses relate to perception of complex *Gestalts*. A complex moving stimulus may well lead to the perception of lines moving in different directions. If cortical synchrony was an indicator of Gestalt integration, it could rather be expected that two moving stimuli invoke more complex Gestalt integration processes and, therefore, more pronounced synchronized activity (von der Malsburg, p.c.). However, it may be argued that the two lines will not be integrated into a gestalt. For this experiment, it appears difficult to decide which cognitive processes are triggered by the differing stimuli.

More or less complex manipulative movements are also correlated with different patterns of synchronized oscillatory brain activity in the gamma range. When a monkey performs a complex motor movement, such as retrieving raisins from slots of a Klüver board, synchronized gamma-band activity can be recorded from motor and somatosensory cortices (Murthy and Fetz, 1992). Coherent gamma-band activity was reduced when monkeys performed simple movements, such as alternating flexion and extension of the wrist. These results were confirmed, in part, by an MEG investigation of human brain responses during complex movements (Kristeva-Feige et al., 1993). During performance of a complex manual task, enhanced gamma-band activity around 30 Hz was found. The enhancement of gamma-band responses around 30 Hz can possibly be attributed to the level of attention or the amount of sensorimotor integration required by complex movements. However, the complexity of the muscle movements could also be critical for the gamma-band response to occur. While these results are consistent with the assumption that cognitive processes, i.e. selective attention to sensorimotor integration, underlie stronger gamma-band responses, they can also be considered a consequence of the complexity of the motor movement to be performed.

Another investigation of MEG responses of the human brain indicates that gamma-band responses reflect cognitive processing. Llinas and Ribary (1993) found reduced 40 Hz activity during delta sleep, while oscillatory activity was stronger during both wakefulness and REM sleep. Because vivid dreaming (which usually occurs during REM sleep) and wakefulness imply cognitive processing, the authors propose that enhanced gamma-band responses are a correlate of cognition. However, it may be argued that various other variables (arousal level, brain activity level etc.) distinguish delta sleep from REM sleep or wakefulness.

In earlier studies, EEG recordings were used for investigating changes of gamma-band activity associated with tasks, such as verbal and visual-spatial problem solving. Sheer and coworkers (Spydel et al., 1979, Sheer, 1984) reported increased spectral power around 40 Hz when subjects engaged in cognitive tasks requiring focusing of attention. However, a possible methodological problem of these studies are EEG artifacts caused by muscle activity. While these authors report no correlation between EEG and EMG spectral responses recorded from temporal and splenius muscles, one may argue that changes in 40 Hz power can be caused by muscles not monitored in these experiments. However, this problem can be solved by investigating spectral responses of even higher frequencies. Spectral power of EMG responses increases with frequency until at least 80 Hz (Cacioppo et al., 1990). Thus, if muscle activity causes differences, for example in the 40 Hz range, the same effects must be present (and they must be even more pronounced) in higher frequency intervals, for example around 60 or 80 Hz. If no differences occur in these bands, muscle activity cannot be the cause of an effect observed in lower frequencies.

In order to decide whether cognitive processing implies specific changes in high frequency brain responses, two conditions must be compared that only differ with regard to the cognitive processes they invoke. Comparing responses to bars and "meaningful" pictures of objects would be one option. However, bars and pictures have very different perceptual complexity and a difference in evoked spectral responses could be a consequence of this physical difference. Two stimulus classes that are of equal perceptual complexity but nevertheless trigger distinct cognitive processes are words and meaningless pseudowords made up of the same letters. A word, such as *moon*, will immediately be comprehended, while a matched pseudoword, such as *noom*, will fail to elicit immediate comprehension. On the cortical level, word presentation should lead to cell assembly ignition, while presentation of pseudowords should fail to ignite specific assemblies. This leads to an obvious prediction: Gamma-band responses to words should be stronger than responses to pseudowords.

DIFFERENT GAMMA-BAND RESPONSES TO WORDS AND PSEUDO-WORDS

The following experiment was carried out in order to investigate EEG responses to words and pseudowords. 15 right-handed German native speakers performed lexical decisions on visually presented words and matched pseudowords (64 stimuli of each category). Each word was presented for 100 ms in the center of a video screen. The inter-stimulus interval varied between 3.5-4.5 s. Words subtended

0.5 degrees of vertical visual angel and a maximum of 2.3 degrees of horizontal visual angel. They were written in black letters on a gray background. The lexical decision was expressed by moving the index finger of the left hand. The EEG was recorded through 17 tin electrodes against linked mastoids. Six of the electrodes were placed close to the perisylvian regions of each hemisphere, respectively. The EEG was recorded for 1.28 s per trial, starting 0.1 s before stimulus presentation (0.1 s baseline). Event-related potentials (ERPs) were calculated for each electrode, condition and subject (artifact rejection and correction as described in Elbert et al., 1985). In order to obtain reference-free data, artifact-free signals obtained from single trials were also submitted to current source density (CSD) analyses. CSD data were filtered in three frequency bands: 25-35 Hz, 35-45 Hz, and 55-65 Hz. These data were normalized by dividing them by the average value obtained in the respective baseline. Finally, mean normalized evoked spectral power was calculated for both conditions (words/pseudowords), for each subject, and for each electrode. For statistical evaluation, data from the 2 x 6 electrodes monitoring the perisylvian cortices were chosen. Average values in three time windows were evaluated: 120-320 ms, 320-520 ms and 520-720 ms post stimulus onset. Three-way analyses of variance were carried out (design: Wordness (word/pseudoword) x Hemisphere (left/right) x Site (six electrodes from anterior to posterior)). Greenhouse-Geisser correction of degrees of freedom was applied when adequate. (For a more detailed description of the methods, see Lutzenberger, Pulvermüller and Birbaumer, 1994.)

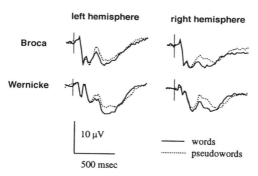

Figure 4. Results from EEG experiment. Event-related potentials (ERPs) after word and pseudoword presentation recorded over the perisylvian cortices of both hemispheres (Broca's and Wernicke's area and homologous areas on the right). Data are averaged over 15 subjects. The late negative-going component around 400 ms after stimulus onset is larger after pseudowords (redrawn from Pulvermüller et al. 1994b)

Figure 4 displays ERPs recorded from two lateral electrodes after word and pseudoword presentation. Around 400 ms post stimulus onset pseudowords elicited more negative ERPs compared to words. There was a significant main effect of Wordness in the 320-520 ms time window, F (1,14) = 12.6, p < 0.003. A larger late negativity after pseudowords compared to words has been reported in an earlier study (Holcomb and Neville, 1990). This larger negativity can be taken as evidence that pseudoword presentation elicited more activity, i.e., a larger number of excitatory post-synaptic potentials, in apical dendrites of numerous neurons (Speckmann and Elger, 1982, Birbaumer et al., 1990).

Figure 5a presents normalized evoked spectral responses in the 30 Hz (25-35 Hz) band recorded from scalp over the left and right perisylvian cortices. Because statistical analysis did not reveal any reliable effect of the factor Site, averaged data from all six electrodes are displayed in this diagram. Spectral responses to words and pseudowords differed around 400 ms past stimulus onset. Statistical analysis revealed a significant Wordness by Hemisphere interaction for the time interval between 320 and 520 ms, $F(1,14) = 8.4$, $p < 0.01$. This interaction is displayed in Figure 5b. Word presentation did not change 30 Hz power compared to the baseline. In contrast, pseudowords elicited reduction of 30 Hz cortical responses over the left hemisphere. This deflection was small (1-2 percent of the baseline values). However, it could consistently be observed in most subjects tested. In contrast, analysis of all other frequency bands did not reveal a similar interaction in any of the time windows.

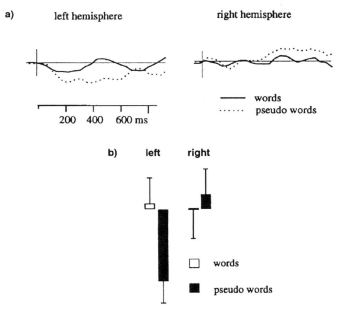

Figure 5. Results from EEG experiment. (a) Normalized evoked spectral responses (arbitrary units) between 25 and 35 Hz elicited by words and pseudowords as a function of time after stimulus presentation. Responses are averaged over 15 subjects. Around 400 ms post stimulus onset, 30 Hz activity is reduced over the left hemisphere after pseudowords ("pseudoword depression"). (b) The significant hemisphere by wordness interaction is displayed. Averaged normalized 30 Hz power recorded from the left and right hemisphere are shown. Redrawn from Lutzenberger et al. 1994.

This difference cannot be the result of an artifact caused by muscle activity. As noted earlier, the power spectrum of potential changes caused by muscle contractions increases with frequencies until around 80 Hz (Cacioppo et al., 1990). The fact that no significant interactions or main effects were observed in the highest frequency band provides (60 Hz band) evidence that muscle potentials did not cause the result.

It may be argued that the reduction of gamma-band power after pseudowords was the consequence of some other recording or evaluation artifact. In addition, we cannot exclude the possibility that this reduction was related to the language used in the experiment (German), to visual stimulus presentation, or to the motor response the participants had to perform. Therefore, we carried out another experiment in which all these features of experimental setting and evaluation procedure were changed. Biomagnetic signals were recorded simultaneously from both hemispheres of five right-handed native speakers of English who heard English words and pseudowords spoken by a professional speaker. In this case, 30 items of each stimulus category (words/pseudowords) were repeated four times, resulting in 120 tokens of each category. The inter-stimulus-interval varied between 2.5 - 3.5 s. Subjects did not have to respond to the stimuli. However, they were asked to memorize all stimuli in order to answer questionnaires which were presented during breaks. In this case, no CSD analysis was necessary, since MEG provides reference-free data. For evaluating spectral responses a method described by Makeig (1993) was used. For a variety of frequency bands (width: 9.6 Hz), spectral power was calculated in overlapping time windows of 200 ms. Two adjacent time windows overlapped by 50 percent. Spectral power values were, again, normalized (0.6 s baseline), and averaged. (For further details, see Pulvermüller, Eulitz, Pantev et al.,1994.)

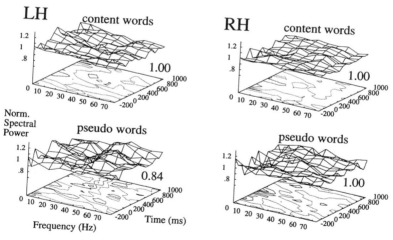

Figure 6. Results from MEG experiment. Normalized evoked spectral responses to words (upper diagrams) and pseudowords (lower diagrams) recorded from one channel over the left hemisphere (diagrams on the left) and the right hemisphere (diagrams on the right) of one individual. Note the "valley" around 30 Hz in the diagram on the lower left.

Figure 6 presents normalized spectral responses of one subject to words and pseudowords obtained from one channel over the left and right hemisphere, respectively. While no pronounced change was elicited by presentation of words, a pronounced deflection of spectral power around 30 Hz followed pseudoword presentation. Reduction of 30 Hz spectral power after pseudowords was observed over the left hemisphere of all five subjects tested. No consistent change of spectral power was present in any of the other frequency bands, or over the right hemisphere. These changes were only seen at anterior channels located above the left inferior frontal lobe. At these channels, magnetic fields evoked by words and pseudowords were larger compared to all other channels, so that a maximal signal-to-noise-ratio can be assumed.

The consistency of results of the EEG and the MEG experiments makes it unlikely that pseudoword-specific gamma-depression is an artifact caused by recording or quantification procedures or that it is affected by parameters such as experimental language, stimulus modality, or response mode.

DIFFERENTIAL GAMMA-BAND RESPONSES: EVIDENCE FOR "COGNITIVE" ASSEMBLIES?

Based on these data, the original prediction of stronger gamma-band responses to words compared to pseudowords, can obviously be verified. A preliminary explanation of this difference is the following: Cell assembly ignition takes place after word presentation. This leads to correlated and periodic activation of large neuron sets which can be observed in EEG and MEG responses. In contrast, pseudoword presentation does not lead to an ignition and, therefore, it causes reduction of 30 Hz power. Activity differences in the gamma-band are primarily seen over the left hemisphere, because most neurons of language assemblies are located in the left hemisphere.

This preliminary hypothesis can be questioned for several reasons: First, the cell assembly theory would predict enhanced gamma-band activity after words compared to the baseline, rather than a reduction after pseudowords. However, it can be assumed that word processing takes place not only immediately after stimulus presentation. Word processing and thinking about these words may last for a few seconds, i.e. throughout the entire inter stimulus interval (which was 2.5-3.5 s in the MEG experiment and 3.5-4.5 s in the EEG experiment). This is consistent with the cell assembly theory. It was postulated that each activated assembly would "hold" for some time. Therefore, word evoked spectral responses may have contaminated the baseline and pseudoword-specific gamma-depression can be interpreted as a power decrease relative to word processing.

A second objection to the reported results concerns the size of the difference between responses to words and pseudowords. These differences are rather small in the EEG experiment (about 1-2 percent of the baseline power) and even the most pronounced differences seen in MEG recordings only amount to some 20 percent of the baseline. This indicates that the differences induced by stimulus presentation was small compared to cortical background activity. However, this is not surprising. Most likely, not all cortical neurons participate in processing information contained in a stimulus word. Perhaps only one cell assembly is activated after a word has been presented. Assemblies are usually assumed to comprise less than 10^6 neurons (Palm, 1993), only a small portion of the 10^{10} to 10^{11} neurons of the cortex. From this perspective, it is rather surprising that a small percentage of neurons generates a change in global activity of one to several percent.

A third possible objection addresses the relationship between spectral responses and event-related potentials (ERPs). Pseudowords lead to larger late negativities around 400 ms post stimulus onset, while 30 Hz power is reduced in this interval compared to words. If both event-related potentials and spectral responses were an indicator of arousal, attention, and/or neuronal mobilization processes, there would be an obvious incompatibility. It appears that ERPs and gamma-band responses reflect different processes. A cognitive and neurobiological model has to account for both 30 Hz depression and larger late negativities evoked by pseudowords.

On the psychological level, 30 Hz activity and late negativities in these experiments may be indicators of lexical access, lexical processing or lexical search. When a word is perceived, comprehended or stored, relatively strong 30 Hz activity is observed. The possible psychological processes underlying gamma depression may be comprehension failure or an unsuccessful lexical search. If no equivalent of a stimulus word is found in the "cortical lexicon", 30 Hz activity decreases. However, there may be an intense lexical search when pseudowords are perceived, whereas for word presentation the search processes terminate much faster, since the stimulus can be matched to a mental representation. In earlier studies it has been found that more intense memory search processes induce larger negativities in the event-related potential (Rockstroh et al., 1989, Rösler et al., 1993). Thus, it may be speculated that 30-Hz power and late event-related potentials indeed mirror distinct cognitive processes. While 30 Hz power indicates whether lexical access and processing take place, late negativities indicate the intensity of lexical search processes. Another view relates the reported data to attention and arousal processes. Pseudowords have never been perceived and are therefore highly unexpected. Thus, their presentation leads to enhanced arousal while word presentation fails to evoke such enhancement. In contrast, word presentation leads to the processing of only one word and to focusing on this stimulus, while pseudoword presentation may activate various associations. According to this latter view, stronger gamma-band responses to words would correspond to the process of selectively attending to words and late negative evoked potentials would covary with general arousal processes. However, such postulates stating correspondence between cognitive processes and psychophysiological measures can be considered unsatisfactory. They have even been called "psychophysio-analytic" by some researchers. It would be highly desirable to specify the neurobiological mechanisms underlying such cognitive processes that also cause the physiological responses. To explain the reported data, brain processes must be referred to.

The following neurobiological model can account for both ERP and 30 Hz differences. After word presentation, *ignition of exactly one assembly* takes place (while competing assemblies are inhibited). This ignition causes a small negative shift in the ERP. Because assembly ignition implies fast periodic and correlated activity of a large number of assembly neurons, strong 30 Hz responses are present. Processes taking place after pseudoword presentation are the following. Not only one, but *several assemblies are preactivated* to some degree. However, no ignition takes place in any of the preactivated assemblies. These assemblies are the neuronal equivalents of words phonologically similar to the pseudoword stimulus (the stimulus *noom* may activate assemblies corresponding to words such as *noon, moon, room* etc.). Because several assemblies are activated, the sum of cortical excitation is larger compared to full ignition of only one assembly. This results in relatively large late negativities in the event-related potential. However, neuronal activity in several assemblies is not correlated, i.e. activation-deactivation cycles differ between assemblies[2]. At a device recording global cortical activity, the effects of fast uncorrelated and periodic activity tend to cancel each other. Therefore, reduced high-frequency responses are observed at recording electrodes and channels. In psychological terms, ignition of a word-specific assembly corresponds to lexical access and processing (which implies focused attention to one particular word), and preactivation of several assemblies without ignition can be considered the correlate of an unsuccessful search process (which may be linked to enhanced global arousal).

Agnowledgement

Research was supported by the DFG (Pu 97/2, Ho 847/6)

REFERENCES

Abeles, M. Local cortical circuits. An electrophysiological study. Berlin: Springer, 1982.

Abeles, M. Corticonics - Neural circuits of the cerebral cortex. Cambridge: Cambridge University Press, 1991.

Ahissar, E. and Vaadia, E. Oscillatory activity of single units in a somatosensory cortex of an awake monkey and their possible role in texture analysis. Proc. Natl. Acad. Sci. ,USA, 1990, 87: 8935-8939.

Ahissar, E., Vaadia, E., Ahissar, M., Bergman, H., Arieli, A. and Abeles, M. Dependence of cortical plasticity on correlated activity of single neurons and on behavior context. Science, 1992, 257: 1412-1415.

Birbaumer, N., Elbert, T., Canavan, A.G.M. and Rockstroh, B. Slow Potentials of the Cerebral Cortex and Behavior. Physiol. Rev., 1990, 70: 1-41.

Bonhöffer, T., Staiger, V. and Aertsen, A.M. Synaptic plasticity in rat hippocampal slice cultures: local "Hebbian" conjunction of pre- and postsynaptic stimulation leads to distributed synaptic enhancement. Proc. Natl. Acad. Sci., USA, 1989, 86: 8113-8117.

Braitenberg, V. Cell assemblies in the cerebral cortex. In: Heim, R. and Palm, G. (Eds.) Theoretical approaches to complex systems. (Lecture notes in biomathematics, vol. 21). Springer, Berlin: 1978: 171-188.

Braitenberg, V. Alcune considerazione sui meccanismi cerebrali del linguaggio. In: Braga, G., Braitenberg, V., Cipolli, C., Coseriu, E., Crespi-Reghizzi, S., Mehler, J. and Titone, R. (Eds.) L'accostamento interdisciplinare allo studio del linguaggio. Franco Angeli Editore, Milano: 1980: 96-108.

Braitenberg, V. and Pulvermüller, F. Entwurf einer neurologischen Theorie der Sprache. Naturwiss., 1992, 79: 103-117.

Braitenberg, V. and Schüz, A. Anatomy of the cortex. Statistics and geometry. Berlin: Springer, 1991.

Bressler, S.L. and Freeman, W.J. Frequency analysis of olfactory system EEG in cat, rabbit and rat. Electroenceph. Clin. Neurophysiol., 1980, 50: 19-24.

Cacioppo, J.T., Tassinary, L.G. and Fridlund, A.J. The skeletomotor system. In: Cacioppo, J.T. and Tassinary, L.G. (Eds.) Principles of psychophysiology. Physical, social, and inferential elements. Cambridge University Press, Cambrigde: 1990: 325-384.

---

[2]Strictly speaking, this implies the following. If periodic activation is caused by inhibitory elements, these inhibitors must be assembly-specific. Otherwise, inhibitors could synchronize activity in distinct assemblies. However, because inhibitory neurons are small and, therefore, do not reach distant pyramidal cells, it can be assumed that each local cluster of assembly neurons activates only its own inhibitors.

Cuffin, B.N. and Cohen, D. Comparison of the magnetoencephalogram and the electroencephalogram. Electroenceph. Clin. Neurophysiol., 1979, 47: 132-146.

Deacon, T.W. Cortical connections of the inferior arcuate sulcus cortex in the macaque brain. Brain Res., 1992a, 573: 8-26.

Deacon, T.W. The neural circuitry underlying primate calls and human language. In: Wind, J., Chiarelli, B., Bichakjian, B.H., Nocentini, A. and Jonker, A. (Eds.) Language origin: a multidisciplinary approach. Kluwer Academic Publishers, Dordrecht: 1992b: 121-162.

Eckhorn, R., Bauer, R., Jordan, W., Brosch, M., Kruse, W., Munk, M. and Reitboeck, H.J. Coherent oscillations: a mechanism of feature linking in the visual cortex? Multiple electrode and correlation analysis in the cat. Biol. Cybern., 1988, 60: 121-130.

Elbert, T. Slow Cortical Potentials reflect the regulaton of cortical excitability. In: V.C. McCallum, S.H. Curry (eds.) Slow Potential Changes in the Human Brain. Plenum Press, New York, 1993, pp 235-251.

Elbert, T., Lutzenberger, W., Rockstroh, B. and Birbaumer, N. Removal of ocular artifacts from the EEG - a biophysical approach to the EOG. Electroenceph. Clin. Neurophysiol., 1985, 60: 455-463.

Engel, A.K., König, P., Gray, C.M. and Singer, W. Stimulus-dependent neuronal oscillations in cat visual cortex: inter-columnar interaction as determined by cross-correlation analysis. Eur. J. Neurosci., 1990, 2: 588-606.

Engel, A.K., König, P., Kreiter, A.K., Schillen, T.B. and Singer, W. Temporal coding in the visual cortex: new vistas on integration in the nervous system. Trends in Neurosciences, 1992, 15: 218-226.

Freeman, W.J. The Physiology of Perception. Sci. Am., 1991, February: 34-41.

Gerstein, G.L., Bedenbaugh, P. and Aertsen, A.M.H.J. Neuronal assemblies. IEEE Trans. Biomed. Engineering, 1989, 36: 4-14.

Gray, C.M., König, P., Engel, A.K. and Singer, W. Oscillatory responses in cat visual cortex exhibit inter-columnar synchronization which reflects global stimulus properties. Nature, 1989, 338: 334-337.

Gray, C.M. and Singer, W. Stimulus-specific neuronal oscillations in orientation columns of cat visual cortex. Proc. Natl. Acad. Sci. ,USA, 1989, 86: 1698-1702.

Gustafsson, B., Wigström, H., Abraham, W.C. and Huang, Y.Y. Long term potentiation in the hippocampus using depolarizing current pulses as the conditioning stimulus to single volley synaptic potentials. J. Neurosci., 1987, 7: 774-780.

Hari, R. and Lounasmaa, O.V. Recording and interpretation of cerebral magnetic fields. Science, 1989, 244: 432-236.

Hebb, D.O. The organization of behavior. A neuropsychological theory. New York: John Wiley, 1949.

Holcomb, P.J. and Neville, H.J. Auditory and visual semantic priming in lexical decision: a comparision using event-related brain potentials. Language and Cognitive Processes, 1990, 5: 281-312.

Kristeva-Feige, R., Feige, B., Makeig, S., Ross, B. and Elbert, T. Oscillatory brain activity during human sensorimotor integration. NeuroReport, 1993, 4: 1291-1294.

Kuhl, P.K., Williams, K.A., Lacerda, F., Stevens, K.N. and Lindblom, B. Linguistic experience alters phonetic perception in infants by 6 months of age. Science, 1992, 255: 606-608.

Llinas, R. and Ribary, U. Coherent 40-Hz oscillation characterizes dream state in humans. Proc. Natl. Acad. Sci. ,USA, 1993, 90: 2078-2081.

Lutzenberger, W., Pulvermüller, F. and Birbaumer, N. Words and pseudowords elicit distinct patterns of 30 Hz EEG responses in humans. Neurosci. Lett., 1994 (in press).

Makeig, S. Auditory event-related dynamics of the EEG spectrum and effects of exposure to tones. Electroenceph. Clin. Neurophysiol., 1993, 86: 283-293.

Mitzdorf, U. Current source density method and application in cat cerebral cortex: investigation of evoked potentials and EEG phenomena. Physiol. Rev., 1985, 65: 37-100.

Mohr, B. Pulvermüller, F. and Zeidel, E. Lexical decision after left, right and bilateral presentation of content words, function words and non-words: evidence for interhemispheric interaction. Neurophysiologia, 1994, 32: 105-124.

Murthy V. N. and Fetz, E.E. Coherent 25- to 35-Hz oscillations in the sensorimotor cortex of awake behaving monkeys. Proc. Natl. Acad. Sci. ,USA, 1992, 89: 5670-5674.

Palm, G. Neural assemblies. Berlin: Springer, 1982.

Palm, G. On the internal structure of cell assemblies. In: Aertsen, A. (Ed.) Brain theory: spatio-temporal aspects of brain function. Elsevier, Amsterdam: 1993: 261-270.

Pandya, D.N. and Yeterian, E.H. Architecture and connections of cortical association areas. In: Peters, A. and Jones, E.G. (Eds.) Cerebral cortex. Vol. 4. Association and auditory cortices. Plenum Press, London: 1985: 3-61.

Pantev, C., Makeig, S., Hoke, M., Galambos, R., Hampson, S. and Gallen, C. Human auditory evoked gamma-band magnetic fields. Proc. Natl. Acad. Sci. ,USA, 1991, 88: 8996-9000.

Pulvermüller, F. Constituents of a neurological theory of language. Concepts Neurosci., 1992, 3: 157-200.

Pulvermüller, F. On connecting syntax and the brain. In: Aertsen, A. (Ed.) Brain theory - spatio-temporal aspects of brain function. Elsevier, New York: 1993: 131-145.

Pulvermüller, F. Syntax und Hirnmechanismen. Perspektiven einer multidisziplinären Sprachwissenschaft. Kongitionswissenschaften, 1994, 4: 17-31.

P, B., Pulvermüller, F. and Zaidel, E. Lexical decision after left, right and bilateral presentation of content words, function words and non-words: evidence for interhemispheric interaction. Neuropsychologia, 1994, 32: 105-124.

Murthy, V.N. anulvermüller, F., Eulitz, C., Pantev, C., Mohr, B., Feige, B., Lutzenberger, W., Elbert, T. and Birbaumer, N. Gamma band brain responses reflect cognitive processing. submitted, 1994.

Pulvermüller, F., Lutzenberger, W. and Birbaumer, N. Electrocortical distinction of vocabulary types. submitted, 1994.

Pulvermüller, F., Mohr, B., Rayman, J., Zaidel, E. and Aertsen, A. Bilateral presentation of words facilitates lexical processing in normals but not in split-brain patients. Soc. Neurosci. Abstr., 1993, 19: 1808.

Pulvermüller, F. and Preißl, H. A cell assembly model of language. Network, 1991, 2: 455-468.

Pulvermüller, F. and Preißl, H. Explaining aphasias in neuronal terms. J. Neurolinguistics, 1994, (in press).

Pulvermüller, F. and Schönle, P.W. Behavioral and neuronal changes during treatment of mixed-transcortical aphasia: a case study. Cognition, 1993, 48: 139-161.

Pulvermüller, F., Eulitz, C., Pantev, C., Mohr, B., Feige, B., Lutzenberger, W., Elbert, T., Birbaumer, N. Gamma-band brain responses reflect cognitive processing, 1994 (submitted)

Rockstroh, B., Elbert, T., Canavan, A., Lutzenberger, W. and Birbaumer, N. Slow cortical potentials and behaviour. Baltimore: Urban & Schwarzenberg, 1989.

Rösler, F., Heil, M. and Glowalla, U. Monitoring retrieval from long-term memory by slow event-related potentials. Psychophysiology, 1993, 30: 170-182.

Schuster, H.G. and Wagner, P. A model for neuronal oscillations in the visual cortex: 1. Mean-field theory and derivation of the phase equations. Biol. Cybern., 1990, 64: 77-82.

Sheer, D.E. Focused arousal, 40-Hz EEG, and dysfunction. In: Elbert, T., Rockstroh, B., Lutzenberger, W. and Birbaumer, N. (Eds.) Self-regulation of the brain and behavior. Springer, Berlin: 1984: 64-84.

Singer, W. Putative functions of temporal correlations in neocortical processing. In: Koch, C. and Davis, J. (Eds.) Large scale neuronal theories of the brain. MIT Press, Boston, MA: 1994.

Speckmann, E.-J. and Elger, C.E. Neurophysiological basis of the EEG and of DC potentials. In: Niedermeyer, E. and Lopes da Silva, F. (Eds.) Electroencephalography - basic principles, clinical applications and related fields. Urban & Schwarzenberg, Baltimore, Munich: 1982: 1-13.

Spydel, J.D., Ford, M.R. and Sheer, D.E. Task dependent cerebral lateralization of the 40 Hz EEG rhythm. Psychophysiology, 1979, 16: 347-350.

Wilson, H.R. and Cowan, J.D. A mathematical theory of the functional dynamics of cortical and thalamic nervous tissue. Kybernetik, 1973, 13: 35-80.

TIME- AND FREQUENCY-DOMAIN ANALYSES OF AUDITORY EVOKED FIELDS

Bernd Lütkenhöner and Christo Pantev

Institute for Experimental Audiology, University of Münster
Germany

INTRODUCTION

The magnetoencephalogram (MEG) represents a superposition of magnetic field contributions from a great number of individual generators with different temporal activation patterns and different frequency characteristics. Thus it is not surprising that the result of a source analysis is generally dependent on the selected time and freqency windows. While the time dependency of the magnetic field distribution is routinely analysed in most MEG studies, investigations of the frequency dependency are still the exception. Two different concepts for a frequency-specific source analysis can be distinguished. First, after narrow-band filtering of the data a time-domain source analysis can be performed in the usual way. Second, special frequency-domain source analysis procedures can be applied to the Fourier transformed data. The first procedure was used e.g. by Williamson et al. (1989), Lü et al. (1992), and Grummich et al. (1992) for the investigation of human parietooccipital alpha activity and by Pantev et al. (1991, 1993) for the investigation of gamma-band auditory evoked fields. The second procedure was proposed already years ago (Lehmann and Michel, 1989, 1990; Michel et al., 1992, 1993) for the analysis of the electroencephalogram (EEG). Inspired by these studies, Lütkenhöner (1992a) formulated a generalized theory and emphasized the close relationship between a source analysis in the time- and the frequency-domain. Recently, Tesche and Kajola (1993) used a frequency-domain analysis to study magnetic alpha activity as well as slow wave activity with superimposed spikes.

As pointed out by Lütkenhöner (1992a) and Tesche and Kajola (1993), a time-domain analysis is in general appropriate for the analysis of transient phenomena (pronounced peaks in the time domain), whereas a frequency-domain analysis can be expected to be superior for the analysis of rhythmic activity (pronounced peaks in the frequency domain). The auditory evoked field (AEF) seems to be composed of both transient and rhythmic activity (Pantev et al., 1991) so that it appears worthwhile to investigate whether the two types of analysis provide complementary information. In this article, time-domain analyses of wideband as well as high-pass filtered AEF are compared with the result of frequency-domain analyses of Fourier transformed AEF.

THE DATA

The data which have been analysed were taken from a study of Pantev and Hampson (1994), primarily devoted to a comparison between the on- and off-response and the sustained field. The experimental procedure was similar to that described by Pantev et al. (1993). Briefly, 1 kHz tone bursts of 2 s duration with a rise and fall time of 15 ms (60 dB normative hearing level, interstimulus interval randomized between 5 and 7 s) were presented to the subject's right ear. Recordings were carried out in a magnetically shielded room (Vacuumschmelze) using a 37-channel biomagnetometer (Magnes™, Biomagnetic Technologies), which was centered over a point lying about 1.5 cm superior to position T3 of the 10-20 system for electrode placement, as close as possible to the subject's head. In each of the 10 investigated subjects, 512 stimulus-related epochs (2.7 s duration, beginning 200 ms before stimulus onset) were recorded with a bandwidth from DC to 100 Hz (slope 60 dB/octave). The sampling interval was 3.36 ms. Fig. 1 shows a typical waveform. The analyses presented in this article were basically confined to the first 100 ms, which are dominated by waves P1m and N1m.

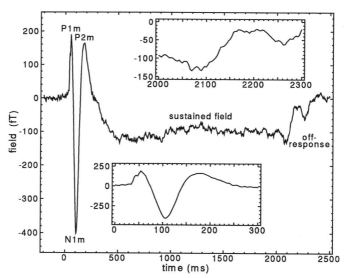

Figure 1. Typical waveform of an AEF elicited by a toneburst of 2 s duration. A high-amplitude transient response (waves P1m, N1m, P2m) is followed by a sustained field. The stimulus offset finally elicits an off-response. The two inserts show on and off response, respectively, in a larger scale. Only the first 100 ms of the response are considered in the present study.

ANALYSIS TECHNIQUES

Signal processing: Time domain

Wideband AEF are dominated by low-frequency components so that phenomena associated with higher frequencies are often hard to detect. Pantev et al. (1991, 1993) therefore used a narrow-band filter (24-48 Hz) to extract oscillatory phenomena in the gamma band (20-90 Hz), which are visible only as small deflections in the wide-band response. Since application of a narrow-band filter inevitably involves the risk that some transient in the original data is transformed into an oscillatory phenomenon, high-pass filtering as a more conservative technique was used in the present study. The most simple high-pass filter operation corresponds to the calculation of the difference between two successive samples. This means that instead of the original time series $u[n\Delta t]$ $(n = 0,1,...)$ the time series

$$\Delta u[n\Delta t] \equiv u[(n+1)\Delta t] - u[n\Delta t] \tag{1}$$

(n=1,2,...) is considered (e.g. Hamming, 1983). For sufficiently small sampling intervals Δt, the operation (1) provides an approximation of the first derivative of the function u(t), except for a factor $1/\Delta t$. Similarly, an approximation of the second derivative (except for a factor $1/\Delta t^2$) is obtained by calculating

$$\Delta^2 u[n\Delta t] \equiv u[(n-1)\Delta t] - 2u[n\Delta t] + u[(n+1)\Delta t] \tag{2}$$

(n=1,2,...). To see the frequency-domain effect of the time-domain operations (1) and (2), it is instructive to investigate how the function sin(ωt) is affected. Simple algebraic operations yield[1]

$$\sin[\omega(t+\Delta t)] - \sin(\omega t) = 2\sin(\omega\Delta t/2)\cos[\omega(t+\Delta t/2)] \tag{3}$$

and

$$\sin[\omega(t-\Delta t)] - 2\sin(\omega t) + \sin[\omega(t+\Delta t)] = -(2\sin(\omega\Delta t/2))^2\sin(\omega t) \tag{4}$$

As to be expected for derivatives of sin(ωt), the right-hand sides of these two equations are proportional to cos(ωt) and -sin(ωt), respectively, except for the time shift $\Delta t/2$ in the case of equation (3). In both cases, the amplitude factor has the value 1 if the frequency of the sine, f=ω/2π, is $1/6\Delta t$. Since the sampling interval is 3.36 ms in the present study, this specific frequency is 49.6 Hz. For lower frequencies the amplitude is reduced, for higher frequencies it is increased by a factor between 1 and 2. In the low-frequency limit, i.e. ω∆t<<1, the amplitude factors on the right-hand sides of

---

[1]The equations (3) and (4) are valid for any value Δt. Thus, it is not absolutely necessary to consider Δt as the sampling interval. An alternative would be, for example, the interpretation of Δt as an integer multiple of the sampling interval.

the equations (3) and (4) can be approximated by $\omega\Delta t$ and $(\omega\Delta t)^2$, respectively. Thus a frequency change by one octave corresponds to amplitude changes by factors of 2 (6 dB per octave) and 4 (12 dB per octave), respectively. In the present case it must be kept in mind that the high-pass filter operations (1) and (2) combined with the 100 Hz anti-aliasing filter yield, of course, a band-pass filter.

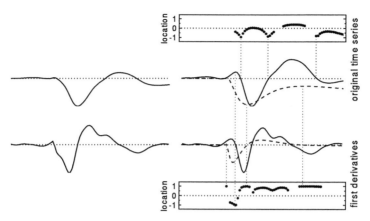

Figure 2. Schematic drawing illustrating that a source analysis of the first derivatives of the original time functions ("field change") can be complementary to a source analysis of the original data. While a one-dipole source analysis applied to the original time function (panel at the top) never provides the correct location of dipole 1 (assumed to have the coordinate value +1), the same analysis applied to the field change provides the correct dipole location at three time instants (panel at the bottom). For more details see text.

Moran et al. (1992, 1993) recently pointed out that situations are conceivable where the difference operation (1) can be quite useful for source analyses of magnetic fields. Their idea shall be explained by means of Fig. 2, in which a two-dipole configuration is considered. The two dipole moments are assumed to have a parallel orientation. The time courses of their amplitudes are represented by the solid and the dashed curve in the second panel on the right. Adding up these two time courses and rescaling yields the upper curve on the left, which can be interpreted as the time course of the magnetic field at locations situated symmetrically with respect to the two dipoles. The panels underneath show the time courses obtained by calculating the first derivatives ("field change"). Suppose now that the dipole represented by the solid curves is located at x=1, whereas the dipole represented by the dashed curves is located at x=-1. Suppose, furthermore, that the dipole location which would be estimated from the magnetic field distribution corresponds to the "center of gravity" of the two dipoles, i.e. $\hat{x} = \left(|P_1| - |P_2|\right) / \left(|P_1| + |P_2|\right)$, where P_1 and P_2 are the amplitudes of the two dipole moments (or their first derivatives).[2] The locations

---

[2]For the calculation of the center of gravity, the absolute value of the moment of a dipole is considered as its "mass". For two parallel dipoles with a distance of a few centimeters the center-of-gravity assumption is quite reasonable, provided that the distance between measurement coils and

defined in this way are shown in the uppermost panel (for the field) and the bottommost panel (for the field change). Times at which the amplitude of the field (or the field change) was smaller than 10% of the maximum amplitude were not considered. In the first case, the location of the second dipole is correctly "estimated" at three time points (corresponding to the zeros of the moment of the first dipole), whereas the location of the first dipole is never "estimated" correctly. In the second case, the second dipole is correctly localized only once (time corresponding to the first significant maximum of moment 1), whereas the first dipole is correctly localized at three time points (corresponding to the onset as well as the two extrema of moment 2). This illustrates that source analyses of original data and first derivatives may yield complementary information: By calculating the first derivatives, contributions of sources with more or less constant activities are eliminated so that the identification of sources with changing activities may be facilitated (Moran et al., 1992, 1993).

Signal processing: Frequency domain

The frequency-domain sampling interval of Fourier transformed data, Δf, corresponds to the inverse of the width of the considered time window, T. Thus, for any application a compromise between frequency specificity and time resolution has to be found. In the frequency-domain analyses described below, time windows of 108 ms (corresponding to 32 samples) were considered. This means that the frequency resolution was 9.3 Hz. Prior to Fourier transformation of data, the mean was removed and a raised-cosine window (von Hann window) was applied (e.g. Hamming, 1983).

A *normalized amplitude* was calculated for all discrete frequencies by dividing the power of the average by the mean power of the single-trial epochs and calculating the square root of the resulting value. The normalized amplitude is closely related to the *signal-to-noise ratio*: Supposed that signal and noise are independent and that x^2 denotes the ratio of signal power and variance of the noise (i.e., x is a measure of the signal-to-noise ratio), then the value to be expected for the normalized amplitude would be $s=x/(1+x^2)^{1/2}$. It follows that a signal-to-noise ratio of x=1 corresponds to a normalized amplitude of $s \cong 0.7$, x=0.5 corresponds to $s \cong 0.45$, and x=0.25 corresponds to $s \cong 0.24$. Since the normalized amplitudes considered in the present study are generally smaller than 0.5, except for the lowest frequencies, it is justified to consider the normalized amplitude as an approximation of the signal-to-noise ratio.

Source analysis

Source analyses were performed using the model of a current dipole in a homogeneous spherical volume conductor. The center of sphere was estimated by means of a least-squares fit algorithm (Lütkenhöner et al., 1990), which served to approximate the scalp region underneath the 37 measurement coils by a spherical surface.

---

dipoles is also of the order of centimeters (Lütkenhöner, 1994). The assumption is inappropriate, on the other hand, for antiparallel dipoles, i.e. for $P_1 P_2 < 0$.

The source of the recorded magnetic field was assumed to be a single dipole with time-varying location and orientation ("moving dipole"). Details of the dipole fit procedure have been described elsewhere (Lütkenhöner et al., 1991, Lütkenhöner, 1992b). Briefly, the original least-squares fit problem was transformed into a minimization problem for the non-linear parameters (dipole coordinates) by replacing the linear parameters (components of the dipole moment) by the algebraic solutions available for their least-squares estimates. The resulting minimization problem was solved iteratively by means of Powell's method (e.g. Press et al. 1986).

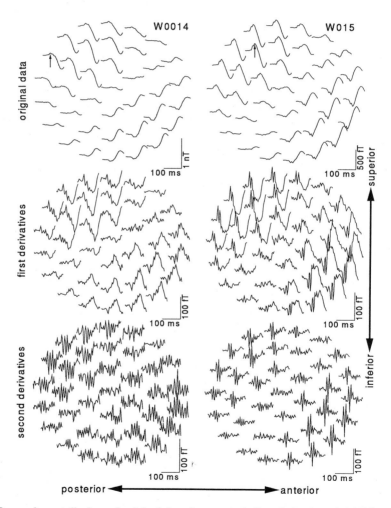

Figure 3. Sensor layout displays of original data (upper row), first derivatives (middle) and second derivatives (bottom row), plotted for subjects W0014 and W0015.

Though the fit algorithm just outlined was developed for time-domain data, it is suitable for frequency-domain data as well: As pointed out by Lütkenhöner (1992a), a minimization for one discrete frequency (complex-valued data) is completely equivalent with a simultaneous minimization for two time points (dipole location and orientation assumed to be invariant).

TIME SERIES ANALYSIS

Time domain

The upper row in Fig. 3 shows for two subjects (W0014 and W0015) how the magnetic field at the 37 measurement positions changes as a function of time during the first 120 ms after stimulus onset. The arrow in the plot for subject W0015 marks the time function already considered in Fig. 1. All the curves have basically the same appearance, apart from the fact that the posterior curves are inverted compared to the anterior ones. Dominating features are two major deflections: around 50 ms (peak P1m) and around 100 ms (peak N1m).

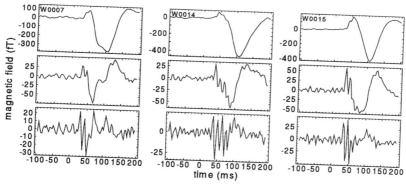

Figure 4. Time course of the magnetic field at a specified measurement location (top), first derivative (middle), and second derivative (bottom), displayed for three subjects. The examples shown in the middle column and on the right correspond to the channels marked by an arrow in Fig. 3.

A closer examination reveals ripples superimposed on peak P1m. A relative enhancement of these ripples can be achieved by calculating first and second derivatives of the original time functions. While in the former case (Fig. 3, middle) the high-frequency ripples still "ride" on a wave with a lower frequency, in the latter case (Fig. 3, bottom) low-frequency components are almost completely missing. The high-frequency phenomenon (ripples in the original time functions) shows a dipolar pattern quite similar to that of the N1m-P1m complex. The different polarity of anterior and posterior curves is most easily recognized in the first derivatives.

To study the relationship between original time functions and first and second derivatives in more detail, single channels are considered in Fig. 4. The curves shown in the middle and on the right (top panel) represent the time functions marked by arrows in Fig. 3, whereas on the left data from a third subject (W0007) are presented. For all three subjects the upper curve shows the original time function, the curve in the middle the first derivative, and the bottom curve the second derivative. The high-frequency phenomenon just described for subjects W0014 and W0015 is obviously present also in the curves from subject W0007. A rough estimate of the frequency of this phenomenon can be derived from the observation that the time required for one cycle is approximately 13 ms. This corresponds to a frequency around 80 Hz, which is obviously different from the 40 Hz "oscillations" observed by Pantev et al. (1991, 1993). It shall be emphasized that the 80 Hz phenomenon is definitely no filter artefact, because the significant peaks found in the first derivatives can be traced back to the small ripples in the original time functions.

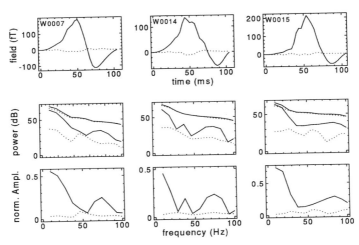

Figure 5. Estimation of the normalized amplitude, exemplified for the three data sets considered in Fig. 4. Upper pannels (solid curves): Cosine-tapered time-domain average. Middle panels (solid curves): Mean power of single-trial epochs (upper curves) and power of cosine-tapered time-domain average (lower curves). Bottom panels (solid curves): Normalized amplitudes, defined as the square root of the ratio obtained by dividing the mean power of the single-trial epochs and the power of the cosine-tapered time-domain average. For comparison, the dotted curves show the corresponding results obtained for a 108 ms pre-stimulus time window.

Quite interesting is the first derivative in the case of subject W0007 (middle panel on the left in Fig. 4): In addition to a low-frequency wave (representing the first derivative of the P1m-N1m complex) and the 80 Hz phenomenon a wave with a negative peak around 60 ms, a positive peak around 80 ms, a second negative peak around 100 ms, and a second positive peak around 120 ms can be observed. The cycle length of 40 ms corresponds to a frequency of 25 Hz. In the original time functions this wave manifests itself as a deformation of peak N1m, whereas it is quite inconspicuous in the second derivative. The latter observation is not surprising, because calculation of a derivative considerably reduces the amplitude of a 25 Hz wave compared to an 80 Hz wave, as explained above.

Frequency domain

The simple qualitative analyses described above have shown that the first 120 ms of AEF in response to tone burst stimuli are composed of a low-frequency wave (P1m-N1m complex) and small deflections associated with much higher frequencies. This finding shall be substantiated now by frequency-domain analyses.

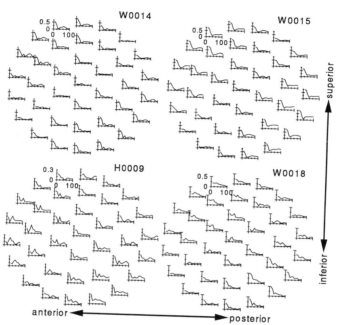

Figure 6. Sensor layout display of the normalized amplitude as a functions of frequency, plotted for four different subjects.

The principle of the analyses carried out is illustrated in Fig. 5, for the three data sets already considered in Fig. 4. The solid curves in the upper panels show the cosine-tapered time-domain average. The panels in the middle show, in an arbitrary decibel (dB) scale as a function of frequency, the mean power estimated for the Fourier transforms of the single-trial epochs (upper solid curves) and the power of the Fourier transform of the average (lower solid curves). Dividing the power of the average by the mean power of the single-trial epochs and calculating the square root of the resulting value yields the *normalized amplitude* (bottom panels), which can be interpreted as an approximation of the signal-to-noise ratio (see section "Methods"). The curves obtained for the normalized amplitude are obviously quite similar to those obtained for the power of the average, but the peak amplitudes have not necessarily the same order of rank. For comparison, the dotted curves show the results obtained by repeating these analyses for a *pre*-stimulus time interval of 108 ms length. The curves obtained for the mean power of the single-trial epochs are almost identical with those obtained for the post-stimulus interval, whereas the values obtained for the power of the average and the normalized amplitude are, as to be expected, generally much smaller than those obtained for the post-stimulus time interval. A comparison with the dotted curves shows that the peaks of the normalized amplitude in the case of the post-stimulus time interval are quite significant.

The results of the frequency-domain analyses presented in Fig. 5. basically confirm the qualitative results obtained by inspecting the time functions shown in Fig. 4: Besides the dominating low-frequency peak, a significant peak is found also around 75 Hz. In the case of subject W0014, an additial peak is found at about 40 Hz. The finding that the 25 Hz wave observed in the time functions from subject W0007 (left column in Fig. 4) does not give rise to a peak in the normalized amplitude (Fig. 5 bottom) is certainly due to the fact that the time window chosen for the frequency-domain analysis was unfavourable for this wave.

In the next two figures the question is investigated as to how representative the results given in Fig. 5 are. For the sake of simplicity, only the normalized amplitude will be considered. Fig. 6 illustrates for four subjects the channel-dependency of the curves obtained for the normalized amplitude. Though the curves from a given subject are generally similar, certain differences cannot be overlooked. In the subjects W0014 and W0018, for example, a peak arround 40 Hz cannot be found in all channels. Most bewildering are the results obtained for subject H0009: A peak around 40 Hz is the dominating feature in some of the channels, but in other channels a dip is found arround that frequency.

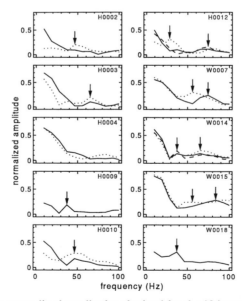

Figure 7. Synopsis of the normalized amplitudes obtained for the 10 investigated subjects. The solid curves always corresponds to the posterior channel with the highest normalized amplitude (i.e. the best signal-to-noise ratio) at 9.3 Hz. For some of the subjects, normalized amplitudes are provided for additional channels (dotted and dashed curves), because in these cases the solid curve was not representative for all channels.

A synopsis of the normalized amplitudes from all subjects is given in Fig. 7. The solid curves represent a standardized location (posterior channel with the best signal-to-noise ratio for the lowest frequency). Since this channel is generally not the channel with the best signal-to-noise ratio at other frequencies, additional curves (dotted or dashed) are given for most of the subjects. Pronounced peaks, marked by arrows, are found between 27 Hz and 83 Hz. In four of the subjects, two different peaks can be observed, though generally not in the same channel.

SOURCE ANALYSIS

In this section, a comparison of time- and frequency-domain source analysis will be presented. The comparison is bascially confined to the data set W0015, which had the highest signal-to-noise ratio and the clearest dipolar pattern in the second derivatives of the data. Only few selected results from the other subjects will be presented.

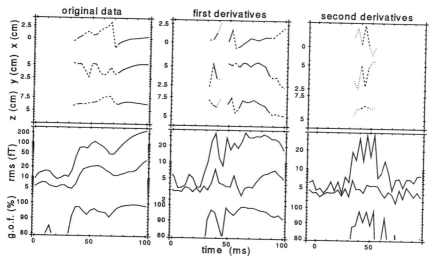

Figure 8. Results of moving-dipole analyses for original data (left column), first derivatives (middle column), and second derivatives (right column). The three traces in the upper panel represent the three Cartesian coordinates of the dipole location. A solid line segment indicates that the goodness of fit (g.o.f.) is better than 95%, a dashed line segment indicates a g.o.f between 90% and 95%, whereas a dotted line segment indicates a g.o.f. between 85% and 90%. The upper two traces in the bottom panel show the field rms value and the residual rms value, whereas the bottom trace represents the goodness of fit.

Time-domain source analysis for subject W0015

Fig. 8 shows the result of a moving dipole analysis for subject W0015. The left column displays the results obtained for the original data, whereas in the middle and on the right the results obtained for the first and second derivatives are presented.

The three traces in the upper panels show, how the three Cartesian coordinates of the estimated dipole change as functions of time. The upper two curves in the bottom panels show the root mean square (RMS) of the magnetic field and the residual RMS, this means the RMS of the differences between measured data and model prediction. The bottom curve, finally, shows the variance accounted for by the moving-dipole model, often called the goodness of fit (g.o.f.). It should be noticed that high RMS values are often associated with high residual RMS values, which indicates an inadequacy of the one-dipole model.

The results obtained for the time points with the highest goodness of fit are compiled in the upper nine rows of Table I. The entries in the first column indicate the type of data analysed (B means original field, whereas $\partial B/\partial t$ and $\partial^2 B/\partial t^2$ means first and second derivatives, respectively). The investigated time point is indicated in the second column. The next three columns provide the estimated dipole coordinates, and in the right-most column the goodness of fit is given. It is remarkable that the coordinate values printed in bold face (corresponding to the results with the best goodness of fit for a given data type) show only differences of the order of millimeters. Slightly greater coordinate differences are found if *all* results are considered. But even in that case the estimated dipoles are located within a relatively small volume (about 1 cm³). The fact that almost the same locations were estimated for the original data and the first and second derivatives suggests that the low-frequency phenomenon (i.e. the classical waves P1 and N1) and the higher-frequency phenomenon (about 30-80 Hz) have a similar origin.

Table I. Overview of estimated source locations.

type of under-lying data	time (ms)	estimated coordinates			variance accounted for (%)
		x	y	z	
B	40	0.44	5.23	6.26	97.4
B	54	0.81	5.22	6.65	96.2
B	91	**0.11**	**5.20**	**6.54**	**99.2**
$\partial B/\partial t$	49	0.17	5.40	6.61	98.4
$\partial B/\partial t$	66	**0.25**	**5.42**	**6.90**	**99.1**
$\partial B/\partial t$	112	0.12	5.42	6.41	96.6
$\partial^2 B/\partial t^2$	40	0.29	4.99	6.20	97.7
$\partial^2 B/\partial t^2$	47	0.20	4.81	6.32	97.5
$\partial^2 B/\partial t^2$	54	**0.03**	**5.42**	**6.27**	**97.8**
9.3 Hz	0-108	0.62	5.30	6.58	87.2
18.6 Hz	0-108	0.46	5.50	6.88	94.0
27.9 Hz	0-108	0.33	5.63	7.35	93.6
83.7 Hz	0-108	0.23	5.54	6.03	91.4

Frequency-domain source analysis for subject W0015

Fig. 9 shows the result of a frequency-domain source analysis for subject W0015. Separate plots are provided for each of the nine discrete frequencies considered in the figure. The complex-valued amplitudes obtained by Fourier transforming the original data are represented by the dots. The thin lines originating in these dots point to the complex amplitudes corresponding to the solution of the inverse problem (the sum of the squared lengths of these lines represents the quantity which was minimized when solving the inverse problem). As to be expected for a one-dipole model, the complex amplitudes corresponding to the solution of the inverse problem are located on a straight line (hatched area in Fig. 9). A goodness of fit better than 90% could be obtained only for 18.6 Hz, 27.9 Hz, and 83.7 Hz. The dipole fit results obtained for these three frequencies as well as for the frequency 9.3 Hz are given in the bottommost rows of Table I. The dipole locations are obviously similiar to those provided by the time-domain analyses: The location obtained at 9 Hz is comparable to that obtained for the original field, whereas the locations obtained at 18 Hz and at 84 Hz are comparable to those obtained for the first and second derivatives, respectively. Only the locations obtained at 28 Hz show larger discrepancies.

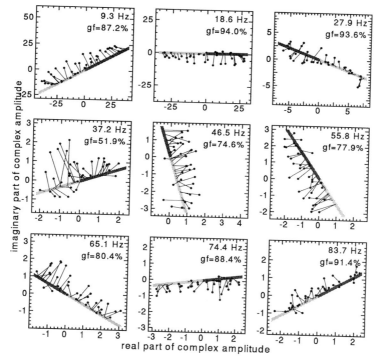

Figure 9. Frequency-domain source analysis for subject W0015. The nine plots collected in this figure show the results obtained for discrete frequencies between 9.3 Hz and 83.7 Hz. The abscissa and the ordinate in each of the panels represent the real and the imaginary part of the complex amplitude obtained by Fourier transforming the time-domain average. Each of the 37 channels is represented by one filled circle. The thin lines connect the measured complex amplitudes (filled circles) with the complex amplitudes corresponding to the best-fitting equivalent current dipole. The latter amplitudes are necessarily located on a straight line through the origin (represented by the hatched field, origin indicated by the transition between the dark hatched field and the light hatched field). The percentage gf refers to the variance accounted for by the model ("goodness of field").

Frequency-domain analyses in eight other subjects

The results presented in Fig. 9 suggest that, for the type of data analysed in this paper, the goodness of fit values obtained in a frequency-domain analysis tend to be much smaller than those obtained in a time-domain analysis. This conjecture is substantiated by Fig. 10 where results from eight other subjects are presented. The frequencies which have been investigated correspond to peaks of the normalized amplitude (marked by arrows in Fig. 7). The highest goodness of fit value obtained in these analyses was only 82.9% (see subject H0003). Dipole localization results associated with such small goodness of fit values cannot be considered as meaningful. Thus we refrained from a comparison between time- and frequency-domain analysis in these cases.

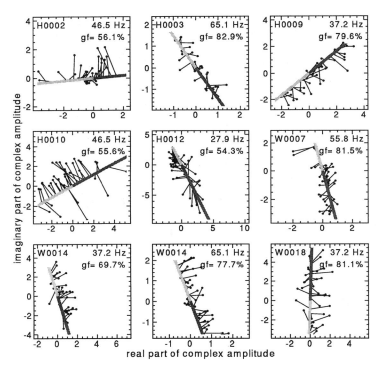

Figure 10. Result of a frequency-domain source analysis for the data corresponding to nine of the peaks marked by an arrow in Fig. 7. The plots collected in this figure are organized analogously to those presented in Fig. 9.

DISCUSSION

The finding of peaks in the power spectrum between 27 Hz and 83 Hz suggests that 40 Hz is not a magic number in the context of tone-burst elicited AEF. Nevertheless, in more than half of the subjects a peak was found between 30 Hz and

50 Hz. In most subjects the frequency-domain peaks corresponded to oscillatory phenomena in the time domain, visible in at least some of the channels. In two subjects these oscillatory phenomena were clearly visible in most of the channels and exhibited a dipolar pattern after high-pass filtering (cf. Fig. 3). It shall be emphasized that neither the peaks observed in the frequency-domain nor the oscillatory phenomena observed in the time-domain can be explained as artefacts produced by signal processing: The frequency-domain analysis was carried out with the unfiltered data, and, as demonstrated in Fig. 4, the high-pass filter operations used in the time-domain analyses did not generate oscillatory phenomena, they only provided a relative enhancement of oscillatory phenomena already present in the unfiltered data.

The question as to where the sources of the oscillatory phenomena are located can be answered only with reservation, because the goodness of fit obtained in the frequency-domain dipole source analyses was quite bad in most cases (cf. Fig. 10). Though in some cases the bad goodness of fit can be explained by a relatively low signal-to-noise ratio, in most other cases the main reason seems to be an inadequacy of the single dipole model.[3] Nevertheless, at least in one subject (W0015) a meaningful source analysis was possible. The result of this analysis suggests that the oscillatory phenomena have a similar macroscopic origin as the main source of wave N1, which is known to be the auditory cortex. Furthermore, the relatively high goodness-of-fit values obtained for the second derivatives, being almost as high as those obtained for the original data (cf. table I), suggest that the generator underlying the oscillatory phenomena has a spatial extent comparable to that of the generator of wave N1. A similar macrosopic origin does not necessarily mean, however, that the generators are the same also in a microscopic sense, because the locations provided by a dipole source analysis can at best be considered as the "center of gravity" of the underlying microscopic source structure. Thus even in the case of a high goodness of fit there are many degrees of freedom for the interpretation of the results. For example, a goodness of fit of 99% is absolutely compatible with the assumption of two parallel dipoles with a separation of up to 4 cm (Lütkenhöner, 1994).

Acknowledgement

Work supported by the Deutsche Forschungsgemeinschaft (Klinische Forschergruppe "Biomagnetismus und Biosygnalanalyse").

REFERENCES

Grummich P., Vieth, J., Kober, H., Scholz, T., 1992, Separation of sources of alpha activity in multi-channel MEG, in: "Biomagnetism: Clinical Aspects," M. Hoke, S.N. Erné, Y.C. Okada, G.L. Romani, eds., Excerpta Medica, Amsterdam, pp. 39-42

Hamming, R.W., 1983, "Digital Filters," second edition, Prentice Hall, Englewood Cliffs.

Lehmann, D., and Michel, C.M., 1989, Intracerebral dipole sources of EEG FFT power maps, Brain Topography 2:155-164

Lehmann, D., and Michel, C.M., 1990, Intracerebral dipole source localization for FFT power maps, Electroenceph. clin. Neurophysiol. 76:271-276

---

[3] If noise would be the main reason for the bad goodness of fit, then a symmetrical distribution with respect to the lines through the origin would be expected for the points plotted in Fig. 10.

Lü, Z.-L., Wang, J.-Z., Williamson, S.J., 1992, Neuronal sources of human parietooccipital alpha rythm, in: "Biomagnetism: Clinical Aspects," M. Hoke, S.N. Ernée, Y.C. Okada, G.L. Romani, eds., Excerpta Medica, Amsterdam, pp. 33-37

Lütkenhöner, B., 1992a, Frequency-domain localization of intracerebral dipolar sources, Electroenceph. Clin. Neurophysiol. 82:112-118

Lütkenhöner, B., 1992b, "Möglichkeiten und Grenzen der neuromagnetischen Quellenanalyse," Lit, Münster/Hamburg

Lütkenhöner, B., 1994, Dipole and multidipole source analysis of magnetic fields: Possibilities and limitations, in: "Proceedings of the 9th International Conference on Biomagnetism," L. Deecke, C. Baumgartner, eds., Elsevier, Amsterdam (in press)

Lütkenhöner, B., Pantev, C., Hoke, M., 1990, Comparison between different methods to approximate an area of the human head by a sphere, in: "Auditory evoked magnetic fields and potentials," F. Grandori, M. Hoke, G.L. Romani, eds. (Advances in audiology, vol. 6), Karger, Basel, pp. 103-118

Lütkenhöner, B., Lehnertz, K., Hoke, M., Pantev, C., 1991, On the biomagnetic inverse problem in the case of multiple dipoles, Acta Oto-Laryngol. (Stockh.) suppl. 491:94-105

Michel, C.M., Lehmann, D., Henggeler, B., Brandeis, D., 1992, Localization of the sources of EEG delta, theta, alpha and beta frequency bands using the FFT dipole approximation. Electroencph. Clin. Neurophysiol. 82:38

Michel, C.M., Koukou, M., Lehmann, D., 1993, EEG reactivity in high and low symptomatic schizophrenics, using source modelling in the frequency domain. Brain Topography 5:389

Moran, J.E., Jacobson, G.P., Tepley, N., 1992, Finite difference field mapping, in: "Biomagnetism: Clinical Aspects," M. Hoke, S.N. Ernée, Y.C. Okada, G.L. Romani, eds., Excerpta Medica, Amsterdam, pp. 801-805

Moran, J.E., Tepley, N., Jacobson, G.P., Barkley, L., 1993, Evidence for multiple generators in evoked responses using finite difference field mapping: auditory evoked fields, Brain Topography 5:229

Pantev C., Makeig S., Hoke M., Galambos R, Hampson S., Gallen C. (1991): Human auditory evoked gamma band magnetic fields. Proc. Natl. Acad. Sci. (USA), 88:8896-9000

Pantev C., Elbert T., Makeig S., Hampson S., Eulitz C., Hoke M. (1993): The auditory evoked sustained field: origin and frequency dependence. Electroenceph. Clin. Neurophysiol. 90:82-90

Pantev C. and Hampson S.(1994): The auditory evoked "off"-response: Sources and comparison with the "on"-and the "sustained"-response (submitted)

Press, W.H., Flannery, B.P., Teukolsky, S.A., and Vetterling, W.T., 1986, "Numerical Recipes: the Art of Scientific Computing," Cambridge University Press, Cambridge

Tesche, C., and Kajola, M., 1993, A comparison of the localization of spontaneous neuromagnetic activity in the frequency and time domains. Electroenceph. Clin. Neurophysiol. 87:408

Williamson S.J., Wang J.-Z., Ilmoniemi R.J., 1989, Method for locating sources of human alpha activity, in: "Advances in Biomagnetism," S.J. Williamson, M. Hoke, G. Stroink, M. Kotani, eds., Plenum, New York, pp. 257-260

ELECTROCORTICAL RHYTHMS IN THE ATTENTIVE CAT: PHENOMENOLOGICAL DATA AND THEORETICAL ISSUES

A. Rougeul-Buser and P. Buser

Institut des Neurosciences
CNRS-Université Pierre & Marie Curie
75005 Paris

INTRODUCTION

The most commonly accepted correlation between spontaneous field potential activity of the neocortex (electrocorticogram, ECoG) and behaviour concerns waking vs slow wave sleep. The first is usually considered accompanied by low amplitude fast activity (called "desynchronized") and slow wave sleep, by rhythmic delta activity and spindles, which imply synchronous firing of neuronal ensembles. More recent data from several groups have now provided evidence that waking can also be accompanied by rhythmic patterns of defined frequencies. The data to be described herein show that in alert cats several sets of characteristic rhythms can be recorded from a variety of cortical and subcortical sites*. Arguments will be developed showing that one of the most interesting behavioural correlates of these rhythms is attention paid by the animal to its surrounding. Our overview here will include, firstly a description of the cortical rhythms, then a summary of our investigations on the thalamic areas presumably also involved in these activities. A third part will be devoted to modulatory actions from deep brainstem structures through aminergic neurotransmitters. In a final section we shall consider some general problems posed by the functional significance of the observed rhythmic activities.

---

* All cortical recordings were performed in monopolar derivation, with electrodes made from isolated wires (diameter .6 mm) introduced into the cortex. The reference electrode was fixed over the frontal sinus. The thalamic gross electrodes consisted of 3 adjacent wires (.3 mm in diameter) whose sections ended at depths differing by 1.5 mm. Recordings were achieved bipolarly using all three combinations between the three leads.

Oscillatory Event-Related Brain Dynamics, Edited
by C. Pantev, Plenum Press, New York, 1994

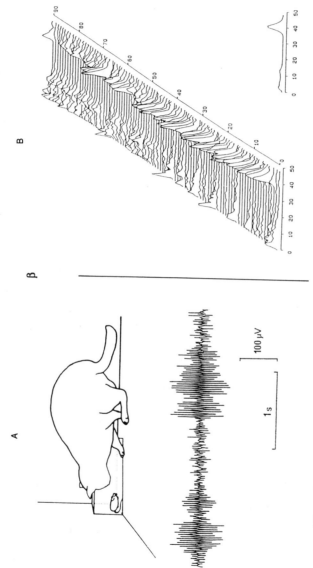

Figure 1 A, B. Beta rhythms in cat parietal cortex and thalamus. Panel A: the animal's attention was focused on a live mouse, installed in a perspex box. Beta rhythms at 40 Hz were recorded from the posterior parietal area 5a (bottom record). B: power spectral display of area 5a ECoG recorded during a 90 mn session in the above situation. Each line corresponds to a spectral analysis of a 1 mn record (0 to 50 Hz) and successive power spectra are lined up on an oblique y axis (average spectrum shown at bottom right). Peaks at 40 Hz correspond to beta rhythms.

CORTICAL RHYTHMS

The fronto-parietal rhythms

Beta rhythms. In the 70s (Bouyer et al., 1974), we described a rhythmic ECoG activity at about 40 Hz in the anterior cortex of the behaving cat. At first, when we grossly evaluated the animal's behaviour, we estimated that this activity was concomitant with "intense wakefulness" (Rougeul-Buser et al., 1975). To pay tribute to the first authors who had previously observed this kind of fast rhythms in man (Jasper and Penfield, 1949), we used the terminology of beta rhythms, a nomenclature that we have so far not abandoned for other terminologies more recently introduced to designate rhythms in the fast frequency band (e.g. "gamma band" rhythms in the olfactory bulb, as described by Freeman, see Freeman and Skarda, 1985). Soon after, we were able to characterize more precisely situations in which these beta rhythms would be enhanced and considerably favoured. We saw them systematically appear while the animal was exploring its surrounding; in such conditions the cat would stop from time to time, and these "arrests" were accompanied by brief episodes of these fast rhythms. We then were lucky to discover another situation when such rhythms could occur as long sustained trains. This situation was easily worked out through installing the animal in the presence of a mouse that he could see, smell but not catch (the mouse being protected by a perspex box). From that time on, we started suggesting that these sustained beta rhythms were concomitants of focused attention of the animal (Rougeul-Buser et al., 1978). We were also able, using arrays of several electrodes 2 mm apart, to identify two foci for beta rhythms in the anterior cortex, one in the motor cortex (area 4 and 6a beta) and one in the posterior parietal area 5 (Bouyer et al., 1981; 1987). As a routine procedure, we from then on used continuous spectral analysis through FFT between 0 and 50 Hz for each minute of tracing, providing a classical 3D "waterfall" spectral display (frequencies as abscissae, recording time as ordinates, spectral power in muV″ as z axis). The recording time was usually 90 mn (Fig. 1 A and B).

Mu rhythms. At about the same period, we also discovered another set of ECoG rhythms, at a much lower frequency i.e. around 15 Hz, whose focus was in the sensory somatic area SI (Rougeul et al., 1972). We first thought that they were also present during focused attention but, much to our surprise, we found them in two different situations, both distinct from that accompanying the beta rhythms. Firstly, as the animal was in a position of "quiet waking", standing still in a "neutral" situation, with no specific item to attract his attention. In these cases, very short episodes of such rhythms at 12-18 Hz usually occurred. We then were again lucky to identify another behavioural situation which markedly favoured these rhythms. This was a particular condition, where the animal waited for an event to occur. This condition of expectation of "something that had a high probability to happen" was practically worked out through placing the animal in front of a small hole in a wall behind which a mouse had been placed. The animal watched the hole, seeing the mouse popping out with her snout from time to time, and hiding again behind the wall. During this little game of hide and seek, the cat usually remained motionless and 15 Hz SI rhythms were sustained, this time for as long as 60 or even 90 mn, with only brief episodes of movements accompanied by desynchronized ECoG activity (Rougeul et al., 1972).

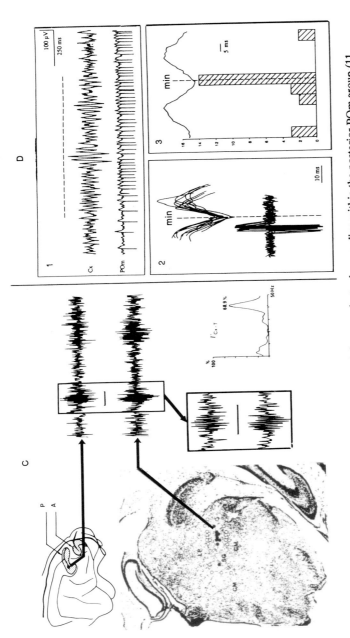

Figure1 C, D. Panel C: at bottom left, gross electrode recording within the posterior POm group (11 cats). Filled circles indicate sites where typical beta rhythms were recorded; blank circles indicate non-rhythmic derivations. At right, samples of beta rhythms simultaneously recorded from the parietal focus (cortical lead P) and the ipsilateral POm. Records in frame are also shown enlarged. Time scales, 0.5 s. Lower right part of panel C: coherence values vs frequency computed between cortex and thalamus, within the 0–50 Hz band. Thalamic nuclei: CM, centrum medianum; GM, medial geniculate body; LP, lateralis posterior nucleus; SG, suprageniculate nucleus. Panel D: time-locking of thalamic POm unit discharges vs corticalparietal beta rhythms. **1**, simultaneous raw records of a POm cell firing (POm) and of the parietal ECoG (Cx). Note close correspondence of firing during part at least of the beta train (the latter indicated by overlying dashed line), **2**, superimposition of the

minima (min) of 12 successive cortical beta waves and of the corresponding spikes. **3**, wave triggered averaging of 10 successive beta waves; the spikes of another cell are now presented as density histogram (number of spikes as ordinates) (After Bouyer et al. 1981).

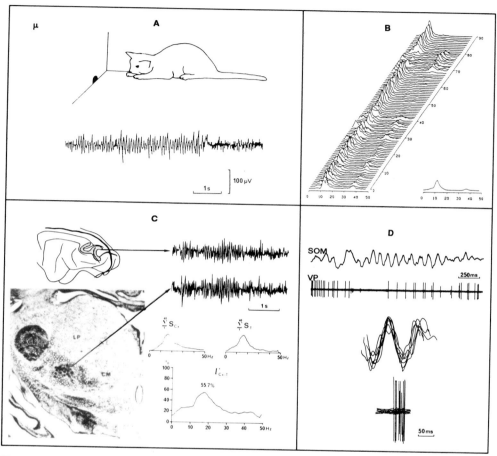

Figure 2. Mu rhythms in cat somatic area SI and thalamus. Same overall presentation as in figure 1. Panel A: cat "expecting" mouse; ECoG from the somatic area SI shows a mu train at 12 Hz. Panel B: power spectral display taken from the SI ECoG recorded in situation A. Peaks at 12 Hz correspond to mu rhythms. Panel C: thalamic location of sites with mu rhythms (filled circles) in the VP nucleus (25 cats). At right, mu rhythms simultaneously recorded from the SI cortex and from the ipsilateral VP. Below: average spectra computed over 40 episodes recorded from these derivations (cortex Cx and thalamus T). Bottom: corresponding coherence values vs frequency between cortex and thalamus, within the 0-50 Hz band. CM: centrum medianum; GL: lateral geniculate body; GM: medial geniculate body; LP: lateralis posterior nucleus; VP: ventralis posterior nucleus. Panel D: thalamic firing compared to cortical activity. Top records: simultaneous recording of a VP cell firing (VP) and of the corresponding mu activity over the SI cortex (SI). Note close correspondence during part at least of the mu train. Bottom: wave-spike correspondence illustrated through superimposition of the minima of 8 successive cortical mu waves and of the corresponding spikes (After Bouyer et al. 1983).

These rhythms pretty much resembled those described by EEGers such as Gastaut (1952), Chatrian et al. (1959), Kuhlmann (1978) and others, as "mu" or "wicket" rhythms. They had also been described previously in cat under the terminology of "sensorimotor rhythms" by Sterman and Wyrwicka (1967). We adopted the human terminology, and decided to call them mu rhythms (Fig. 2 A and B).

Table 1. Type of dominant sensorimotor rhythms, depending on the external situation, "neutral" or presence of a visible target (inducing focused attention) or of a hidden mouse ("expectation"). Three cases are considered: normal animal; animal after lesion of the dopaminergic system; animal after lesion of the noradrenergic system. Notice that in all animals, normal or lesioned, no dissociation was ever observed between behaviour and accompanying rhythms. On the other hand, the behaviour and corresponding rhythms were no more adjusted to the external situation in lesioned animals.

NORMAL ANIMAL

Neutral situation: transient beta then mu then drowsiness then sleep
Focused Attention on visible target (mouse): dominance of beta
Expectation (hidden mouse): dominance of mu

AFTER LESION OF VTA (CONTAINING DOPAMINERGIC A10 CELL GROUP)

Neutral situation: transient mu then drowsiness then sleep
Focused attention: no beta; restless animal
Expectation: dominance of mu

AFTER LESION OF THE NORADRENERGIC SYSTEM

Neutral situation: dominance of mu
Focused Attention: dominance of mu
Expectation: dominance of mu

To sum up this first section, it then turned out that, using the ECoG indices, we were able to separate two distinct features which were both belonging to what is generally considered as attention, focused attention on a target, and expectation of an event. The two situations were characterized by the dominance of distinct rhythms on the sensorimotor cortex, motor and parietal beta rhythms in focused attention, somatic mu rhythms during expectation.

More generally, one may consider three situations, in which to put the animal; one, called neutral, is just placing the cat in his experimental room. Despite the fact that all our animals were tame and perfectly habituated to their recording space, they always showed a first period of exploration, with arrests and concomitantly, the appearance of some beta activity. Then the animal passed a short episode of quiet waking with some mu, and then quickly went to drowsiness then to sleep with slow waves and spindles. A second situation was that of focused attention, when a visible (perspex protected) mouse was put in the recording chamber. In such case, an almost complete dominance of the beta rhythms was noticed. The third situation was that of expectation, with the wall-and-hole paradigm; it was accompanied this time by the presence of a high amount of mu rhythms. Table 1 (top) summarizes these three situations for the normal animal.

The bulk of the above data had been collected in quasi-normal situations that any domestic cat can experience in his daily life. But they were very far from the much more sophisticated conditions under which attentional processes are usually explored in experimental animals, especially monkeys. We therefore elaborated another protocol in which the animal's head was painlessly fixed. He faced a video-screen which was kept blank during a period of 10 s. After these 10 s, a grating would appear, either stable or moving at a certain constant speed. In positive trials, this grating was presented during 5 s, a food delivery occurring at the 4th second. In negative trials, the grating was different (vertical stripes moving horizontally vs horizontal stripes moving vertically) and was maintained during 5 s without any food delivery. This paradigm did not imply any movement of the animal, the food reward (milk) being brought to his mouth through a moving coil operated device (Buser and Buser, to be published).

The data thus obtained have indicated that a clear difference exists between the period of expectation of the signal (that when the screen is blank) or the period of what we considered as attentive observation of the grating until food delivery. As shown on Fig. 3 A, the mu activity was high during the period preceding the signal, and significantly decreased when the positive signal appeared; on the other hand, the beta rhythms behaved in the opposite way, with an increase in the period from onset of the positive signal until food delivery. Fig. 3 A also shows that with negative signals the changes were much less pronounced.

The posterior rhythms

Alpha-like rhythms. More recently, we investigated the ECoG rhythms existing in the posterior parts of the cortex, in the visual area (Chatila et al., 1992). These rhythms were of two types. Firstly, the most easy to record were rhythms at a frequency at about 10 Hz. They often appeared within about the same time epochs as the anterior mu rhythms, although not simultaneously. In particular they occurred when the animal was waiting for the signal in the grating protocol (Fig. 3 B). On the other hand, they disappeared, just as did the mu rhythms, in the period between appearance of the grating and food delivery. In other cases, the two rhythms showed independence, with the anterior mu rhythms being dominant while the posterior 10 Hz were absent, thus showing that the two activities were not linked together. Using several other tests (not detailed herein, cf Chatila et al., 1992), we finally considered that the 10 Hz visual rhythms were equivalent of the human alpha rhythm. They were recorded from parts of areas 17 and 18 and the 17/18 limit, corresponding to the projection of the inferior hemifield.

Visual 40 Hz rhythms. We also recorded a 40 Hz activity from the visual cortex. This also appeared within the same time segments as the anterior beta, at least in the grating training protocol: these rhythms increased just as did the beta in the period following presentation of the grating (Fig. 3 B). These data can be considered preliminary, though, since we have not, so far, indications on the topography of these rhythms over the cortex. Whether these global rhythmic field potentials are equivalent to those recorded through intracortical microelectrodes (Eckhorn et al., 1988; Gray and Singer, 1989) remains to be determined.

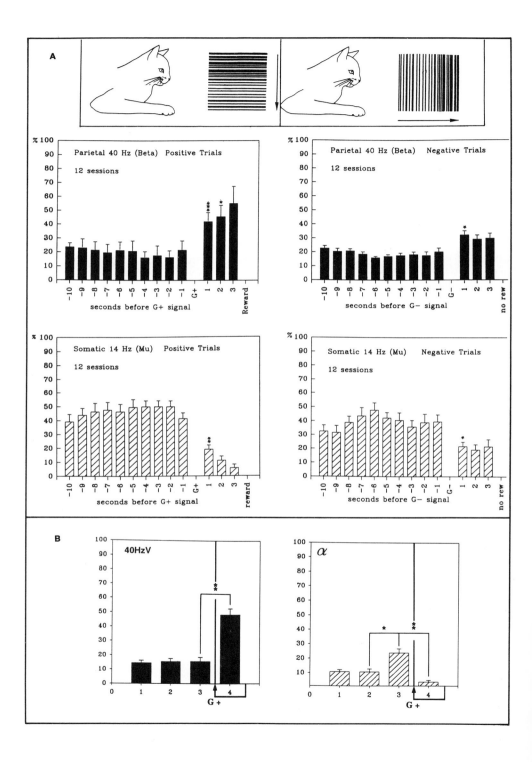

THALAMIC RHYTHMS

A next step in our investigations was reached when we found rhythmic activities in the thalamus, that very closely corresponded to one or other of the rhythms recorded on the cortex. For at least two sets of rhythms, namely the fast beta recorded from the parietal area 5 and the slower mu in somatic area SI, we performed a series of very parallel studies including: i) gross recording (see footnote on page 1) from a thalamic nucleus to circumscribe the active area, together with computation of the coherence function between the activities recorded at both sites, cortex and thalamus (coherence values > 60% being considered as strongly indicative of a close relationship between the two activities, that in the thalamus and that in the cortex). ii) lesioning the identified thalamic structure and observing the disappearance of the corresponding activity on the cortex. iii) micropipette extracellular unitary recording from the thalamus to identify cells that fired at the frequency of the cortical rhythm. This part of the analysis was completed in part at least of our studies by several computations such as spike triggered averaging, wave triggered averaging, comparing autocorrelation function of the cortical rhythms to data of the autocorrelation analysis of the thalamic firing etc (Bouyer et al., 1981; Canu and Rougeul, 1994).

The results have been as follows: i) the parietal beta rhythms seemed to be strongly correlated with an identical rhythmic activity recorded from an area belonging to the posterior group of the thalamus, called POm; this area is well known from anatomical studies to project with long-axoned neurons to the cortical area 5 (Avendano et al., 1985); ii) the somatic mu was instead correlated with rhythms simultaneously occurring in a part of nucleus ventralis posterior thalami, that corresponding to the projection of the forepaw area (Bouyer et al., 1983). Figs 1 (C and D) and 2 (C and D) summarize some of the data thus obtained, for both sets of explorations, those of the parietal beta and those of the somatic mu rhythms. Finally, in a more recent exploration of the lateral geniculate (GL) nucleus of the

---

Figure 3. Mu and beta ECoG rhythms, alpha and 40HzV rhythms undergo opposite changes in a "passive" attention paradigm. Evolution of the amount of rhythms during a visual conditioning (grating appearing on a computer screen in front of the cat, after a period of expectancy of several seconds with blank screen). A. Amount of simultaneously recorded beta (black columns) and mu rhythms (hatched columns) calculated by blocks of 1 s, over 10 s of the waiting period before the appearance of the signal pattern (columns -10 to -1), then in the presence of the signal, either positive (G+) followed by reward after 3 s (left), or negative, not reinforced (G-) (right). Each column is the mean and SEM of the number of epochs when the rhythm was present during the considered second, calculated over the 12 sessions, each with 30 presentations. Significance levels for comparison of a column with the preceding one: *, $p < .05$; **, $p < .01$; ***, $p < .005$. Notice that: beta rhythms were significantly more abundant at time 1 than at -1 and also at 2 than at 1; a significant ($p < .05$) increase occurred after the negative signal only at 1. Simultaneously the amount of mu rhythms was strongly reduced ($p < .01$) at times 1 vs -1 and at 3 vs 2 during the positive signal. After onset of the negative signal, a less significant decrease (at $p < .05$) occurred at time 1 vs -1. Otherwise, changes were not significant. B. Changes in visual rhythms observed in another cat at the end of the waiting period (blocks 1, 2 and 3) and during presentation of the positive stimulus G+ until reward (block 4); black blocks: visual 40 Hz (40HzV); hatched blocks: alpha rhythms. In each block the data from 3 successive s of records [(-9, -8, -7), (-6, -5, -4) and (-3, -2, -1)] were this time pooled together. At left, significant ($p < .01$) increase of 40HzV. At right, significant decrease ($p < .01$) of the alpha rhythms (block 4) vs presignal rate in block 3 ($p < .01$) and also vs presignal rates in blocks 2 and 1 ($p < .05$).

thalamus (Chatila et al., 1993) we could identify sites which were displaying clear rhythms at the "alpha" frequency with a high correlation with the cortex. We did not so far, however, carry out the complementary microelectrode study in this area. The active zone in GL was located in the antero-dorsal part of the nucleus.

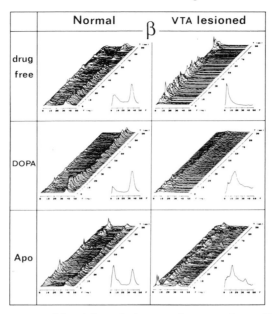

Figure 4. Beta rhythms are modulated through the ventral tegmental area (VTA) dopaminergic system. Power spectral displays established from the area 5a ECoG during 90 mn sessions. Upper row: in drug-free conditions, cat was placed in the presence of a mouse (as in Fig. 1). Note the 35 Hz (beta) peaks. After bilateral VTA lesion, the same cat showed total absence of beta rhythms. Peaks from 0 to 4 Hz are movement artifacts, due to hyperactivity of the cat continuously moving around the mouse box. Middle row: at left, after IP injection of DOPA (30 mg/kg) in a normal animal but *in absence of mouse*. Even so, notice beta peaks throughout the recording time. Right, same cat after VTA lesion; no more beta peak under DOPA. Bottom row left: after IP injection of apomorphine (8 mg/kg); no mouse present. Notice presence of a double peak, at 0-4 Hz and at 35 Hz, corresponding to episodes of stereotyped movements and of immobility with beta rhythms, respectively. Right: after 4 mg/kg of apomorphine in the same cat after VTA lesion; occurrence of some rare beta peaks alternating with mu peaks (14 Hz). No stereotyped movements were observed at this low dose (higher doses were lethal to the lesioned animal) (After Montaron et al. 1982).

Several issues could be raised regarding this set of data. Firstly, we have no irrefutable proof that a thalamic nucleus is acting as the pace-maker for the corresponding cortical rhythms, although several very valuable indications tend to favour this view, in line with the most current opinion expressed in the literature. The microelectrode data may perhaps not be the best argument; on the other hand, the existence of "macrorhythms" in the thalamus, with strong coherence with those on the cortex, and the disappearance of the cortical rhythms after thalamic lesions are we think very solid indications. Moreover, rhythms at the mu frequency were still observed in the cat thalamic VP nucleus after cortical ablation (Andersson and Manson, 1971). Given the presence of the thalamic rhythms, the alternative hypotheses would evidently be that the thalamus is entrained by the cortical activity through the activation of corticothalamic neurons. For a very long time no real proof had been given of the existence of descending actions from the cortex down to the thalamus, despite the abundance of corticothalamic neurons. Recent data have shed a new light on this issue, which may perhaps introduce a new evaluation of the mechanisms of how the thalamo-cortical oscillations are generated (Mc Cormick and Van Krosigk, 1992; Funke and Eysel, 1992). Second, another question was whether nucleus reticularis thalami (RET) was involved in the genesis of mu or beta rhythmicity. We explored the RET while studying the parietal beta and could not find any cell that was homorhythmic to the cortical beta, at variance with opinions expressed elsewhere (Pinault and Deschênes, 1992). But we also considered slow wave sleep, in particular the typical spindles, which are quite distinct from the focal attention rhythms and accompany normal slow wave sleep. As opposed to the lack of correlation with beta rhythms, we found very clear correlations between the cortical spindles and the single unit spiking in the RET (Canu and Rougeul, 1992), indicating that this nucleus participates in the sleep spindles (see e.g. Steriade, 1993).

To conclude this second section, we tend to consider that the focal attention rhythms, at least the parietal beta, the somatic mu and the posterior alpha are due to the activity of distinct thalamo-cortical channels, with a major role given to the thalamus in rhythmogenesis. The question remains open for the 40 Hz visual rhythms for which we have only preliminary evidence that they can be found in the thalamus. It should however be mentioned that stimulus dependent oscillations were recently described in the lateral geniculate body of lightly anaesthetized paralyzed cat (Podgivin et al., 1992). Finally, the case with the anterior beta focus in the motor cortex awaits definite proof that it exists in the thalamus. Our data so far are indicative of the role of a fringe area between n. ventrolateral and mediodorsal. Rather surprisingly though, the thalamic area has so far revealed to be very small.

BRAINSTEM AMINERGIC MODULATION OF ATTENTION AND CORRELATIVE FOCAL RHYTHMS

A further step was reached when we showed that the above thalamo-cortical focal rhythms were undergoing some modulation from structures known to be the origin of aminergic pathways. That attentive states can be modulated by a variety of bioamines has been known for a long time, with two main landmarks in these studies: the "noradrenergic hypothesis", lending importance to the noradrenergic pathway from the locus coeruleus (especially the dorsal bundle, see Mason and Iversen, 1979), also recently assessed by Devauges and Sara (1990) and a "dopaminergic hypothesis" derived from observations that rats with lesions of the VTA (ventral tegmental area) displayed hyperactivity and lack of capacities of attentive behaviour (Simon et al., 1980). With these previous findings and/or hypotheses in mind, we performed various studies to test the importance of the dopaminergic or the noradrenergic systems, using both the animal's behaviour and its concomitant synchronized activities.

Dopaminergic control of the beta rhythms

Let us first consider the beta activities. A variety of converging data have now indicated that the dopaminergic pathway, originating from the A10 cell group in the ventral tegmental area (VTA) exerts a permissive effect on the thalamocortical beta rhythm system. In the neutral situation, the animal with bilateral VTA lesion exhibits some transient mu, no beta and soon goes to sleep. In the focused attention paradigm, the cat is restless, turning around the mouse with no more beta (Fig. 4): lesioning the VTA thus elicits in these conditions a "VTA syndrom" similar to that previously described in the rat. In the expectation paradigm, mu rhythms are completely dominating (Montaron et al., 1982).

Among other proofs, single unit recording from the VTA revealed cells that increased their firing rate 1 s before the appearance of each beta train and slowed down at about the end of the train, this strongly suggesting a permissive effect of the VTA on the thalamo-cortical beta system (Montaron et al., 1984). Other indications came from pharmacological investigations: administration of DOPA, the precursor of dopamine to the normal cat was shown to enhance the amount of beta up to a sustained high level, with a corresponding change in the cat's attitude, which pretty much recalled that of focused attention, even in the absence of any target. After lesioning the VTA, this effect of DOPA was completely abolished. Another characteristic effect was that of apomorphine. In the normal animal, this agonist of postsynaptic D1 receptors also elicits a persistence of beta rhythms, only interrupted by episodes of stereotyped movements, as classically described in the rat. After VTA lesion, the apomorphine remained of a moderate efficacy, indicating that some postsynaptic receptors had been left and could still act on the thalamocortical system (Fig. 4).

It is interesting to notice that, while in the normal animal, the rhythms and behaviour were absolutely concomitant, with fixation accompanying beta rhythms, but both being contingent on the presence of a target, drug administration changed completely the picture with respect to the external cause of induction of the rhythms (presence of a target): the behaviour and rhythms now became completely non-adjusted to the external situation, while the concomitance between rhythms and behaviour was perfectly preserved (this is also summarized in Table 1).

Noradrenergic control of the rhythms in the 10-18 Hz band

Now turning to the noradrenergic system, we naturally considered the role of the locus coeruleus. We did not record from that structure (as we had from the VTA), but we lesioned it and could as a result show a considerable increase in mu activity, in all three situations, neutral, focused attention and expectation (Delagrange et al., 1993). The same remark as above holds true in this case: the animal lost adjustment to the external situation, while the behaviour and accompanying rhythms remained coherent and concomitant with each other (Fig. 5 B). Another lesion was achieved using the neurotoxic drug DSP4, which destroys noradrenergic pathways originating from the locus coeruleus (Delagrange et al., 1989). Pretty much as after coeruleus lesions, the animal displayed an extreme abundance of mu rhythms together with a persistent attitude of expectation, whatever the surrounding situation (Fig. 5 A and Table 1). To sum then, mu rhythms seem to be negatively correlated to the activity of the noradrenergic system: a high noradrenergic activity depresses the expectation system. This observation is not in contradiction with data indicating an increase of focused attention under high noradrenergic activation. We have had various data (not detailed herein) showing that the two systems, beta and mu can in some cases be antagonistic. It is therefore possible that a strong depression of the mu system favours the system of focused attention.

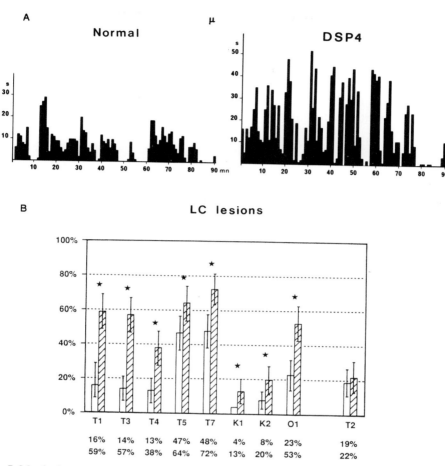

Figure 5. Mu rhythms are modulated by a noradrenergic system originating from locus coeruleus LC. A. Effect of a unique injection of neurotoxic DSP4. Before (normal) and 5 days after injection (DSP4), the cat was placed during 90 mn in the situation depicted on Fig. 2 (waiting near a mouse hole). The two graphs present the total time (in s) occupied by ECoG mu rhythms on the SI area, for each minute of the 90 mn session. Note considerable increase after DSP4. This change accompanied an exaggeration of the time spent by the animal in a state of immobile expectation (significant difference at p < .005, Mann-Whitney U tests). B. Effect of lesioning LC on % of cortical mu rhythms during the 90 mn recording time. In these experiments, the animals were placed in a "neutral" situation (mouse neither present nor hidden). Open bars designate preoperative values and hatched bars, postoperative ones. All animals with LC lesions displayed a significant increase in mu rhythm (at p < .05), together with a prolonged state of expectation. Cat T2 at extreme right was a control animal. (After Tadjer et al. 1988; Delagrange et al. 1989, 1993).

We more recently tried to determine whether the posterior visual rhythms were also under some DAergic or NAergic control. In this line our results are only preliminary and only concern the alpha band. We tested one drug, namely prazosin, known as a specific noradrenergic postsynaptic alpha1 blocker (Rogawski and Aghajanian, 1982). We were able to notice, on this occasion, a simultaneous increase of both the mu and the alpha rhythms, indicating that in cat, the posterior alpha may probably also be under NAergic control, in the same line as the somatic mus.

That both NAergic effects, on the anterior mu, or the posterior alpha, could be due to a direct modulation at the thalamic level, was suggested by the existence of NAergic endings, as demonstrated in the VP nucleus through immunocytochemical labeling (Delagrange et al., 1990) and, for the GL, by the existence of alpha1 NAergic receptors (Rogawski and Aghajanian, 1982). This does not hold true for the above described dopaminergic control of the beta system, which almost certainly passes through intermediate relays.

MORE ON RHYTHMS AND SOME THEORETICAL ISSUES RAISED

The above presented data raised several questions that deserve some comments.

Slow rhythms and their possible significance

Firstly, let us consider more broadly the frequency spectrum of the field potentials that could be recorded from the cat cortex by gross electrodes. We described above two sets of rhythms, those in the 10-18 Hz band, that we called alpha or mu, depending on their location and reactivity, and others, in the 35-45 Hz band, consisting of beta in the anterior cortices and "40HzV" in the visual area(s). We tried to show that the two sets were considerably reinforced in two distinct states of attention paid by the alert animal to its surrounding, expectation and focused attention respectively. Further data exist though, indicating that other types of rhythms exist on the cat cortex, which belong to a lower frequency band, that of 5 to 8 Hz and are distinct in their topography and conditions of occurrence from both the above discussed faster rhythms and the slow delta pattern and spindles characterizing "slow wave sleep". We haven't so far identified all conditions in which these 4-8 Hz rhythms occur. Their most evident correlate is that of a state of drowsiness (Rougeul et al., 1974), occurring during the transition between quiet waking and slow wave sleep. Many years ago, we trained animals to a standard operant conditioning of pressing a lever for food delivery to a positive (light) signal, also with a differentiation i.e. no-response to a negative stimulus (buzzer). The interesting observation was that at delivery of the negative non-reinforced signal, long trains of 4-8 Hz rhythms appeared on the somatic, visual and auditory cortex, at a latency that was roughly of the same order of magnitude as the response time of the animal to the positive stimulus. We interpreted this activity as underlying a state of "internal inhibition" as described by Pavlov. Actually the cat's attitude was markedly similar to that of the dogs, considered by Pavlov under internal inhibition after repetition of the negative stimulus. These two situations are probably not the only ones inducing slow rhythmic patterns. Speculatively, it is not excluded that both belong to a more general behavioural repertoire, which may be termed "withdrawal of attention to the outside world", comprising drowsiness, internal inhibition, and possibly also situations of "despair" (animal attached and ceasing to struggle) etc. These slow neocortical rhythms are quite distinct from the well documented hippocampal "theta activities" which appear in a variety of situations (locomotion, exploration, REM sleep), none of them corresponding to the above described ones.

Functional ambivalence of neuronal "synchronization"

Now regarding the various frequencies and their behavioural correlates, it turns out that if we would consider the "cognitive performance with respect to the surrounding" vs the rhythm frequency, a kind of surprising proportionality would show up, with high frequency bands (beta) corresponding to focused attention and low frequencies (drowsiness) to a low level of awareness. Going even lower in the performance scale, one encounters the stage of loss of consciousness in slow wave sleep. This time the electrophysiological correlate is somewhat more complex since sleep displays both slow waves (delta band, 2 to 3 Hz) and spindles (at about 14 Hz).

From this kind of observation, one is inclined to conclude that cortical and thalamic synchronizations of the neuronal activities have several distinct functional correlates. This of course, if and only if we assume (as a sensible hypothesis) that a causality link exists in all cases between the state of rhythmicity of the neuronal network and this network's functional performance, an alternative hypothesis being that rhythms are epiphenomena with no functional bearing on cortical integration processes. A more heuristic and fruitful assumption (the one that we tend to favour here even if we do not so far have all desirable proofs) is that neuron ensembles that undergo rhythmicity work differently than neurons not involved in such forced oscillations : oscillations and synchronization may influence the way in which these neurons handle information at the thalamic and/or the cortical level, thus strongly influencing the resulting perceptual processing. Two opposite views have been expressed: the rhythmic state may either facilitate such processing or may on the contrary impede it. The first assumption is definitely the one nicely developed by the studies on the visual 40 Hz rhythms, showing how a narrow correlation between several neuron ensembles, achieved by the rhythmic oscillations contributes to "binding" between separate parallel channels (Freeman and Van Dijk, 1987; Eckhorn et al., 1988; Sporns et al., 1989; Gray and Singer, 1989; Eckhorn, Engel in this Symposium). The functional importance of rhythms in this frequency band, as an essential component of auditory perceptual processing was also stressed by the observation of their presence during the auditory evoked potentials (see e.g. Pantev et al., 1991). The second assumption is indeed what probably leads to decrease of awareness and loss of consciousness, which occurs during slow wave sleep. But other data are also available (Delagrange et al., 1987), showing that under mu rhythmicity, processing of sensory information may be modified in nucleus ventralis posterior in the sense that the signal-to-noise ratio is decreased during mu episodes, as opposed to episodes when the cortical activity is "desynchronized", the dependent variable being the VP response to an air-puff. It is interesting to notice that in some recent microelectrode explorations of the monkey VP nucleus some state-related modulations of the thalamic somatosensory responses were observed by one group (Morrow and Casey, 1992), but not by another one (Tremblay et al., 1993); no particular emphasis was placed though, in these studies, on the recording of specific rhythmic field activities (except in the first one, just as a means to monitor the animal's general state of arousal).

In their now classical studies on guinea-pig thalamic slices, Jahnsen and Llinas (1984) suggested that all thalamic neurons can undergo two distinct states, either the "oscillatory mode", or the "transfer mode" (see also Steriade and Llinas, 1988). It is only in the latter one that the cell can transfer information to the cortex, which means that in this view, oscillations are impeding integration rather than facilitating it. Clearly, given these various and sometimes contradictory inferences, future studies on the information processing in thalamo-cortical channels should contribute to a more thorough and better insight into one of the major issues raised by higher integrative activities.

How is the oscillating thalamic network organized ?

Going back for a while to our own thalamic explorations, one observation has been that, in the VP as well as in the POm nucleus, the number of oscillating cells was rather low; those were intermingled with others, non rhythmic ones. In case of the VP, part of the oscillatory cells were characterized as long-axoned thalamocortical (TC) neurons; very surprisingly, these TC cells could not be activated by any of the natural stimulations applied during the recording, tactile (such as air-puff) or pressure. On the other hand, these rhythmic TC cells were surrounded by classical TC relay cells (TCR), which were easily driven by peripheral stimuli. The POm was not as easy to explore in this line, since cells were generally not driven by any peripheral stimulus, and because we failed to clearly identify the thalamocortical nature of the oscillating cells as long-axoned ones through the usual collision test. On the other hand, the proportion of oscillating cells was of the same order of magnitude as that in the VP, i.e. about 7% of the encountered isolated neurons (our explorations in both nuclei were carried out with glass micropipettes and were therefore strictly of the single unit type). Provisionally then, one may hypothesize that the thalamic oscillating system consists in each particular nucleus (VP, POm and maybe others) of a network made of a limited number of cells interconnected either directly or through interneurons (?) and scattered throughout the nucleus. At least in the VP, these cells would be distinct from the standard sensory relay cells.

High frequency anterior rhythms and movement

Among other problems that may be raised, one is that of the significance of the anterior fast, motor and parietal beta rhythms. The question to ask is whether these rhythms are in some way correlated with the preparation of a movement (the close relationship between area 5 and the motor areas is a well documented finding, see Avendaño et al., 1988). From our data with the training of animals to grating presentation, we take that beta rhythms can occur without any overt movement following them: in our paradigm, the only movement that the cat had to perform was laping milk. In other situations, beta rhythms were observed during prolonged immobility of the cat, with no identifiable preparation to move. In still other conditions though, rhythms were described in the monkey, very closely related to performance of an exploratory movement (Murthy and Fetz, 1992 and this symposium). In man, Pfurtscheller & Neuper recently (1992) reported recording 40 Hz from the fronto-parietal human scalp before performing a movement (see also Giannitrapani, 1967; De France and Sheer, 1988). The question in these latter cases is thus posed concerning the precise timing of the phasic movement vs rhythmic activity: does the latter precede immediately or accompany the movement ? The problem is a general and almost theoretical one, in so far as in the first case, rhythms mean contribution to preparation to move, while in the second case, they accompany some episodes of execution of the movement by the cortical network. The issue is, we think, of importance and should also be explored very carefully in future studies.

Aknowledgments

Collaborators who participated in one or other part of the herein reported experiments from our laboratory are greatly acknowledged, namely J.J. Bouyer, M.H. Canu, M. Chatila, P. Delagrange and M.F. Montaron (among others). Computations of spike and wave triggered averagings were recently achieved with the kind cooperation of Drs Y. Fregnac and V. Bringuier. The very efficient technical help of C. Durand throughout the years is also warmly acknowledged.

We wish to thank Dr T. Bullock for his helpful comments on an earlier version of this paper.

This study was supported by DRED, CNRS, DRET (Contract N⁻ 92/69) and by the Fondation pour la Recherche Médicale.

REFERENCES

Andersson S.A. and Manson J.R., 1971, Rhythmic activity in the thalamus of the unanaesthetized decorticate cat, *Electroenceph. Clin. Neurophysiol.* 31:21-34.

Avendaño C., Rausell E., Perez-Aguilar D. and Isorna S., 1988, Organization of the association cortical afferent connections of area 5: a retrograde tracer study in the cat, *J. Comp. Neurol.* 278:1-33.

Avendaño C., Rausell E. and Reinoso-Suarez F., 1985, Thalamic projections to areas 5a and 5b of the parietal cortex in the cat: a retrograde horseradish peroxidase study, *J. Neurophysiol.* 5:1446-1470.

Bouyer J.J., Dedet L., Konya A. and Rougeul A., 1974, Convergence de trois systemes rythmiques thalamocorticaux sur l'aire somesthésique du chat et du babouin, *Rev. EEG Neurophysiol.* 4:397-406.

Bouyer J.J., Montaron M.F. and Rougeul A., 1981, Fast frontoparietal rhythms during combined focused attentive behaviour and immobility in cat: cortical and thalamic localizations, *Electroenceph. Clin. Neurophysiol.* 51:244-252.

Bouyer J.J., Montaron M.F., Vahnee J.M., Albert M.P. and Rougeul A., 1987, Anatomical localization of the cortical beta rhythms in cat, *Neuroscience* 22:863-869.

Bouyer J.J., Tilquin C. and Rougeul A., 1983, Thalamic rhythms in cat during quiet wakefulness and immobility, *Electroenceph. Clin. Neurophysiol.* 55:180-187.

Canu M.H. and Rougeul A., 1992, Nucleus reticularis thalami participates in sleep spindles, not in beta rhythms concomitant with attention in cat. *C.R. Acad. Sci. (Paris)* 315:513-520.

Canu M.H. and Rougeul A., 1994, Relationship between posterior thalamic nucleus (POm) unit activity and parietal cortical rhythms (beta) in the waking cat, *Neuroscience*, in press.

Chatila M., Milleret C., Buser P. and Rougeul A., 1992, A 10 Hz "alpha-like" rhythm in the visual cortex of the waking cat, *Electroenceph. Clin. Neurophysiol.* 83:217-222.

Chatila M., Milleret C., Rougeul A. and Buser P., 1993, Alpha rhythm in the cat thalamus, *C.R. Acad. Sci. (Paris)* 316:51-58.

Chatrian G., Petersen M. and Lazarte J., 1959, The blocking of the rolandic wicket rhythm and some central changes related to movement, *Electroenceph. clin Neurophysiol.* 11:497-510.

De France J. and Sheer D., 1988, Focused arousal, 40 Hz EEG and motor programming. *In* The EEG of Mental Activities, Giannitrapani and Muri (Eds), Karger, Basel, 153-168.

Delagrange P., Canu M.H.., Rougeul A., Buser P. and Bouyer J.J., 1993, Effects of locus coeruleus lesions on vigilance and attentive behaviour in cat, *Behav. Brain Res.* 53:155-165.

Delagrange P., Conrath M., Geffard M., Tadjer D., Bouyer J.J. and Rougeul A., 1990, Noradrenaline-like terminals in the cat nucleus ventralis posterior of the thalamus, *Brain Res. Bull.* 26:533-537.

Delagrange P., Tadjer D., Bouyer J.J., Rougeul A. and Conrath M., 1989, Effect of DSP4, a neurotoxic agent, on attentive behaviour and related electrocortical activity in cat, *Behav. Brain Res.* 33:33-43.

Delagrange P., Tadjer D., Rougeul A. and Buser P., 1987, Activité unitaire de neurones du noyau ventral postérieur du thalamus pour divers degrés de vigilance chez le chat normal, *C.R. Acad. Sci. (Paris).* 305:149-155.

Devauges V. and Sara S., 1990, Activation of the noradrenergic system facilitates an attentional shift in the rat, *Behav Brain Res.* 39:19-28.

Eckhorn R., Bauer R., Jordan W., Brosch M., Kruse W., Munk M. and Reitboeck H.J., 1988, Coherent oscillations: a mechanism of feature linking in the visual cortex, *Biol. Cybern.* 60:121-180.

Freeman W. and Skarda C., 1985, Spatial EEG patterns, non-linear dynamics and perception: the neo-Sherringtonian view, *Brain Res. Rev.* 10:147-175.

Freeman W. and Van Dijk B., 1987, Spatial patterns of visual cortical fast EEG during conditioned reflex in a rhesus monkey, *Brain Res.* 422:267-276.

Funke K. and Eyse U.T., 1992, EEG-dependent modulation of response dynamics of cat dLGN relay cells and the contribution of corticogeniculate feedback, *Brain Res.* 573:217-227.

Gastaut H., 1952, Etude électrocorticographique de la réactivité des rythmes rolandiques, *Rev. Neurol.* 87:176-182.

Giannitrapani D., 1969, EEG average frequency and intelligence, *Electroenceph. clin. Neurophysiol.* 27:480-486.

Gray C.M. and Singer W., 1989, Stimulus specific neuronal oscillations in orientation columns of cat visual cortex, *Proc. Natl Acad. Sci. USA* 86:1698-1702.

Jahnsen H. and Llinas R., 1984, Ionic basis for electroresponsiveness and oscillatory properties of guinea pig thalamic neurones *in vitro, J. Physiol.* 349:227-247.

Jasper H.H. and Penfield W., 1949, Electrocorticogram in man: effect of voluntary movement upon the electrical activity of precentral gyrus, *Arch. Psychiat. Z. Neurol.* 183:163-174.

Kuhlmann W., 1978, Functional topography of the human mu rhythm, *Electroenceph. clin. Neurophysiol.* 44:88-93.

Mason S.T. and Iversen S.D., 1979, Theories of the dorsal bundle extinction effect, *Brain Res. Rev.* 1:107-137.

Mc Cormick D.A. and Krosigk M., 1992, Corticothalamic activation modulates thalamic firing through glutamate metabotropic receptors, *Proc. Natl Acad. Sci. USA* 89:2774-2778.

Montaron M.F., Bouyer J.J. and Rougeul A., 1979, Relations entre l'attention et le rythme mu chez le chat et le singe, *Rev. EEG Neurophysiol.* 33:333-339.

Montaron M.F., Bouyer J.J., Rougeul A. and Buser P., 1982, Ventral mesencephalic tegmentum (VMT) controls electrocortical beta rhythms and associated attentive behaviour in the cat, *Behav. Brain Res.* 6:129-145.

Montaron M.F., Bouyer J.J., Rougeul, A. and Buser P., 1984, Activité unitaire dans l'aire tegmentale ventrale et état d'attention focalisée chez le chat normal éveillé *C.R. Acad. Sci.(Paris)* 298: 229-236.

Morrow T.J. and Casey K.L., 1992, State-related modulation of thalamic somatosensory responses in the awake monkey, *J. Neurophysiol.* 67:305-317.

Murthy V.N. and Fetz E.E., 1992, Coherent 25 to 35 Hz oscillations in the sensorimotor cortex of awake behaving monkeys, *Proc. Natl Acad. Sci. USA* 89:5670-5674.

Pantev C., Makeig S., Hoke M., Galambos R., Hampson S. and Gallen C.,1991, Human auditory evoked gamma-band magnetic fields, *Proc. Natl Acad. Sci. USA* 88:8996-9000.

Pfurtscheller G. and Neuper Ch., 1992, Simultaneous EEG 10-Hz desynchronisation and 40-Hz synchronization during finger movements, *NeuroReports* 3:1057-1060.

Pinault D. and Deschénes M., 1992, Voltage-dependent 40 Hz oscillations in the rat reticular thalamic neurons *in vivo, Neuroscience,* 51:245-258.

Podvigin N.F., Jokeit H., Pöppel E., Chizh A.N. and Kiselyeva N.B., 1992, Stimulus-dependent oscillatory activity in the lateral geniculate body of the cat, *Naturwissenschaften,* 79:428-431.

292

Rogawski M.A. and Aghajanian G.K., 1982, Activation of lateral geniculate neurons by locus coeruleus or dorsal noradrenergic bundle stimulation: selective blockade by the alpha1 adrenoceptor antagonist prazosin, *Brain Res.* 250:31-39.

Rougeul A., Corvisier J. and Letalle A., 1974, Rythmes électrocorticaux caractéristiques de l'installation du sommeil naturel chez le chat. Leurs rapports avec le comportement moteur, *Electroenceph. Clin. Neurophysiol.* 37:41-57.

Rougeul A., Letalle A. and Corvisier J., 1972, Activité rythmique du cortex somesthésique primaire en relation avec l'immobilité chez le chat libre éveillé, *Electroenceph. clin. Neurophysiol.* 33: 23 39.

Rougeul-Buser A., Bouyer J.J. and Buser P., 1975, From attentiveness to sleep. A topographical analysis of localized "synchronized" activities on the cortex of normal cat and monkey, *Acta Neurobiol. Exp.* 35:805-819.

Rougeul-Buser A., Bouyer J.J. and Buser P., 1978, Transitional states of awareness and specific attention; neurophysiological correlates and hypotheses, *In* Int. Symp. on Cerebral Correlates of Conscious Experience, A. Rougeul-Buser and P. Buser (Eds) Elsevier, Amsterdam.

Simon H., Scatton B. and Le Moal M., 1980, Dopaminergic A10 neurons are involved in cognitive functions, *Nature*, 286:150-151.

Sporns O., Gally J.A., Reeke G.N. and Edelman G.M., 1989, Reentrant signaling among simulated neuronal groups leads to coherency in their oscillatory activity, *Proc. Natl Acad. Sci. USA* 86:7265-7269.

Steriade M., 1993, Central core modulation of spontaneous oscillations and sensory transmission in thalamocortical systems, *Current Opinion in Neurobiology* 3:619-625.

Steriade M. and Llinas R.R., 1988, The functional states of the thalamus and the associated neuronal interplay, *Physiol. Rev.* 68:649-742.

Sterman M. and Wyrwicka W., 1967, EEG correlates of sleep: evidence for separate forebrain substrates, *Brain Res.* 6:143-163.

Tremblay N., Bushnell M.C. and Duncan G.H., 1993, Thalamic VPM nucleus in the behaving monkey. II. Response to air-puff stimulation during discrimination and attention tasks, *J. Neurophysiol.* 69:753-763.

MODELS FOR THE NEURONAL IMPLEMENTATION OF SELECTIVE VISUAL ATTENTION BASED ON THE TEMPORAL STRUCTURE OF NEURAL SIGNALS

Ernst Niebur and Christof Koch

Computation and Neural Systems Program, California Institute of Technology, Pasadena CA 91125, USA

INTRODUCTION

Selective attention has been studied quantitatively at least since the time of Helmholtz. With the availability of the necessary technology, these early psychophysical studies were later complemented by the observation of the electrical activity which accompanies the functions of the brain. While EEG recordings make visible only the average activity of very large neural populations, more and more selective methods became available over time. Modern micro-electrode studies allow the observation of attentional effects at the level of single neurons. Selective attention can thus be observed quantitatively over the whole range of neural populations, from a single neuron (by micro-electrodes) over several scales of larger populations (Electrocorticogram and EEG) to the performance of the whole system (psychophysics). We believe that this makes selective attention a prime candidate for the elucidation of the higher mental functions and their neurobiological basis.

Since the term attention is often used by different authors with slightly different meanings, we start by defining the term for the purposes of the present paper. By selective attention we understand a process whereby a particular piece of information is selected from a sensory array for further processing. Visual selective attention usually manifests itself within a single, spatially circumscribed part of the visual field that can vary in size and that scans objects in the visual field at a rate of about 30-50ms per object. As Posner first explicitly showed, movements of the "focus" of attention can be divorced from visible eye movement (Julesz, 1991; Kanwisher and Driver, 1992; Posner and Driver, 1992; Posner and Petersen, 1990; Treisman, 1988).

How attention acts at the level of single neurons in extrastriate cortex has been elucidated by Desimone and colleagues (Moran and Desimone, 1985; Desimone et al., 1991). When two different objects, say a red and a green bar, are both located

Oscillatory Event-Related Brain Dynamics, Edited
by C. Pantev, Plenum Press, New York, 1994

within the receptive field of a V4 neuron selective for red, the neuron will respond vigorously if the monkey attends to the red stimulus, but respond much less if the monkey is attending to the green stimulus. The stimulus is identical in both cases (a red and a green bar); the difference is only in the internal state of the monkey. The Moran and Desimone (Moran and Desimone, 1985) effect is mainly suppressive, since the response of the cell to the attended stimulus does not increase significantly over its response if the monkey is attending to a stimulus outside the receptive field.

In the present work, we presume the existence of a "saliency map" à la Koch and Ullman(1985) (see also Treisman, 1988) which encodes information on where salient (conspicuous) objects are located in the visual field, but not on what these objects are. Saliency is here meant to be understood in terms of simple operations, implemented by center-surround type of operations, i.e. a green object among many red ones or a moving stimulus in an otherwise stationary scene would constitute very salient objects. After a short time, the location of the presently most salient object becomes inhibited in the saliency map, and attention switches to the next most conspicuous location. In the present work, we assume that salient objects have been selected in the visual field by such a mechanism and that they are "tagged" by modulating the temporal structure of the neuronal signals corresponding to attended stimuli.

We are concerned here with the mechanisms by which the result of the selection process is communicated to the information-processing occipito-temporal pathway. It will be our assumption that subcortical structures modulate the neural signals generated in the various visual cortices (for instance by subthreshold synaptic input to cortical cells). In order to simplify matters somewhat, we here assume that this modulation only occurs at the level of primary visual cortex V1. The main candidate for such subcortical structures are the efferent connections from the different visual maps present in the pulvinar nuclei of the thalamus (inferior, lateral and medial) into striate and extrastriate cortex (for the neuroanatomy of the pulvinar and its relationship to saliency see Garey, Dreher and Robinson, 1991; Robinson and Petersen, 1992). On the basis of inactivation and PET studies, parts of the pulvinar are known to be involved in the control of attention (LaBerge and Buchsbaum, 1990; Petersen et al., 1987; Desimone et al., 1989; Kubota et al., 1988).

How can such a signal be used to lead to the changes in V4 cell activity observed by Desimone and colleagues? Following Crick and Koch(1990) and Desimone (1992), we assume the existence of competition between cells in area V4 which is biased in favor of cells representing attended stimuli. Furthermore, we assume that this is implemented using different temporal structures (but identical average spike rates) of the spike trains generated by V1 neurons inside and outside the focus of attention. At the level of V4 and beyond, signals along the tagged pathway compete with signals in the untagged pathway, leading to an inhibition or reduction in the response of neurons in the untagged pathway.

In principle, the modulation imposed on the spike trains corresponding to attended stimuli could be of any form, provided that (i) this modulation can be detected, i.e., decoded, in higher cortical areas, and (ii) it does not change the mean spike rate in V1 (Moran and Desimone, 1985). In this paper, we discusss two particular modulation schemes and we show that they can be implemented using standard "neural hardware," without the need to postulate any novel biophysical mechanisms.

Both models have identical anatomical substrates; they differ in the temporal fine structure of the modulation at the single-cell level. In the first case, we assume oscillatory modulation of the spike trains with a frequency in the γ-range (referred to as "40Hz" oscillations). This oscillatory structure should be visible in the autocorrelation function of the spike trains. In contrast, for the second model we assume random time structure in any given spike train. The attentional modulation is manifested only in the cross-correlation between spike trains, which is at chance level outside the focus of attention but has a detectable structure inside. We will refer to the first model as the "oscillatory model" and to the second as the "synchronicity model."

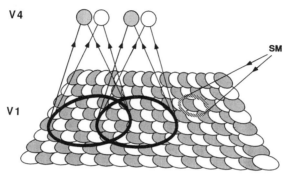

Figure 1A. Architecture of the model. Receptive fields of V1 cells are represented by overlapping circles arranged in a two-dimensional array (actual receptive fields in the model are square). White and gray circles represent cells receptive to the two features considered in the text). Actual overlap is larger than shown in the figure: every point on the retina is represented in the receptive fields of 4 cells of each feature type (i.e., by 4 white and 4 gray cells). Unfilled black circles denote the receptive fields of V4 cells, i.e., all cells in a stack in V4 receive input from all V1 cells in the corresponding circle (arrows) with the same feature selectivity (see Fig. 1B for details of the connectivity). The shaded unfilled circle indicates the focus of attention; the activity of all V1 cells inside this circle is subject to temporal modulation by the saliency map (SM).

MODEL ANATOMY

The gross architecture of our models is shown in Figure 1A. Input from the two-dimensional retina is fed via the lateral geniculate nucleus (LGN; not shown) into area V1, where the attentional modulation-originating in the saliency map at a subcortical site-is added. We assume that cells in V1 are only selective to one of two different features, here the colors red or green (our results do not depend on the simplification of using only two features). The output of our model V1 projects into neuronal "stacks" in our model V4 area (see Fig. 1B), where it excites pyramidal cells as well as inhibitory interneurons.

For the sake of simplicity of language, we will sometimes refer to neurons whose receptive fields are inside the focus of attention as "attended neurons" and those outside as "unattended."

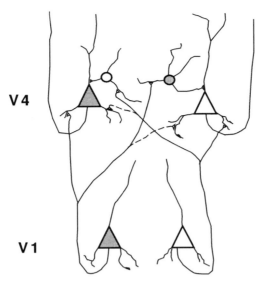

Figure1B. Schematic connectivity of the model. The two types (white and gray) of V1 pyramidal cells (triangles) project to a stack of cells in V4. Strong excitatory synapses are made to pyramidal cells (triangles) and smooth interneurons (circles) with the same preferred color (full lines), weaker connections are made to pyramidal cells of the other color (dashed lines). The smooth interneurons project with inhibitory synapses on pyramidal cells of the other color.

Attentional Modulation

The visual input into each V1 cell is provided by a 10 by 10 array of "pixels." The output of any V1 cell are binary action potentials, generated using a Poisson process of mean firing rate λ in combination with a refractory period, which is chosen randomly from a uniform distribution with values between 2ms and 5ms. Bair, Koch, Newsome and Britten (Bair et al., 1993) analyzing 216 cells recorded from extrastriate area MT in the awake and behaving monkey, showed that the stochastic properties of non-bursting cells-firing at high discharge frequencies-can be described on the basis of a Poisson process in combination with a short refractory period (see also Softky and Koch, 1993).

The total firing rate λ of any neuron is the sum of the spontaneous firing rate λ_{spont} and the stimulus-dependent rate λ_0. If no stimulus or a non-preferred stimulus is present in the receptive field of the cell under study, we have $\lambda_0=0$. If a preferred stimulus is present, λ_0 is chosen to be proportional to the degree of spatial overlap of the receptive field of the cell with the stimulus. The overlap varies between zero (no overlap) and unity (complete overlap). More complex models of receptive fields can easily be incorporated.

The action of attention is to modulate this discharge *without* affecting its mean rate λ, since this is determined by the stimulus conditions. We achieve this by using an inhomogeneous Poisson process, whose instantaneous firing rate $\lambda(t)$ varies around λ, such that the temporal mean of $\lambda(t)$ is λ. In the oscillatory model, the instantaneous firing rate of cells in the focus of attention is given by

$$\lambda(t)=\lambda_0(1+A \sin \omega t) +\lambda_{spont} \tag{1}$$

where A (with $0 \leq A \leq 1$) is proportional to the overlap between the receptive field and the focus of attention, and $\omega = 2\pi \times 40 \ s^{-1}$.

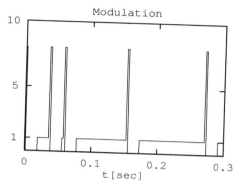

Figure 2. Synchronicity model: Stochastic modulation of neuronal activity in attended cells. Shown is the modulation imposed on cells with receptive field inside the focus of attention. At random times (subject to a Poisson distribution), P(t) deviates from its average value (P=1) during events which consist of a 2.5ms long elevation of the firing rate (in the case shown, to a level of P=8), followed by a subsequent depression (to P=0) of a duration of 17.5 ms. The mean activity, averaged over any period of time which includes only entire events (i.e. no incomplete events) is unity. i.e. the same as for unattended cells.

In the synchronicity model, we assume that the rate $\lambda(t)$ of the Poisson process is determined itself by another stochastic process termed the synchronicity process. The events of this process do not correspond to action potentials of V1 neurons but to a sequence of synchronized elevations and depressions of the spiking rates of **all** attended neurons. More concretely, we assume that the mean rate of an attended neuron is given by

$$\lambda(t)=\lambda_0 P(t)+\lambda_{spont} \tag{2}$$

where P(t)=1, except during a modulation event. If a synchronicity event occurs at $t=t_0$, P(t) instantaneously increases for 2.5 ms to 8, and then drops for 17.5 ms to 0. This assures that the mean rate is always constant, independent of the level of attentional modulation (see Fig. 2). This approximately eightfold increase of the instantaneous spike rate occurs for all the neurons in the "spotlight" of attention, assuring that their firing is highly correlated during the 2.5 ms period of enhanced firing.

In both models, selective attention activates competition within a population of V4 neurons with overlapping receptive fields. In the presence of multiple stimuli, the neurons responding to the different stimuli will compete against each other and attentional modulation will bias this competition in favor of attended stimuli. In our model, there are two classes of neurons in every stack: inhibitory interneurons which detect the temporal structure in their input, and excitatory pyramidal neurons, which provide the output to higher cortical areas. Members of both classes receive overlapping input from 100 V1 cells which makes the receptive fields of V4 cells much larger than those of V1 cells. We model the pyramidal neurons as integrate-and-fire neurons which generate an action potential whenever the membrane potential reaches the cell's spiking threshold (and the cell is not refractory). The neuron receives synaptic input in the form of conductivity changes whose reversal potentials correspond to Na^+ for excitatory synapses and Cl^- for inhibitory synapses, respectively. Synapses also have different time courses: Since we assume the excitatory synapses to be glutamate synapses of the AMPA (amino-3-hydroxy-5-methyl-4-isoxazolepropionic acid) type, they were modelled by an exponential with rapid decay time (1.5 ms). The activation time constants for inhibitory $GABA_A$ (γ-Aminobutyric acid A) synapses are much longer and these synapses were modelled as decaying exponentials with time constants of 60 ms. Neurons in V4 also receive input from other neurons in V4 as well as other cortical areas. This background activity is represented by two stochastic (Poisson) spike trains, one conveying excitatory input and the other inhibitory input. The rates of these processes were determined by the requirement of obtaining a spontaneous firing rate of a few spikes per second in the absence of stimulation. All neurons have a refractory period, chosen randomly with uniform distribution from the interval [2ms, 5ms].

In the following two sections, we discuss the dynamics of the inhibitory interneurons (stellate cells) for the two models we studied in detail.

Detection of Temporal Structure in V4: Synchronicity Model

In this model, the V4 interneurons need to detect coincident spikes in their input. This is implemented by providing them with a dynamic threshold which is essentially a gliding average with a short time constant (10 ms). Therefore, the response of the neurons will be largely independent of the average input rate (which provides information about the stimulus properties) and instead will be dominated by the temporal fine structure at a time scale of about 10 ms, which reflects the attentional state of the animal. Synaptic input from V1 and from other areas is simulated by conductance changes in postsynaptic cells, as in the model described above for the input to the pyramidal V4 cells. Neurons have a refractory period and noise is added whose level is sufficient to generate low spontaneous activity.

Thus, correlated synaptic input from V1 optimally excites these cells, while uncorrelated input causes a smaller response. Interneurons which detect such coincidences inhibit the pyramidal cells of opposing selectivity (Fig. 1B). For instance, if the output of V1 cells responding to red are correlated, their postsynaptic interneurons in V4 will inhibit the reponse of the green pyramidal cell in V4. Because green cells in V1 are not correlated (they are outside the spotlight of attention), the associated V4 inhibitory interneurons will respond only little. Thus, in the competition among red and green pyramidal cells in V4, red will win out.

Detection of Temporal Structure in V4: Oscillatory Model

In this model, we presume the existence of frequency-selective inhibitory interneurons in V4. These interneurons are assumed to act like bandpass filters, selective to spikes arriving every 25 ms or so. Cells with bandpass characteristic in the 10-50Hz range have been identified by Llinas, Grace and Yarom in a cortical slice preparation (Llinas et al., 1991). Because the neurons are located in layer IV and are small, with smooth, aspiny dendrites, Llinas and colleagues (1991) argue that they correspond to inhibitory interneurons. There are numerous other examples of cells which can be described as bandpass filters, for instance the frequency-selective haircells of the bullfrog, turtle or lizard cochlea (Crawford and Fettiplace, 1981; Fettiplace, 1987).

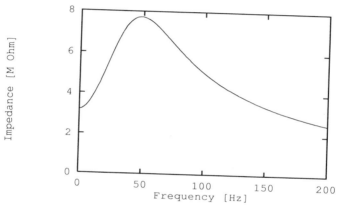

Figure 3. Oscillatory model: Impedance (in MΩ) of a model inhibitory V4 neuron as a function of the frequency (in Hz) of the applied current. The resonance frequency is close to 40Hz. The behavior shown corresponds to that of linearized Hodgkin-Huxley membrane (Koch, 1984).

A bias in favor of attended stimuli is generated by the frequency modulation imposed on the V1 activity in the present model. V4 inhibitory cells respond stronger to 40 Hz synaptic input than to higher or lower input frequencies. The frequency selective interneurons then inhibit the pyramidal cells of opposing response selectivity (Fig. 1B) and the response to the unattended stimulus is suppressed.

The electrical behavior of frequency-selective cells is usually modeled by describing the different ionic currents underlying their behavior using a Hodgkin-Huxley like formalism (Yamada et al., 1989). This approach is not only computationally expensive, it also requires detailed knowledge of the relevant channel kinetics. A simpler method consists in linearizing the system of equations around the resting potential. One now obtains a linear system of ordinary differential equations with a bandpass behavior (Fig. 3), with the resonance frequency varying from a few to several hundred Hertz, depending on the density and kinetics of the various currents (Crawford and Fettiplace, 1981; Hudspeth and Lewis, 1988). Such bandpass behavior, albeit at lower frequency, has also been observed in neocortical neurons

(Hutcheon and Puil, 1992) and is sometimes termed quasi-active (Koch, 1984). As always in our simulations, neurons have a refractory period (chosen randomly from the interval [2ms, 5ms]) and synaptic noise is added to generate spontaneous activity (Niebur et al., 1993).

RESULTS

Figures 4 and 5 illuminate the characteristic properties of the V1 spike trains which are used by the inhibitory interneurons in area V4 to distinguish attended from unattended stimuli. In the case of the oscillation model, the instantaneous firing frequency is modulated periodically. Figure 4 shows the power spectra of two model V1 cells (for the oscillatory model), one outside the focus of attention ($A_{max}=0$; Fig. 4A) and one inside ($A_{max}=0.75$; Fig. 4B). It is helpful to recall that the power spectrum of a Poisson process of mean rate λ is given by:

$$S(f) = \lambda + 2\pi\lambda^2\delta(f) \tag{3}$$

i.e., no particular nonzero frequency is preferred.

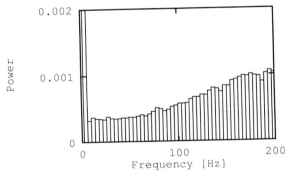

Figure 4A. The power spectra of the firing frequency response of V1 and V4 cells. The Fourier-transform of the power spectrum is identical to the autocorrelation function commonly used by electrophysiologists (e.g. Gray and Singer, 1989). All data are averaged over 64 runs of 1,024ms each; the spectra are computed using the FFT routine; the frequency imposed on attended V1 neurons was varied across trials in equal steps between 35 and 45 Hz. The panel shows the power spectrum of a V1 cell while attention is outside it's receptive field ($A_{max}=0$).

The spectra shown in Figs. 4A,B exhibit two departures from this behavior. First, the power increases as a function of frequency. It has been shown (Bair et al., 1993) that this effect is due to the presence of the refractory period: Poisson distributed spike trains with a refractory period and a mean interspike interval substantially longer than the refractory period will show a dip at low frequencies. The second deviation from the Poisson behavior, present only in the case of the attended stimulus (Fig. 4B), is the peak around 40Hz, caused by the imposed frequency modulation. The spectra of unattended and attended V4 interneurons are shown in Figs. 4C and D, respectively. The most prominent feature is here the presence of a very strong peak in the spectrum of the attended neurons (Fig. 4D) which is completely absent in the case of the unattended neurons (Fig. 4C). Figure 4E shows the power spectrum of an unattended V4 pyramidal neuron and Figure 4F that of an attended V4 pyramidal neuron. A small peak around 40Hz is visible in the spectrum of the attended neuron, but is nearly completely absent in the case of the unattended neuron.

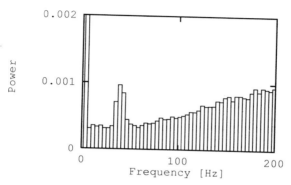

Figure 4B. The power spectrum while the spotlight of attention overlaps with its receptive field (A_{max}=0.75).

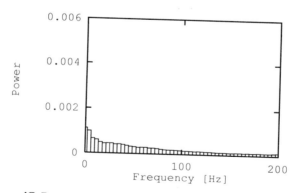

4C. Power spectrum of the inhibitory V4 cells without attention.

Note that for all cells except for the inhibitory V4 interneurons, only a small fraction of the total power is concentrated around the modulation frequency and that this modulation is not readily detected except when explicitly looking for it (as by computing the frequency spectrum). In the case of the inhibitory V4 interneurons, the appearance of a strong peak in their spectrum is consistent with the observation of such cells by Llinas and collaborators (Llinas et al., 1991).

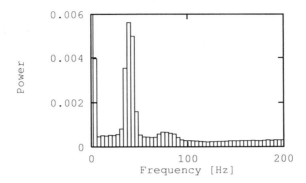

Figure 4D. Power spectrum of the inhibitory V4 cells with attention. Our model predicts that these cells show a very large peak in the 40 Hz range (a second harmonic around 80 Hz is present, too).

The importance of the 40 Hz peak in the power spectrum of the attended V4 neuron is that it allows the system to continue to use "temporal tagging" via frequency modulation to effect the same attentional gating at the next cortical site, inferotemporal cortex (IT). Thus, if two objects are located within an IT receptive field (which can cover a substantial fraction of the entire visual scene), and the observing subject is attending to one of them, interneurons in IT can pick up the frequency modulation of the V4 afferent neurons and inhibit the IT cell selective to the non-attended stimulus.

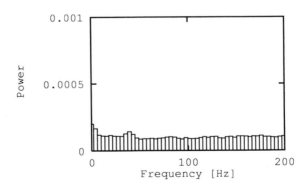

4E. Power spectrum of excitatory V4 cells without attention.

In the case of the synchronicity model, the distinction between attended and unattended stimuli rests on the properties of the cross-correlation functions between spike trains of attended and unattended cells in V1 and V4, as shown in Fig. 5. While the correlation function between unattended V1 cells is flat (not shown), attentional modulation generates a peak between attended V1 spike trains (Fig. 5). The central peak is accompanied by valleys with a minimum at about 15 ms which are caused by the suppression of activity for 17.5 ms after each 2.5 ms long period of elevated firing (see after eq.2).

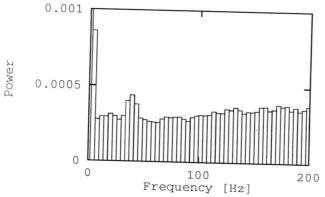

Figure 4F. Power spectrum of excitatory V4 cells with attention. Notice the threefold increase in the mean level of the spectrum when going from the non-attended to the attended situation (i.e. from Fig 4E to Fig. 4F). The 40 Hz peak is approximately as strong here as in an attended V1 cell, allowing the same temporal "tagging" mechanism to be repeated between V4 and its next cortical target IT, in agreement with the experiments of Moran and Desimone (1985).

A similar peak is observed between attended V4 cells (Fig. 5B). In contrast, there is nearly no correlation between unattended V4 cells (Fig. 5C). We also found that the rate of the unattended V4 cells is reduced, comparable to the reduction in spike rates observed by Moran and Desimone (1985). In contrast, and again in agreement with these experiments, this suppression is contingent on the attended stimulus being in the same receptive as the unattended. If the attended stimulus is outside the receptive field, or if there are two unattended stimuli in the receptive field, no suppression occurs.

As it was the case in the oscillation model, we find that the temporal tag imposed on the activity of the V1 cells is projected to the V4 pyramidal neurons. In the case of the synchronicity model, the tag is represented by the correlation between spike trains. Comparison between Figs. 5A and B shows that about the same level of synchronicity is found between spike trains of the output neuron in area V4 as between those of area V1. Therefore, we may again infer that the system can continue to use temporal tagging in IT. Such hierarchies can possibly extend throughout the entire visual system.

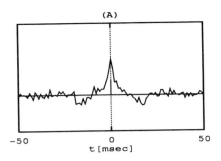

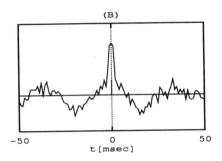

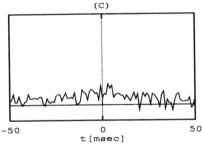

Figure 5. Cross-correlation functions between (A) two V1 cells with receptive fields in the focus of attention, (B) two excitatory V4 cells in the focus of attention, (C) two excitatory cells outside the focus of attention. All functions are averaged over 1000 s of activity, and a shift predictor is subtracted. The modulation-induced peak in V1 (A) is reproduced approximately between attended V4 neurons (B), while only a very weak correlation is observed between unattended V4 neurons(C). Valleys in (A) and (B) are due to the 17.5 ms long suppression of activity after each 2.5 ms long period of elevated firing.

DISCUSSION

One of the hypotheses underlying the present work is the idea that the nervous system might make better use of the bandwidth available to spiking neurons by using not only the average spike rate but also the details of the temporal structure of the spike trains (at a resolution of a few milliseconds). Here, stimulus properties are coded in terms of the average spike rate of neurons in primary cortical areas, whereas information about the attentional state of the animal (i.e, which stimulus is at any given time attended to) is provided in the fine temporal structure of the spike trains.

For both models discussed, we have shown that the key aspect of this "temporal tagging" hypothesis can be implemented by plausible single-cell models connected in a realistic architecture: the firing rate is modified by attention only in area V4, not in V1. We find (Niebur et al., 1993; Niebur and Koch, 1994) that a suppression of the unattended stimulus in V4 can be achieved which is comparable to that found experimentally in awake monkey (Moran and Desimone, 1985). While, in principle, this result can always be obtained in a trivial way, given sufficiently high levels of modulation and arbitrary sensitivity of the detectors, we have shown that low levels of temporal modulation of the spike trains are sufficient to allow detection by generic neural mechanisms. Furthermore, the required modulation levels are low enough to escape casual observation. Therefore, only careful experimental study of the fine temporal structure of V1 and V4 spike trains will reveal the validity of our models.

If the implementation of selective attention is based on oscillatory mechanisms, periodic modulation should be visible in the recordings of cells with attended stimuli in their receptive fields (but their amplitude can be expected to be relatively weak). In contrast, modulation by synchronization would not be visible in any single spike train, but only in multi-unit recordings. Furthermore, both models predict correlations between spike trains in different cortical areas, receiving the same input from V1 as long as the spike trains are generated by attended objects. The discovery of stimulus-induced (but not stimulus-locked), long-distance correlations between cells in visual cortex of cat (Eckhorn et al., 1988; Engel et al., 1991; Gray and Singer, 1989) and visual and sensorimotor cortex of monkey (Kreiter and Singer, 1992; Livingstone, 1991; Murthy and Fetz, 1992) is certainly compatible with this prediction, insofar as it shows that the neural substrate is capable of producing such synchronization. Oscillatory phenomena have been observed to occur in correlation with states of focused attention in cats (Bouyer et al., 1987), monkeys (Rougeul et al., 1978) and humans (Tiitinen et al., 1993). However, the electrodes used in these experiments were too large to allow any fine differentiation between attended and unattended regions. Furthermore, since the subjects were allowed to move around during the experiment in most of these studies, no control of the location and extension of the focus of attention was possible in these experiments.

Keele and collaborators used psychophysical methods in humans to study temporal mechanisms for attentional binding (Keele et al., 1988). They found that binding of different features of an object is closely related to the common location of these features, whether the features are presented simultaneously or not. These findings provide little support for models in which external temporal information is imposed on neurons for attentional binding (e.g. Horn et al., 1991, von der Malsburg and Schneider, 1986), but they are consistent with models in which the temporal patterns are generated internally, as in our models. Indeed, the strong correlation of attentional binding with the spatial location of the object is consistent with both of our models, since they are based on a spatially defined focus of attention in which the modulation takes place.

Acknowledgements

We thank Francis Crick for helpful discussions, Wyeth Bair for writing the routines for computing power spectra and correlation functions, and Jochen Braun for a critical reading of the manuscript. This work was supported by the Office of Naval Research, the Air Force Office of Scientific Research, and the National Science Foundation.

REFERENCES

Bair W., Koch C., Newsome W., and Britten K. (1993). Power spectrum analysis of MT neurons in the awake monkey. In Bower J. and Eeckman F., editors, Computation and Neural Systems. Kluwer, Norwell, MA.

Bouyer J.J., Montaron M.F., Vahnee J.M., Albert M.P., and Rougeul A. (1987). Anatomical localization of cortical beta rhythms in cat. Neuroscience, 22(3):863-869.

Crawford A.C. and Fettiplace R. (1981). An electrical tuning mechanism in turtle cochelar hair cells. Journal of Physiology (London), 312:377-412.

Crick F. and Koch C. (1990). Towards a neurobiological theory of consciousness. Seminars in the Neurosciences, 2:263-275.

Desimone R. (1992). Neural circuits for visual attention in the primate brain. In Carpenter G. and Grossberg S., editors, Neural networks for vision and image processing. MIT Press, Cambridge.

Desimone R., Wessinger, M., Thomas L., and Schneider W. (1989). Effects of deactivation of lateral pulvinar or superior colliculus to selectively attend to a visual stimulus. Soc. Neurosci. Abstr., 15:162.

Desimone R., Wessinger, M., Thomas L., and Schneider W. (1991). Attentional control of visual perception: cortical and subcortical mechanisms. Symp. Quant. Biol., 55:963-971.

Eckhorn R., Bauer R., Jordan W., Brosch M., Kruse W., Munk M., and Reitboeck H.J. (1988). Coherent oscillations: a mechanism of feature linking in the visual cortex? Biol. Cybern., 60:121-130.

Engel A.K., König P., Kreiter A.K., and Singer W. (1991). Interhemispheric synchronization of oscillatory neuronal responses in cat visual cortex. Science, 252:1177-1179.

Fettiplace R. (1987). Electrical tuning of hair cells in the inner ear. Trends in Neurociences, 10:421-425.

Garey L.J., Dreher B., and Robinson S.R. (1991). The organization of the visual thalamus. In Dreher B. and Robinson S.R., editors, Neuroanatomy of the visual pathways and their developement, pages 176-234. CRC Press, Boca Raton.

Gray C.M. and Singer W. (1989). Stimulus-specific neuronal oscillations in orientation columns of cat visual cortex. Proc. Nat. Acad. Sci., USA, 86:1698-1702.

Horn D., Sagi D., and Usher M. (1991). Segmentation, binding and illusory conjunctions. Neural Computation, 3:510-525.

Hudspeth A.J. and Lewis R.S. (1988). Kinetic analysis of voltage- and ion-dependent conductances in saccular hair cells of the bull frog, rana catesbeiana. Journal of Physiology (London), 400:237-274.

Hutcheon B. and Puil E. (1992). A low-frequency subthreshold resonance in neocortical neurons generated mainly by I_h. Soc. Neurosci. Abstr., 18(2):1344.

Julesz B. (1991). Early vision and focal attention. Rev. Mod. Physics, 63:735-772.

Kanwisher N. and Driver J. (1992). Objects, attributes and visual attention: which, what, and where. Current Directions in Psychological Science, 1:26-31.

Keele S.W., Cohen A., Ivry R., Liotti M., and Yee P. (1988). Tests of a temporal theory of attentional binding. J. Experimental Psychology: Human Perception and Performance, 14:444-452.

Koch C. (1984). Cable theory in neurons with active, linearized membranes. Biol. Cybern., 50:15-33.

Koch C. and Ullman S. (1985). Shifts in selective visual attention: towards the underlying neural circuitry. Human Neurobiol., 4:219-227.

Kreiter A.K. and Singer W. (1992). Oscillatory neuronal responses in the visual-cortex of the awake macaque monkey. Europ. J. Neurosci., 4(4):369-375.

Kubota T., Morimoto M., Kanaseki T., and Inomata H. (1988). Visual pretectal neurons projecting to the dorsal lateral geniculate nucleus and pulvinar nucleus in the cat. Brain Res. Bull., 20:573-579.

LaBerge D. and Buchsbaum M.S. (1990). Positron emission tomographic measurements of pulvinar activity during an attention task. J. Neurosci., 10:613-619.

Livingstone M.S. (1991). Visually-evoked oscillations in monkey striate cortex. Soc. Neurosci. Abstr., 17(1):176.

Llinas R.R., Grace A.A., and Yarom Y. (1991). In vitro neurons in mammalian cortical layer 4 exhibit intrinsic oscillatory activity in the 10- to 50-Hz frequency range. Proc. Natl. Acad. Sci. USA, 88:987-901.

Moran J. and Desimone R. (1985). Selective attention gates visual processing in the extrastriate cortex. Science, 229:782-784.

Murthy V.N. and Fetz E.E. (1992). Coherent 23- to 35hz oscillations in the sensorimotor cortex of awake behaving monkey. Proc. Nat. Acad. Sci., USA, 89:5670-5674.

Niebur E. and Koch C. (1994). A model for the neuronal implementation of selective visual attention based on temporal correlation among neurons. J. Comput. Neuroscience (in press)

Niebur E., Rosin C., and Koch C. (1993). An oscillation-based model for the neural basis of attention. Vision Research, 1993:2789-2802.

Petersen S.E., Robinson D.L., and Morris J.D. (1987). Contributions of the pulvinar to visual spatial attention. Neuropsychologia, 25:97-105.

Posner M.I. and Driver J. (1992). The neurobiology of selective attention. Current Opinion in Neurobiology, 2:165-169.

Posner M.I. and Petersen S.E. (1990). The attention system of the human brain. Ann. Rev. Neurosci., 13:25-42.

Robinson D.L. and Petersen S.E. (1992). The pulvinar and visual salience. Trends in Neurociences, 15(4):127-132.

Rougeul A., Bouyer J.J., Dedet L., and Debray O. (1978). Fast somato-parietal rhythms during combined focal attention and immobility in baboon and squirrel monkey. Electroenchephalography and Clinical Neurophysiology, 46:310-319.

Softky W. and Koch C. (1993). The highly irregular firing of cortical-cells is inconsistent with temporal integration of random EPSPs. J. Neurosci., 13(1):334-350.

Tiitinen H., Sinkkonen J., Reinikainen K., Alho K., Lavikainen J., and Näätänen (1993). Selective attention enhances the auditory 40-Hz transient response in humans. Nature, 364:59-60.

Treisman A. (1988). Features and objects: the fourteenth Bartlett memorial lecture. Quant. J. Exp. Psychol., 40A:201-237.

von der Malsburg C. and Schneider W. (1986). A neural cocktail party processor. Biol. Cybern., 54:29-40.

Yamada W., Koch C., and Adams P.R. (1989). Multiple channels and calcium dynamics. In Koch C. and Segev I., editors, Methods in Neuronal Modeling: From Synapses to Networks, pages 97-134. MIT Press, Cambridge.

20 HZ BURSTS OF ACTIVITY IN THE CORTICO-THALAMIC PATHWAY DURING ATTENTIVE PERCEPTION

Andrzej Wróbel, Marek Bekisz and Wioletta Waleszczyk

Department of Neurophysiology, Nencki Institute of Experimental
Biology, 3 Pasteur St. 02-093 Warsaw, Poland

INTRODUCTION

It is well established that thalamo-cortical fibers of the visual pathway of the cat send out collaterals to two recurrent loops: inhibitory, via GABAergic interneurons of the perigeniculate nucleus (PGN) and excitatory, relayed by the pyramidal cells of layer 6 of the striate cortex. Both of these loops terminate on principal cells of lateral geniculate nucleus (LGN). There are data indicating the possible role of PGN neurons in synchronization of thalamo-cortical rhythmic activity (see Steriade and Llinas, 1988 for a review), whereas hypotheses concerning the functions of the cortico-geniculate pathway lack clear experimental support. This pathway should be important in view of the fact that cortical axons outnumber all other excitatory inputs to LGN principal cells (Wilson et al., 1984). One of the main reasons for the lack of understanding of the role of cortical input may be the poor responsiveness of layer 6 pyramidal cells in anaesthetized cats. This was demonstrated by Livingstone and Hubel (1981) who showed further that when layer 6 cells become active, after the cat recovered from the anaesthesia, also their specific responses to visual stimuli were noticeably enhanced. These effects were also accompanied by more vigorous responses of LGN principal cells to specific stimulation.

It has been recently shown (Lindström and Wróbel, 1990) that synapses of the cortico-geniculate pathway have a built-in frequency amplification. At low frequencies of electrical stimulation of the cortex the excitatory postsynaptic potentials, recorded intracellularly from LGN principal cells, were barely detectable but at 20 Hz they often exceeded in amplitude those of retinal origin. It was postulated that frequent activation of layer 6 cells can change, by means of such a potentiation mechanism, the membrane polarization of geniculate cells and can control the input-output gain of the geniculate relay. A prediction derived from this hypothesis is that the cortico-geniculate pathway is activated according to the needs of the animal, e.g.,

Oscillatory Event-Related Brain Dynamics, Edited
by C. Pantev, Plenum Press, New York, 1994

during attentive perception. The present experiment was designed to investigate the possible changes of activity in the cortico-thalamic part of the visual system in two behavioral situations which required attention to either visual or nonvisual (acoustic) cues.

METHODS

Four cats were trained to respond to two different stimuli, a visual and an acoustical, which were presented at random during the same session (12 times each). The animals were placed in a small (20x45x45 cm) wooden cage and faced two translucent doors situated 5 cm from each other. The cat and the doors were separated by a transparent removable screen. The visual stimulus was a small (0.5x2 deg) slit of light of 3 cd/m2 intensity which was projected on the front wall at the level of the cat's eyes and moved along a sinusoidal pathway from left to right and backwards (linear speed of about 10 cm/s). After about ten to twenty seconds the slit was stopped on one of the two doors, indicating that a piece of meat was hidden behind that door. Then it was switched off. After a 1-3 second delay the transparent screen was raised and by pressing the correct door the cat could reach for the reward. The incorrect door was locked and the animal could not open it. The cat was not allowed to correct an error.

The acoustic stimulus, produced by a pocket-radio loudspeaker, was a noise with a fundamental frequency of 5 kHz and with an intensity modulated with a 2.5 Hz frequency between 50 dB and 55 dB. It was switched on behind the part of the wall between the doors and consecutively moved around the corner of the cage behind the left or right side-wall. After ten to twenty seconds the stimulus was switched off, and there was a 1-3 second delay before the cat was allowed to press one of the doors. The rewarded door was the one on the side on which the auditory stimulus had been turned off.

The learning procedure started with the visual task. The acoustic stimulus was introduced after stabilization of the response to the visual stimulus. For one cat (Cat 1), the visual stimulus was also present and oscillated in the vertical plane between the doors (neutral position) during the acoustic task to provide similar light-flux input in both visual and auditory modality tasks. The recordings obtained from this cat did not differ from those of the other animals. Training was considered complete when animals reached 90% performance accuracy during 3 successive experimental days. Ten to 15 sessions (one session a day, each lasting about half an hour) were sufficient for all cats to learn the task.

After the completion of training, surgery was performed under Nembutal anaesthesia. Tungsten recording electrodes were inserted under electrophysiological control: two in the left lateral geniculate and/or perigeniculate nucleus, and one in the contralateral hippocampus. The LGN electrode was lowered to the central position of the nucleus, aiming for representation of area centralis. A row of three chromonickel electrodes, about 1.5 mm apart, was placed in the left primary visual cortex of all animals. In two cats an additional electrode was inserted into the ipsilateral auditory cortex. The electrodes and a plug were mounted on the animal's head by means of dental cement. The recordings started 5 to 7 days after surgery with one session a day, four sessions a week. The EEG signals (1 Hz - 0.5 kHz) were amplified and stored on a magnetic tape recorder (Racal V-store). After the experiment, the cats were anesthetized and perfused for histological verification of the recording sites.

The data analysis was performed off-line by an IBM-386 compatible, personal computer. The EEG signals were filtered up to 100 Hz and digitized with a 200 Hz sampling rate. Activity contaminated by movement artifacts was not analyzed. The amplitude spectra were calculated from 2.5 s epochs by the Fast Fourier Transform (FFT) procedure and averaged for the same modality of stimulus and the same performance (correct or incorrect) of the task. Further elaboration of the "20 Hz band" was performed after filtering with cut-off frequencies of 16-24 Hz by means of a digital filter.

To obtain the cumulative amplitude function the mean value of the signal was calculated. The cumulative function was then calculated by adding absolute values (moduli) of the differences between signal measured at each successive sample and mean value.

In order to find a relation between the EEG activity recorded from different sites we finally calculated the directed transfer functions (Kaminski and Blinowska, 1991). This method will be described in detail below.

RESULTS

The amplitude spectra of cortical activity as recorded from two cats in different behavioral situations are presented in Fig. 1. Most of the power in the EEG was concentrated at lower frequencies but this part of the spectrum was not analyzed in the present report. For most recording sites in the visual cortex (VCx) and lateral geniculate nucleus (LGN) a prominent peak in the beta range appeared during periods when the animals attended to the visual stimulus in order to perceive the location of its disappearance. These peaks were shown to differ from the corresponding parts of the spectra obtained during acoustic trials (indicated by stars in Fig. 1 A, C) by means of t-test. During the acoustic trials the amplitude of the EEG activity in the visual centers decreased smoothly with increasing frequency (Fig. 1 A, C). To assure that the observed increase of beta activity was not due only to the visual stimulation, we allowed Cat 1 to see the same (but in this case nonsignalling) slit of light in between the doors during acoustic trials. The difference between the spectra obtained in visual and auditory situations for this cat (Fig. 1 A) was similar to that observed in the remaining animals (e.g., Fig. 1 C). Consequently, we doubt that the observed beta peak was due to stimulus induced activation of the visual centers but we believe that it results from setting the proper mechanism for the "attentive state".

The characteristic beta peak appeared only in trials which ended with a correct response. The spectra obtained during erroneous trials did not show such a peak (Fig. 1 B). This supports the notion that the observed beta component of an EEG power spectrum is related to attentional processes in the thalamo-cortical system. The enhanced frequency band encompassed the 16 to 24 Hz range for different animals and remained characteristic for each cat during the experiment. For one cat (Cat 2), the difference in amplitude spectra at 20 Hz as obtained for the two different modalities did not reach statistical significance at the 0.05 level. This cat had learned the task very quickly and performed with no apparent behavioral manifestation of attention directed towards the stimuli.

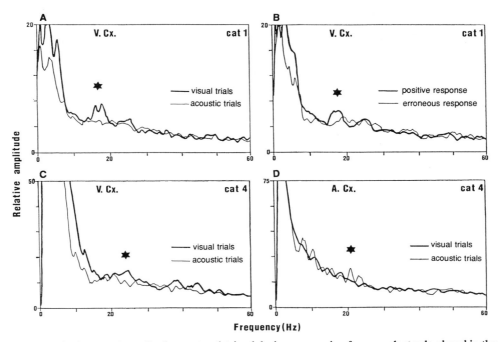

Figure 1. A. Averaged amplitude spectra obtained during one session from an electrode placed in the primary visual cortex during seven positive trials with visual (thick line) and ten with acoustic (thin line) cues. Each averaged spectrum was calculated from 14 independent, 2.5 s long EEG epochs. An irrelevant visual stimulus accompanied the acoustic trials; B. Averaged spectra obtained from epochs recorded during the same session with the same electrode as in A. The EEG epochs were chosen for averaging from five incorrect trials (5 epochs) and six correct trials (8 epochs); C. Another cat. Averaged spectra calculated from recordings obtained from the same primary visual cortex site during seven acoustic (14 epochs) and six visual trials (12 epochs); D. Similar spectra obtained from simultaneously recorded epochs from auditory cortex of the same cat. The statistically significant differences in the 20 Hz band have been marked by stars. See text for details. (This figure has been reprinted from Bekisz and Wróbel 1993, by permission of Acta Neurobiol. Exp.)

The hypothesis that the sensory channels are activated according to the needs of the animal is furthermore supported by the finding of a similar increase of the same frequency (beta) component in the EEG spectrum obtained from activity in the auditory cortex during attentive listening in one cat (out of two) which had an additional electrode implanted in ectosylvian gyrus (Fig. 1 C, D). However, we did not see such a component in the spectrum of the auditory cortex activity of the other cat nor did it appear in individual spectra obtained from the visual cortex of different animals. It should be noticed that the increase in the amplitude in the 20 Hz band, though observed in all cats, was usually most prominent from only one of the electrodes implanted in the visual cortex (e.g., most posterior recording site in Fig. 5). The set of only three electrodes gave a poor resolution but we could observe that the most prominent beta peaks were often recorded from electrodes close to the representation of the area centralis. This is congruent with the observation that in most of the trials cats were following the moving stimulus with their eyes. The magnitude of the observed 20 Hz peak could vary with experimental days, as the performance of the animals did. The full analysis of this effect is not yet complete.

From trial to trial we have also observed significant, higher frequency peaks (more than 30 Hz) in the spectra calculated for epochs encompassing presentation of the visual stimulus (cf. Fig. 1 C). We have not analyzed these systematically in the present experiment.

To better characterize the additional activity underlying the enhancement of the 20 Hz band we used a digital filter to filter out signals in the range of 16 - 24 Hz. It appeared that this activity consisted of short bursts of oscillations, lasting about 100 - 1000 ms. From the raw data of Fig. 2 it can be seen that the introduction of the visual stimulus (the cue for visual differentiation) enhanced the amplitude and frequency of appearance of these bursts in visual cortex. A similar picture was obtained when beta activity from the lateral geniculate nucleus was analyzed and also, during acoustic trials in the auditory cortex, from the site where the enhanced beta peak was observed. No changes in amplitude or frequency of appearance were noted for the prominent bursts observed in the hippocampus. Thus, within the range of our recording sites, the increase of this specific activity seemed to be limited to the geniculo-cortical system.

The increase in beta burst activity shown in Fig. 2 was consistently found throughout the visual trials. To better analyze the dynamics of this activity we have calculated cumulative amplitude functions throughout the trials as shown in Fig. 3. The analysis of these functions showed that in both LGN and VCx activity was roughly stable during a given trial. The slopes of the cumulative amplitudes were, however, different as measured for visual and acoustic trials (Fig. 3). The difference between mean slopes of the two experimental conditions were significant (Mann-Whitney U-test, $p < 0.05$) for most recording sites in LGN and VCx for all animals except the one which did not show the enhanced beta activity during the experimental trials. In general, the conclusions drawn from comparing the cumulative functions agreed well with the results obtained by Fourier analysis.

The detailed analysis of the cumulative functions revealed synchronous changes in the activity measured in cortical and geniculate sites (not shown) suggesting that both of them might be driven by a common mechanism. To study this possibility further we have built up crosscorrelation functions between the bursts appearing at the two sites. A "burst event" was counted when signal amplitude exceeded an arbitrary threshold. The midpoint of each above-threshold period was approximated as the moment of appearance of the burst. The crosscorrelations obtained between burst events recorded simultaneously during one day's session for selected recording sites are presented in Fig. 4.

The uppermost left correlogram in Fig. 4 presents the crosscorrelation function calculated between bursts occurring in the LGN and the most posterior cortical recording site. We recorded from both electrodes before fixing them in the two structures in such a way that the corresponding strongest multiunit responses were elicited from retinal regions placed not further than about two degrees of visual angle from each other. When electrode locations were aligned this way, the oscillatory bursts recorded during visual trials tended to occur simultaneously. The control cor-

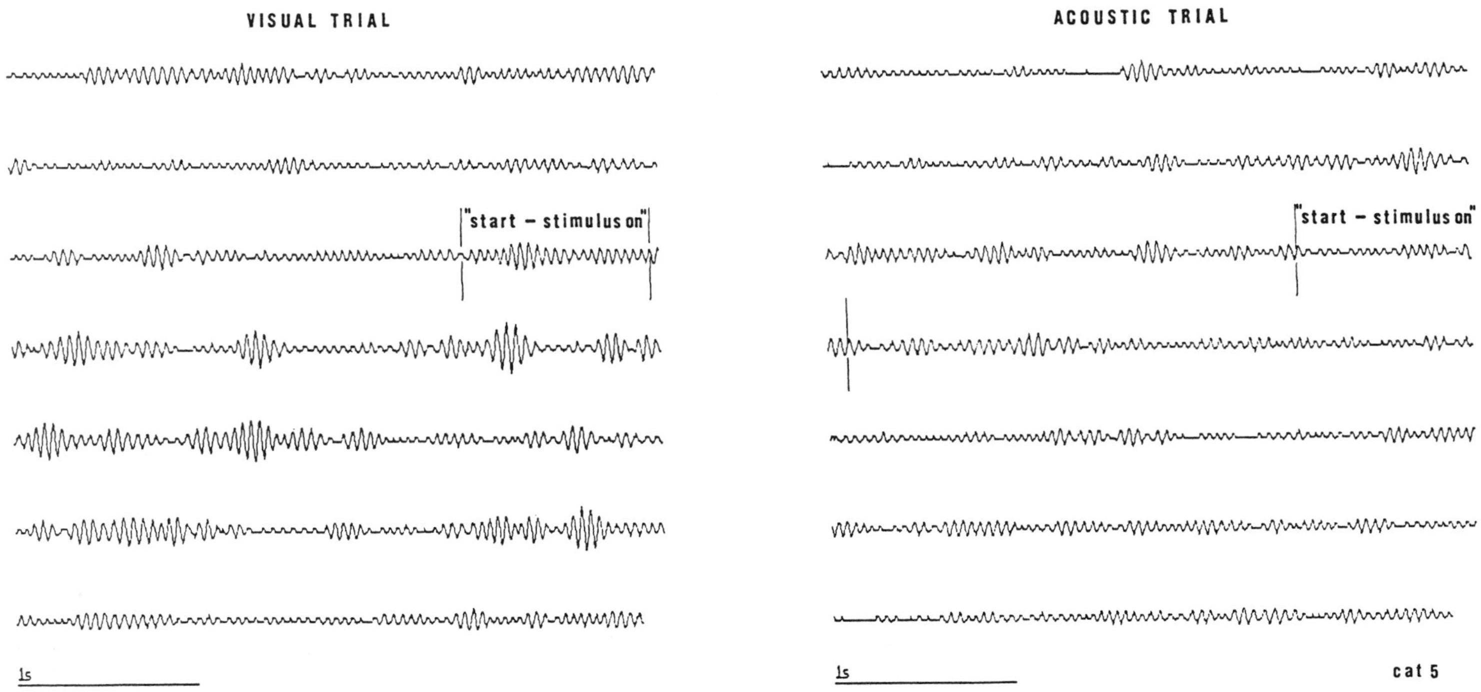

Figure 2. The filtered (16-24 Hz) activity recorded in successive visual (left traces) and acoustic (right traces) trials in one experiment. Each column represent continuous recording divided into seven traces. "Start - stimulus on" marks the words of the experimentator registered on the auditory channel of the recorder. Note the enhanced amplitude and frequency of bursts after the appearance of the relevant visual stimulus.

316

relogram (Fig. 4, left column, middle) was obtained by shifting the traces of cortical activity in relation to the signals from the geniculate by a number of visual trials. The burst activity recorded from the same sites during acoustic trials was much lower since the amplitude of the bursts rarely reached the specified threshold (cf. Fig. 2). Nevertheless, during acoustic trials a central peak (lowermost, left correlogram) still seemed to be present, indicating that a common mechanism was potentially operative. In contrast to the recording from the LGN, the numerous 20 Hz bursts recorded from the hippocampus during visual trials were not correlated with events occurring in the visual cortex (lowermost, right correlogram).

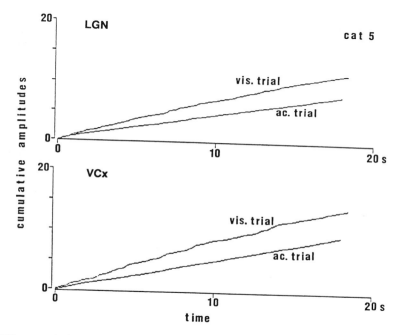

Figure 3. The cumulative amplitudes of the filtered signals as presented in the Fig. 2 recorded from Cat 5. The cumulative curves (see methods) have been obtained from successive visual and acoustic trials in one session. The simultaneous activity recorded from LGN and VCx recording sites are shown in separate graphs. Note that the traces representing visual trials are steeper.

The correlation between the burst activities at LGN and VCx sites, which were further apart in the visual field (about five degrees of visual angle), is presented in the uppermost right corner of Fig. 4. The central peak of this correlogram is barely visible indicating less correlated burst activity at these two sites which were not aligned.

317

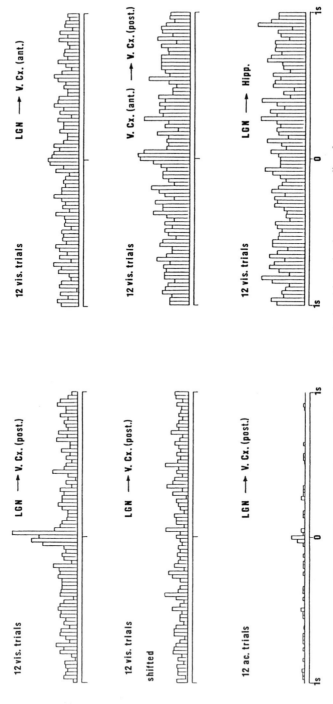

cat 5

Figure 4. The crosscorrelograms of bursting activity calculated from simultaneous recordings between different recording sites during different behavioral situations. The session consisted of 12 visual (vis.) and 12 acoustic (ac.) trials, all ending with correct responses. The right sides of the correlograms are labelled. The unlabelled left sides show the correlation values for the opposite direction. One second (1 s) window, 25 ms bin width. Bursts were filtered by a 16-24 Hz digital filter. See text for details.

It could also be that the correlation between activity recorded in LGN and the anterior electrode of VCx resulted from an indirect pathway and was coursed due to quite strong coincidental burst appearance in both (posterior and anterior) cortical sites (right column, middle correlogram of Fig. 4).

Coincidental bursting activity in the cortico-thalamic system was a consistent finding for all cats. Although our spatial resolution was poor, resulting from a limited number of electrodes, we typically found the strongest correlations between the well aligned sites in visual cortex and lateral geniculate nucleus. The retinotopic locations of all such correlated pairs were at or in the vicinity of the area centralis. More accurate studies, however, are needed to confirm these preliminary observations.

Were our hypothesis of cortical control of thalamic transmission true, the source of the observed correlated activity should be at the cortical rather than the geniculate site. To distinguish these possibilities, we calculated the directed transfer functions (DTFs), a method developed to measure the direction and frequency content of the flow of EEG activity between different brain locations (Kaminski and Blinowska, 1991). The method based on a multichannel autoregression model calculates DTF values which are meant to describe the percentage of activity at a given frequency recorded from one electrode, which then appears with some delay in the recording registered by a second electrode at the target point.

Figure 5 shows DTFs for activity measured at six recording sites during one experimental day in Cat 5. The left matrix contains the functions obtained during successful visual trials and the right matrix the corresponding DTFs calculated during periods of successful acoustic trials. On the diagonals of both matrices we present the power spectra of activity recorded by each of the corresponding electrodes, calculated by means of the autoregression method. Unfortunately, the program we used scales all the power spectra on the diagonal to the maximal amplitude value measured in any of them (in Fig. 3 it corresponds to the low frequency end of the spectrum measured for the hippocampal activity) and, therefore, the corresponding spectra on the diagonals of the left and right matrices cannot be compared (the right diagonal spectra are overestimated). Nevertheless, on the spectra measured for geniculate and cortical sites in the left matrix diagonal one can trace humps at about 20 Hz, correspondingly to the peaks observed in the amplitude spectra obtained by means of Fourier analysis as presented in Fig. 1.

Each box outside the diagonals of the two matrices in Fig. 5 contains the directed transfer function calculated between two EEG signals recorded at the corresponding sites which are indicated above and to the side of the matrix. The largest flow of activity was found between the posterior-most cortical and geniculate electrodes during visual trials (fourth column, first row of the left matrix in Fig. 5). The other cortical sites also seemed to be targets for the 20 Hz activity from the most posterior electrode (two last boxes in the same column).

No corresponding reciprocal flow was observed at the same time from geniculate nucleus towards the posterior cortical site (empty box in the forth row, first column). It should be mentioned here that at the 200 Hz sampling frequency used in our calculations, the DTF method would only exhibit time shifts between signals that are longer than five milliseconds (Kaminski and Blinowska, 1991). Thus, it is possible that such a dead period

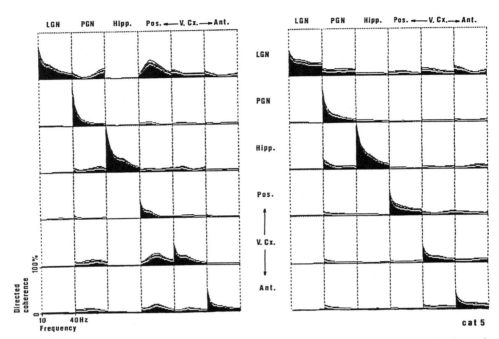

Figure 5. Directed Transfer Functions (DTFs) obtained between signals recorded from six electrodes placed under physiological control as follows: LGN (azimuth 2°/ elevation 2° PGN (5°/ 0°); VCx (from posterior to anterior: 0°/ 2°, 2°/ 0°, -1°/ -2°). The hippocampal electrode was lowered stereotactically and later confirmed to reach dentate gyrus. DTFs directed from LGN to other recording sites are shown in the first column below the diagonal and those directed towards the LGN site in the first row, above diagonal, etc. Each box on the diagonal contains the power spectrum obtained with autoregression method from appropriate site. The power spectra are normalized within each matrix taking the maximal peak among all spectra from a diagonal as 100%. The left matrix exhibits functions obtained from the signals recorded during visual trials ending with correct response. The right matrix shows data from acoustic trials. The elevation of the DTF shows the amount of signal flow at the given frequency (see text for details). Variability shown by corridors of SEM. Note much more "cross-talk" between visual centers during performance of the visual task than during the auditory task.

of time is long enough to allow transient (nonoscillatory) visual signals to be transferred from LGN to the cortex.

The DTF method suggests that the 20 Hz rhythmic activity is propagated from its place of appearance to other parts of the primary visual cortex and towards the lateral geniculate nucleus. It is remarkable that we observed the 20 Hz component transfer only during visual trials (Fig. 5, left matrix) for which we registered the excess of 20 Hz activity as demonstrated above by FFT and cumulative amplitude function methods. The differences in appropriate DTF values between the visual and acoustic trials were significant as shown by variation (SEM) of the curves in the two matrices. Notice also that, in general, much more of the activity seems to be related inbetween visual brain areas during the experimental situation requiring visual attention (Fig. 2, left part) than during the acoustic task (Fig. 2, right part).

It is worth mentioning that only during visual trials the dip in the 20 Hz frequency band appears in the transfer function directed from the PGN towards the lateral geniculate nucleus as compared to acoustic trials (first row, second column boxes of the appropriate matrices). This suggests that at least the recorded part of PGN activity (at least in the posterior part of the nucleus, where our electrodes were placed) might not be a source of the 20 Hz oscillations registered in other electrodes. Furthermore, in none of the experiments we have observed evidence that either the hippocampal (third column of each matrix in Fig. 5) or the auditory cortical recording sites (not shown) could be a source of the 20 Hz signal for visual centers.

DISCUSSION

In the present experiment we have shown that a specific 20 Hz frequency band (average of 16-24 Hz range) of the EEG spectrum recorded from the visual cortex and lateral geniculate nucleus of the cat grew significantly when the animal was attending to a visual cue used to solve a simple task. This increase in power was due to the increase in amplitude and frequency of appearance of relatively short (0.1-1 s) bursts of oscillations which appeared coincidentally in VCx and LGN. The enhancement of this specific beta frequency component was observed mainly in one of the cortical recording sites close to retinotopical representation of area centralis and an aligned site in the lateral geniculate nucleus. The calculation of directed transfer functions suggested that the beta activity was propagated by a descending cortico-geniculate pathway.

Although electrophysiological correlates of behavior have been intensively studied for years, there are only a few findings in this area related to the analysis of beta frequency oscillations in the discussed range. In complex studies of visual and acoustic evoked potentials, Basar (1980) has shown a stimulus induced stabilization of certain frequency bands in the sensory systems of the awake cat. Interestingly, the 10 and 20 Hz bands were best represented in the frequency spectra of responses evoked from lateral/medial geniculates and visual/auditory cortices. In this experimental situation no behavioral activity was required from the animal. This could be the reason why the EEG recorded between stimuli did not contain any specific, enhanced frequencies. In one cat with an electrode in the auditory cortex, we also saw the increase in the 20 Hz rhythm content during attentive perception of the acoustic stimulus. Thus, this beta band might be similarly enhanced during specific activation of any of the sensory systems.

A frequency of 10 Hz has been shown to dominate the EEG recorded from area 18 (Chatila et al., 1992) and 17 (Rougeul-Buser, this volume) of cats quietly waiting for a hidden mouse to become visible. These authors report that the rhythm vanished during attentive fixation. Since we recorded both 10 and 20 Hz oscillations from the striate cortex of attending animals, the difference between our observations and theirs might be due to different mechanisms being involved in the different behavioral situations.

Interestingly, the increased 20 Hz EEG content has been observed in many behavioral situations in different sensory systems of various mammals. Lopes da Silva et al. (1970) have shown a 20 Hz rhythm occurring in the visual cortex of a dog "focusing his attention on a target". At the same time, however, only an 11 Hz peak appeared in the Fourier spectrum of the signal recorded from LGN. It should be emphasized here that the 20 Hz peak in geniculate activity observed in our experiment was always weaker and more difficult to notice then the cortical one. In the experiment of Lopes da Silva, closing the dogs eyes stopped the beta rhythm in the cortical EEG. Similarly, we found that the 20 Hz rhythm disappeared in the activity of the visual centers when the cat started to attend to the acoustic stimulus.

Quite similar bursts of about 20 Hz frequency were recorded by Rougeul et al. (1979) in the S1 area of baboon, directly preceding a manipulatory movement. These authors related such activity to the maintenance of immobility in the somaesthetic channel though the observed bursting may also have been due to increased sensitivity of the thalamic cells (for a recent discussion see Chatila et al., 1992). Notice also that in the awake rat suppression of cortical activity in area S1 reduces the responsiveness of ventrobasal neurons most severely in the 20 Hz band (Yuan et al., 1986).

It has been recently demonstrated that excitatory postsynaptic potentials (EPSPs) evoked in geniculate cells by cortical fibers are frequency potentiated and, with repetitive cortical activation reaching 20 Hz, they exceed by far the optic tract evoked EPSP (Lindström and Wróbel, 1990). Consequently, the hypothesis was put forward that the potentiation mechanism might be used during attentive perception. This hypothesis required frequent activation of layer 6, cortico-thalamic pyramidal cells to increase the gain of retino-cortical transmission through the geniculate relay. In other experiments it was further observed that the tonic phase of a focal seizure evoked in the cortico-thalamic loop had a resonance frequency of about 20 Hz (Lindström and Wróbel, unpublished). Subthreshold oscillations of similar and higher frequencies have also been reported in many thalamo-cortical cells of the sleeping cat (Steriade et al., 1991). The present observation that the 20 Hz rhythm originating in the visual cortex during visual attention reaches the geniculate cells agrees with the proposed role for cortico-geniculate pathway. Short synchronized bursts of beta frequency in the cortico-thalamic system might thus be used for gain setting during attentive perception. Though there are good reasons to believe that the EEG signal reflects the activity of the surrounding local neuronal network, the present findings should be confirmed by recording from single cells in behaving animals.

The idea that the rich cortico-thalamic projection modulates the activity of the lateral geniculate nucleus was put forward long ago (see Frigyesi et al., 1972, for a review) but lacked experimental support. It has been proposed that the descending system might be activated during attention (Hernandez-Peon, 1966; Ahlsen et al., 1985) and that this activation correlates with EEG desynchronization observed in vigilant animals (see Lindsley, 1960; Steriade and Llinas, 1988, for reviews). Recent investigations suggest that the "desynchronized" EEG is comprised of fast oscillations of small amplitudes and different frequencies, serving as a carrier for co-ordination between cortical areas (see Bressler, 1990). It is not known whether these oscillations influence the thalamic cells' activity (Steriade et al., 1991). There are findings suggestting that at least part of the beta rhythms appearing in the cortex might have their source in the thalamic structures (Lopes da Silva, 1991; Boyer et al., 1992). It can be argued, however, that when beta and gamma rhythms have anything to do with directed attention or other higher order information processing, their sources should be located in the cortical structures rather than the thalamus.

Acknowledgements

We are indebted to M. Kaminski and K. Blinowska for providing the DTF program and helpful methodological discussions. We would like to thank D. Krakowska for excellent technical assistance and E. Kublik for preparing the histological verification. This study has been supported by grants from the State Committee for Scientific Research: 0430/P2/92/02 and statutable to the Nencki Institute.

REFERENCES

Ahlsen, G., Lindström, S., Lo, F.-S., 1985, Interaction between inhibitory pathways to principal cells in the lateral geniculate nucleus of the cat, Exp. Brain Res. 58: 134-143.

Basar, E., 1980, "EEG - Brain Dynamics. Relation Between EEG and Brain Evoked Potentials", Elsevier/North-Holland Biomedical Press, Amsterdam-New York-Oxford, 411p.

Bekisz, M., Wróbel, A., 1993, 20 Hz rhythm of activity in visual system of perceiving cat, Acta Neurobiol. Exp. 53: 175-182.

Bouyer, J.J., Montaron, M.F., Buser, P., Durand, C., Rougeul, A., 1992, Effects of mediodorsalis thalamic nucleus lesions on vigilance and attentive behaviour in cats, Behav. Brain Res. 51: 51-60.

Bressler, S.L., 1990, The gamma wave: a cortical information carrier? TINS. 13: 161-162.

Chatila, M., Milleret, C., Buser, P., Rougeul, A., 1992, A 10 Hz "alpha-like" rhythm in the visual cortex of the waking cat, Electroenceph. clin. Neurophysiol. 83: 217-222.

Frigyesi, T.L., Rinvik, E., Yahr, M.D., eds., 1972, "Corticothalamic Projections and Sensorimotor Activities", Raven Press, Publishers, New York, 585p.

Hernandez-Peon, R., 1966, Physiological mechanisms in attention, in: "Frontiers in Physiological Psychology", R.W. Russell, ed., Academic Press, New York.

Kaminski, M.J., Blinowska, K.J., 1991, A new method of the description of the information flow in the brain structures, Biol. Cybern. 65: 203-210.

Lindsley, D.B., 1960, Attention, consciousness, sleep and wakefulness, in: "Handbook of Physiology", Sect. 1: Neurophysiology III, Amer. Physiol. Soc., Washington. p. 1533-1593.

Lindström, S., Wróbel, A., 1990, Frequency dependent corticofugal excitation of principal cells in the cat's dorsal lateral geniculate nucleus, Exp. Brain Res. 79: 313-318.

Livingstone, M.S., Hubel, D.H., 1981, Effects of sleep and arousal on the processing of visual information in the cat, Nature. Lond. 291: 554-561.

Lopes da Silva, F., 1991, Neural mechanisms underlying brain waves: from neural membranes to networks, Electroenceph. clin. Neurophys. 79: 81-93.

Lopes da Silva, F.H., Van Rotterdam, A., Storm van Leeuven, W., Tielen, A.M., 1970, Dynamic characteristics of visual evoked potentials in the dog. II. Beta frequency selectivity in evoked potentials and background activity, Electroenceph. clin. Neurophysiol. 29: 260-268.

Rougeul, A., Bouyer, J.J., Dedet, L., Debray, O., 1979, Fast somato-parietal rhythms during combined focal attention and immobility in baboon and squirrel monkey, Electroenceph. clin. Neurophysiol. 46: 310-319.

Steriade, M., Curro Dossi, R., Pare, D., Oakson G., 1991, Fast oscillations (20-40 Hz) in thalamocortical systems and their potentiation by mesopontine cholinergic nuclei in the cat, Proc. Natl. Acad. Sci. USA. 88: 4396-4400.

Steriade, M., Llinas, R.R., 1988, The functional states of the thalamus and the associated neuronal interplay, Physiol. Rev. 68: 649-742.

Wilson, J.R., Friedlander, M.J., Sherman S.M., 1984, Fine structural morphology of identified X- and Y-cells in the cat's lateral geniculate nucleus, Proc. R. Soc. Lond. B 221: 411-436.

Yuan, B., Morrow, T.J., Casey, L., 1986, Corticofugal influences of S1 cortex on ventrobasal thalamic neurons in the awake rat, J. Neurosci. 6: 3611-3617.

SSR-MODULATION DURING SLOW CORTICAL POTENTIALS

Matthias Müller[1], Brigitte Rockstroh[1], Patrick Berg[1],
Michael Wagner[1], Thomas Elbert[2], Scott Makeig[3]

[1]Department of Psychology, University of Konstanz,
78434 Konstanz, Germany
[2]Institute of Experimental Audiology, University of Münster
48194 Münster, Germany
[3]Naval Health Research Center, San Diego, CA USA

INTRODUCTION

The present series of studies examines the relationship between slow event-related changes in electrical brain activity and higher-frequent activity evoked by a series of stimuli presented in rapid succession (around 40Hz).

The repetitive delivery of brief stimuli at a high enough rate causes event-related responses to individual stimuli to overlap, eliciting a so-called steady-state response (SSR; Galambos et al., 1981; Galambos and Makeig 1988; Makeig, 1989; Makeig, 1993a; Rohrbaugh et al., 1990). The acoustically evoked SSR has been shown to have a primary generator in the auditory cortex (Mäkelä and Hari, 1987; Mäkelä et al., 1990; Hari et al., 1989; Pantev et al., 1991; Pantev at al., in press; Makeig et al., 1992) and to be a "hybrid" of middle latency responses (Galambos et al., 1981). In a series of studies, Makeig (for summary see Makeig and Galambos, 1989; Makeig, 1993a,b) demonstrated systematic and relatively long-lasting variations in the amplitude and phase of the SSR following presentation of foreground auditory stimuli. Perturbations in the amplitude and phase of an ongoing SSR time-locked to a distinct stimulus, such as occasional omissions of a pulse interspersed within the continuous 40-Hz stimulus train, were labeled the "complex event-related potential" (CERP, Makeig, 1985). By varying attentional demands in an oddball paradigm, Makeig (1993) recently demonstrated a characteristic CERP phase advance following rare tones regardless of whether the subject was attending to the stimuli, while an above-baseline peak in SSR amplitude with a 400-msec latency occurred when subjects directed their attention to the target stimuli. Rohrbaugh et al.

Oscillatory Event-Related Brain Dynamics, Edited
by C. Pantev, Plenum Press, New York, 1994

(1990) also found a phase[1] advance in the SSR during a 200-msec period following a foreground stimulus in an orienting paradigm. Since this phase advance occurred parallel to the development of a negative slow wave the authors interpreted both phenomena to reflect a transient sensitization during orienting to the foreground stimuli.

In the studies reported here, the relationship between slow brain potentials and the SSR with its stimulus- and task-related perturbations was investigated. The SSR was regarded as a "tool" to explore the functional meaning of slow cortical potentials, such as the P300, the Contingent Negative Variation (CNV) and the Bereitschaftspotential. Elbert and coworkers (Elbert and Rockstroh, 1987; Rockstroh et al., 1989; Elbert, 1992, 1993) have previously suggested that such slow potentials indicate the modulation of excitation and excitability in cortical neuronal networks. This hypothesis is based on a model which may account for certain features observable in neuronal networks: information may be coded by an increase or a decrease in firing rates of neurons; it may equally well be coded through distinct spatial patterns of activation, or most likely by a combination of both (Van der Malsburg and Schneider, 1986), but certainly not through synchronous activation of many or all neuronal elements. Therefore, overexcitation must somehow be controlled by the brain's intrinsic mechanisms, and excitability must be regulated by imposing limits on the dynamic patterns of neural mass action (Elbert, 1993). It is well known that depolarization in the apical dendritic trees of cortical pyramidal cells results in surface negative potentials such as the CNV (Speckmann et al., 1984). These cells are the most likely candidates for enabling and "tuning" cortical excitability. The modulation of excitability, in turn, may be at the basis of attentional mechanisms, reflecting a preparatory state for cerebral processing in the underlying networks (Rockstroh et al. 1989). In contrast, slow positive shifts like the P300 imply a "disfacilitation" of cortical neural networks (Elbert, 1993). This hypothesis has been examined in two previous studies using a "probe" technique (Rockstroh et al., 1992; Rockstroh et al., 1993). In these experiments, occasional probe stimuli were presented during surface-negative potentials induced by a forewarned reaction time paradigm, or during an acoustic oddball paradigm, inducing a P300. Probe-evoked responses (reaction time, probe-evoked potential), indeed, varied with the slow potential shifts induced by the primary task, being facilitated during the development of the CNV but inhibited parallel to the development of the P300 (see also Woodward et al., 1991). It is important to note that both evoked and slower cortical potentials were partially generated within the same cortical regions.

However, a trade-off between the two tasks in dual-task paradigms has been shown to reduce the amplitude of cognitive slow potentials (Strayer and Kramer, 1990), a result which is consistent with the type of regulation described by Elbert & Rockstroh (1987, Elbert, 1993). By using the continuously driven SSR to probe excitability instead of responses to distinct and occasional probe stimuli, the impact of the dual task of responding to probes and responding to warning or oddball stimuli may be circumvented.

---

[1] "Phase" refers to the relationship or time interval between the onset of stimuli in the steady state train and the waveform of the SSR, defined as evoked potential bandpass-filtered at the probe stimulus repetition rate.

Results of three experiments, in which the presentation of a continuous train of stimuli (driving an SSR) was combined with an oddball task or a voluntary response task, will be reported. The modulation of SSR-amplitude following target stimuli (brief changes in the frequency of the SSR stimulus train) or following a voluntary button press was considered a measure of the modulation of cortical excitability during the oddball task and the voluntary response task. We hypothesized that if the continuous stimulus train was comparable to probes used in the previous experiments, then the probe train presented to excitable brain regions during the development of a negative-going potential shift in the respective regions should be processed more efficiently. In this case, we should expect a parallel increase in SSR-amplitude, and perhaps even a phase advance signifying an SSR latency reduction. If the train of probes would be presented to less excitable neuronal networks such as during a P300 in the oddball task and the positive-going motor potential complex in the voluntary response task, we might expect a relative suppression of central processing of the probes and, hence, a reduction in SSR amplitude and an accompanying phase retard. If, on the other hand, every stimulus and task, irrespective of their nature and accompanying slow potentials, cause similar perturbations of the SSR amplitude and phase, similar CERP amplitude shifts and/or phase advance would be found in both paradigms, and we must conclude that the time course of the CERP amplitude/phase does not follow the slow potential.

METHODS

Subjects

In all three experiments, healthy, right-handed student volunteers received course credit or monetary compensation for participation. Forty-five male subjects (mean age 22.2 years) participated in Experiment 1, ten subjects (5 male, 5 female, mean age 26.3 years) in Experiment 2, which was comprised of two sessions, and seventeen subjects (7 male, 10 female, mean age 24.8 years) in Experiment 3.

SSR-Generation, Apparatus and Electrophysiological Recordings

A steady-state stimulus train (SSR) was created by presenting 5-msec pulses (Gaussian envelope) of 1000 Hz (65 dB SPL (A)) at a rate of 40 per second in Experiments 1 and 2, and 39.25 per second in Experiment 3, with a zero rise and fall time of the stimuli. The SSR was presented via earphones monaurally to the right ear in Experiments 1 and 2 and binaurally in Experiment 3.

During the experiment the subject sat in a partially electrically shielded, dimly lit and sound-attenuated subject chamber. After being prepared for the physiological recordings, the subject received written instructions informing him/her about the stimuli and the task. The instructions also included a request to avoid blinks and eye or head movements by adopting a relaxed position and fixating a spot on the opposite wall. In order to familiarize the subject with the continuous stimulus train, the first two minutes of the experiment were comprised of only 1000 Hz frequency deviations.

An ASYST program running on an AT 386 computer controlled timing of the experimental stimuli and the storage of electrophysiological responses. Coulbourn Instruments modules were used to produce the acoustic stimuli.

EEG was recorded along the mid-sagittal line (Fz, Cz, Pz) in Experiments 1 and 2, and, in addition, from C3 and C4 in Experiment 3. Recordings were obtained with a DC-amplifier (MES, Munich) using a time constant of 5 seconds in Experiments 1 and 2, and DC in Experiment 3. In all experiments, the reference electrode was affixed to the left earlobe. Nonpolarizable silver-silver/chloride electrodes (ZAK) were used, and Grass EC2 electrolyte served as the conducting agent. The skin under the electrodes was prepared by cleansing with alcohol and rubbing with an abrasive paste (OMNIPREP). The vertical EOG was recorded using Ag/AgCL-electrodes (ZAK) centered about 1 cm above and below the left eye.

Data were filtered with a low-pass filter at 100 Hz and were digitized at a rate of 400 Hz (Experiments 1 and 2) or 312.5 Hz (Experiment 3). In Experiment 3 EMG was recorded by a pair of silver/silver-chloride electrodes (Beckman) placed on the forearm (musculus flexor pollicis longus). After amplification (gain 1000, time constant 0.025 sec) the EMG signal was rectified and integrated with a time constant of 20 msec. The integrated EMG was digitized at the same rate as the EEG recordings (312.5 Hz).

In order to separate SSR and slow potential responses, subject-averages[2] were separately filtered with a 25-Hz high-pass filter (2nd order) and with a low-pass filter of 15-Hz (Experiments 1 and 2) or 8 Hz (Experiment 3), respectively. Perturbations in the SSR time-locked to experimental events were analyzed by complex demodulation using a 5-sec smoothing window. The amplitude of the resulting time series correspond to the peak-to-baseline amplitude of the SSR, and its phase corresponds to the latency shifts of the SSR peaks relative to stimulus onsets[3].

Effects of foreground stimuli and tasks were evaluated by analyses of variance. For within-subject comparison of recording sites, p-values were obtained after adjusting the degrees of freedom by the Greenhouse-Geisser-Epsilon. Means ± standard errors are presented.

EXPERIMENT 1

In the first study, an oddball task was superimposed on the series of 1000-Hz pulses by shifting their frequency for 100 msec at a constant interval of 1.5 sec. Of the total of 1200 frequency-shift events, changes to 2000 Hz on 70% of the trials (840) constituted standard or non-target events, while changes to 500 Hz on 30% of the trials (360) constituted the target stimuli. Thus, the SSR shifted briefly every 1.5

---

[2] Prior to data analysis, artifacts were corrected using a computer program (Berg, 1986) that first classified blinks, muscle potentials, drifts or large DC-shifts on the basis of various templates. The influence of eye blinks on the EEG was then corrected using a regression analysis. The artifact correction procedure led to the rejection of five subjects in Experiment 1, two subjects in Experiment 2 and one subject in Experiment 3. For the remaining subjects, data analysis was based on 83.3% of the data in Experiment 1, 83.1% in Experiment 2, and 78.8 % in Experiment 3.

[3] This method can be considered comparable to the moving-average discrete Fourier transform applied by Makeig and Galambos (1989).

sec to either higher or lower frequency tone pulses. The relationship of frequency and standard/ target stimulus was counterbalanced across subjects[4]. Subjects were asked to press a button with their dominant hand as quickly as possible following every target.

Results

The frequency-shift stimulus in the oddball task elicited the expected ERPs including a sequence of N100 and P300 features (see Fig. 1a). As found for other auditory oddball stimuli (Segalowitz and Barnes, 1993), the frontally pronounced N100 reached a later peak and a larger amplitude following targets than following standards , differences being most pronounced at the frontal and central leads (p< .01 for all effects). The P300, peaking after 349 msec on average, was largest at the parietal recording in response to targets (p< .01 for all effects).

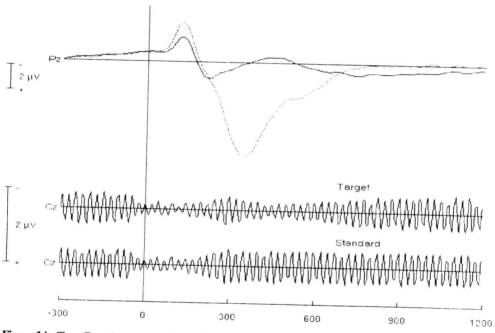

Figure 1A. Top: Grand average of the parietal ERP (in μV, negativity up) to oddball stimuli (targets: dotted line, standards: solid line) during 300 msec prior to and 1.2sec following the stimuli. The vertical line marks the time point of oddball stimulus presentation. Bottom: Examples of SSR perturbation at the central electrode parallel to the ERP, averaged across 12 subjects separately for targets and standards.

---

[4] No effects were obtained between the subgroups in an ANOVA accounting for the frequency/target relationship.

The 40-Hz stimulus train also drove an SSR[5] with a mean amplitude of 0.51±0.03 µV at Fz, 0.50±0.03 µV at Cz and 0.38±0.02 µV at Pz (baseline-to-peak, F(2,72)= 63.1, p< .001). Following the oddball stimuli SSR-amplitude was reduced by 0.32±0.02 µV (to 37 % of baseline) at Fz and Cz at 45.4±4.8 msec peak latency [6], and by 0.23±0.02 µV (to 39 %) at Pz (ELECTRODE, F(2,36)= 66.2, p< .01; see Fig.1a for the evoked responses, and Fig. 1b for the results of the complex demodulation). For the sake of brevity and clarity, this first amplitude reduction during the first 100 msec will hereafter be referred to as R100 . Following the R100, which did not differ between standard and target stimuli, SSR-amplitude increased to above-baseline levels. This amplitude augmentation above baseline near 400 msec (referred to as A300), described by Makeig as a regular feature of the CERP, was significant at all electrodes and under all conditions (F(1,36)= 168.0, p< .001). A300 was more pronounced following standard than following target stimuli (F(1,36)= 7.1, p< .05) and was reached earlier at Fz and Cz than at Pz (390±15 msec versus 434±15 msec, F(2,72)= 6.9, p< .01).

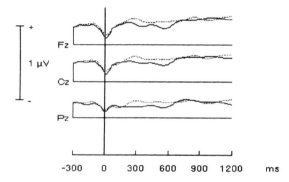

Figure 1B. Modulation of SSR amplitude as determined by complex demodulation, plotted separately for the recording sites and perturbations following targets (solid lines) and standards (dotted lines). Minus signifies amplitude reduction, plus signifies amplitude increase relative to baseline values. Time scale as in Fig.1a.

A second amplitude reduction, R400, ocurred at a mean latency of 422.1±7.6 msec by 0.16±0.02 µV at Fz and Cz and by 0.12±0.01 µV at Pz which corresponded to a reduction to 68% of baseline values at all three electrode sites; F(2,72)= 34.9,

---

[5] Data of 3 Ss who did not show any measurable synchronisation were not included in the analysis. Reports are based on 37 Ss.

[6] Because of a programming error, the interval between two successive 1000 Hz pulses was increased prior to every oddball stimulus from 25 to 44 msec. Although this delay was not subjectively noticeable, the SSR modulation around the oddball stimuli must have been affected by this "compound" stimulus, and phase shifts could not be analysed. However, effects of stimulus meaning such as differences between target and standard stimuli cannot be attributed to this error, which equally affected both types of stimuli.

p< .001. R400 was more pronounced following target than following standard stimuli; $F(1,36)= 1.9$, p< .01. The differences between stimulus types were larger at Fz (reduction to 50% and 86% of baseline values, resp.,) and at Cz (reduction to 50% and 84% of baseline levels, resp.) than at Pz (reduction to 55% and 81% resp.; ELECTRODE x STIMULUS; $F(2,72)= 16.9$, p< .001; see Fig.1). It is interesting to note that the peak latency of the R400 was shorter at Pz relative to Fz, while the latencies of the R100 exhibited a fronto-parietal gradient (p< .05). This shift in latencies suggests some connection of the second minimum with the P300 complex.

Mean SSR amplitude during the time segment 270 to 650 msec confirmed these results for the R400: an amplitude reduction by 0.11 ± 0.01 µV (to 76%) relative to baseline followed target stimuli, while the average amplitude returned to baseline levels following standard stimuli, this difference being more pronounced at Fz and Cz (reduction to 74% of baseline levels) than at Pz (to 80% of baseline levels; ELECTRODE x STIMULUS; $F(2,72)= 23.9$, p< .01; STIMULUS; $F(1,36)= 41.4$, p< .01).

The first study confirms earlier results: a CERP followed each stimulus, beginning within the first 100 msec, with a fronto-central predominance, and is followed by an augmentation above baseline in amplitude during the subsequent 300-400 msec. Furthermore, another perturbation of SSR-amplitud was observed following target stimuli only. A differentiation of SSR-amplitude by the present stimulus- and task-conditions began with the A300. It could be tempting to associate the following second SSR-amplitude reduction with the target-evoked P300, as it became pronounced in the same latency range. However, with every target-evoked P300 there was also a motor response; thus, either the combination of the two or one of the factors alone could have influenced the SSR near 400 msec following target stimuli. Experiment 2 was designed to separate these contributions, by comparing SSR perturbations when subjects pressed a button to targets, in one condition, or silently counted the targets in another.

EXPERIMENT 2

Under the same experimental conditions as described in Experiment 1, ten subjects participated in two experimental sessions on successive days. In one session their task was to respond with a motor response to every target (as in Experiment 1), whereas in the other session they were asked to silently count the number of targets. The order of these tasks was counterbalanced across subjects. Motivation to count was assured by asking the subject at irregular time intervals during the session for the number of targets counted so far. A frequency change to 500 Hz[7] constituted the target.

---

[7] In order to evaluate effects of tone frequency on SSR modulation, two of the ten subjects participated in two further sessions, in which a frequency change to 2000 Hz constituted the targets.

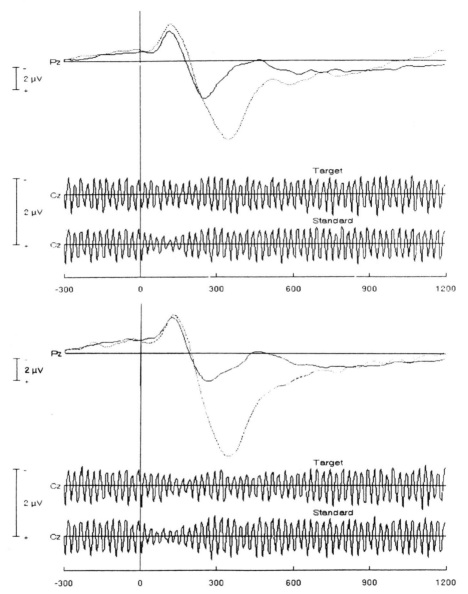

Figure 2A. Grand averages of the parietal ERP (in μV, negativity up) to oddball stimuli (targets: dotted line, standards: solid line) during 300 msec prior to and 1.2sec following the stimuli. Top: session with motor response required to targets, bottom: session in which targets were to be counted. The vertical line marks the time point of oddball stimulus onset. Below each ERP, an example of SSR perturbations at the central electrode is given, averaged across subjects separately for targets and standards.

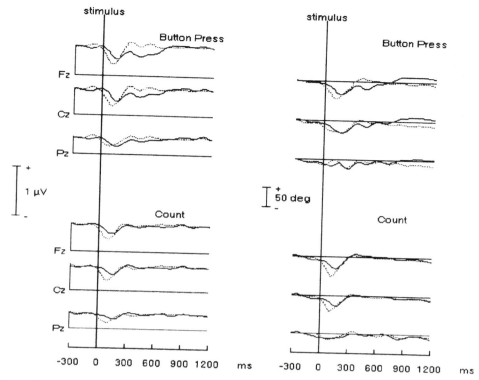

Figure 2B. Modulation of SSR amplitude (left) and phase (right) as determined by complex demodulation, plotted separately for the recording sites, perturbations following targets (solid lines) and standards (dotted lines), and the sessions with a button press required to targets (top) and the session with targets to be counted. Ordinate: Amplitude modulation in μV (minus indicating amplitude reduction) or phase modulation in degrees (minus indicating phase advance relative to baseline). Time scale as in Fig.2a.

Results

As in Experiment 1, subjects[8] developed the expected ERP to the oddball stimuli[9] (see Fig. 2a). While N100 did not differ between tasks or conditions, P300 was larger (F(1,7)= 19.0, p< .01) and peak amplitude was later (F(1,7)= 33.7, p< .001) after targets than after standards, in particular under conditions in which subjects indicated target detection by a motor response (STIMULUS x TASK; F(1,7)= 7.3, p< .05, STIMULUS x ELECTRODE; F(2,14)= 28.6, p< .01).

As reported for the first study, the 40-Hz stimulus train drove an SSR (mean amplitudes 0.59±0.03 μV at Fz, 0.55±0.03 μV at Cz and 0.37±0.02 μV at Pz, baseline-to-peak, F(2,7) = 91.3, p< .001). Oddball stimuli elicited an R100 with an average

---

[8] Data of two subjects, who did not show an SSR were excluded from further analyses, thus, results reported here are based on the data of eight subjects.

[9] including an N100 (mean peak latency 121.31±2.5 msec, mean amplitude -4.82±0.45 μV, frontal maximum, F(2,14)= 50.5, p< .001) and P300 (latency 325.35±7.3 msec, amplitude 5.16±0.5 μV, centro-parietal dominance, F(2,14)= 140.3, p< .001.

minimum at 97.6±7 msec (see Fig. 2a) to 52 % of baseline level at Fz, 51% at Cz and 49% at Pz (no significant difference between sites). The peak latency of the R100 was 55 msec later following target relative to standard stimuli; $F(1,7)= 10.9$, $p< .05$. The peak R100 amplitude was more pronounced following standards than targets (ELECTRODE x STIMULUS; $F(2,14)= 4.5$, $p< .05$)[10].

The phase component of the R100 was a phase advance (see Fig.2b) by 48° (of 3.3 msec) at Fz, 42° (2.9 msec) at Cz and 34° (2.4 msec) at Pz, peaking 126.8 msec after the oddball stimulus onset, with a pronounced fronto-parietal gradient ($F(2,14)= 4.8$, $p< .05$) which did not differ between stimulus conditions or tasks.

The A300 (SSR amplitude augmentation) in the latency range of 340 - 410 msec was smaller following targets (2 % of baseline values) and larger following standards (by 17 %) when subjects indicated target detection by button press; smaller differences (17 % versus 16 %) occurred when subjects silently counted the targets (TASK x STIMULUS, $F(1,7)= 5.4$, $p< .06$[11]). A slight phase retard of 11.5° (or 0.8±.01 msec) occurred later following targets (490±17 msec) than following standards (410±13 msec, $F(1,7)= 6.1$, $p< .05$), but was not affected by task.

Figure 2 also shows the presence of an R400 amplitude reduction. For mean SSR amplitude during the time interval 270 to 650 msec following oddball stimulus onset, a TASK x STIMULUS interaction indicates a more pronounced R400 when subjects pressed the button following targets (TASK x STIMULUS, $F(1,7)= 9.2$, $p< .05$[12]). Following target stimuli, mean amplitude at the R400 peak was a reduction to 88% of baseline, the modulation being larger at Fz (reduction to 86%) than at Cz and Pz (reduction to 91%). Following standard stimuli there was no change in mean amplitude from the baseline level for any of the electrodes (ELECTRODE x STIMULUS; $F(2,14)= 8.4$, $p< .01$). The R400 was more pronounced at Fz (to 89 % at Fz) than at Cz (93 %) and Pz (92%) when subjects pressed a button, wheras there was no difference between electrode sites when subjects counted the targets (reduction to 96%, 98%, 94%, resp. , TASK x ELECTRODE, $F(2,14)= 7.8$, $p< .01$). Parallel to the R400, a phase advance was larger (17°, 1.2 msec) in response to targets than in response to standards (10° or 0.7 msec; $F(1,5)= 7.9$, $p< .05$), while no significant task effects could be found. Subsequent to this R400, mean amplitude modulations (650-1450 msec) by stimulus types and tasks were reversed: targets for which a button press was required produced a sustained amplitude augmentation of 5.8%, while an amplitude reduction by 4 % of baseline was observed following standard stimuli. Targets to be counted, however, produced only a slight amplitude

---

[10] This effect has to be attributed to the tone frequency: When – in two Ss – the target stimulus was a higher frequency (2000 Hz), the relationship between amplitudes to targets and standards was reversed, i.e. the higher tones, whether targets or standards, whether counted or answered with a button press, induced SSR amplitude reductions of about 0.3 μV, lower tones relative to the 1000-Hz stimulus train induced SSR amplitude reductions of about 0.1 μV.

[11] Post-hoc comparisons confirmed the significant stimulus difference under conditions of button press ($t= 2.8$, $p < .05$), and the larger A300 following counted targets ($t=2$ 2.4, $p< .05$), while differences between stimulus types under counting conditions did not reach significance ($t= 0.5$).

[12] Post-hoc t-tests confirmed a significantly larger R400 following a button press relative to targets to be counted ($t= 3.2$, $p< .05$), and relative to standards of both types ($t= 4.1$, $p< .05$), while R400 did not differ significantly between targets and standards under the counting condition.

reduction, while no change relative to baseline levels followed standard stimuli ($F(1,7)= 5.5$, $p< .05$).

The results of this second study closely replicated the findings of Makeig (1993). The interaction of stimulus and task suggests that both motor responding and stimulus processing, with their related slow potentials, may contribute to the observed SSR perturbations. The R100 was equally pronounced at all recording sites (as percent reduction from baseline), suggesting that a common set of generators or equally effective modulation of all generators contributes to the response at all three sites. The R100 was sensitive to physical (tone) frequency but not to psychological (task-related) aspects of the stimuli. The subsequent CERP features (A300 and R400) revealed an impact of stimulus and task relevance.

Experiment 3 was designed to explicitly examine SSR perturbations related to slow potentials that develop prior to and following motor responses. To our knowledge, SSR perturbations parallel to the development of a surface-negative Bereitschaftspotential (BP), the negative motor potential and the positive-going motor potential complex has not been studied previously. As described in the introduction, the slow postive shifts like the P300, or the slow positive changes following a motor response, have been viewed as widespread disfacilitatory responses (Elbert, 1993). A parallel reduction in SSR-amplitude and a phase retard is constistent with this model. The Bereitschaftspotential seems to be restricted to generators in the somatomotor cortex and areas in the frontal lobe (Rockstroh et al., 1989). A facilitation of neuronal activity in these regions should not have much of an effect on the SSR generators in the auditory cortex.

EXPERIMENT 3

In order to elicit movement-related potentials, sixteen right-handed subjects were instructed to press a button with their dominant hand approximately every 10 seconds while the stimulus train was delivered to both ears through headphones at 65 dB (1000 Hz tone frequency, 39.25 Hz stimulation rate). A total of 250 button presses were collected per subject[13].

The course of the slow potential prior to and following the voluntary button presses was described by the following components (named according to Neshige et al., 1988): mean potential during 1.5 - 0.5 sec prior to the button press[14], the Bereitschaftspotential (BP); mean potential during the last 500 msec prior to the button press, the negative slope (NS) of the BP; the negative peak surrounding the button press, the motor potential (MP). Post-response components were described by the maximum positive peak (positive motor potential, PMP), as well as mean potential during the 500 msec interval following the button press (motor potential complex, MPC); mean potential during the interval 0.5 - 2 sec following the button

---

[13] Prior to the experimental period, subjects were trained to estimate the time interval to avoid their using a distracting strategy of silently counting to ten between every button press.

[14] In this analysis, components are related to the time point of the button press (closure of the microswitch). Analysis relative to the first EMG response will be reported elsewhere.

press, slow motor potential, MPS). SSR perturbations during similar time intervals were related to these scores.

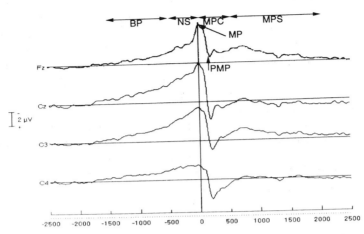

Figure 3A. Top: Grand average of slow potentials 2.5 sec preceding and 2.5 sec following the button press (in µV, negativity up), plotted separately for the frontal and three central recording sites. The vertical line marks the time point of microswitch closure. BP, NS, MP, PMP, MPC, MPS: scores determined to describe the course of the slow potentials (see text).

Results

The ERP revealed a Bereitschaftspotential prior to the voluntary button press in all subjects except one. Mean potential was -1.56±0.2 µV during the interval 1.5 - 0.5 sec prior to the button press (BP), - 3.1±0.3 µV during the last 500 msec before the button press (NS), and -4.3±0.4 µV during the motor potential (MP) at the time point of the button press (mean latency 8.4±5.5 msec relative to the switch closure). For each component, negativity (relative to baseline) was most pronounced at Fz, followed by Cz and C3 (see Fig. 3a)[15]. As expected, the pre-movement negativities (NS and MP) were larger over C3 than C4 (t(15)= 1.9, p< .1 and 5.1, p< .01), i.e., contralateral to the moving hand. Button presses were followed by a positive peak at 161.6±8.5 msec (PMP, mean amplitude +1.85±0.5 µV), followed by a slower positive deflection (MPC). Both post-motor potentials exhibited negative values at the frontal, and positive values at centro-parietal recording sites, with less pronounced amplitudes over the brain areas for which larger pre-response negative shifts had been observed (for the difference C3-C4: t(15)= 2.3, p< .05). Only the PMP differed

---

[15] The development of these negative SCPs was confirmed by an ANOVA including baseline values and the three components (COMPONENT x ELECTRODE, F(12,180)= 16.97, p< .001; COMPONENT: F(3,45)= 24.4, p< .001; ELECTRODE: F(4,60)= 13.2, p< .001) as well as by an ANOVA comparing difference values (baseline - BP, NS, and MP: COMPONENT x ELECTRODE: F(8,120)= 19.2, p< .001; COMPONENT: F(2,15)= 20.6, p< .001; ELECTRODE: F(4,60)= 13.2, p< .001). The increase in negativity relative to baseline was 1.2±0.1 µV during the BP, 2.6±0.3 µV during the last 500 msec before the button press and 3.9±0.4 µV during the MP.

336

significantly from baseline values (F(1,15)= 6.1, p< .05), while differences between baseline and MPC reached significance only at Fz, Pz and C3 (F(4,60)= 7.1, p< .01). The post-movement positivities (PMP and MPC) were smaller at C3 than at C4 (t(15)= 2.3 and 3.6, p< .05).

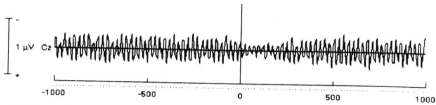

Figure 3B. Example of SSR averaged across 16 subjects driven by the 39.25 Hz stimulus train in μV, from the central recording site during an interval of 1 sec before and 1 sec following the button press (marked by the vertical line).

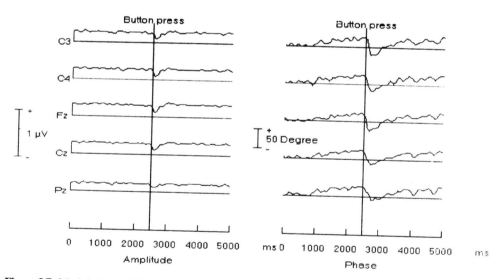

Figure 3C. Modulation of SSR amplitude (left) and phase (right) as determined by complex demodulation, plotted separately for the five recording sites. Ordinate: SSR-amplitude change in μV (minus indicating amplitude reduction) and phase modulation in degrees (minus indicating phase advance relative to baseline). The vertical line marks the time point of switch closure.

All 16 subjects showed SSR-synchronization at 39.25 Hz, SSR amplitude being 0.29±0.03 μV at Fz, 0.27±.02 at Cz and at Pz (0.21±0.02 μV, baseline-to-peak; see Fig. 3b for an example of the central SSR). This gradient was observed not only under baseline conditions but for all response-related time periods (p< .01). SSR amplitude did not change significantly during the period preceding the button press. The button press was followed by a marked reduction in SSR amplitude (R100; see Fig. 3b and c), which was significant for both measures, the peak amplitude change following the button press, and mean amplitude in the 500 msec post-response (parallel to the MPC): Relative to baseline values, SSR amplitudes were reduced to a peak minimum of 46.7 % of baseline after an average latency of 105.0±8.2 msec following the button press ($F(1,15)= 47.9$, $p< .001$ for the R100, $F(1,15)= 13.8$, $p< .01$ for the MPC-related SSR perturbation). Although SSR amplitudes prior to and following the button press were (non-significantly) smaller at C3 than at Cz and C4, amplitude changes from baseline did not differ significantly between recording sites. Following the R100, SSR amplitudes returned to baseline values [16]. (Ten out of the 16 subjects demonstrated an ongoing slow negative shift, i.e., negative MPS relative to baseline.)

CERP phase parallel to the development of the negative slow potential preceding the voluntary movement showed a retard of 1.21±0.5 msec (16.8°) beginning 1.5 sec before the button press and of 1.83±0.5 (25.4°, $p< .01$) msec during the negative slope, resp. (see Fig. 3c). Phase modulation did not differ between recording sites. A phase advance by 1.99±0.6 msec (27.8°, $p< .01$) relative to baseline values at a mean peak latency of 161.6±19.3 msec following the button press vanished during the subsequent 500 msec (no significant difference from baseline for the phase perturbation parallel to the MPC).

In Experiment 3, SSR-amplitude was smaller than the SSR-amplitudes obtained in the preceding experiments. A major difference in the designs for the two experiments was that subjects had to pay attention to the continuous stimulus train in the first two experiments in order to detect changes in tone frequency, while no attention to the stimulus train was required in the last study. Possibly, the boring task of pressing a button for long periods of time may have distracted subjects from attending the stimulus train or may have induced drowsiness, a condition known to affect SSR amplitude (Galambos and Makeig, 1988). The voluntary movement clearly induced a CERP with features in many respects comparable to those described in the previous studies (% amplitude reduction and phase advance). The reduction in SSR amplitude during the execution of a motor response is in line with the results of Böcker et al. (1993), who observed smaller SEP amplitudes to distinct somatosensory probes during the movement (250 msec after the motor response), compared to SEPs to probes prior to the response or following the response at a later point in time. However, the SSR in this experiment is thought to be generated mainly in the auditory and not in the somatomotor cortex. While the size of the phase advance during the CERP is in line with the results obtained in Experiment 2, as well as results reported by Makeig (1993), the phase retard parallel to the development of the negative Bereitschaftspotential contrasts with findings of Rohrbaugh et al. (1990), who described a phase advance parallel to the negative Slow wave following an auditory stimulus, which was attributed to an orienting response.

---

[16] No significant differences from baseline were found for the mean SSR amplitude during 0.5 - 2 sec following the button press (parallel to the MPS in the SCP) at Cz and Pz, while at Fz, SSR amplitude exceeded the baseline SSR amplitude by 5.1% (BL-MPS x ELECTRODE, $F(4,60)= 6.1$, $p< .01$, ELECTRODE, $F(4,60)= 11.1$, $p< .01$).

DISCUSSION

The series of experiments reported in this chapter clearly demonstrate perturbations of the auditory SSR following auditory stimuli, and particularly in conditions that require stimulus evaluation and motor performance. Under these conditions, auditory stimuli also evoked slow potential shifts typically found in the oddball and in the voluntary response paradigm. As we know of only very few studies (other than Makeig, 1993) systematically exploring the relationship between SSR perturbations and slow potentials, the interpretation of the present results must remain speculative.

In the present experiments, a CERP amplitude reduction and phase advance, relative to SSR baseline – was consistently observed following a foreground stimulus, such as the oddball frequency shifts, but also following voluntary motor responses. This finding replicate results by Makeig (1993). However, varying the stimulus meaning (non-target and target events) provoked an additional SSR perturbation during the time period, during which evaluation of and response to a meaningful stimulus is supposed to take place.

Two hypotheses guided the present series of experiments and provide a framework for discussion of the results: as outlined in the introduction, SSR modulation is an appealing approach to studying the functional meaning of slow potentials. Within the framework of our model (Elbert and Rockstroh, 1987; Elbert, 1993), which considers slow negative potentials to indicate enhanced excitability, and surface positive slow potentials to indicate "disfacilitation" of underlying neuronal networks, the processing of the SSR stimulus train should vary accordingly: if increased excitability corresponds to facilitated input processing, a larger SSR might be expected when negative slow potentials are generated in temporal regions, in the same way as we found evoked responses to occasional probe stimuli to be larger (Rockstroh et al., 1993). If reduced excitability during positive slow potentials corresponds to disfacilitated input processing, smaller SSR amplitudes might be expected during positive slow potentials, such as the P300 and the positive motor potential, in a similar way as we found evoked responses to occasional probes to be smaller during the development of a P300 (Rockstroh et al., 1992). The present results support the latter relationship. No correspondence between slow negative potentials preceding the voluntary response and SSR amplitudes was found. These results are consistent with the assumption that the positive potentials indicate widespread disfacilitation while nagtive shifts are generated specifically in task-related brain regions (Rockstroh et al., 1989, Elbert, 1993). The regulation of slow negativities is therefore closely linked to attentional regulation.

Makeig (1993) links SSR perturbations to shifts in attention. If processing of a foreground stimulus as well as focusing on the execution of a cued or voluntary button press binds attentional capacities, as well as their neuronal substrates, similar perturbations might be expected for different tasks. This hypothesis is supported by the similarity of SSR perturbations following an event in the three experiments. We may even attribute the phase retard during the preparation of the voluntary response to such shifts in attention, if we assume that during this interval, the subjects' attention was focused on the motor response and distracted from the high-frequency stimulus train.

The finding of a CERP induced by a voluntary motor response seems remarkable and highly interesting. It suggests an influence of the preparation and execution of a motor response, known to be generated in the motor cortex and motor-related areas, on the SSR, generated in the auditory cortex. A suppression of auditory input prior to, during or just following a voluntary movement, which certainly represents controlled processing, should have an impact not only on the conceptualization of the SSR, but also on hypotheses about attentional processes.

It remains to be explored in further experiments, as to what extent the SSR perturbations are global, task-specific and/or modality-specific. If the intensity of SSR perturbations vary systematically with slow potential amplitudes, relating the two phenomena in a functional sense. On the other hand, the conclusion reached by Eckhorn (this volume) that stimulus-dominated synchronization inhibits cortex-dominated synchronization might provide a framework for an interpretation of the present findings.

Acknowledgment

Research was supported by the Deutsche Forschungsgemeinschaft (Ro 805). We gratefully acknowledge the assistance of A. Heinz, S. Schrader, K. Oberhauser, and W. Formhals in data collection and analysis.

REFERENCES

Berg, P. , 1986, The residual after correcting event-related potentials for blink artefacts. *Psychophysiol.* 23: 354-364.

Böcker, K., Forget, R. and Brunia, C.H.M. , 1993, The modulation of somatosensory evoked potentials during the foreperiod of a forewarned reaction time task. *J.Electroenc. Clin. Neurophysiol*

Elbert, T., 1992, A theoretical approach to the late components of the event-related brain potential. In: Information Processing in the Cortex. A. Aertsen, V. Braitenberg (eds.) Heidelberg, Springer-Verlag, pp. 225-245.

Elbert, T., 1993, Slow cortical potentials reflect the regulation of cortical excitability, *in*:: Slow Potential Changes of the Human Brain, W.C. McCallum, S.H. Curry, (eds) New York, London, Plenum Publishing Corp., pp 235-252.

Elbert, T. and Rockstroh, B. , 1987, Threshold regulation – a key to the understanding of the combined dynamics of EEG and event-related potentials. *J. Psychophysiol.* 4: 317-333.

Galambos, R., Makeig, S., and Talmachoff,P., 1981, A 40 Hz auditory potential recorded from the human scalp, *Proc. Natl. Acad. Sci, USA*, 78: 2643-2647.

Galambos, R., and Makeig, S., 1988, Dynamic changes in steady-state potentials. *in*: Dynamics of Sensory and Cognitive Processing of the Brain, E. Basar, ed., Springer , Berlin/Heidelberg, 178-199.

Hari, R., Hämäläinen, M. and Joutsiniemi, S. , 1989, Neuromagnetic steady-state responses to auditory stimuli. *J. Acoust. Soc. Am.* 86: 1033-1039.

Makeig, S. Studies in Musical Psychobiology, 1985, University Microfilms, Ann Arbor.

Makeig, S., and Galambos, R. , 1989, The CERP: Event-related perturbations in steady-state responses, *in*: Brain Dynamics: Progress and Perspectives, E. Basar and T. Bullock, eds., Springer , Berlin/Heidelberg, 373-400.

Makeig, S. , 1993a, Effects of attention and stimulus probability on the auditory complex event-related potential, (submitted).

Makeig, S., and Inlow, M. 1993b, Lapses in alertness: coherence of fluctuations in performance and EEG spectrum,*J. Electroenc. Clin. Neurophysiol.*, 86: 23-35.

Makeig, S., Pantev, C., Schwartz, B., Inlow, M. and Hampson, S. , 1992, The auditory compoex event-related field to omitted stimuli. Amsterdam: *Excerpta Medica* , 165-170.

Mäkelä, J.P. and Hari, R. , 1987, Evidence for cortical origin of the 40 Hz auditory evoked response inman. *Electroenceph. clin. Neurophysiol.* , 66: 539-546.

Mäkelä, J.P., Karmos, G., Molnar, M., Csepe, V. and Winkler, I. , 1990, Steady-state responses from the cat auditory cortex. *Hear. Res.* 45: 41-50.

Neshige,R., Lüders,H., Shibasaki,H., 1988, Recording of movement-related potentials from scalp and cortex in man. *Brain*, 111: 719-736.

Pantev, C., Makeig, S., Hoke, M., Galambos, R., Hampson, S. and Gallen, C. , 1991, Human auditory evoked gamma band magnetic fields. *Proc. Natl. Sci. USA* , 88: 8996-9000.

Rockstroh, B., Elbert, T., Canavan, A., Lutzenberger, W., Birbaumer, N., 1989, Slow Cortical Potentials and Behavior, Urban and Schwarzenberg, Munich.

Rockstroh, B., Müller, M., Elbert, T. and Cohen, R. , 1992, Probing the functional brain state during P300-evocation. *J. Psychophysiology*, 6: 175-184.

Rockstroh, B., Müller, M., Wagner, M., Cohen, R. and Elbert, T. , 1993, "Probing" the nature of the CNV. *J. Electroenc. Clin. Neurophysiol*, (in press).

Rohrbaugh, J., Varner, J., Paige, S., Eckardt, M. and Ellingson, R. , 1990, Event-related perturbations in an electrophysiological measure of auditory sensitivity: Effects of probability, intensity and repeated sessions. *Int. J. Psychophysiology*, 10: 17-32.

Segalowitz, S.J. and Barnes, K.L., 1993, The reliability of ERP components in the auditory oddball paradigm. *Psychophysiology*, 30: 436-450.

Speckmann, E.-J., Caspers, H. and Elger, C.E. , 1984, Neuronal mechanisms underlying the generation of field potentials, *in*: Self-Regulation of the Brain and Behavior, T.Elbert, B.Rockstroh, W.Lutzenberger, N.Birbaumer, eds., Springer, Berlin, 9-25.

Strayer, D.L. and Kramer, A.F. , 1990, Attentional requirements of automatic and controlled processing. *J.Exp. Psychology*, 16: 67-82.

Van der Malsburg, C. and Schneider, W. , 1986, A neural cocktail-party processor. *Biological Cybernetics* , 54: 29-40.

Woodward, S.H., Brown, W.S., Marsh, J.T. and Dawson, M.E. 1991, Probing the time course of the auditory oddball P3 with secondary reaction time. *Psychophysiol.*, 28: 609-618.

SYNCHRONOUS OSCILLATIONS IN SENSORIMOTOR CORTEX OF AWAKE MONKEYS AND HUMANS

Venkatesh N. Murthy, Fumi Aoki and Eberhard E. Fetz

Dept. of Physiology & Biophysics, and
Regional Primate Research Center
University of Washington
Seattle, WA 98195, USA

INTRODUCTION

Oscillatory neural activity in the frequency range of 20-70 Hz has been observed in various sensory and motor cortical areas in awake and anesthetized mammals (Freeman, 1978; Eckhorn et al., 1988; Gray and Singer, 1989; Engel et al., 1991; Murthy and Fetz, 1992; Kreiter and Singer, 1992; Singer, 1993). High-frequency oscillations in the visual cortex have been suggested to play a role in associating stimulus features and in segmentation of objects in the visual scene (Eckhorn et al., 1988; Gray et al., 1989). Synchronous 20-40 Hz oscillations of local field potentials (LFP) have also been observed in the sensorimotor cortex of behaving monkeys (Rougeul et al., 1979; Murthy and Fetz, 1992; Murthy et al., 1992; Sanes and Donoghue, 1993) and humans (Sheer, 1984; Ribary et al., 1991; Pfurtscheller and Neuper, 1992). In awake, behaving monkeys, synchronous LFP oscillations were found during performance of trained motor tasks (Sanes and Donoghue, 1993) as well as during less constrained exploratory movements of the arm and hand (Murthy and Fetz, 1992). This suggests that the oscillations may play a role in attention or facilitating interactions during sensorimotor integration.

Previously, we showed that the LFP oscillations occurred more frequently during exploratory hand and arm movements that involved ongoing sensorimotor coordination, than during periods of repetitive constrained wrist movements (Murthy and Fetz, 1992). Moreover oscillations occurred synchronously over wide areas (Murthy et al., 1992), suggesting the hypothesis that the oscillations facilitate associations between cells involved in the same task. This hypothesis predicts that synchronization of LFP oscillations would tend to occur between coactivated cortical sites and be specific to particular behaviors. For instance, oscillations in pre- and post-central sites could become synchronized during a task that involved integration of somatosensory signals with movements. To test this hypothesis we quantitatively estimated the coherence between paired sites in pre- and post-central cortex during different behaviors. We also recorded bilaterally in the hand areas of the motor cortex when the monkey performed a manipulatory task with either hand alone and with both hands together. The associational hypothesis predicts that the LFP oscillations at the two sites would be synchronized more often during the performance of bimanual manipulations.

To determine if oscillations observed in monkeys could also occur in humans under similar behavioral conditions, we obtained subdural recordings from the sensorimotor cortex of human patients while they performed a visuomotor task. We report here the presence of clearly identifiable oscillations in electrocorticograms (ECoG) over particular sites in sensorimotor cortex which were correlated with the preparatory and execution phases of a two-dimensional tracking task. Recently, similar results were reported for electroencephalograms (Pfurtscheller and Neuper, 1992) and magnetoencephalograms (Kristeva-Feige et al., 1993).

METHODS

Two monkeys (*Macaca mulatta*) were trained to generate alternating flexion and extension torques about the wrist to track a visual target on a screen with a cursor. During this task both hands were restrained and the active hand generated isometric torque to control the cursor. The monkeys also readily performed unconstrained reaching movements to obtain pieces of food offered either to the side of their heads, in the slots of a visible Klüver board, or from an unseen box.

After the monkeys were trained, they underwent aseptic surgery in which stainless steel chambers were implanted over the sensorimotor cortical area for chronic recordings. Recordings were made from the sensorimotor cortex with two or more glass-coated tungsten microelectrodes (1-2 MΩ impedance) simultaneously while the monkeys performed the above tasks. The signal from each electrode was filtered at two band-pass settings to record LFPs (10-100 Hz) and unit activity (0.3-3 kHz). In some experiments, electromyograms (EMGs) were recorded from forearm muscles with pairs of stainless steel wires. In a few experiments behavior was recorded on videotape simultaneously with LFP traces and audio clicks generated from oscillatory cycles to correlate movements with oscillatory episodes. Data were recorded on FM tape for off-line analysis.

In collaboration with Dr. George Ojemann, epicortical recordings were made from the sensorimotor cortex of human patients treated for epilepsy at the University of Washington Medical Center. A week before surgical resection of cortical tissue, subdural grid electrodes were implanted over the central and temporal cortex to provide estimates of the location of epileptic foci and to map the sensory and motor sites. The location of the central sulcus was estimated from motor responses evoked by stimulation and from somatosensory evoked potentials recorded at various sites (Ojemann and Engel, 1987). After the subject recovered from implantation surgery, recordings were made from electrodes overlying the sensorimotor cortex while the subject performed a 2-dimensional tracking task. This consisted of following a moving target on a video screen with a cursor controlled by a hand-held joystick. Surface EMGs were monitored over the forearm and digit muscles. ECoG, EMG and cursor movements were recorded on magnetic tape for subsequent analysis.

Data Analysis

To monitor changes in oscillatory activity, power spectra of LFPs were calculated for 256-msec windows which were shifted successively by 100 msec. Average power in different bands was plotted as a time series along with LFPs (and EMGs if available), allowing changes in power to be correlated with behavior. To document the activity associated with oscillatory cycles, we calculated cycle-triggered averages (CTA) of LFPs and EMG activity, and histograms of unit activity aligned on the cycles of LFP oscillations (Murthy and Fetz, 1992). Since oscillatory episodes in the LFP occurred during a small proportion of total recording time, the CTAs captured the patterns of unit firing and associated LFPs specifically during LFP oscillations.

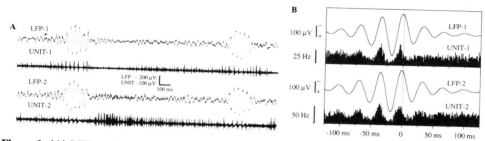

Figure 1. (A) LFPs and unit activity recorded in the motor cortex with two electrodes separated by ~800 mm. The monkey was retrieving a raisin from a Klüver board. (B) Cycle-triggered averages of LFPs and unit activity aligned on the cycles of LFP-2. (From Murthy and Fetz, 1992.)

Pairwise cross-correlations of LFPs (digitized at 1 kHz) from different sites were also performed to quantify the extent of synchronization. Correlations were calculated for time windows of 200 msec, successively shifted by 100 msec. For each window, the peak correlation closest to 0 delay and the corresponding time shift were obtained. A probability distribution of the correlation peak amplitudes (one point from each 200-msec segment) was made for recording epochs of up to 50 sec. This distribution was integrated to yield a cumulative probability distribution. Since high correlations could occur purely by chance, especially if the two correlated signals had similar frequencies, we determined a significance level by the following procedure. For the same recording epochs, cross-correlations were recalculated with one of the channels shifted by a time interval much larger than the oscillatory episodes, usually 1 sec. From the cumulative distribution of the "shifted" correlations, the significance level for the regular correlations was taken as the level below which 95% of the "shifted" correlations lay.

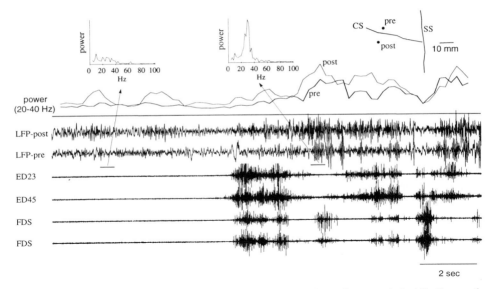

Figure 2. Pre- and post-central LFPs and EMGs of contralateral muscles recorded while the monkey was reaching to the side of its head to wrest a raisin from the experimenter's hand. For both LFPs the power in the 20-40 Hz band (calculated for 256-msec windows that were shifted successively by 100 msec) is plotted above the LFPs. After an initial period when the monkey was sitting quietly without any overt movement, a raisin was offered to the side. When the monkey reached for the raisin, oscillations increased at both cortical sites. Insets show the power spectra for precentral LFP at times indicated by the horizontal bars. Right top indicates the approximate location of the recorded sites.

RESULTS

Properties of Oscillations

Figure 1 shows examples of LFP and unit oscillations in the hand area of the motor cortex of a monkey retrieving a raisin from a Klüver board. During oscillatory episodes, the power in the 20-40 Hz band increased by more than 10 times above the baseline. When clear oscillations appeared in the LFP traces, the units fired in bursts during the negative deflections in LFP; these units were also active in the absence of LFP oscillations. CTAs of LFPs and units shown in Fig. 1b confirm the synchrony between the LFP oscillations and the firing of units. In order to correlate the occurrence of oscillations with movement, we recorded EMGs from forearm muscles simultaneously with LFPs from pre- and post-central sites. During the premovement period, when the monkey was waiting with its arm unrestrained (and the EMGs were silent), there was little oscillatory activity (Fig. 2). When the monkey was presented with a raisin to the side of its head, it attempted to wrest it from the experimenter's hand. Oscillations appeared at both pre- and post-central recording sites, although these episodes were not time-locked to the onset of EMG. During this period of exploratory movements, oscillations occurred much more frequently than in the quiet period. The following parameters of the oscillations were compared under different behavioral conditions: the average number of cycles per oscillatory episode, the mean frequency of oscillations within episodes, and the number of oscillatory episodes per second. The mean number of cycles per episode (4.5) and the mean frequency of oscillations (27 Hz) remained similar across behaviors, but the number of oscillatory episodes per second was different (Murthy and Fetz, 1992). When the monkey was quiet or performing the overtrained flexion-extension task, oscillations occurred at a rate of 0.2 per second; during unconstrained exploratory movements their occurrence increased to almost 1 per second (Fig. 3). There was no consistent relation between the bursts of EMG and LFP oscillations, but CTAs of rectified EMGs often revealed an oscillatory modulation synchronized with the LFP oscillations (Murthy and Fetz, 1992).

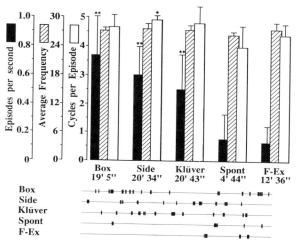

Figure 3. Parameters of LFP oscillations for different behaviors. The number of cycles per episode, episodes per second, and average cycle frequency were calculated as described (Murthy and Fetz, 1992). The amount of recording time for each behavior is indicated below each set of bars. Shown below are triggers from cycles of oscillations for a typical 20-sec period for each behavior. Behaviors were: Box, retrieving food pieces from a box at the level of the hip outside the range of vision; Side, retrieving pieces of food from the experimenter's hand at the side of the monkey's head; Klüver, retrieving pieces of food from the slots of a Klüver board; F-Ex, alternating flexion and extension of wrist; Spont, quiet sitting, no overt movement. (From Murthy and Fetz, 1992.)

Synchronization Between Sensory and Motor Cortical Sites

The LFP oscillations could occur synchronously over a wide area of the sensorimotor cortex. Ipsilaterally, the oscillations could become synchronized between pre- and post-central sites (Fig. 4). There was no detectable change in the phase of oscillations from anterior to posterior sites. However, traveling waves of oscillations similar to those suggested to occur in humans (Ribary et al., 1991) would have a velocity too large to generate significant phase shifts for spatial separations of our electrodes (time shifts would have been less than 1 msec). Synchronization of LFP and multiunit activity in the postcentral and precentral sites suggests that coherent oscillations affect large populations of neurons and may play a role in associating their activity.

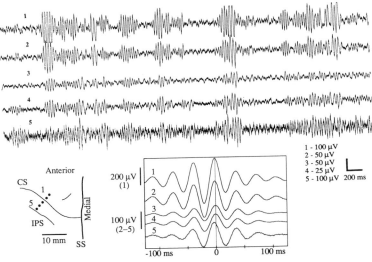

Figure 4. LFPs recorded simultaneously at five sites in pre- and post-central cortex (indicated on the sketch of cortex at bottom left) when the monkey was reaching for a raisin offered to the side of its head. Oscillations occurred synchronously at all sites, with negligible phase shift, as confirmed by cycle-triggered averages aligned on the cycles of LFP-1 (lower right). (From Murthy and Fetz, 1992.)

Synchrony and Behavior

If the oscillations play a role in associating activity at specific sensory and motor areas, the synchronization of oscillations at different sites could depend on behavioral context. This associational hypothesis would predict that different cortical sites may exhibit independent oscillations, which become synchronized preferentially during those particular behaviors that require coordination of neural responses at these sites. To test this hypothesis we recorded LFPs at two independent sites simultaneously, in pre- and post-central sites in the distal arm representation area, while the monkey (1) rested quietly without any overt movement, (2) performed the over-trained wrist flexion and extension task, and (3) performed reaching movements to retrieve raisins from unseen locations. As previously described, the oscillations occurred more often during unconstrained reaching movements than during repetitive trained movements. For the same recordings, cross-correlations between pairs of LFPs and the autocorrelations for each LFP were calculated for each behavior. If either site or both sites exhibited LFP oscillations, as judged by the magnitude of the secondary peak in the autocorrelations, the corresponding cross-correlation peak was measured; the average values are plotted in Fig. 5. More correlation peaks were included in the average for periods of reaching movements than for the other two periods because oscillations were much more likely during periods of reaching.

However, the mean amplitudes of the correlation peaks and the proportion of oscillatory episodes that were synchronized were similar for the different behaviors (Fig. 5).

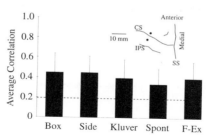

Figure 5. Amplitudes of peaks of cross-correlations between LFP oscillations recorded at a pre- and post-central site in the hand representation, during different behaviors. For each 200-msec window cross-correlations and autocorrelations of both signals were calculated. If either or both autocorrelation had a secondary peak above 95% significance level (usually 0.5), the magnitude of the cross-correlation peak for that window was included in the average. The significance level calculated from shifted correlations was 0.19 and is indicated by the dashed line. Behaviors indicated are the same as in Fig. 3.

Bilateral Synchrony

Recent experiments in the visual cortex of cats have indicated that activity of neurons in the two hemispheres could become synchronized, and that the synchronization was mediated at least in part by callosal connections (Engel et al., 1991; Munk et al., 1992). Primates commonly perform synergistic bimanual movements, which probably involve interhemispheric cortical interactions. To test whether oscillations in the sensorimotor cortex are involved in mediating this interaction, we recorded bilaterally in a monkey while it made manipulatory movements with either hand alone and with both hands together. LFPs and multiunit activity were recorded simultaneously at two homologous sites bilaterally in the hand representation of the motor cortex. The monkey manipulated unshelled peanuts with either the right hand alone (left hand restrained), or vice versa, and with both hands. EMGs were also recorded from a flexor and extensor muscle in each forearm. The occurrences and parameters of the LFP oscillations at both sites were documented, as was the synchrony between them.

As shown in Fig. 6, oscillations occurred in both hemispheres when the monkey made bimanual movements. However, oscillations occurred in both hemispheres even when the monkey made movements with one hand alone. The frequency of occurrence of oscillatory episodes in either hemisphere when the monkey used both hands was the same as during unimanual activity. Synchronization of LFPs was determined by calculating cross-correlations for 200-msec windows shifted by 100 msec. The cumulative distributions of correlation peaks over a period of 40 sec each when the monkey was actively making manipulatory movements with the left hand, the right hand, and both hands were very similar (Fig. 6b).

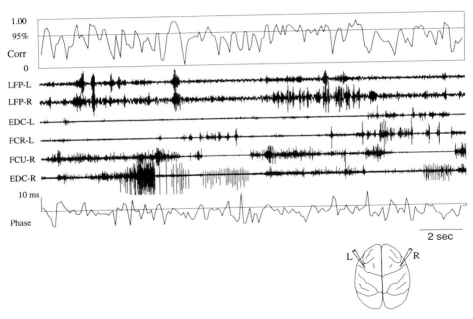

Figure 6 A. LFPs recorded at homologous sites in the left and right motor cortex (indicated in the sketch of the cortex) and EMGs from left and right hand muscles while the monkey manipulated an unshelled peanut with both hands. Cross-correlation peaks (Corr, top) and the corresponding time delays from zero (Phase, bottom) were calculated for 200-msec windows (100-msec shift between successive windows). Oscillations occurred at both sites and could become synchronized with negligible phase-shift. The 95% significance level determined from "shifted" correlations was 0.65 and is indicated by the horizontal line labeled 95% level.

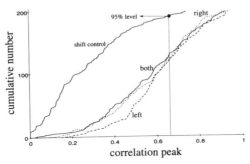

Figure 6 B. Cumulative distributions of the amplitudes of correlation peaks for 40-sec recordings under each condition of uni- or bimanual manipulations. The shift control distribution was obtained by shifting the left LFP forward by 1 sec, recalculating the correlations for all three conditions, and taking the average.

If synchronization of LFPs had increased preferentially during the bimanual task, the distribution of correlations during this task would have shifted to the right in comparison with those for either hand alone. The lack of such a shift suggests that bilateral synchronization of LFP oscillations may not be preferentially correlated with overt bimanual movements. This does not rule out the possiblity of more specific synchronization of smaller groups of neurons at a finer temporal resolution, which may not be reflected in LFPs. Cross-correlating simultaneously recorded single- and multi-unit activity at different sites would be required to resolve this.

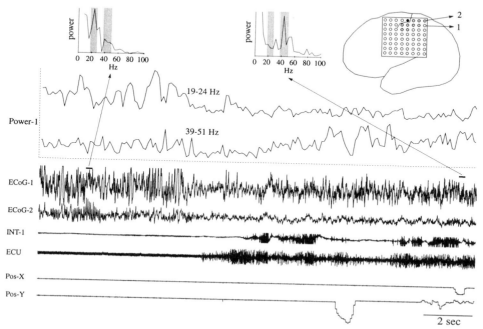

Figure 7. Subdural recordings from a patient while he was tracking a visual target on a video screen with a hand-held joystick. Right top shows the approximate location of the electrode grid. The black dot indicates a site whose stimulation evoked movement of the contralateral hand and the grey shaded dots indicate sites that showed somatosensory evoked potentials from the contralateral hand and arm. Recordings include: ECoGs from sites 1 and 2 (ECoG-1 and -2); surface recordings of EMGs from thumb (INT-1) and wrist extensor (ECU); and the 2-dimensional position of the joystick. For ECoG-1 the power in the two bands indicated was calculated for 256-msec windows (shifted successively by 100 msec), and is plotted above the ECoGs. Power spectra are illustrated for two windows taken during the waiting and tracking periods with the same relative scale. Shaded areas indicate the frequency bands for which power is plotted. After the onset of tracking, power diminished in the 20-Hz range and increased in the 39-51 Hz band.

In order to confirm that the recordings at bilateral sites were independent and not merely volume-conducted signals from distant sources, we calculated CTAs of neurons recorded at both sites aligned on the cycles of LFP oscillations at one site. In 34 of 43 pairs of recordings, averages indicated oscillatory modulation of units at both sites, confirming the presence of local generators of synchronized rhythmic activity in each hemisphere.

Oscillatory Activity in Human Electrocorticograms

Techniques for recording ECoG directly from surface of sensorimotor cortex (Ojemann and Engel, 1987) provide an opportunity to determine whether high-frequency oscillations similar to those described above occur in humans during movements requiring fine sensorimotor control. The existence of 20-30 Hz rhythms in scalp EEG over sensorimotor areas has been documented (Jasper and Penfield, 1949; Pfurtscheller, 1981), but their correlation with motor behavior has not been studied extensively. In collaboration with Dr. George Ojemann, we were able to monitor subdural ECoG at multiple sites, with better spatial resolution of signals than obtained with EEG. Multichannel ECoG recordings from the sensorimotor cortex were obtained from two patients when they performed a visuomotor tracking task using the hand contralateral to the recorded cortical hemisphere. The results from the two subjects were quite similar and detailed analysis is presented for one subject (Fig. 7). When the subject was prepared to begin the task, but made no overt movement of the hand, high-amplitude rhythmic activity around 20 Hz was observed in the channels overlying the sensory cortex. When he began the tracking task, the 20-Hz activity was suppressed over the sensorimotor areas. Simultaneously, higher frequency oscillations around 40 Hz appeared in one channel posterior to the estimated location of the central sulcus. The 20-Hz reduction was also observed when the subject made a forceful pinching movement with his thumb and forefinger after a period of waiting. However, the pinching task produced no significant increase in the 40-Hz oscillations. This suggests that the reduction in the 20-Hz rhythm may be a general effect spread over the entire sensorimotor area, but the increase in the 40-Hz rhythm may be more specific to the motor task performed.

DISCUSSION

Although there is now ample evidence of widespread oscillatory activity in the cortex, there is little agreement about its possible function. High-frequency cortical oscillations in the range of 20-70 Hz occurring in sensory areas are thought to play a role in perception. Synchronous rhythmic activity in the visual cortex was postulated to bind different stimulus features of a single object and to represent it as distinct from other objects in the visual field (Eckhorn et al., 1988; Engel et al., 1992; Singer, 1993). In support of this hypothesis, oscillations have been shown to be synchronized between different visual cortical areas and between two hemispheres of the cat in a stimulus-specific manner (Engel et al., 1991). In the sensorimotor cortex, 20-40 Hz oscillations occur across a wide area and have been correlated with certain behaviors (Murthy and Fetz, 1992). In overtrained tasks, LFP oscillations were relatively rare during the actual movements (Murthy and Fetz, 1992; Sanes and Donoghue, 1993). However, in a delayed response task oscillations were evident during the waiting period before movement began (Sanes and Donoghue, 1993). When the monkeys made less constrained exploratory movements to retrieve food, oscillations occurred more frequently (Murthy and Fetz, 1992). Oscillations were not reliably associated with every movement, indicating that they are not essential for execution of movements. Rather, their general increased rate of occurrence during interactive movements suggests a higher-order role in sensorimotor integration.

Oscillations could become synchronized over wide areas in the primary motor cortex, between motor and sensory cortex, and between the motor cortices of the two hemispheres. Synchronization of LFP and multiunit oscillations in pre- and postcentral sites could be a possible substrate for sensorimotor integration. During exploratory tasks, sensory feedback would be essential to make precise movements (Asanuma and Arissian, 1984). Under these conditions, rhythmic synchronous activity could lead to efficient excitation of neurons that are part of the same functional network. There are abundant anatomical connections between pre- and post-central cortex (DeFelipe et al., 1986). A close functional link between the two areas is also supported by close similarities in response properties of precentral and postcentral neurons, with regard to sensory input and motor-related activity (Fetz, 1984). Our findings suggest that synchronization of spike discharges of neurons in the two areas could be a further substrate for functional interaction.

Synergistic bimanual movements most likely involve neural interactions between the two hemispheres through callosal or commisural connections. While the supplementary motor area has been suggested to be important for mediating bimanual synergies (Brinkman, 1984; Wiesendanger et al., 1992), there is clear evidence that the primary motor cortex also plays an active role (Tanji et al., 1988; Aizawa et al., 1990). Some motor cortex neurons respond with ipsilateral movements (Tanji et al., 1988), and stimulating certain sites in the motor cortex elicits bilateral hand movements (Aizawa et al., 1990). We found that LFP oscillations occurred at bilateral sites for unimanual as well as bimanual movements. LFP oscillations at bilateral sites could become synchronized with negligible phase shifts during bimanual manipulations. Further, single- or multi-unit activity could become synchronized with the LFP oscillations. Interestingly, synchronization of LFP oscillations was just as likely and just as strong for unimanual manipulations as for bimanual ones, suggesting that synchronization was not specific for bimanual coordination. However, since LFPs are complex averages of underlying neural activity, they may not be appropriate indicators of synchronization of small groups of neurons. Smaller groups of neurons may exhibit transient and specific synchronization for bimanual tasks which may not be detected by average measures such as LFPs.

Widespread synchronization of LFP oscillations in the sensorimotor cortex appears to be too nonspecific and episodic to be involved directly in motor control, particularly when compared with the specificity of synchronization in the visual cortex (Engel et al., 1992). However, the complexity of motor cortical organization with its convergent and divergent cortico-cortical and corticofugal projections could account for the extent of synchronization. Low-threshold intracortical microstimulation of multiple disjoint sites in the primary motor cortex can evoke similar digit-muscle responses (Sato and Tanji, 1989). Correlational evidence indicates that single corticomotoneuronal cells can have postspike effects on multiple muscles, both agonists and antagonists (Fetz and Cheney 1980). More directly, single axons of pyramidal tract cells have been shown to arborize among different motoneuron pools in the spinal cord (Shinoda et al., 1981). Sites that evoke movement across different joints are connected by intrinsic axon collaterals in a complex manner (Huntley and Jones, 1991). These connections could provide an anatomical basis for the synchronization of LFP oscillations over the entire arm representation. Since most natural movements of the hand and the arm involve coordination of many muscles, widespread synchronization of LFP oscillations may provide a general increase in excitability to allow recruitment of more specific neurons for an upcoming movement. Other observations further support the idea that oscillations may play a general role in attention, rather than a specific role in movement control.

Single units in the motor cortex discharge reliably at the onset of and during particular movements. In contrast, LFP oscillations were not consistently correlated with occurrence of EMG bursts. However, CTAs of EMGs revealed oscillatory modulation of EMGs synchronized with LFPs (Murthy and Fetz, 1992). This suggests that even if the oscillatory episodes are not reliably timed to the execution of movements, they nevertheless involve output cells that can influence muscle activity. Precisely timed synchronous inputs to spinal circuits might allow subtle yet fast modification of ongoing muscle activity, especially during fine finger movements. Further, during attentive periods before movement begins synchronous oscillations could help bring functionally related neurons close to firing threshold and facilitate their response to subsequent inputs.

Oscillations could also play a role in modulating synaptic efficacy among neurons that are active simultaneously. The synaptic inputs that arrive during the depolarizing phase of the oscillations could be potentiated selectively. In the hippocampus, in-phase stimulation of afferents to CA1 during carbachol-induced theta rhythm can lead to long-term potentiation (Huerta and Lisman, 1993). Preliminary data from our lab indicate that evoking excitatory postsynaptic potentials (EPSPs) in cortical neurons in phase with the negative deflections of naturally occurring LFP oscillations can transiently potentiate the EPSP (Chen and Fetz, 1993).

We found that oscillations similar to those described in monkeys occurred in the sensorimotor cortex of humans during performance of sensorimotor tasks. In light of the results from the monkey experiments, we asked the human patients to perform a task that involved ongoing sensorimotor integration instead of repetitive movements. We found that the power in the 20-Hz band decreased at the onset of both finely controlled movement and simple forceful pinching of the digits; however, 40-Hz power increased robustly only during the tracking task. This is consistent with a recent report that in addition to a decrease in the 20-Hz rhythm, there was an increase in power near 40 Hz in the EEG at the onset of movement (Pfurtscheller and Neuper, 1992). Although the occurrence of these oscillations during sensorimotor tasks suggests some similarities with oscillations in monkey cortex, there were clear differences. In humans, high-frequency oscillations were most prominent over the postcentral cortex, whereas in monkeys oscillations were usually more robust in the precentral cortex. Second, ECoG in human sensorimotor cortex exhibited a clear shift in frequency from around 20 Hz during the preparatory phase to around 40 Hz during actual movement, but such shifts were not obvious in monkeys (Sanes and Donoghue, 1993). Further, the 40-Hz oscillations in humans were spatially restricted to one or a few electrode sites; oscillations in monkeys were more widespread. In this respect, the 20-Hz oscillations in humans were similar to the 20-40 Hz oscillations in monkeys because: (1) they both occurred over a wide area of cortex, and (2) they both occurred during preparatory phases of movements and were suppressed at the onset of movement in a delay task (Sanes and Donoghue, 1993). Further experiments are required to determine the differences and parallels between oscillatory activity in monkeys and humans.

In summary, we have considered three possible roles for the synchronous 20-40 Hz oscillations in sensorimotor cortex. First, they could play a role in associating neurons that are coactivated during a task and facilitate interactions in functional networks of neurons. Second, they could simply reflect an increased excitability of neurons caused by a general increase in the attentional or arousal state of the animal. Third, they could play a role in altering synaptic transmission. Our evidence from multiple site recordings in monkeys does not appear to support the first hypothesis since we have been unable to find preferential synchronization of oscillations during behavior involving neurons at the relevant sites. The second possibility is more compatible with our data since oscillations occur more frequently over a wide area of the cortex during exploratory movements requiring attention. The third hypothesis is supported by preliminary data (Chen and Fetz, 1993), but remains to be further explored.

Acknowledgements

We thank Dr. George A. Ojemann for access to recordings from patients and Mr. Ettore Lettich for technical help with ECoG recordings. We also thank Mr. Jonathan Garlid for technical help in monkey experiments and Mr. Larry Shupe for computer programming. This work was supported in part by NIH grants NS 12542 and RR 00166.

REFERENCES

Aizawa, H., Mushiake, H., Inase, M. and Tanji, J. 1990. An output zone of the monkey primary motor cortex specialized for bilateral hand movement. Exp. Brain Res. 82:219-221.

Asanuma, H. and Arissian, K. 1984. Experiments on functional role of peripheral input to motor cortex during voluntary movements in the monkey. J. Neurophysiol. 52:212-227.

Brinkman, C. 1984. Supplementary motor area of the monkey's cerebral cortex: short- and long-term deficits after unilateral ablation and the effects of subsequent callosal section. J. Neurosci. 4:918-929.

Chen, D.-F. and Fetz, E.E. 1993. Effect of synchronous neural activity on synaptic transmission in primate cortex. Soc. for Neurosci. Abstr. 19:781.

DeFelipe, J., Conley, M. and Jones, E.G. 1986. Long-range focal collateralization of axons arising from corticocortical cells in monkey sensory-motor cortex. J. Neurosci. 6:3749-3766.

Eckhorn, R., Bauer, R., Jordan, W., Brosch, M., Kruse, W., Munk, M. and Reitboeck, H.J. 1988. Coherent oscillations: a mechanism for feature linking in the visual cortex? Biol. Cybern. 60:121-130.

Engel, A.K., König, P., Kreiter, A.K., Schillen, T.B. and Singer, W. 1992. Temporal coding in the visual cortex: new vistas on integrtion in the nervous system. Trends in Neurosci. 15:218-226.

Engel, A.K., König, P., Kreiter, A.K. and Singer, W. 1991. Interhemispheric synchronization of oscillatory neuronal responses in cat visual cortex. Science 252:1177-1179.

Fetz, E.E. and Cheney, P.D. 1980. Postspike facilitation of forelimb muscle activity by primate corticomotoneuronal cells. J. Neurophysiol. 44:751-772.

Fetz, E.E. 1984. Functional organization of motor and sensory cortex: symmetries and parallels, in: "Dynamic Aspects of Neocortical Function," G.M. Edelman, W.E. Gall and W.M. Cowan, eds., John Wiley.

Freeman, W. 1978. Spatial properties of an EEG event in the olfactory bulb and cortex. Electroencephal. Clin. Neurophysiol. 44:586-605.

Gray, C.M. and Singer, W. 1989. Stimulus-specific neuronal oscillations in orientation columns of cat visual cortex. Proc. Natl. Acad. Sci. USA 86:1698-1702.

Gray, C.M., König, P., Engel, A.K. and Singer, W. 1989. Oscillatory responses in cat visual cortex exhibit inter-columnar synchronization which reflects global stimulus properties. Nature 338:334-337.

Huerta, P.T. and Lisman, J.E. 1993. Heightened synaptic plasticity of hippocampal CA1 neurons during cholinergically induced rhythmic state. Nature 364:723-725.

Huntley, G.W. and Jones, E.G. 1991. Relationship of intrinsic connections to forelimb movement representations in monkey motor cortex: A correlative anatomic and physiological study. J. Neurophysiol. 66:390-413.

Jasper, H. and Penfield, W. 1949. Electrocorticograms in man: effect of voluntary movement upon the electrical activity of the precentral gyrus. Arch. Psychiat. Nervenkr. 183:163-174.

Kreiter, A.K. and Singer, W. 1992. Oscillatory neuronal responses in the visual cortex of the awake macaque monkey. Eur. J. Neurosci., 4:369-375.

Kristeva-Feige, R., Feige, B., Makeig, S., Ross, B. and Elbert, T. 1993. Oscillatory brain activity during a motor task. NeuroReport 4:1291-1294.

Munk, M.H.J., Nowak, L.G., Chouvet, G., Nelson, J.I. and Bullier, J. 1992. The structural basis of cortical synchronization. Eur. J. Neurosci. Suppl. 5: 21.

Murthy, V.N. and Fetz, E.E. 1992. Coherent 25-35 Hz oscillations in the sensorimotor cortex of awake behaving monkeys. Proc. Natl. Acad. Sci. USA 89:5670-5674.

Murthy, V.N., Chen, D.-F. and Fetz, E.E. 1992. Spatial extent and behavioral dependence of 20-40 Hz oscillations in awake monkeys. Soc. Neurosci. Abstr. 18:847.

Ojemann, G. A. and Engel, Jr., J. 1987. Acute and chronic intracranial recording and stimulation, in: "Surgical Treatment of Epilepsies," J. Engel, Jr., ed., Raven Press, New York. pp. 263-288.

Pfurtscheller, G. 1981. Central beta rhythm during sensorimotor activities in man. Electroenceph. Clin. Neurophysiol. 51:253-264.

Pfurtscheller, G. and Neuper, C. 1992. Simultaneous EEG 10 Hz desynchronization and 40 Hz synchronization during finger movements. NeuroReport, 3:1057-1060.

Ribary, U., Joannides, A.A., Singh, K.D., Hasson, R., Bolton, J.P.R., Lado, F., Mogilner, A. and Llinas, R. 1991. Magnetic field tomography of coherent thalamocortical 40 Hz oscillations in humans. Proc. Natl. Acad. Sci. USA 88:11037-11041.

Rougeul, A., Bouyer, J.J., Dedet, L. and Debray, O. 1979. Fast somato-parietal rhythms during combined focal attention and immobility in baboon and squirrel monkey. Electroenceph. Clin. Neurophysiol. 46:310-319.

Sanes, J.N. and Donoghue, J.P. 1993. Oscillations in local field potentials of the primate motor cortex during voluntary movement. Proc. Natl. Acad. Sci. USA 90:4470-4474.

Sato, K.C. and Tanji, J. 1989. Digit-muscle responses evoked from multiple intracortical foci in monkey precentral motor cortex. J. Neurophysiol. 62:959-970.

Sheer, D.E. 1984. Focused arousal, 40 Hz EEG and dysfunction, in: "Self Regulation of the Brain and Behavior," T. Elbert, B. Rockstroh, W. Lutzenberger and N. Birbaumer, eds., Springer, Berlin. pp. 64-84.

Shinoda, Y., Yokota, J-I. and Futami, T. 1981. Divergent projection of individual corticospinal axons to motoneurons of multiple muscles in the monkey. Neurosci. Lett. 23:7-12.

Singer, W. 1993. Synchronization of cortical activity and its putative role in information processing and learning. Ann. Rev. Physiol. 55:349-374.

Tanji, J., Okano, K. and Sato, K.C. 1988. Neuronal activity in cortical motor areas related to ipsilateral, contralateral and bilateral digit movements of the monkey. J. Neurophysiol. 60:325-343.

Wiesendanger, M., Corboz, M., Hyland, B., Palmeri, A. Maier, V., Wicki, U. and Rouiller, E. 1992. Bimanual synergies in primates, in: "Control of Arm Movement in Space," R. Caminiti, P.B. Johnson and Y. Burnod, eds., Exp. Brain Res. Series 22 (Springer, Berlin).

EVENT-RELATED DESYNCHRONIZATION (ERD) AND 40-HZ OSCILLATIONS IN A SIMPLE MOVEMENT TASK

Gert Pfurtscheller

Ludwig Boltzmann-Institute for Medical Informatics and
Neuroinformatics and
Department of Medical Informatics Institute of Biomedical
Engineering, University of Technology
Brockmanngasse 41
A-8010 Graz, Austria

INTRODUCTION

The brain has the ability to generate rhythmic activities in a broad range, whereby components within the alpha, beta and gamma bands play an important role in sensory processing and motor behavior. One of the first groups to record EEG from the intact skull during sensory stimulation was Jasper and Andrew (1938), who reported on alpha components from 8-13 Hz, beta components from 17-30 Hz and possible gamma components from 35-48 Hz. Their main finding was that precentral beta potentials were independent of the occipital alpha potentials in response to sensory stimulation.

From Jasper and Penfield (1949) there is first evidence from electrocorticograms in man that finger movement results in a localized and circumscribed blocking of pre and post central beta rhythms in the hand area but not in the face area. This finding is very important because it shows that the blocking or desynchronization of rhythmic components is highly specific and localized.

Desynchronization or blocking of central alpha components (mu rhythms) is found not only during finger or hand movement, but also before movement in the preparatory state. Such a desynchronized mu rhythm over the contralateral central region in man was reported by Gastaut et al. (1952) and Chatrian et al. (1959).

These early observations of brain oscillations show that one important feature is the reactivity in form of an alpha amplitude attenuation or desynchronization to externally or internally paced events. We have therefore introduced the term "Event-

Related Desynchronization" (ERD) to describe the amplitude attenuation or blocking response of rhythmic EEG components (Pfurtscheller and Aranibar, 1977) and "Event-Related Synchronization" (ERS) to describe the induced and non-phase locked oscillations in close connection with an event (Pfurtscheller, 1992).

ERD of alpha and beta components were reported during sensory stimulation (Van Winsum et al., 1984; Boiten et al., 1992; Klimesch et al. 1990) and voluntary movement (Pfurtscheller and Aranibar, 1979; Pfurtscheller and Berghold, 1989; Derambure et al., 1993). 'Event-related desynchronization' of magnetoencephalic 10-Hz rhythms was also demonstrated in the human visual cortex by Williamson and Kaufman (1989) and in the human auditory cortex by Tiihonen et al. (1991).

The aim of this paper is to give some examples of ERD (ERS) time courses and spatial distributions (maps) obtained in man, to show the importance of EEG reactivity measurements and to get more insight in complex functional interactions of the brain in a movement task.

Interpretation of ERD and ERS

The amplitude of a wave measured with scalp electrodes results from the spatial averaging effect over a limited scalp area. Large 10-Hz waves on the scalp electrode can only be measured when the electrical activity within an area of some cm2 is coherent or synchronized (Cooper et al., 1965; Lopes da Silva, 1991). This means that ERS can be interpreted as a rhythmic cooperative behavior of relatively large numbers of cortical neurons. ERD can be interpreted as the interruption of such cooperative or synchronized behavior of cortical neurons and be seen as a reliable correlate of increased cellular excitability in thalamocortical systems (Steriade and Llinas, 1988).

An important aspect is the frequency dependence of the transmission of the bioelectrical activity from the cortex to scalp. Attenuation factors between 10 and 10,000, depending on the cortical area involved, have been reported (Cooper et al. 1965). Because of phase-shifts measured between closely spaced subdural electrodes, the attenuation due to spatial averaging is larger for beta compared with alpha components (Pfurtscheller and Cooper, 1975) or in other words, the beta components are of much higher amplitude in electrocorticograms in comparison to scalp recordings. Therefore, it is not suprising that Jasper and Penfield (1949) and recently Murthy and Fetz (1993) reported on dominant central beta components in electrocorticograms of man blocked by movement.

METHODS

Experimental paradigm and data acquisition

ERD (ERS) was studied during simple finger, toe and tongue movements in man. The experimental paradigm is shown in Fig. 1. The EEG was recorded with 2 coupled 32-channel amplifier systems (BEST, Fa. Grossegger, Austria) with digital output and an optical link to a 486-PC. The EEG was amplified between 0.15 Hz and 60 Hz (3dB points, 16 dB/octave) and sampled at 128 Hz. Fifty-six electrodes arranged in an approximate rectangular array with interelectrode intervals of 2.5 cm and located in a cap, were placed over the left and right central, pre- and post-central areas. All silver silver-chloride scalp electrodes were referred to an electrode on the tip of the nose. In addition, horizontal and vertical electrooculograms (EOG) were recorded to control eye movements. A scheme of the electrode array and a picture of the cap is shown in Fig. 2.

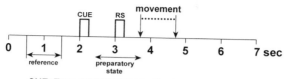

CUE: Tone (1200/800Hz) indicating which movement (finger, toe, tongue) should be made.
RS: Reaction stimulus (indication to start a brisk movement)

Figure 1. Experimental paradigm used for data recording.

For the experiments a CNV-paradigm was used. The first stimulus (CUE) was a sinus tone delivered via loudspeaker (200 msec duration, 70 dB (A)) with a frequency of either 800 Hz or 1200 Hz. The pitch of this auditory cue indicated which of 2 possible movements should be made. The second stimulus (RS), a 50 msec 1000 Hz tone burst, was presented 1 sec after the CUE and indicated to execute the movement. This means that the time between the CUE and the RS was related to the planning, and the time after the RS was related to the execution of a specific movement. The stimuli were presented at an inter-trial interval of 12 sec. The following movements were investigated:
- right thumb vs. right index finger (pressing of a microswitch)
- right finger (dorsal flexion) vs. tongue
- right toes vs. left toes (dorsal flexion)

In experimental runs with contingent finger and toe movements, the subjects were asked to make one brisk but distinct movement in each trial, without any speed instruction. In the case of the tongue movement the instruction was to touch the upper gums as if speaking the letter "d".

RESULTS

Measurement of ERD (ERS)

Each single trial with a length of 7 sec was digitally band pass filtered with finite impulse response (FIR) filters with a width of 2 Hz (in some cases also 4 Hz) centered between 8 Hz and 42 Hz. Examples of such filtered data in the range 36-40 Hz are presented in Fig. 3.

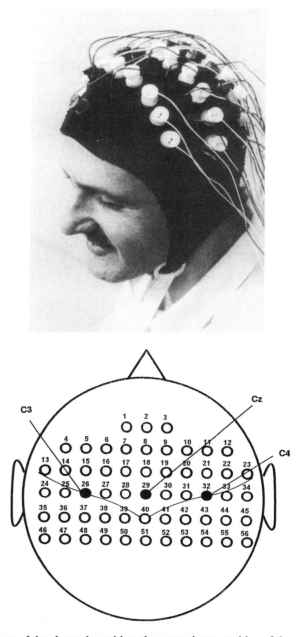

Figure 2. (Top): Schema of the electrode position, the approximate position of the central sulcus is indicated. (Bottom): electrode cap with inserted electrodes.

After filtering, each sample value was squared and squared samples were averaged over all trials. To reduce the variance, 16 consecutive power samples were averaged. Using this procedure, 8 power estimates per second were obtained. The mean band power (R) in the reference interval (second 0.5-1.5 of each trial) was determined and relative power changes were computed according to the equation (R-A)/R*100, whereby A are the power estimates of the 7 sec epochs. Further details about data processing and computation of maps are reported elsewhere (Pfurtscheller and Aranibar, 1979; Pfurtscheller and Berghold, 1989).

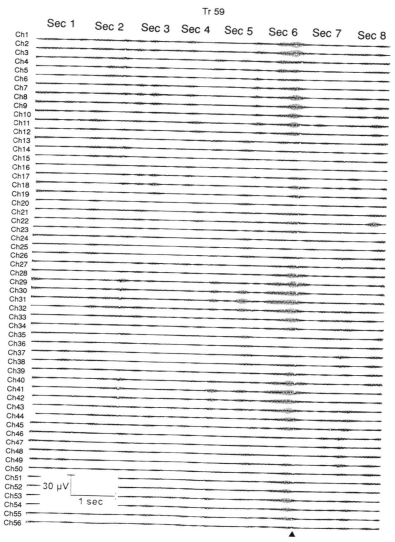

Figure 3. Example of filtered 40-Hz data recorded from 56 electrodes during a single right index finger movement. Recording from top to bottom corresponds to electrodes 2-56 as marked in Fig. 1. Finger movement-onset is indicated by a black triangle.

There are also other ways to calculate ERD time courses. One is to calculate the envelope of band pass filtered data using the Hilbert transformation (Seri and

Cerquiglini, 1993). Another method is to calculate the inter-trial variance in parallel with the event-related potential (ERP). The latter is proposed by Kaufmann in this volume. A comparison of ERD time courses calculated by band pass filtering and squaring and the intertrial variance revealed similar results although in the former case the power of the ERP can disturb the time course.

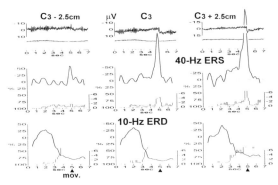

Figure 4. Examples of movement-related potentials (inclusive intertrial standard deviation) and ERD time courses calculated for 40-Hz and 10-Hz bands. In addition to the power changes in relation to the reference period (0.5 to 1.5 sec) the statistical significance (scale from p=0 to p= -6) is plotted. The data from 3 electrodes arranged in a transverse montage with intervals of 2.5 cm are displayed (C3 position, -2.5 cm in lateral direction, +2.5 cm in medial direction). Note the short bursts of 40-Hz activity on the electrode C3 and the more medial electrode and the relative long-lasting 10-Hz ERD on all 3 electrodes.

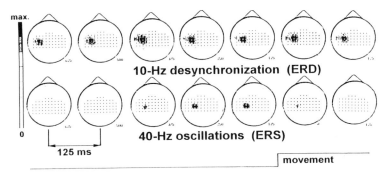

Figure 5. Series of maps displaying 10-Hz ERD and 40-Hz ERS during right finger movement. 'Black' marks the location of maximal 10-Hz power decrease (ERD) and 40-Hz power increase (ERS), respectively. Each map represents an interval of 125 msec.

Finger movement

Some data from right index finger movements are displayed in Fig. 4. Beside the time course of 10-Hz ERD and 40-Hz ERS, the ERP with the Bereitschafts-potential are shown. It can be seen, that the 10-Hz ERD was largest on electrode C3 and smaller on the neighboring electrodes. In contrast to this, the 40-Hz ERS was largest on the most medial electrode and suppressed on the lateral electrode. With movement-onset (marked by a triangle), the Bereitschaftspotential shows a reset, the 40-Hz oscillations are suppressed and only the 10-Hz desynchronization is sustained.

The topographical localization of 10-Hz ERD and 40-Hz ERS can be seen in Fig. 5. This figure shows a series of maps calculated in intervals of 125 msec before and after movement-onset. As can be seen in Fig. 5 the 10-Hz desynchronization became bilateral symmetrical with movement, whereas the 40-Hz ERS is focused to the contralateral side and limited to the time interval prior to movement-onset.

Tongue and toe movement

The topographical display of tongue and left and right toe movement are shown in Fig. 6. The foci of maximal 40-Hz oscillations are localized close to the primary sensorimotor areas. Similar maps are obtained with left and right toe movement with 40-Hz oscillations close to the vertex.

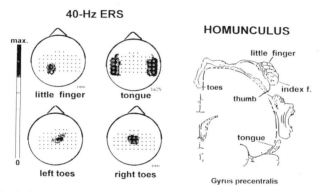

Figure 6. Maps displaying 40-Hz ERS during different types of movements. Scale from 0 to 150%. 'Black' marks areas with most prominent 40-Hz oscillations.

DISCUSSION AND CONCLUSION

The investigation has shown that finger movement is accompanied not only by a desynchronization of central mu rhythms (10-Hz ERD) but also by brief oscillations in the frequency range around 40 Hz (40-Hz ERS). Both phenomena have slightly different topographical patterns and different time courses and can be recorded on the intact scalp in man at the same moment in time.

It can be speculated that the observed desynchronization of 10-Hz components involves cortical modules or assemblies distributed over at least some cm2 of senso-rimotor areas. The relatively early start of the 10-Hz ERD of more than 1 second before finger movement-onset (Pfurtscheller and Berghold, 1989) and the relatively widespread distribution in the order of several cm2 can be interpreted as follows. The central 10-Hz ERD displays a cortical mechanism of preparation of all the cortical structures in pre- and post-central areas that are needed to perform a self-paced movement.

In the case of finger movement, brief 40-Hz oscillations were embedded in the 10-Hz ERD. In contrast to the 10-Hz ERD, which becomes bilateral symmetrical with movement-onset, these 40-Hz oscillations were found exclusively over the contralateral hand area during finger movement and were suppressed during movement. The 40-Hz oscillations are, therefore, more directly related to the programming of a specific motor act than to the desynchronization of 10-Hz components. There is also evidence from recordings in monkeys that the 40-Hz oscillations are not only focused to sensorimotor areas but are also suppressed during movement (Gaal et al. 1993). In addition, 40-Hz oscillations were reported during focal attention and immobility in cats (Bouyer et al., 1980); there, oscillations were blocked by the slightest body movement.

The amplitude of these 40-Hz oscillations are about one tenth below the 10-Hz waves and therefore, reflect a smaller area of coherent cortical activity compared with the alpha-band waves. It can be speculated that for the recording of 40-Hz oscillations on the scalp at least a cooperative behavior or synchronization of neural activity over still many mm2 of cortical tissue is necessary. In this respect, it is of interest to note that synchronized 40-Hz oscillations were recorded in the monkey's motor and somatosensory cortex on 5 electrodes separated by 2 mm. This means that synchronized oscillations occur across the central sulcus over at least 8 mm (Murthy and Fetz, 1992).

The specificity of the 40-Hz oscillations is documented by the different topographical patterns during finger, toe and tongue movements. The data reported are preliminary and need further intensive research work. The maxima of the 40-Hz bursts were found on electrodes overlying approximately the primary areas. It is of interest that tongue movements resulted in bilateral symmetrical 40-Hz oscillations lateral to C_3 and C_4 respectively, close to the cortical tongue representation area. The 40-Hz oscillations during toe movements were localized slightly posterior to the vertex electrode C_z close to the foot-toe representation area (Pfurtscheller et al. 1993). These observations of the localization of 40-Hz oscillations close to the primary sensorimotor areas are partially supported by measurements of regional cerebral blood flow (CBF) using PET during tracking movement of finger, toes and tongue (Grafton et al., 1991); here a CBF increase over the primary areas was reported.

While studying 40-Hz oscillations on the scalp it is very important to investigate whether or not muscle activity is present (DeFrance and Sheer, 1988). The influence of muscle activity can be ruled out in our experiments because:

(i)   The oscillations found around 40-Hz were only present in a narrow frequency band. For example, during finger movement oscillations were maximal in the 36-40 Hz band and suppressed in the frequency bands below and above.

(ii)   The 40-Hz oscillations displayed a circumscribed focus close to primary sensorimotor areas.

(iii) The 40-Hz oscillations were found before finger movement onset and were suppressed during movement.

In conclusion, it can be stated that brief 40-Hz oscillations are not only found after different types of sensory stimulation (Bressler and Freeman, 1980; Eckhorn et al. 1988; Gray et al., 1988; Sheer 1989 and others) but also during motor activity in man. This gives further evidence that 40-Hz oscillations are a basic phenomenon in the entire neocortex and may serve as an important cortical information carrier (Bressler, 1990).

Acknowledgements

I would like to thank Dr. Christa Neuper, Silvia Gölly, Doris Petz and Wolfgang Mohl for data acquisition, Hannes Berger and Joachim Kalcher for technical assistance and Britta Holzner for typing the manuscript. This research is supported by the "Fonds zur Förderung der wissenschaftlichen Forschung" project S49/02 and the "Oesterreichische Nationalbank" project 4534.

REFERENCES

Basar, E., Rosen, B., Basar-Eroglu, C. and Greitschus, F., The associations between 40-Hz EEG and the middle latency response of the auditory evoked potential. Intern. J. Neuroscience. 33 (1987) 103-117.

Boiten, F., Sergeant, J. and Geuze, R. Event-related desynchronization: the effects of energetic and computational demands. Electroenceph. clin. Neurophysiol. 82 (1991)302-309.

Bouyer, J.J., Montaron, M.F., Rougeul-Buser, A. and Buser, P., A thalamo-cortical rhythmic system accompnying high vigilance levels in the cat., In: Pfurtscheller, G., Buser, P., Lopes da Silva, F.H. and Petsche, H. (Eds.) Rhythmic EEG Activities and Cortical Functioning. Elsevier/North Holland Biomedical Press Amsterdam, New York, Oxford 1980 pp. 63-77.

Bressler, S.J., Freeman, W.J., Frequency analysis of olfactory system EEG in cat, rabbit and rat. Electroenceph. clin .Neurophysiol. 50 (1980) 19-24.

Bressler, S.L. The gamma wave: a cortical information carrier? TINS 13(5) (1990) 161-162.

Chatrian, G. E., Petersen, M.C. and Lazarete, J.A., The blocking of the rolandic wicket rhythm and some central changes related to movement. Electroenceph. clin. Neurophysiol. 11 (1959) 497-510.

Cooper, R., Winter, A.L., Crow, H.J. and Grey, Walter W., Comparison of subcortical, cortical and scalp activity using chronically indwelling electrodes in man . Electroenceph. clin. Neurophysiol. 18 (1965) 217-228.

De France, J. and Sheer, D.E., Focused arousal, 40-Hz EEG and motor programming. In: Giannitrapani, Murri (eds.) The EEG of Mental Activities, Karger, Basel 1988, pp. 153-168.

Derambure, P., Defebvre, L., Dujardin, K., Bourriez, J.L., Jacquesson, J.M., Destee, A. and Guieu, J.D. Effect of aging on the spatio-temporal pattern of event-related desynchronization during a voluntary movement. Electroenceph. clin. Neurophysiol. 89 (1993) 197-203.

Eckhorn, R., Bauer, R., Jordan, W., Brosch, M., Kruse, W., Munk, M. and Reitboeck, H.J. Coherent oscillations: a mechanism of feature linking in the visual cortex? Biol. Cybern. 60 (1988) 121-130.

Gaal, G., Sanes, J.N. and Donoghue, J.P., Motor cortex oscillatory neural activity during voluntary movement in macaca fasicularis, Neurosci. Soc. Abs. (1992), 848.

Gastaut, H., Etude electrocorticographique de la reactivite des rythmes rolandiques. Rev. neurol., 87 (1952) 176-182

Grafton, S.T., Woods, R.P., Mazziotta, J.C. and Phelps, M.E., Somatotopic mapping of the primary motor cortex in humans: Activation studies with cerebral blood flow and positron emission tomography, J. of Neurophysiol., 66 (1991) 735-743.

Gray, C.M., König, P., Engel, A. and Singer, W. Oscillary responses in cat visual cortex exhibit inter-columnar synchronization which reflects global stimulus properties. Nature 338 (1988) 334-337.

Jasper, H.H. and Andrew, H.L. Electro-encephalography III. normal differentiation of occipital and precentral regions in man. Arch. Neurol. Psychiat. 39 (1938) 96-115.

Jasper, H.H. and Penfield, W. Electrocorticograms in man: effect of the voluntary movement upon the elctrical activity of the precentral gyrus. Arch. Psychiat. Z. Neurol., 183 (1949) 163-174.

Klimesch, W., Pfurtscheller, G., Mohl, W. and Schimke, H. Event-related desynchronization, ERD-mapping and hemispheric differences for words and numbers. Int. Journal of Psychophysiology, 8 (1990) 297-308.

Lopes da Silva, F. Neural mechanisms underlying brain waves: from neural membranes to networks. Electroenceph. clin. Neurophysiol. 79 (1991) 81-93.

Murthy, V.N. and Fetz, E.E., Coherent 25- to 35-Hz oscillations in the sensorimotor cortex of awake behaving monkeys, Proc. Natl. Acad. Sci. USA, 89 (1992) 5670-5674

Pfurtscheller, G. Event-related synchronization (ERS): an electrophysiological correlate of cortical areas at rest. Electroenceph. clin. Neurophysiol. 83 (1992) 62-69.

Pfurtscheller, G. and Cooper, R. Frequency dependence of the transmission of the EEG from cortex to scalp. Electroenceph. clin .Neurophysiol. 38 (1975) 93-96.

Pfurtscheller, G. and Aranibar, A. Event-related cortical desynchronization detected by power measurements of scalp EEG. Electroenceph. clin. Neurophysiol. 42 (1977) 817-826.

Pfurtscheller, G. and Aranibar, A. Evaluation of event-related desynchronization (ERD) preceding and following voluntary self-paced movement. Electroenceph. clin .Neurophysiol. 46 (1979) 138-146.

Pfurtscheller, G. and Berghold, A. Pattern of cortical activation during planing of voluntary movement. Electroenceph. clin. Neurophysiol. 72 (1989) 250-258.

Pfurtscheller, G. and Neuper, Ch. Simultaneous EEG 10-Hz desynchronization and 40-Hz synchronization during finger movements. NeuroReport 3 (1992) 1057-1060.

Pfurtscheller, G., Neuper, Ch. and Kalcher, J. 40-Hz oscillations during motor behavior in man. Neuroscience letters. 164 (1993) 179-182.

Seri, S. and Cerquiglini, A. Event-related desynchronization as a probe of linguistic processes. Brain Topography, 5 (1993) 310.

Sheer, D.E. Sensory and cognitive 40-Hz event-related potentials: Behavioral correlates, brain function and clinical application. In: Basar, E. and Bullock, T.H. (Eds.) Springer Series in Brain Dynamics 2, Springer Verlag, Berlin, Heidelberg, (1989) 339-374.

Steriade, M. and Llinas, R.R. The functional states of the thalamus and the associated neuronal interplay. Physiol. Rev., 68(3), (1988) 649-742.

Tiihonen, J., Hari, R., Kajola,M., Karhu, J., Ahlfors, S. and Tissari, S. Magnetoencephalographic 10-Hz rhythm from the human auditory cortex. Neuroscience Letters, 129 (1991) 303-305.

Van Winsum, W., Sergeant, J. and Geuze, R. The functional significance of event-related desynchronization of alpha rhythm in attentional and activating tasks. Electroenceph. clin. Neurophysiol. 58 (1984) 519-524.

Williamson, S.J. and Kaufman, L. Advances in neuromagnetic instrumentation and studies of spontaneous brain activity. Proc. Topographic EEG Analysis and Brain Mapping, Valle d'Aosta, Italy, Sept. 7-10, 1989.

SUBCORTICAL P300-40-HZ RESPONSE OF THE CAT BRAIN

Canan Basar-Eroglu[1] and Martin Schürmann[2]

[1]Institute of Psychology and Cognition Research
University of Bremen
28359 Bremen, Germany
[2]Institute of Physiology
Medical University of Lübeck
23538 Lübeck, Germany

INTRODUCTION

The study of the 40-Hz activity of the brain was begun by Adrian (1942) who analyzed the relationship of electrical events in the olfactory bulb to the olfactory process. Freeman (1975) emphasized that the 40-Hz wave packet has a key function in percepts of models in the olfactory bulb of the rabbit. Several papers handled the importance of 40-Hz activity also in states of attention and motivation (Montaron et al., 1982; Sheer, 1989). The latter processes are also thought to contribute to the genesis of cognitive event-related potentials (ERPs), the most popular of the ERPs being the P300 wave. It is widely accepted that this late positive ERP component reflects the human cognitive processing associated with detection of a target signal. These correlations between human ERP and cognitive processes have generated great interest in the search for neural sources underlying these electrophysiological events. Therefore, conventional paradigms for eliciting P300 waves in humans have recently been adapted for studies in animals in order to develop animal models for ERPs, in spite of some limitations.

40-Hz responses in intracellular recordings

Following recent publications concerning 40-Hz neuronal oscillatory waveforms in the visual cortex of the cat brain (Gray and Singer, 1987), and the confirmation of these results by Eckhorn et al. (1988) research on 40-Hz potentials gained further importance. The results by Gray and Singer (1989), Gray et al. (1989), and Eckhorn et al. (1988, 1989a, b) gave rise to an important question: Can the oscillatory damped waveforms of 40-Hz activity be related clearly to given functional states of the brain?

Oscillatory Event-Related Brain Dynamics, Edited
by C. Pantev, Plenum Press, New York, 1994

Cortical 40-Hz response

Cortical oscillatory activity in 40-Hz range has been observed in man during cognitive tasks and following sensory stimulation, as analyzed by electroencephalographic (EEG) and magnetoencephalographic (MEG) means (Galambos et al., 1981; Mäkelä and Hari, 1987; Weinberg et al., 1988). Pantev et al. (1991) reported magnetic gamma band response from supratemporal auditory cortex, whereas Ribary et al. (1991) illustrated magnetic field tomography of coherent thalamo-cortical 40-Hz oscillations and Tiitinen et al. (1993) demonstrated a physiological correlate of selective attention in the 40-Hz transient response in humans.

40-Hz response – a universal operator?

In preliminary investigations of our research group on 40-Hz activity we stated that 40-Hz damped oscillations are part of every single evoked potential independent of the type of sensory stimulation used (Basar, 1980; Basar et al., 1987; Basar-Eroglu and Basar, 1991): 40-Hz waveforms could be considered as an universal operator in brain function which might be related to various functions. Furthermore, a series of our studies concluded that the 40-Hz responses might be nearly as important as the alpha activity. (See also Pfurtscheller and Neuper, 1992).

Event-related potentials (ERPs): P300 waves and cognitive functions

For investigations in humans, several types of ERP paradigms have been introduced. According to the view of several authors, the family of P300 waves are of special importance to describe cognitive functions of the brain, motivation and attention (Galambos and Hillyard, 1981). A typical P300 paradigm is based on randomly omitting stimuli from an otherwise regular train of stimuli ("omitted stimulus paradigm", Sutton et al., 1965). The omitted stimulus is referred to as "target". The existence of P300-like waves in the cat and monkey brain was already demonstrated by several groups (Buchwald and Squires, 1982; Harrison et al., 1986; Arthur and Starr, 1984; Neville and Foote, 1984; Katayama et al., 1985; Paller et al., 1988; Pineda et al., 1987; Basar-Eroglu and Basar, 1987).

The present study deals with two main questions:
1. Is the 40-Hz response only a cortical pattern or can it also be found in subcortical structures?
2. Is the 40-Hz response correlated with sensory and/or cognitive functions of the brain?

METHODS

Recording electrodes

Mono-electrodes Our experiments were performed on 9 freely moving female cats with chronically implanted electrodes. The stainless steel electrodes of 100 µm diameter were located while the cats were under Nembutal anesthesia (35mg/kg) in the auditory cortex (GEA; gyrus ectosylvianus anterior), left dorsal hippocampus (HI), mesencephalic reticular formation (RF). The coordinates and details of the surgical procedure were described in previous publications (Basar-Eroglu et al.,

1991a). The derivations were against a common reference which consisted of three stainless steel screws in different regions of the skull.

Multi-electrodes Also in the right hippocampus we placed a multielectrode array having four tips in 8 cats. The diameters of the electrodes were 25 µm. The distance between the tips of electrodes was 0.7 mm. The electrodes were labelled as HI1, HI2, HI3, and HI4: the first electrode was located in the upper pyramidal layer of the hippocampus (CA1), the second electrode between upper pyramidal layer and gyrus dentatus, the third and fourth electrode on or in the pyramidal layer (CA3, CA4), respectively.

Experimental procedure

The cats were sitting in a cage in a soundproof and echo-free room which was dimly illuminated and the EEG was monitored during all the experimental sessions. Movement artefacts and stages in which the cats had shown sleep spindles or slow waves in the cortex were eliminated off-line after recording sessions. The cats had not been previously trained or conditioned. Long and tiring experimental sessions were avoided to eliminate the effects of adaptation.

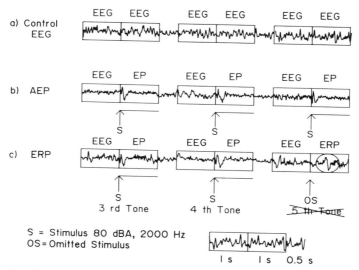

Figure 1. Paradigm. Sequence of experiments in order to elicit event related potentials (ERPs) in cats. Every experimental session consists of 3 experiments: (a) control-EEG (b) auditory EPs (c) ERPs with omitted stimuli. Every 5th stimulus is omitted.

Figure 1 shows our paradigm to elicit ERP by means of an omitted stimulus. Every experimental session consists of three experiments (Figure 1) in the following order:

1. *Recording of EEG* for control and comparison with an ERP experiment with omitted stimulus;

2. *Recording of auditory evoked potentials.* The tones were presented repetitively in order to give the cats the possibility to anticipate for the ERP experiment in marking roughly the time of omitted stimulus. We used for stimulation a 2 kHz tone with an intensity of 80 dB SPL with a duration of 1 s and a stimulus interval of 2.5 s;

3. *Recording of ERPs.* Our systematical investigations with five cats showed that the optimal rate for target is the paradigm with every fifth tone omitted (Basar-Eroglu et al., 1989; 1991).

Data acquisition and analysis

In order to analyze the ERPs in the time domain, we used again a "combined analysis procedure" of EEG and ERPs. This methodology and its relevant theoretical considerations were described in several previous publications (Basar, 1980). The method we used was basically the same except minor modifications. Therefore we describe the methods shortly as follows:

1. Recording of EEG-EP/ERP epochs: With every stimulus presented or omitted segments of 1s EEG activity preceding and 1 s EP/ERP following the stimulus/omitted stimulus were digitized, labelled and stored on computer disc memory (Hewlett-Packard 1000 F Data Acquisition System). This operation was repeated about 100 times for auditory EPs. For ERPs we had to record about 300 trials since in our paradigm every fifth stimulus is omitted (i.e., 240 EEG/EP epochs and 60 EEG/ERP epochs, see Figure 1).

2. Selective averaging of ERPs: The stored raw single EEG-EP/ERP epochs were selected with specified criteria after the recording session: Movement artefacts were eliminated with double check; the stages in which the cat showed sleep spindles or slow waves in the cortex were also eliminated. (Such an a posteriori approach permits more efficient artefact rejection than various direct online methods; Basar, 1980).

3. The selectively averaged EP/ERPs were transformed to the frequency domain with the Fourier Transform (FFT) in order to obtain the amplitude frequency characteristic of the studied brain structure. Details of the transient response frequency characteristics method (TRFC) are given in references (Basar, 1980; Basar-Eroglu et al., 1992).

4. The ERPs were filtered with digital filters with no phase shift as described earlier (Basar et al., 1984).

RESULTS

P300-like waveforms in cats

Figure 2 presents results for the auditory cortex, hippocampus and reticular formation (grand average curves across the 9 cats). At the left side of the figure the transient ERPs and at the right side the amplitude frequency characteristics - computed from the transient ERPs given - are presented. Although amplitudes and latencies of P300 were different in the curves presented, all the three structures showed a clear positive wave at 300 ms after omitted stimulus. The structures showed marked differences in frequency contents: whereas the hippocampus showed resonance properties (dominant maxima) at the theta frequency range, reticular formation and auditory cortex responded in the 10 Hz frequency range.

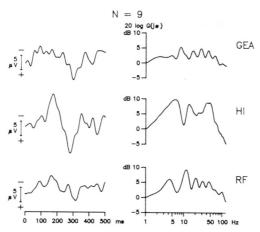

Figure 2. Averaged ERPs recorded from auditory cortex, hippocampus and reticular formation of 9 cats (left column), amplitude frequency characteristics computed from ERP (right column). Abscissa is frequency in Hz, ordinate is amplitude in relative units and decibels (dB).

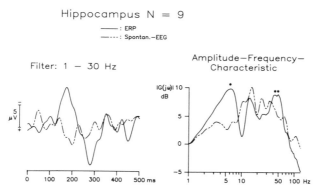

Figure 3. Grand average of ERPs/EEG of hippocampus across the nine cats (left side). Solid line represents averaged responses to omitted stimuli, dashed line represents averaged control-EEG. Corresponding amplitude frequency characteristics computed from ERPs/EEG (right side).

In order to be certain that these typical ERPs cannot be produced by averaging EEG segments, we computed the average control EEG (hippocampal recording) in nine cats (Figure 3, left column). The dashed curve in the illustration shows the grand average of the averaged EEGs. We compared the amplitudes of the positive waves between 250 ms and 400 ms. These amplitudes were significantly greater in ERP recordings than in the averaged control EEG ($p < 0.05$, Wilcoxon Test). The right side of Figure 3 shows the amplitude frequency characteristics computed from grand average EEG and ERP of Hippocampus. We observed a marked change in the amplitude frequency characteristics at around 5 Hz and 40 Hz ($p < 0.05$ and p < 0.01, Wilcoxon Test), whereas the 12 Hz maximum for ERP was less distinct than its counterpart for EEG.

Compound P300-40-Hz responses recorded in the hippocampus

In Figure 4 grand averages of wide band filtered ERPs and the 40-Hz responses are shown together. This illustration (grand averages, N=8, 30-50 Hz band) demonstrates again that the slow wave activity (the so-called N200-P300 response) is most marked in the CA3 region, whereas the response in the first upper hippocampal electrode (HI1) is less significant (cf. Basar-Eroglu et al., 1991b). Moreover, the waves recorded in the first and second hippocampal locations (HI1 and HI2) do not show a prominent N200 response. Also the wave at around P300 with a frequency of 40 Hz is in these locations not as marked as in the CA3 region. In other words the appearence of P300 response is usually combined with a wave packet of 40-Hz frequency. Comparing the 40-Hz responses (at P300 latency) recorded from the four electrode locations we mostly found responses of highest amplitude in CA3 (HI3 and HI4)). This is supported by statistically significant differences between amplitudes of HI3 and HI4 responses on the one hand and amplitudes of HI1 responses on the other hand (Wilcoxon-Wilcox-Test, see Basar-Eroglu and Basar, 1991). It is also to be emphasized that in a large number of sweeps, just at the moment of the occurrence of omitted stimulation (the first 100 ms), a small 40-Hz wave packet is seen, but usually it is not as high as the 40-Hz wave at the P300 location. We know from our earlier results that the hippocampus depicted significant 40-Hz responses to sensory stimulation reaching responses up to 200 µV in single sweeps (Basar, 1980, 1988).

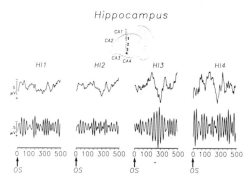

Figure 4. Grand averages (mean values from 8 cats) of ERPs in various layers of hippocampus; Top: Location of multielectrode; CA1 and CA3 corresponding to HI1 and HI3/HI4 respectively. Middle: Unfiltered ERPs. Bottom: Filtered ERPs (30-50 Hz) (After Basar-Eroglu and Basar, 1991, with permission).

The question whether the appearance of a slow wave, in this case a N200-P300 complex, is usually combined with a wave package of 40-Hz activity cannot be answered in this study. This would require another type of statistical evaluation, where significance of correlation between a) the envelope of 40-Hz packet and b)

ERP should be evaluated. Such a project is planned and will be subject for future publications.

P300-40-Hz responses observed in single-trial ERPs

The grand averages given above globally demonstrate the occurrence of P300-40-Hz responses: Figure 5 presents single trials recorded in one cat during an experimental session. It shows about 10 sweeps of single ERP epochs recorded during a P300 experimental session by using digital filters in the range of 30 to 50 Hz.

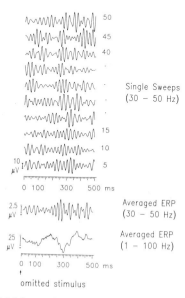

Figure 5. ERPs of lower pyramidal layer (CA3) of hippocampus (one cat). Top: Single ERP sweeps (epochs) also filtered 30-50 Hz. Middle: Averaged ERP filtered between 30 and 50 Hz. Bottom: Unfiltered ERP, average of about 50 artifact-free epochs (After Basar-Eroglu and Basar, 1991, with permission).

At the bottom of the illustration the average of 50 sweeps (filtered 30-50 Hz) from the same experimental session is shown together with the wide band filtered ERP. The wide band filtered averaged curve shows a marked ERP with peaks at around N200 and P300. Without filtering it is not possible to recognize the 40-Hz component which is probably masked by the low frequency activity. However, the filtered single epoch and the averaged curve show a 40-Hz burst around 270-300 ms following the omitted stimulation (target). In a large number of single sweeps, in fact, a 40-Hz wave packet can be seen. The maximal peak-to-peak amplitudes of 40-Hz wave packets usually do not exceed 50 µV; as a rule they were in the range of 20 µV. The time-locking is weak. However, we observed an alignment of 40-Hz wave-packets around 300 ms after omitted stimulus.Usually in the first 10-20 sweeps at the beginning of the experiments single sweeps depict weak phase locking, or no phase locking. As a rule a better and sometimes almost strong time-locking was observed against the end of the recording.

P300-40-Hz responses in the reticular formation

How does reticular formation behave concerning the P300-40-Hz response? Figure 6 shows the grand average of ERPs of reticular formation again from the same nine cats as already presented above.The dashed curves in the illustration show the grand average of averaged EEG (at the left side). By comparing the amplitudes of the positive waves between 250 ms and 400 ms we found that the amplitudes were significantly greater in ERP (p < 0.05, Wilcoxon Test). At the right side of Figure 6 we present amplitude frequency characteristics computed from averaged EEG as well as from ERP of reticular formation. The differences in amplitudes of maxima at 10 Hz and 40 Hz were significant (p < 0.05; Wilcoxon Test).

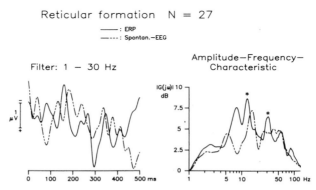

Figure 6. Grand average of ERPs/EEG of reticular formation of nine cats, three experiments each (left side). Solid line represents averaged responses to omitted stimuli, dashed line represents averaged control-EEG. Corresponding amplitude frequency characteristics computed from ERPs/EEG (right side).

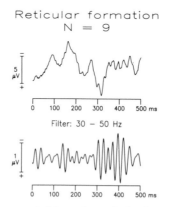

Figure 7. Grand average of ERPs recorded from the reticular formation (nine cats). Top: unfiltered ERPs. Bottom: Filtered ERPs (30-50 Hz).

Figure 7 presents grand averages of wide band filtered ERP and the 40-Hz responses together. 40-Hz wave packet mostly shows a longer duration in reticular formation than in hippocampus. In most of the cats investigated, the beginning of the 40-Hz wave packet is time locked with a maximum of P300 waves, and it shows long sustained oscillations.

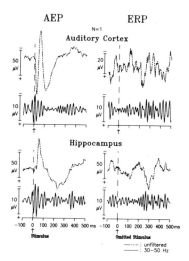

Figure 8. Representative averaged auditory EPs and ERPs recorded from auditory cortex and hippocampus in a single cat. Dashed lines represent unfiltered auditory EPs/ERPs and solid lines represent filtered auditory EPS/ERPs (30-50 Hz).

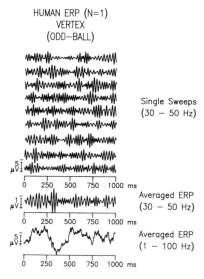

Figure 9. ERPs recorded from Vertex (Cz) by means of acoustical odd-ball paradigm. Top: Single ERP sweeps (Filtered. 30-50 Hz). Middle: Averaged ERP (Filtered. 30-50 Hz). Bottom: Unfiltered ERP, average of about 50 artifact-free sweeps.

In Figure 8 we present comparative results from two experiments (auditory EP and ERP) as well as two different structures (auditory cortex and hippocampus) of one cat. As can be easily seen, the 40-Hz response timely correlated either with the onset of auditory stimuli or with the P300 response (about 300 ms after omitted stimulation). Also similar results concerning P300-40-Hz response of human frontal, vertex and parietal locations by acoustical odd-ball paradigm were reported by Basar et al. (1993). Figure 9 presents ERPs recorded from human vertex. As can be easily seen, in unfiltered averaged ERP (bottom) 350 ms following task-relevant stimuli a distinct positive wave occurred. Similar to hippocampal results (cf. Figure 5) an enhanced 40-Hz response accompanied the P300 wave of the ERP. We may add to this point that in our earlier reports we could also show that the grand average of 15 human subjects had a 40-Hz enhancement at around 300 msec following the omitted stimulation (Basar et al., 1989).

DISCUSSION

The present study describes the 40-Hz response of subcortical structures with about 300 ms latency ("P300 latency"). We also refer to this response as a "P300-40-Hz response". With the cognitive role of the P300 response being firmly established we hypothesize that in the paradigm used the 40-Hz response also has cognitive functions (Basar-Eroglu and Basar 1991).

40-Hz activity in earlier studies

Adrian (1942) reported that the application of odorous substances to the olfactory mucosa gave rise to a train of sinusoidal oscillations lasting for the duration of stimulation. This response was termed "induced activity" to differentiate it from the intrinsic higher frequency activity of the bulb. This induced activity was usually between 40 and 60 Hz and appeared only with stimuli of sufficient intensity to supress the intrinsic activity. Later, Lavin et al. (1959) and Hernandez-Peon et al. (1960) stated that the 40-Hz activity could be elicited by a wide range of nonolfactory stimuli and that they reflect more the state of sensory stimuli rather than the activity triggered by olfactory processes. Further, Gault and Leaton (1963) have been able to measure the appearance of a burst of 40-Hz in the olfactory bulb of the cat when animals sniff. A neural mode for the generation of 40-Hz activity in attention has been detailed by Freeman (1975, 1983). The ongoing 40-Hz rhythm has been interpreted further as indexing a focused state of cortical arousal (Spydell et al., 1979; Sheer, 1984). Reasons for the recently increased interest in 40-Hz responses have been pointed out in the introduction; cortical 40-Hz responses will be dealt with below.

40-Hz activity and the concept of brain dynamics

The rationale of a recent study (Basar et al., 1987)was to further our long-standing investigation of 40-Hz resonance phenomenon induced by auditory stimulation. Our first observation of "40-Hz resonance" to acoustical stimuli has been reported in the cat auditory cortex (Basar, 1972) and in the cat hippocampus (Basar and Özesmi, 1972; Basar and Ungan, 1973). The 40-Hz human auditory response was first published by our research group (Basar et al., 1976b). The 40-Hz resonance phenomenon (or 40-Hz enhancement) was further analyzed as a distributed property of various brain structures including cerebellar cortex and has been indicated as one of the main components of the middle latency response of the auditory pathway evoked potentials (Basar, 1980). One of our previous studies had the aim to show the enhancement of the 40-Hz EEG-activity elicited with auditory stimuli and to demonstrate the important relation between 40-Hz spontaneous activity and the middle latency peaks of the human evoked potentials (Basar et al., 1987).

The brain is a dynamic system. In a dynamic system, at any instance, each state variable has a rate of change which depends on the current state of the system. The same proposition applies to the various bands of EEG-activity which have been taken to reflect different aspects of brain activity. Therefore, if such a dynamic system is stimulated one can expect that for every state (EEG-stage) a different excitability or different response susceptibility should exist. This was the main reason for introducing in our earlier works single brain responses together with the brain spontaneous electrical activity prior to stimulation (Basar et al., 1976a, b; Basar, 1980). The studies demonstrated that the amplitude of single-trial EPs was dependent on the amplitude of the EEG-activity prior to stimulation.

Are 40-Hz responses related to focused attention and perception?

Freeman (1975; Freeman and Skarda, 1985) has shown that the EEG of the olfactory bulb and cortex in awake, motivated rabbits and cats shows a characteristic temporal pattern consisting of bursts of 40-80-Hz oscillations, superimposed on a surface-negative baseline potential shift synchronized to each inspiration. He has since interpreted this finding as follows:

"The neural activity which is induced by an odour during a period of learning provides the specification for a neural template of strength connections between the neurons made active by that odour. Subsequently when the animal is placed in the appropriate setting, the template may be activated in order to serve as a selective filter for search and detection of the expected odour" (Freeman, 1979).

"A key step in information processing depends on an orderly transition of cortical activity from a quasi-equilibrium to a limit-cycle (synchronized oscillation state and back again). In this interpretation, the synchronized 40-Hz EEG activity, or the 40-Hz limit-cycle activity, serves as an operator on sensory output, to abstract and generalise aspects of the input into pre-established categories, thereby creating information for further central processing" (Freeman, 1983).

In other words, the limit cycle may represent an image of the input that the animal expects to identify or is searching for. The results of the present study and those of Basar-Eroglu and Basar (1991) support this interpretation and the possibility of its generalization to a wider range of EEG frequencies. Expectancy and selective attention, associated with regular, frequent target stimuli, result in highly synchronized EEG activity. This regular "limit cycle" activity occurs in various frequency ranges between 1 and 40 Hz. In this study we have focused our analysis on the 40-Hz range. Accordingly, we now conclude tentatively that the 1-4-Hz, 4-7-Hz, and 8-13-Hz activities serve as "operators" in the selective filtering of expected target stimuli. Basar et al. (1989) suggest that Freeman's concept can be generalized to various sensory systems and to EEG frequencies such as delta (around 1-4 Hz), theta (around 4-7 Hz), and alpha (around 8-13 Hz).

40-Hz sensory response

The 40-Hz evoked response can be elicited by using auditory or visual stimuli (Basar et al., 1987). According to these results it can be argued that the 40-Hz activity might represent an important general information processing of the brain similar to 10-Hz activity. The emphasis and analysis remained underestimated till now. The spontaneous 40-Hz activity has small amplitudes in comparison to alpha activity or theta activity and is therefore almost masked by these large magnitude activities. The use of digital filtering, however, makes it possible to analyze 40-Hz activity separately. Another scope, which would be interesting, but is not analyzed in this study, is the relation between 40-Hz and 10-Hz activities. Do these brain activities exist in competition or in mutual enhancement? This consideration might open new avenues of research, and various paradigms can be developed for experiments on cognitive and sensory information processing.

The important conclusion drawn from earlier findings is that EEG operators appear to modulate the response characteristics as a function of learning. Basar et al. (1984) used the expression "operative stage" for degrees of brain synchronization in delta, theta and alpha activity during brain information processing associated with sensory and cognitive inputs. In contrast to the statements made by Freeman (1983), Basar (1980) and Basar et al. (1984) suggested that EEG activity also contains response fragments to internal sensory and cognitive inputs (more details and examples concerning the EEG template pattern are analyzed by Basar et al., 1989).

Resonance phenomena at the cellular level

Recently Gray and Singer (1987, Gray et al., 1989) have reported that neurons in the cat visual cortex exhibit oscillatory responses in the range of 40-60 Hz. These oscillations occur in synchrony for cells located within a functional column and are tightly correlated with local oscillatory field potentials. This led Gray and Singer to the working hypothesis that the synchronization of oscillatory responses of spatially distributed, feature-selective cells might be a way to establish relations between features in different parts of the visual field. Later, Gray and Singer (1989) provided evidence that neurons in spatially separate columns synchronize their oscillatory responses. This synchronization occurs on the average with no phase difference and depends on the spatial separation and orientation preference of the cells. The discovery made by Singer's group was later confirmed by Eckhorn et al. (1988, 1989a, b), who raised the important question whether coherent oscillations do reflect a mechanism of feature linking in the visual cortex. They also found stimulus-evoked resonances of 35-85 Hz throughout the visual cortex when the primary coding channels were activated by their specific stimuli. The highly relevant experimental results by the groups of Singer and later Eckhorn and their interpretation was commented upon by Stryker (1989): "Is Grandmother an oscillation? Is it possible that the neurons in visual cortex activated by the same object in the world tend to discharge rhythmically and in unison? Such a one-note neural harmony could, in principle at least, provide the neurons at higher cortical levels with stronger inputs so that they associate the activities of lower-order neurons with one another." Stryker further notes, "Exploring the rhythms of the brain, revered by the pioneers of

Electroencephalography but now mostly dismissed as irrelevant to neural information processing, may even come back into fashion."

Cortical P300-40-Hz responses in humans

As already mentioned above, in earlier studies we have shown that the so-called "middle latency response" of the human auditory EP reflects the 40-Hz wave packet elicited in the first 100 ms following stimulus (Basar et a., 1987). We also extended our analysis by searching for 40-Hz cognitive potentials in human ERP recordings by means of acoustical odd-ball and omitted stimulus paradigms (Basar et al., 1993).

Our preliminary results showed that P300-40-Hz responses occurred 250-400 ms following task-relevant stimuli. We should be emphasized that, this response showed enormous intraindividual and interindividual differences concerning amplitude and latency values in frontal, vertex and parietal derivations in humans. The delayed 40-Hz response is a weak signal, often masked by slow frequencies. However, the similar results of ERPs elicited from cat and human brain is encouraging us to develop new paradigms for animal model.

The EEG-activity as universal operator of ERPs in general

Is the use of the expression "EEG activity as a functional operator in CNS a too general and strong assumption, or is the expression "operative states" for degrees of synchronization in various EEG frequencies a too courageous step? Since 1980 we have used the terms "alpha response" or "theta response" for evoked-potential components and published a general working hypothesis that all brain rhythms are linked to brain responsiveness. Furthermore, different oscillatory responses in human visual evoked potentials were described: (a) damped oscillatory responses of 5-6 Hz and (b) damped oscillatory responses of 10 Hz (Basar, 1988) and 40-Hz response oscillations. In a highly important recent review Llinas (1988) discussed the relevance of 6 and 10-Hz oscillations in the CNS. Llinas asks the following question:" How do the oscillatory properties of central neurons relate to the information-carrying properties of the brain as a whole?" Llinas (1988) described the thalamic neurons oscillating at two distinct rhythms: if the cell is depolarized, it may oscillate at 10 Hz; if the cell is hyperpolarized, it tends to oscillate at 6 Hz. Llinas concludes that oscillation and resonance allow single elements in the CNS to be woven into functional states capable of representing and embedding external reference frames into neural connectivity. In addition to these embedding properties, oscillation and resonance generate global states such as sleep-wakefulness rhythms and probably emotional and attentive states. This conclusion, which was reached by using results of the resonance concept at the neuronal level, has parallels to our conclusion that various cognitive tasks or sensory communications result in a specific combination of various "resonant modes", and that the ensemble of resonant modes could achieve an important function in the sensory communication of the brain (Basar 1980, 1988; see also Pfurtscheller et al., 1988).

The statement of Llinas (1988) that oscillation and resonance allow single elements in the CNS to be woven into a functional state supports the view of Freeman on 40-Hz activity (1975) and that of Basar (1980) on general resonance phenomena in theta and 10-Hz ranges. Although the designs of experiments were completely different in the studies mentioned, the coincidence of functional interpretation is highly striking.

40-Hz response: universal building block of responses to sensory and cognitive inputs

The general discussion of results in the previous studies and present paper (Basar-Eroglu et al., 1991a, b) led us to formulate the tentative working hypothesis as follows: it has been shown that the theta and 10 Hz-EEG components were enhanced upon omitted stimuli. In hippocampal P300 the 4-5-Hz enhancement is the dominating response whereas in cortex and reticular formation the 10 Hz response component had more weight (Basar-Eroglu et al., 1991). The significant 40-Hz response in the CA3 region of hippocampus, in the reticular formation and in the auditory cortex adds an important aspect to our working hypothesis that the resonating EEG is an important building block of the ERP elicited by omitted stimuli.

Concluding Remarks

Our results show that the 40-Hz component of EP/ERP is not only a cortical pattern. Besides that, the 40-Hz response is to our opinion a multimodal (sensory and cognitive) reaction of the brain. This component can be recorded in different brain structures and upon application of various sensory and cognitive inputs.

REFERENCES

Adrian, E.D., 1942, Olfactory reactions in the brain of the hedgehog. Journal of Physiology, 100:459-473.

Arthur, D., and Starr, A., 1984, Task relevant late positive component of the auditory event-related potential in monkeys resembles P300 in humans. Science, 223:186-188.

Basar, E., 1972, A study of the time and frequency characteristics of the potentials evoked in the acoustical cortex. Kybernetik, 10:61-64.

Basar, E., 1980, "EEG-Brain Dynamics. Relation between EEG and Brain Evoked Potentials", Elsevier/ North-Holland, Amsterdam.

Basar, E., 1988, EEG-dynamics and evoked potentials in sensory and cognitive processing by the brain, in: "Dynamics of Sensory and Cognitive Processing by the Brain", E. Basar ed., Springer, Berlin Heidelberg New York (Springer Series in Brain Dynamics, vol. 1).

Basar, E., Basar-Eroglu, C., Demiralp, T., Schürmann, M., 1993, The compound P300-40 Hz response of the human brain. Electroencephalography and Clinical Neurophysiology, 87:R76.

Basar, E., Basar-Eroglu, C., Röschke, J., and Schütt, A., 1989, The EEG is a quasi-deterministic signal anticipating sensory-cognitive tasks, in: "Brain Dynamics", E. Basar, and T.H. Bullock eds., Springer, Berlin Heidelberg New York (Springer Series in Brain Dynamics, vol. 2).

Basar, E., Basar-Eroglu, C., Rosen, B., and Schütt, A., 1984, A new approach to endogenous event-related potentials in man: Relation between EEG and P300-wave. International Journal of Neuroscience, 24:1-21.

Basar, E., Gönder, A., and Ungan, P., 1976a, Important relation between EEG and brain evoked potentials. I. Resonance phenomena in subdural structures of the cat brain. Biological Cybernetics, 25:27-40.

Basar, E., Gînder, A., and Ungan, P., 1976b, Important relation between EEG and brain evoked potentials. II. A system analysis of electrical signals from the human brain. Biological Cybernetics, 25:41-48.

Basar, E., and ôzesmi, C, . 1972, The hippocampal EEG activity and a systems analytical interpretation of averaged evoked potentials of the brain. Kybernetik, 12:45-54.

Basar, E., and Ungan, P., 1973, A component analysis and principles derived for the understanding of evoked potentials of the brain: Studies in the hippocampus. Kybernetik, 12:133-140.

Basar, E., Rosen, B., Basar-Eroglu, C., and Greitschus, F., 1987, The associations between 40 Hz-EEG and the middle latency response of the auditory evoked potential. International Journal of Neuroscience, 33:103-117.

Basar-Eroglu, C. and Basar, E., 1987, Endogenous components of event related potentials in hippocampus. An analysis with freely moving cats, in: "Current Trends in Event-Related Potential Research", R. Johnson, Jr., J.W. Rohrbaugh and R. Parasuraman, eds. (Electroencephalography and Clinical Neurophysiology, 40, 440-444).

Basar-Eroglu, C. and Basar, E., 1991, A compound P300-40Hz response of the cat hippocampus. International Journal of Neuroscience, 60:227-237.

Basar-Eroglu, C., Basar, E., Demiralp, T., and Schürmann, M., 1992, P300-response: possible psychophysiological correlates in delta and theta frequency channels. A review. International Journal of Psychophysiology, 13:161-179.

Basar-Eroglu, C., Basar, E., and Schmielau, F., 1991a, P300 in freely moving cats with intracranial electrodes. International Journal of Neuroscience, 60:215-226.

Basar-Eroglu, C., Schmielau, F., Schramm, U. and Schult, J. (1991b). P300 response of hippocampus with multielectrodes in cats. International Journal of Neuroscience, 60:239-248.

Buchwald, J.S., and Squires, N.S., 1982, Endogenous auditory potentials in the cat, in: "Conditioning: Representation of Involved Neural Function, " C.D. Woody, ed., Plenum Press, New York.

Eckhorn, R., Bauer, R., Jordan, W., Brosch, M., Kruse, W., Munk, M., and Reitboeck, H.J., 1988, Coherent oscillations: a mechanism of feature linking in the visual cortex? Biological Cybernetics, 60:121-130.

Eckhorn, R., Reitboeck, H.J., Arndt, M., and Dicke, P., 1989a, Feature linking via stimulus-evoked oscillations: experimental results from cat visual cortex and functional implications from a network model. Conference on Neural Networks, Washington, abstract volume.

Eckhorn, R., Bauer, R., and Reitboeck, H.J., 1989b, Discontinuities in visual cortex and possible functional implications: relating cortical structure and function with multielectrode/correlation techniques, in: "Brain Dynamics", E. Basar and T.H. Bullock, eds., Springer, Berlin Heidelberg New York (Springer Series in Brain Dynamics, vol. 2).

Freeman, W.J., 1975, "Mass Action in the Nervous System". Academic Press, New York.

Freeman, W.J., 1979, Nonlinear dynamics of paleocortexs manifested in the olfactory EEG. Biological Cybernetics 35:21-37.

Freeman, W.J., 1983, Dynamics of image formation by nerve cell assemblies, in: "Synergetics of the Brain" E. Basar, H. Flohr, H. Haken, and A.J. Mandell, eds., Springer, Berlin Heidelberg New York.

Freeman, W.J., and Skarda, C.A., 1985, Spatial EEG-patterns, non-linear dynamics and perception: the neo-Sherringtonian view. Brain Research Reviews, 10:147-175.

Galambos, R., and Hillyard, S.A., 1981, Electrophysiological approaches to human cognitive processing. Neuroscience Research Program Bulletin, 20.

Gault, F.P. and Leaton, R.N., 1963, Electrical activity of the olfactory system. Electroencephalography and Clinical Neurophysiology, 15:299-304.

Gray, C.M., and Singer, W., 1987, Stimulus specific neuronal oscillations in the cat visual cortex: a cortical function unit. Society of Neuroscience Abstracts, 404:3.

Gray, C.M. and Singer, W., 1989, Stimulus-specific neuronal ascillations in orientation columns of cat visual cortex. Proceedings of the National Academy of Sciences of the USA, 86:1698-1702.

Gray, C.M., König, P., Engel, A.K., and Singer, W., 1989, Oscillatory response in the cat visual cortex exhibit inter-columnar synchronization which reflects global stimulus properties, Nature, 338:334-337.

Harrison, J.B., Buchwald, J.S., and Kaga, K., 1986, Cat P300 present after primary auditory cortex ablation. Electroencephalography and Clinical Neurophysiology, 63:180-187.

Hernandez-Peon, R., Lavin, A., Alcocer-Cuaron, C., and Marcelin, J.R., 1960, Electrical activity of the olfactory bulb during wakefulness and sleep. Electroencephalography and Clinical Neurophysiology, 12:41-58.

Katayama, Y., Tsukiyama, T., and Tsubokawa, T., 1985, Thalamic negativity associated with the endogenous late positivity component of cerebral evoked potentials (P300): recordings using discriminative aversive conditioning in humans and cats. Brain Research Bulletin, 14:223-226.

Lavin, A., Alcocer-Cuaron, C., and Hernandez-Peon, R., 1959, Centrifugal arousal in the olfactory bulb. Science, 129:332-333.

Llinas, R.R., 1988. The intrinsic electrophysiological properties of mammalian neurons: insights into central nervous system function. Science, 242:1654-1664.

Mäkelä, J.P., and Hari, R., 1987, Evidence for cortical origin of the 40 Hz auditory evoked responsein man. Electoencephalography and Clinical Neurophysiology, 66:539-566.

Montaron, M.P., Bouyer, J.J., Rougeul, A., and Buser, P., 1982, Ventral mesencephalic tegmentum (VMT) controls electrocortical beta rhythms and associated attentive behavior cat. Behavioural Brain Research, 6:129-145.

Neville, H.J., and Foote, S.L., 1984, Auditory event-related potentials in the squirrel monkey: parallels to human late wave responses. Brain Research, 298:107-116.

Paller, K.A., Zola-Morgan, S., Squire, L.R., and Hillyard, S.A., 1988, P3-like brain waves in normal monkeys and monkeys with medial temporal lesions. Behavioral Neuroscience, 102:714-725.

Pantev, C., Makeig, S., Hoke, M., Galambos, R., Hampson, S., and Gallen, C., 1991, Human auditory evoked gamma-band magnetic fields. Proceedings of the National Academy of Sciences of the USA, 88:8996-9000.

Pfurtscheller, G., and Neuper, Ch., 1992, Simultaneous EEG 10 Hz desynchronization and 40 Hz synchronization during finger movements. NeuroReport, 3:1057-1060.

Pfurtscheller G., Steffan, J., and Maresch, H., 1988, ERD mapping and functional topography: temporal and spatial aspects, in: "Functional Brain Imaging", G. Pfurtscheller and F.H. Lopes da Silva, eds., Huber, Toronto.

Pineda, J.A., Foote, S.L., and Neville, H.J., 1987, Long latency event-related potentials in squirrel monkeys: further characterization of wave form morphology, topographic and functional properties. Electroencephalography and Clinical Neurophysiology, 67, 77-90.

Ribary, U., Ioannides, A.A., Singh, K.D., Hasson, R., Bolton, J.P.R., Lado, F., Mogilner, A., and Llinas, R., 1991, Magnetic field tomography of coherent thalamocortical 40-Hz oscillations in humans. Proceedings of the National Academy of Sciences of the USA, 88: 11037-11041.

Sheer, D.E., 1984, Focused arousal, 40 Hz EEG, and dysfunction, in: "Self-regulation of the brain and behavior", Th. Elbert, B. Rockstroh, W. Lutzenberger, and N. Birbaumer, eds., Springer, Berlin Heidelberg New York.

Sheer, D.E., 1989, Sensory and cognitive 40-Hz event-related potentials: Behavioral correlates, brain function, and clinical application, in: "Brain Dynamics, " E. Basar, and T.H. Bullock, eds., Springer, Berlin Heidelberg New York (Springer Series in Brain Dynamics, vol. 2).

Spydell, J.D., Ford, M.R., and Sheer, D.E., 1979, Task dependent cerebral lateralization of the 40 Hertz EEG rhythm. Psychophysiology, 16:347-350.

Stryker, M.P., 1989, Is grandmother an oscillation? Nature, 338:297-298.

Sutton, S., Braren, M., John, E.R., and Zubin, J., 1965, Evoked potential correlates of stimulus uncertainty. Science, 150:1187-1188.

Tiitinen, H., Sinkkonen, J., Reinikalnen, K., Alho, K., Lavikainen, J., and Näätänen, R., 1993, Selective attention enhances the auditory 40-Hz transient response in humans. Nature, 364:59-60.

Weinberg, H., Cheyne, D., Brickett, P., Gordon, R., and Harrop, R., 1988, An interaction of cortical sources associated with simultaneous auditory and somesthetic stimulation, in: "Functional Brain Imaging", Pfurtscheller, G., Lopes da Silva, F.H., eds., Huber, Toronto.

STIMULUS-RELATED OSCILLATORY RESPONSES IN THE AUDITORY CORTEX OF CATS

Valéria Csépe[1], Georg Juckel[2], Márk Molnár[1], and George Karmos[1]

[1]Institute for Psychology of the Hungarian Academy of Sciences
Department of Psychophysiology, Budapest, Hungary
[2]Laboratory of Clinical Psychophysiology
Department of Psychiatry, Freie Universität, Berlin, Germany

INTRODUCTION

The electroencephalographic oscillations in the higher frequency band have been the subject of intense scientific interest in the past few years. The higher frequency synchronous oscillation, called gamma band, has been extensively studied in the olfactory, visual, somatosensory and auditory system.

Both in man and animals repetitive acoustic stimuli elicit steady-state responses, which increase in amplitude when the stimulus rate approaches 40 Hz. In a previous study of Mäkelä et al. (1990) steady-state type responses (SSR) were measured in cats above the auditory cortex to clicks presented in trains of 350 ms. The SSR was enhanced at 30-40 Hz stimulation and were abolished by barbiturate anesthesia. The auditory cortical origin of SSR was confirmed by further studies (Karmos et al., 1993) using intracortical recordings.

In the present study we focussed on the auditory gamma band response (GBR), a transient response time locked to the stimulus onset. The transient GBR has been identified in man by Makeig (1990). He showed that the GBR increases in amplitude as the interstimulus interval increases. The neuromagnetic studies of Pantev and his colleagues (1991) revealed the auditory cortical origin of the transient GBR. These studies gave an impetus to our own studies in cats.

The main purpose of our recent study was to compare the intensity dependence of the auditory evoked potentials and the transient gamma band response. An increase in amplitude of several deflections of the auditory evoked potentials (AEP) caused by increased stimulus intensity have been found in humans (for review see Carrillo-de-la-Pena, 1992). The intensity/amplitude relationship of P1, N1 and P2 components has shown nonlinearity and pronounced interindividual variability.

Oscillatory Event-Related Brain Dynamics, Edited
by C. Pantev, Plenum Press, New York, 1994

The intensity dependence of visual evoked potentials has extensively been studied in cats. However, no systematic studies have been performed in the auditory modality in cats. In our present study the following questions were addressed: (1) Do the different components of the AEPs of cats show different sensitivity to the stimulus intensity? (2) Does the GBR show intensity dependence? (3) Are there differences between the intensity dependence of the primary and secondary auditory cortical AEPs and GBRs?

METHODS

Six adult cats were used as subjects. The cats were housed in individual cages, and their daily care met international standards. The recording electrodes were implanted under i.p. pentobarbital anesthesia. Surface electrodes, made from 0.23 mm diameter enamel-coated stainless steel were implanted epidurally above the A I and A II areas of the auditory cortex. The recording surface was a 1 mm uninsulated tip on the wire. Stainless steel bipolar twisted electrodes (0.23 mm o.d.) were placed into the ventral region of the medial geniculate body (MGB). A stainless steel screw placed 10 mm anterior to the bregma served as reference electrode. All electrodes were connected to a small 26-pin Winchester socket mounted on the skull. The localization of the electrodes was verified histologically using Nissl staining.

The experiments were carried out in a sound-attenuated room. The auditory stimuli were given through a 14x16x8 mm bone conductor (OTICON 10380) fixed to the animals' forehead during the experiments (for details see Karmos et al., 1970). The stimuli were 4 kHz tone bursts 5 ms duration (rise-fall time 1 ms). Four intensity levels (50, 60, 70, 80 dB peak equivalent SPL), calibrated with an artificial mastoid, were used. Forty two stimuli of each intensity were presented in an experimental block, in random order with a randomized ISI (1.8-2.2 s). Not more than two stimuli of the same intensity came in sequence. The first 8 stimuli (2 of each intensity) were rejected in order to avoid short-term habituation effects. Since attention effects on the intensity dependence are described in humans (Baribeau and Laurent 1987), all animals were habituated to the stimuli and the experimental conditions for at least one week.

Auditory stimulation and recording were performed by an IBM computer (486 processor). The brain electric activity was led to high-impedance preamplifiers through a light low-noise cable allowing the cats to move freely in the cage. The signals were amplified with a bandpass of 1.6-1000 Hz with a rolloff of 12 dB. Neocortical and hippocampal activity, electrooculogram and electromyogram were monitored on paper for classifying the animal's vigilance level. The behavior of the cats was continuously monitored via closed circuit television and documented in the experimental protocol. After a one week habituation the stimuli did not elicit behavioral or EEG arousal reactions during the experiments.

The data were corrected for movement, technical and ocular artifacts. 35-40 responses per intensity level were obtained in each recording. Averaging across animals were avoided because of the non-identical electrode locations within the analyzed areas. Repeated measures analysis of variance were used to evaluate the intensity effect on the amplitudes of the A I and A II components.

RESULTS

The main components of the AEPs obtained in the A I and A II area are presented in Figure 1. In the A I responses, the first positive component with a sharp peak (P12) is followed by a sharp negativity (N19) and broader waves like P34 and N55. The positive component around 110 ms in the A I responses was not present in all recordings. In the A II responses, the first positive component (P13) was followed by a broader second positive component (P25) and by the N50.

In the AEPs recorded over the A I area (left panel) the amplitude of the positive components (P12, P34, P110) increased significantly in all animals with an increase in intensity. The amplitude of the N19-component showed no significant increase, while that of the N55 showed only a trend of change with increasing intensity. The amplitude increase of P12 and P34 has reached its plateau at 70 dB. This implies a saturation of the amplitude increase at higher intensities in the A I region.

In all cats, the P25 wave of the A II responses exhibited a linear amplitude increase as a function of stimulus intensity enhancement. Different patterns of intensity modulation of the two early positive components were observed. While the P13 exhibited a small amplitude increase with increasing intensity, the amplitude of P25 showed a robust (statistically significant) increase, suggesting independence of the generating mechanisms of these waves. The N50 wave of the A II responses expressed small amplitude changes to the stimulus intensity increase, similar to the N55 of A I responses. The effect of increasing stimulus intensity on the amplitude of investigated AEP components was reproducible during different sessions over a four weeks measuring period. The intensity dependence of the component amplitudes showed small (not significant) fluctuations over time.

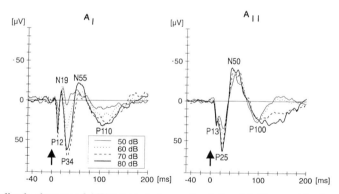

Figure 1. Amplitude changes of AEPs recorded over the A I and A II areas (single session, one cat).

The GBR was filtered out from the AEPs by using a symmetric filter around the peak frequency of the short-time frequency spectrum. Figure 2 and Figure 3 show the AEPs (left panels) and GBRs (right panels) for the four investigated intensities of one cat (filter: 28-48 Hz) recorded in quiet wakefulness. The overplotted responses represent the results of two sessions (time difference of one week).

According to the visual investigation and to the statistical analysis both the AEPs and GBRs show reasonable stability over time. The AEPs are recorded over six recording sites of the acoustic cortex (A I: AC1-AC4, A II: AC5-AC6), the association cortex (ASC) and the medial geniculate body (MGB).

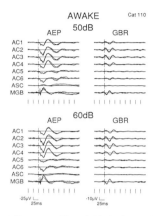

Figure 2. AEPs and GBRs at 50 and 60 dB intensities in awake cats.

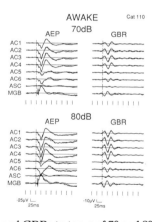

Figure 3. AEPs and GBRs to tones of 70 and 80 dB intensities.

Figure 2 compares the AEPs and GBRs elicited by the 50 and 60 dB tones. The amplitude of the second positive wave of auditory cortical responses increases with the stimulus intensity change from 50 to 60 dB. The consecutive negative-positive wave does not express amplitude increase. The ASC responses do not show any changes for this intensity increase. The early sharp negative wave (peak latency: 8 ms / N8) of the MGB response shows a smaller amplitude increase than the following negativity (peak latency: 25 ms / N25). The GBR responses are time-locked to the stimulus, but not phase-locked over the auditory cortex (AC1-AC6). Phase differences can be seen between GBR waves when comparing A I (AC1-4) and A II (AC5-6) responses. There is an expressed phase-shift between the GBRs recorded from the A I and the MGB. The peak-to-peak amplitude of the GBR recorded over the A I and from the MGB increases as the stimulus intensity changes from 50 to 60 dB.

Figure 3 represents the AEPs and GBRs recorded during the same sessions as the responses shown in Figure 2 but elicited by the two highest intensities, 70 and 80 dB. The P25 wave of the auditory cortical AEPs and the N25 wave of the MGB responses show further increase compared to the 50 and 60 dB. At 70 dB, GBR can be detected over the A II as well. All detectable peaks of the auditory cortical GBR increase with increasing stimulus intensity (the prestimulus waves can be due to some filter ringing). However, the intensity level of the first significant peak-to-peak amplitude increase of the GBR differs according to the recording site (60/70 dB A I, 70/80 dB for A II and MGB).

DISCUSSION

This study demonstrated a systematic stimulus intensity effect on amplitudes of evoked responses recorded epidurally over the primary and secondary auditory cortices in behaving cats. With increasing tone intensity the amplitudes of the P12/P34 (A I) and the P25 (A II) components increased linearly. This finding corresponds to previous AEP results (Starr and Farley, 1983) found in paralyzed cats. Our results are also in line with those of Phillips and Kelly (1992) indicating a strong intensity dependence of the first AEP components of the primary auditory cortex in the behaving cat.

The intensity dependence of AEPs recorded over the primary and secondary auditory cortices of behaving cats seems to reflect a rather complex process. Intensity function of the evoked response is a widespread phenomenon along the auditory pathway. The first negative deflection recorded from the MGB (N8) exhibited a pronounced intensity dependence of the amplitude.

The observed complexity of the intensity function in the cat primary and secondary auditory cortices leads to the assumption that the underlying processes generating the individual AEP components differ between the two areas. The GBR amplitude and phase differences also suggest different sensitivity in the stimulus intensity processing. Schreiner and Cynader (1984), studying single neurons in the auditory cortex of cats have found that the sensitivity of A I responses to tonal stimuli was greater than that in A II. This finding could explain the stronger intensity dependence of the first positive component in the A I than in the A II area.

The different sensitivity of the two auditory areas is also suggested by the differences in the intensity dependence of the GBR responses. The phase shift observed between the A I and A II GBRs suggests that the oscillatory activity may have a particular function in synchronizing the different regions of the auditory cortex. The time-locked but not exactly phase reversed MGB-GBR indicates probably multiple sources.

Acknowledgement

This study was supported by the Hungarian Research Fund (OTKA I/3-2595). We would like to acknowledge the technical assistance of G. Ujvari.

REFERENCES

Baribeau, J.C. and Laurent J-P., 1987, The effect of selective attention on augmenting/intensity function of the early negative waves of AEPs, In: R. Johnson, Jr., J.W. Rohrbaugh and R. Parasuraman, eds., Current Trends in Event-Related Potential Research, Electroenceph. clin. Neurophysiol., Suppl 40, Elsevier, Amsterdam, 68-75.

Carrillo-de-la-Pena, M.T., 1992, ERP augmenting/reducing and sensation seeking: a critical review. Int. J. Psychophysiology, 12: 211-220.

Karmos, G., Mäkelä, J.P., Ulbert, I. and Winkler, I., 1993, Evidence for intracortical generation of the auditory 40-Hz response in the cat, in: New Developments in Event-Related Potentials, H.-J. Heinze, T.F. Münte and G.R. Mangun, eds., Birkhaeuser, Boston-Basel-Berlin, 87-93.

Karmos, G, Martin, J., Kellényi, L. and Bauer, M., 1970, Constant intensity sound stimulation with a bone conductor in the freely moving cat, Electroenceph. clin. Neurophysiol., 28: 637-638.

Makeig, S., 1990, A dramatic increase in the auditory middle latency response at very slow rates, in: C.M. Brunia, A.K. Gaillard and A. Kok, eds., Psychophysiological Brain Research, Tilburg University Press, Tilburg, 56-60.

Mäkelä, J.P., Karmos, G., Molnár, M., Csépe, V. and Winkler, I., 1990, Steady-state responses from the cat auditory cortex. Hearing Research, 45: 41-50.

Pantev, C., Makeig, S., Hoke, M., Galambos, R., Hampson, S. and Gallen, C., 1991, Human auditory evoked gamma-band magnetic fields, Proc. Natl. Acad. Sci.,88: 8896-9000.

Phillips, D.P. and Kelly, J.B., 1992, Effects of continuous noise maskers on tone-evoked potentials in cat primary auditory cortex. Cerebral Cortex, 2: 134-140.

Schreiner, C.E. and Cynader, M.S., 1984, Basic functional organization of second auditory cortical field (A II) of the cat. J. Neurophysiol., 51: 1284-1304.

Starr, A. and Farley, G.R., 1983, Middle and long latency auditory evoked potentials in cat. II. Component distributions and dependence on stimulus factors, Hearing Res., 10.

DETECTION OF UNAVERAGED SPONTANEOUS AND EVENT-RELATED ELECTROPHYSIOLOGICAL ACTIVITIES FROM FOCAL REGIONS OF THE CEREBRAL CORTEX IN THE SWINE

Yoshio C. Okada and Chibing Xu

Magnetophysiology Laboratory
V A Medical Center
Albuquerque, NM 87108 and
Departments of Neurology and Physiology
University of New Mexico School of Medicine
Albuquerque, NM 87131

INTRODUCTION

Today interests seem to be rapidly increasing in the analysis of event-related electrophysiological responses of the brain that are not necessarily well time-locked to external events, but exhibit some variability in their temporal patterns of response (Pfurtscheller and Sranibar, 1977; Pfurtscheller et al., 1988; Kaufman, et al., 1990, 1992; Makeig, 1993; Makeig and Inlow, 1993). This class of responses includes as a subset those that are time-locked to external events, but the whole class is presumably vastly larger. Attention to this type of data seems to be fueled by the perceived need in this community to develop methods to detect activities related to higher functions of the brain so that the area of application can be widened for the MEG and EEG research.

There are certain basic requirements for a method to successfully detect the presence of MEG or EEG signals that are related to a certain event but are variable in latency or temporal waveform. Since averaging tends to reduce amplitudes of such responses, it is necessary to devise some method to avoid this cancellation (Pfurtscheller et al., 1988; Klismesch et al., 1988; Kaufman, et al., 1990, 1992) or to detect responses unaveraged (Robinson and Rose, 1992). Such a method must be also capable of selectively detecting activities from regions of interest in the cortex, discriminating them from unrelated activities in other cortical areas.

In this contribution we describe a simple method for unaveraged detection of on-going activities located in focal regions of the cerebral cortex. We have used an *in*

Oscillatory Event-Related Brain Dynamics, Edited
by C. Pantev, Plenum Press, New York, 1994

vivo animal preparation to evaluate the feasibility of studying such activities in real time with MEG. Our study shows that it is possible with a high-resolution MEG system to clearly detect on-going cortical activities from a specific, identified cortical area. We detected MEG signals associated with spontaneous cortical activities located within or near the somatosensory cortex during a rest and a pre-stimulus period. We also observed long-lasting synchronized oscillatory activities in the same cortical area continuing for several seconds after the termination of a somatic afferent stimulation. These responses were clearly observable in unaveraged records, but they could not be seen in averaged responses since they were not time-locked to the stimulation.

CHARACTERISTICS OF SPONTANEOUS MEG SIGNALS

We first assessed some general characteristics of spontaneous MEG signals from a swine preparation detectable with a high-resolution sensor called a microSQUID (Buchanan et al., 1989). Figure 1A shows spontaneous signals measured from a juvenile swine under sodium pentobarbital anesthesia (11 mg/kg/hr i.v.). The animal was prepared as in our previous studies (Okada et al., 1992), i.e. the anesthetized animal was placed in a non-ferromagnetic head-holder and positioned below the MEG sensor in a magnetically shielded room. The sensor consisted of four channels of detectors; each detector was an asymmetric, first-order gradiometer with a 4 mm-diameter pickup coil and a 16 mm baseline between the pickup coil and the upper cancellation coil (Okada, 1992). The four pickup coils were located at the corners of an imaginary 6 mm x 6 mm square. The equivalent field sensitivities of the detectors were about 50 fT/√Hz. The detecting coils were placed just above the exposed dura mater of the cortex near the vertex; the pickup coils were about 1.2 mm above the dural surface. The MEG signals within a bandwidth of 0.3 Hz to 100 Hz were as much as 20 pT in amplitude with strong positive correlations across the four detectors. The sensor outputs were flat, reflecting the instrumentation noise, after the animal was euthanized with barbiturate overdose (Fig. 1B). Figure 1C shows the Fourier spectra of the top traces in Figs. 1A and 1B. The signals from the active brain were as much as 30 times stronger than the control condition at low frequencies and above the control even at frequencies as high as 50 Hz. Thus, most of the variance in records of Fig. 1A was not due to some instrumentation or environmental noise. Nor did the fields seem to be due to muscle activities. The measurements in Fig. 1A were made above the vertex after removing the entire scalp and the dorsal portion of the skull. The fields were weaker (about 0.1 pT at 1 Hz) than the signals in Fig. 1A even when the detecting coils were placed directly above and touching the neck muscles. The high degree of positive correlations among the MEG channels indicates that the generators of the fields were predominantly away from the detector array, since the fields measured at the four sensing coils would be similar in magnitude and polarity for distant sources. Nevertheless, the waveforms were not exactly the same, and at some time instances the fields were opposite in direction across the four channels, suggesting that cortical activities directly below the detector array might have also contributed to the spontaneous MEG signals.

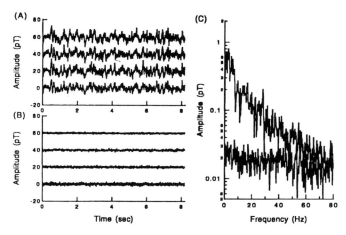

Figure 1. Spontaneous MEG signals from a juvenile swine. (A) MEG signals between 0.3 Hz and 100 Hz recorded with 4 channels of the microSQUID. (B) MEG signals with the same bandwidth recorded above the cortex of an euthanized swine. (C) Fourier spectra of the top trace of Figs. 1A and 1B.

A METHOD OF ANALYSIS OF SPONTANEOUS MEG SIGNALS

If an array of detectors has a sharp sensitivity for cortical sources, it should be possible to aim such a detector array to a particular region of interest such as the somatosensory cortex or some associative area and study how the activity in that area is modulated by various tasks presented to an animal or to human subjects. One simple method to increase the sharpness of the sensitivity function is to use the bipolar recording method as is commonly done in EEG research. Our microSQUID sensor has two pairs of channels arranged along two orthogonal directions on a plane. As shown by Cohen et al. (1980) and Cuffin and Cohen (1989), one can use the difference in the outputs of each pair of such detectors to maximize the sensitivity of the detector array for a dipolar current source located just below the center of such an array and oriented tangential to the plane of measurements. This bipolar detection method is now being used in displaying the data from a 122-channel whole-head system (Ahonen et al., 1992), since in this system there are two orthogonally oriented detectors at each measurement location with each detector having the geometry of an off-axis, planar first-order gradiometer, i.e. each detector consists of a pair of rectangular pickup coils connected in series and located on a plane nearly tangential to the surface of the head.

Figure 2 illustrates the bipolar method for MEG. On the first row of Fig. 2 a focal cortical source is represented by a tangential current dipole oriented along the x-axis and having a rectangular amplitude variation. The bottom row shows the same source oriented along the y-axis. The component of the magnetic field normal to the x-y plane (B_z) is shown at each of the four channels of the microSQUID for the two cases. Using the right-hand rule of electromagnetism, it can be easily seen that the B_z component is directed upward from the plane of this paper at channel 3 and downward at channel 1 for the dipole source oriented toward the direction of the positive

391

x-axis. The field pattern is shifted in orientation by 90° for the dipole Q_y oriented along the y-axis. Also shown are the fields that would be sensed by two bipolar pairs of channels, channels 3 and 1 and channels 4 and 2.

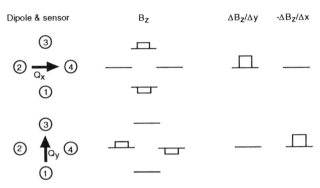

Figure 2. Unipolar (B_z) and bipolar ($\Delta B_z/\Delta y$ and $-\Delta B_z/\Delta x$) outputs from four channels of the microSQUID for a tangential current dipole oriented along the x- and y-axis. The temporal waveform of these dipole moments is rectangular.

From these examples of extreme cases one can intuit that the unipolar recordings at four probe locations of the microSQUID should show polarity reversals if the source is below the detector array and within the 8.5 mm x 8.5 mm rectangular area defined by the boundaries of the pickup coils. Furthermore, the outputs of the bipolar recordings should be maximum for a source located at the center of the detector array. To obtain a concrete idea of the selectivity of our method to cortical sources we computed the sensitivity functions of unipolar and bipolar MEG recordings for cortical sources.

Sensitivity Function of Unipolar MEG

The unipolar sensitivity function is presented first for one gradiometer channel of our microSQUID in detecting a tangential current source $\mathbf{Q} = \{Q_x, Q_y\}$ in a homogeneous half-space. The magnetic field $\mathbf{B}$ detected by an ideal magnetometer with an infinitely small pickup coil is given by Biot-Savart Law: $\mathbf{B} = \mathbf{L}\mathbf{Q} = (\mu_o/4\pi)\{y/r^3, -x/r^3\}\cdot\{Q_x, Q_y\}$, where $r = (x^2+y^2+z^2)^{1/2}$, and μ_o is the permeability of free space. The vector $\mathbf{L}$ is called the lead field and its magnitude, $|\mathbf{L}| = (\mu_o/4\pi)\sqrt{(y^2+x^2)}/r^3$, can be viewed as a sensitivity function of the particular detector. It is circularly symmetric. A more complex but similar expression can be derived, as was done here, for a realistic gradiometer with a pickup coil and a cancellation coil of finite sizes using the vector potential technique (Jackson, 1975).

Figure 3A shows the magnitude $|\mathbf{L}|$ of the sensitivity function for one of the micro-SQUID channels at a distance of 5.2 mm between the pickup coil and the tangential dipole source. This distance applies to our experimental situation, since we typically measure MEG signals with the pickup coil 1.2 mm above the exposed dura and depths of the sulci are commonly about 2-6 mm. Since we were interested in the sensitivity functions for shallow cortical sources, the brain was approximated by a flat homogeneous conductor. Although the volume conductor was homogeneous in our calculation, our result also applies to the case of planar inhomogeneous conductors or radially inhomogeneous spherical volume conductors since the layers of differing conductivities do not contribute to the normal component (for the planar case) and radial component (for the spherical case) of the magnetic field outside such volume conductors (Grynszpan and Geselowitz, 1973). The unipolar sensitivity function was computed only for a tangential dipole since MEG is not sensitive to a vertical dipole in cylindrically symmetric conductors.

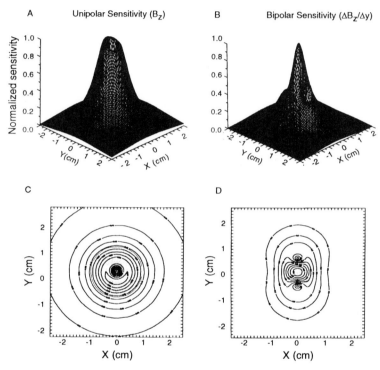

Figure 3. Magnitudes of the MEG sensitivity functions for (A) a unipolar recording and (B) a bipolar recording. Iso-sensitivity contour maps corresponding to the sensitivity functions of (A) and (B). Distance between the pickup coil and a tangential source = 5.2 mm. A semi-infinite volume conductor.

The sensitivity function is not completely symmetrical, unlike for an ideal magnetometer, since the cancellation coil is asymmetrically located relative to the pickup coil. However, it exhibits the well known basic features for a magnetometer (Williamson and Kaufman, 1981). The gradiometer is not sensitive to a dipole located near the center of the gradiometer pickup coil; i.e. there is a hole in the middle of the unipolar sensitivity function. But, it is maximally sensitive to dipoles slightly away from the center, at the location approximately $d/\sqrt{2}$(approximate because of the finite size detector), where d is the distance between the pickup coil and the dipole source.

Sensitivity Function of Bipolar MEG

The bipolar sensitivity function was computed by taking the difference of the lead fields of two detectors, channel 1 and 3 or channels 2 and 4. Figure 3B presents the magnitude of this bipolar sensitivity function for a pair of detectors aligned along the y-axis. It is sharper than the unipolar sensitivity and maximal just below the bipolar pair of detectors as already pointed out above. This difference in sharpness can be clearly seen in the iso-sensitivity contour maps for the unipolar and bipolar MEG recordings (Fig. 3C and 3D, respectively). The iso-sensitivity contour at half amplitude is shaded to indicate the difference in sharpness of the uni- and bipolar sensitivity functions. The bipolar function ($\Delta B_z/\Delta x$) for the other pair has the same shape but its orientation is shifted by 90°.

Both the unipolar and bipolar sensitivity functions are sharper for detecting dipole sources that are closer to the detector and more spread for deeper sources. Also, the sensitivity declines for detecting deeper sources. We quantified these relationships for the bipolar sensitivity in order to obtain a concrete notion of the usefulness of the bipolar technique for focusing on active tissues at different depths. The decline of the peak sensitivity with source depth was found simply from the extremum of the bipolar sensitivity function. The size of sensitive area was determined from the width of the sensitivity function at half amplitude for varying source depths.

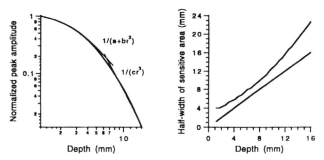

Figure 4. (Left) Normalized peak amplitude of bipolar sensitivity function. (Right) Half width of the MEG bipolar sensitivity function at its half maximum amplitude for tangential dipole sources (upper curve). For comparison, the lower curve shows a linear function with a slope of unity.

Figure 4 shows the peak amplitude of the bipolar sensitivity function normalized to unity at a depth of 1.2 mm from the pickup coil and the half width of the bipolar sensitivity at half amplitude as a function of depth of a tangential dipole. The sensitivity declines slowly for sources between 4-5 mm of the pickup coils because of the averaging produced by the finite-size (4 mm diameter) pickup coils. The function within this distance is well approximated by an inverse square function. At distances between about 4 mm and 16 mm, which is the baseline of each gradiometer, the peak sensitivity declines with the cube of the distance as it should be for a bipolar detection of a dipolar field. As the depth of the source becomes deeper than the baseline of the gradiometers (not shown in the figure), the peak sensitivity is influenced by the cancellation coil of the gradiometer and the rate of decline is faster than $1/r^3$, approaching $1/r^4$. Thus, the bipolar recording using our microSQUID seems to be differentially sensitive to shallow sources, i.e. to cortical sources within about 1 cm of the pickup coils.

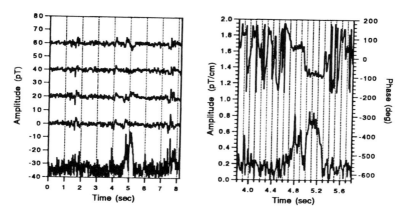

Figure 5. Spontaneous MEG signals under deep barbiturate anesthesia. (Left) spontaneous MEG signals recorded at four channels of the microSQUID together with the magnitude of vector $\mathbf{Y} = (\Delta Bz/\Delta y, -\Delta Bz/\Delta x)$. (Right) Magnitude and phase of $\mathbf{Y}$ during 3.8-5.8 sec after start of the epoch.

The right panel of Fig. 4 shows the half width of the bipolar sensitivity, defined as that of the iso-sensitivity contour at half amplitude along the x-axis. The half width is about 4 mm for detecting sources within about 2 mm of the pickup coil. It increases linearly with depth for sources as deep as about 8 mm, then faster for deeper sources. For our study with the juvenile swine, the diameter of the focal sensitivity can be expected to be about 1 cm or less even when fields are measured outside the skull, since the pickup coils can be placed as close as 1.2 mm from the surface of the skull and the skull is about 2 mm thick.

CORTICAL ACTIVITIES DETECTED WITH MEG

Next, some results will be presented from the swine to demonstrate spontaneous activities and event-related oscillatory activities within focal regions of the cortex detected with our MEG sensor. In displaying the bipolar MEG data, we combined the outputs of the two orthogonal bipolar derivations to produce a vector, $\mathbf{Y} = (\Delta B_z/\Delta y, -\Delta B_z/\Delta x)$, whose magnitude is maximum independently of source orientation for a tangential current dipole located beneath the center of the detector array. We eliminated the orientation dependence of the bipolar recordings by using this vector $\mathbf{Y}$. We inferred that MEG signals detected with the microSQUID were most likely due to activities within a cortical area with a diameter of about 1 cm when there were polarity reversals across the four channels of the sensor and the magnitude of the vector **Y** was well above its baseline. This working hypothesis seemed useful to us, although it was not possibe with this type of simple method to exclude the possibility of the source being located below the cortex, in a subcortical region, or outside of the detector array, since in both cases the magnetic field could be opposite in sign at two channels of the sensor and the field could be strong if the underlying activity was synchronized within a large volume of tissue. We felt that one could always validate the inferences concerning properties of specific cortical areas once some interesting properties are suggested by this simple analysis.

Spontaneous Cortical Activities Under Deep Barbiturate Anesthesia

Cortical activities can be seen when spontaneous neuronal activities become sporadic under a deep level of barbiturate anesthesia. The 8-sec record of spontaneous MEG (0. 3 - 30 Hz) (Fig. 5 left) was obtained after injecting a 2 cc/100 mg bolus of sodium pentobarbital i.v. to a juvenile swine weighing 8 kg and anesthetized under a continuous infusion of the same barbiturate at a rate of 11mg/kg/hr i.v.

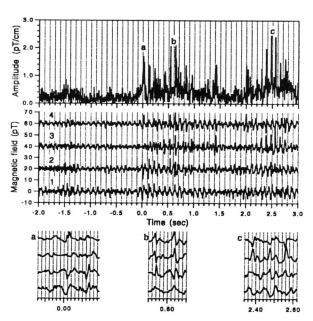

Figure 6. Single-trial MEG during a 5 sec epoch. (Top) Magnitude of vector Y. (Middle) Raw MEG signals before, during and after the presentation of 1 sec-long, 10 Hz, 50 μs, 1.0 mA electric stimulations of the snout. Channel numbers are indicated. (Bottom) Expanded records of the traces shown in the middle panel. Bandwidth = 10 - 100 Hz.

In comparison to Fig. 1 the on-going level of spontaneous activity was reduced except for the sporadic spike-wave complexes. At 1.5 sec after the start of recording period a spike-wave complex was detected at four probe locations. However, the waveforms were similar in amplitude and magnitude. Thus the magnitude of vector **Y** (|**Y**|) derived from the bipolar derivations, shown at the bottom of the figure, was not large relative to the baseline (this function is shown in an arbitrary unit with a zero offset of -40). The spike-wave complex occurring at 3.9 sec had a large value of |**Y**| during the initial spike, but the polarity and magnitude of the wave following this spike were similar across the four channels and consequently the |**Y**| did not deviate remarkably from the baseline. In contrast, the wave of the third spike-wave complex was biphasic with opposite polarities across the channels. Therefore, the magnitude |**Y**| was visibly large. The magnitude and phase of **Y** are plotted in the right panel of Fig. 5 for a time period between 3.8 and 5.8 sec after the start of MEG recording; the left axis now shows the calibrated unit of this magnitude function. Interestingly, the phase of **Y** was constant during each half of the biphasic wave, indicating that the underlying source was stationary during each segment of the wave. The phase shift after the transition was less than 180∘ (about 160∘), indicating that the biphasic wave was not due to a reversal of the direction of current of the source. This examples clearly demonstrates that a combination of unipolar and bipolar MEG recordings can be useful in discriminating signals arising from a particular area of the cortex against signals from a larger, surrounding region.

Event-related Oscillatory Cortical Activities

In the next experiment we obtained evidence for spontaneous and event-related oscillatory activities localized within the primary somatosensory cortex. After a juvenile swine (9 kg) was anesthetized with ketamine (15 mg/kg/hr i.v.) and xylazine (6 mg/kg/hr i.v.) and placed in a non-ferromagnetic head-holder, we measured the magnetic field evoked by stimulating a portion of the snout with a bipolar electrode. The MEG signals were measured on a plane 2 mm above the exposed dura, tangential to the area of the cortex receiving the projection from the snout. After mapping the magnetic field over the field plane, the 4-channel sensor was placed over the projection area where the MEG signals showed polarity reversals across the four detectors, i.e. directly above the primary cortical projection area. Then, the magnetic field was recorded unaveraged for several seconds: (1) to detect spontaneous activities arising from the somatic cortex and (2) to obtain evidence for some modulation of on-going spontaneous activities within the specific somatic projection area by a stimulation of the somatic afferents.

In the results to be reported, the level of spontaneous activity was reduced by increasing the dosage of ketamine and xylazine by a factor of three (i. e. ketamine - 45 mg/kg/hr and xylazine - 18 mg/kg/hr). During a set of 27 replications, we recorded MEG signals for 5 sec with a bandwidth of 0.3 - 200 Hz. Two seconds after the start of each recording, a somatic stimulus (a 1 sec-long, 10 Hz burst of 100 μs, 5 mA electrical pulses) was applied to the same area of the snout used for localizing the particular projection area. About 15 minutes after increasing the dosage, the MEG signal started showing the interesting features presented in Figs. 6 and 7. In both figures the raw data were filtered between 10 Hz and 100 Hz.

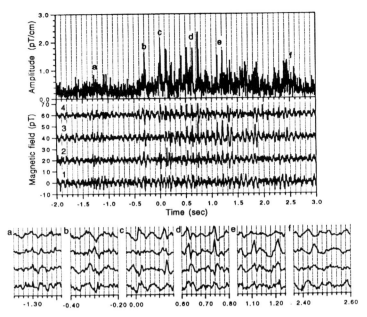

Figure 7. Records of another set of single epoch MEG signals from the same study as in Fig. 6. The 1 sec-long, 10 Hz train of 50 μs, 1 mA stimuli was delivered at time 0 to the same area of the snout as in Fig. 6.

In Fig. 6 the MEG signals were relatively quiet during the pre-stimulus period. During the 1-sec stimulus presentation period the MEG signals time-locked to the stimulus train increased in amplitude. As shown in detail at the bottom (a and b), the MEG signals were opposite in direction at channels 2 and 4, starting about 15 ms after the start of the stimulus train. Although the responses during the stimulus train were time-locked, there was some evidence that the pattern of cortical activity was not stationary since the relative magnitudes of the field at the four probe locations varied over the 1-sec period, as can be seen in the records during periods a and b. Interestingly, the magnetic field after the stimulus train did not return immediately to the pre-stimulus baseline. Strong oscillatory activities were detected for several seconds after the termination of the stimulus train. The bipolar MEG signals during the post-stimulus period were strong in magnitude, comparable to or even more than during the stimulus presentation in some cases. As shown in detail in c, these strong oscillatory activities exhibited polarity reversals across the probe locations of the microSQUID, although the reversals occurred across channels 3 and 4 rather than channels 2 and 4 during the stimulus period. We, therefore, infer that the oscillatory activities were induced after the termination of a stimulus train in a cortical area close to the area receiving the projection from the snout.

Figure 7 shows the long-lasting cortical activity found in Fig. 6, but also contains two additional features. During the pre-stimulus period, there were some strong oscillatory activities during two brief periods (a and b). Oscillatory activities were present during period a, but the fields detected at the four probe locations (shown in detail at the bottom) did not show any clear-cut polarity reversal. Correspondingly, the magnitude of the difference vector **Y** was not remarkably larger than the background level. During period *b*, however, the magnitude of **Y** was quite large. The MEG signals showed polarity reversals across channels 1, 3 and 4. We, thus, interpreted this result as being due to sychronized population activities within the neighborhood of the primary projection site, occurring without a stimulus, analogous to spontaneous discharges detectable with a field potential electrode implanted within the cortex. Thus, it appears that bipolar MEG can be used to selectively detect activities arising from a focal region of the cortex, even during a rest.

As in Fig. 6, the event-related oscillatory activities continued to exist until at least 1.5 sec after the termination of the stimulus train. During the stimulus presentation polarity reversal was present across channels 2 and 4 as indicated by insets *c* and *d*. The strong oscillatory activities during period *e* showed a large field gradient across the four probe positions, but did not show any clear polarity reversal. Polarity reversal was, however, present across channels 3 and 4 during period *f.* Thus, it appears that the stimulus train modulated activities of the cortical area receiving the projection from the stimulated snout area. These modulated activities were not well time-locked to the stimulus train, since the post-stimulus activities were cancelled out when ten consecutive epochs were averaged. The amplitudes of the post-stimulus MEG signals were comparable to those during the pre-stimulus period in the averaged record.

DISCUSSION

Our results demonstrate that it is possible using MEG to detect spontaneous cortical activities arising from a specific area and event-related oscillatory events arising from an identified area of the cortex. This demonstration was feasible because the array of four detecting coils could be placed within a few millimeters from the active cortical tissue. The close distance of measurement increases the spatial resolution of the measurements, that is the width of the sensitivity function is relatively small. Also close measurements increase the signal amplitude since the amplitude of a magnetic field due to a current dipole is inversely related to the square of distance of measurement.

One important implication of the present results is that a high-resolution MEG sensor such as our microSQUID may possibly be used to study modulations of spontaneous activities in different parts of the cortex, including the associative areas, that are elicited by external stimulations, various pharmacological manipulations and tasks given to unanesthetized as well as anesthetized animals.

In human studies one must consider improving the sharpness of the sensitivity function for MEG measurements since the pickup coils are commonly 3-4 cm away from the surface of the cortex. It may be necessary to combine the outputs from more than four channels to narrow the zone of sensitivity. One such method may be the spatial filtering technique developed by Robinson and Rose (1992). In this technique the weights for all the channels of a detector such as a whole-head detector are adjusted to focus the sensitivity of an entire array to a given region of the brain, including subcortical regions. We expect that as techniques such as the spatial filtering method are developed and validated for detecting activities from focal regions of the brain it may become possible to use MEG to detect on-going activities from specific regions of the human cerebral cortex including its associative areas.

Acnowledgements

This research was supported by NIH grant RO1-NS30968. We wish to thank Karen Nishimura for her help with data analysis and Anna Barrios for her help with the experiment.

REFERENCES

Ahonen, A.I., Hämäläinen, M.S., Kajola, M.J., Knuutila, J.E.T., Laine, P.P., Lounasmaa, O.V., Simola, J.T., Tesche, C.D., and Vilkman V.A., 1992, A 122-channel magnetometer covering the whole head. in: "Proc. Satellite Symp. Neurosci. & Tech., XIV Ann. Int. Conf. IEEE EMBS", A. Dittmar, J.C. Froment, eds. , Lyon, pp. 16-20.

Buchanan, D.S., Crum, D.B., Cox, D., and Wikswo, J.P. Jr., 1989, MicroSQUID: A close-spaced four channel magnetometer. in: S.J. Williamson, M. Hoke, G. Stroink and M. Kotani, eds., "Advances in Biomagnetism", Plenum, New York, pp. 677-679.

Cohen, D., Palti, Y., Cuffin, B.N., and Schmid, S.J., 1980, Magnetic fields produced by steady currents in the body, *Proc. Natl. Acad. Sci., USA*, 77:1447-1451.

Cuffin, B.N., and Cohen, D., 1989, Comparison of the magnetoencephalogram and electroencephalogram. *Electroenceph. clin. Neurophysiol.*, 47:132-146.

Grynszpan, F. and Geselowitz, D.B., 1973, Model studies for the magnetocardiogram. *Biophys. J.*, 13:911-925.

Jackson, J.D., 1975, "Classical Electrodynamics", Wiley, New York, pp. 177-180.

Kaufman, L., Curtis, C., Wang, J.-Z., and Williamson, S.J., 1992, Changes in cortical activity when subjects scan memory for tones. *Electroenceph. clin. Neurophysiol.*, 82: 266-284.

Kaufman, L., Schwartz, B., Salustri, C., and Williamson, S.J., 1990, Modulation of spontaneous brain activity during mental imagery. *Cogn. Neurosci.*, 2:124-132.

Klismesch, W., Pfurtscheller, G., and Mohl, W., 1988, ERD mapping and long-term memory: The temporal and topographic pattern of cortical activation. in: G. Pfurtschuller and F.H. Lopes da Silva, eds., "Functional Brain Imaging", Huber, Toronto, pp. 131-141.

Makeig, S. and Inlow, M., 1993, Lapses in alertness: coherence of fluctuations in performance and EEG spectrum, *Electroenceph. clin. Neurophysiol*, 86: 23-35.

Makeig, S., 1993, Auditory event-realted dynamics of the EEG spectrum and effects of exposure to tone, *Electroenceph. clin. Neurophysiol.*, 86:283-293.

Okada, Y.C., 1992, Applications of a μ-SQUID in studies of the physiological basis of MEG. *Biomed. Res.*, 13:29-33.

Okada, Y.C., Kyuhou, S., Lähteenmäki, A. and Xu, C., 1992, A high-resolution system for magneto-physiology and its applications. in: M. Hoke, S.-N. Erné, Y.C. Okada and G.L. Romani, eds., "Biomagnetism: Clinical Aspects". El Sevier, Amsterdam, pp. 375-383.

Pfurtscheller, G., and Sranibar, A., 1977, Event-related cortical desynchronization detected by power measurements of scalp EEG. *Electroenceph. clin. Neurophysiol*, 42:817-826.

Pfurtscheller, G., Steffan, J. and Maresch, H., 1988, ERD mapping and functional topography: Temporal and spatial aspects. in: G. Pfurtschuller and F.H. Lopes da Silva, eds., "Functional Brain Imaging", Huber, Toronto, pp. 117-130.

Robinson, S.E., and Rose, D.F., 1992, Current source image estimation by spatially filtered MEG. in: "Biomagnetism: Clinical Aspects," M. Hoke, S.N. Erné, Y.C. Okada, and G.L. Romani, eds., Excerpta Medica, Amsterdam, pp. 761-765.

Williamson, S. J. and Kaufman, L., 1981, Magnetic fields of the cerebral cortex. in: S, N, Erné, H.-D. Hahlbohm, and H. Lübbig, eds., "Biomagnetism", Berlin, de Gruyter, pp. 353-402.

COGNITION AND LOCAL CHANGES IN BRAIN OSCILLATIONS

Lloyd Kaufman

New York University, New York, USA

INTRODUCTION

Event related desynchronization (ERD) refers to the systematic diminution of level of spontaneous brain activity that accompanies the performance of many different tasks. Pfurtscheller and his colleagues (Klimesch et al.1988; Pfurtscheller, 1988; Pfurtscheller & Aranabar, 1977;Pfurtscheller et al., 1988) introduced the term "ERD" to distinguish this phenomenon from conventional event related potentials (ERPs). These are voltage changes that are time- locked to some observable event, such as a visual stimulus or onset of a motor act. By contrast, voltage changes in, say, the band of alpha frequencies need not be time-locked to the observable event. Even so, the power in that band may be modulated in-step with the event. This modulation is not apparent in the straightforward average of the scalp-detected activity. However, the envelope of the alpha activity recovered by rectification reveals an amplitude modulation during motor acts and also when subjects perform some cognitive tasks.

It is noteworthy that the scalp topography of ERD varies, depending upon the modality involved and the nature of the task. This suggests that the modulation of the intrinsic brain oscillations occurs locally, and is not merely a global phenomenon related to changes in nonspecific levels of arousal. Also, as we shall describe later, the modulation of intrinsic (spontaneous) brain activity is not correlated with the so-called "cognitive" components of the ERP. Apparently, ERD is an independent measure of brain activity which complements the ERP. Therefore, it is of considerable interest to inquire into the fundamental nature of the ERD.

The term "desynchronization" implies a theory as to the origin of alpha activity and its blockage. This implicit theory presumes that macroscopic alpha rhythms and the slow waves of sleep are due to the summed activity of many neurons acting in concert. When they are no longer active in-step with each other, their individual effects tend to be self-cancelling. Thus, self-cancellation of fields or potentials arising from independently active neurons is the hypothetical basis for phenomena such as alpha blockage and ERD. However, a computer simulation (Kaufman et al., 1991) raised serious questions about the general validity of this hypothesis. In the

Oscillatory Event-Related Brain Dynamics, Edited
by C. Pantev, Plenum Press, New York, 1994

simulation a large number of current dipoles oriented normal to the surface of a "cortex" were the generators of extracranial magnetic fields. The current dipoles can be likened to macrocolumns of cortical cells. Their fields were summed at the scalp. It was found that synchronization of these current elements (defined as currents of random magnitude, but all flowing in the same direction at the same time) does not necessarily lead to a large amplitude magnetic field or electric potential at the scalp. In fact, synchronization of these elements can actually result in weaker scalp fields or potentials than is obtained when the current elements are desynchronized, i.e., where both the magnitudes and directions of current flow are randomly selected. Two factors contribute to this non-intuitive result. The first is the geometry of the cortex itself. If on average all of the randomly selected current elements are of equal strength, then the determining factor is the geometry of the cortex. Synchronized current elements populating a symmetrical fold will produce less average field power than do the same elements populating an asymmetrical cortical fold. Furthermore, the average field power of synchronized elements populating a symmetrical cortical fold is less than that of desynchronized elements populating the same fold. These effects are due to the breaking of symmetry due either to differences in structure, or momentary imbalances in amounts of activity across the walls of a sulcus. Therefore, without an explicit model that takes into account both the statistics of neuronal activity and the underlying anatomical structure, it is not possible to ascribe alpha blockage, cortical "activation" or ERD to desynchronization. For this reason we prefer the theoretically neutral term "suppression" to "desynchronization". For us suppression signifies activation of cortical networks. This type of activation reveals aspects of cognition that are apparently unavailable in the classic evoked responses.

In several magnetoencephalographic (MEG) studies alpha suppression was found to occur in different regions of the brain, depending upon the nature of the task. For example, if subjects responded to a flashed word on a screen either by forming a mental image of the object represented by the word or by performing a verbal task, the alpha suppression had different durations at different scalp locations, depending upon the task (Kaufman et al., 1989;1993; Cycowicz et al., 1994). Similarly, in the classic Sternberg (1966; 1969) experiment, reaction time increases linearly with the size of the memory set. In our experiment on memory search for tones (Kaufman et al., 1991) the duration of alpha suppression detected over right temporal region also increased linearly with size of the memory set, and was commensurate with the simultaneous measured reaction time. It is interesting to note that Zatorre et al. (1993) used PET scanning techniques to confirm that this region is involved in memory search. Also, in a related study of responses to visual forms (Kaufman et al., 1990), the suppression effect correlated with reaction time was observed over the visual areas.

In their study of short-term memory for tones Kaufman et al. (1991) examined how the amplitude of the electric N100 varied with set size. No systematic effect was found. However, the vertex N100 reflects activity of generators in both hemispheres (Naatanen & Picton, 1987). To identify any hemispheric differences that might be related to the memory search task, the amplitudes of the magnetic counterpart to N100 over each hemisphere were compared with reaction times obtained with different set sizes. Although the two hemispheres were found to respond differently, neither was systematically related to the reaction time data.Thus we must conclude that N100 does not reflect the process of searching short- term memory. Also, the relation between P300 latency or amplitude and reaction time in the memory search task is widely considered in the literature. Neither the latency nor the amplitude of P300 appears to be a consistent index of memory search time. For example, in older subjects P300 does not change with set size, but reaction time increases with set size (Marsh, 1975). Also, where more than two set sizes are used, in most studies neither latency nor amplitude of P300 changes linearly with set size (Pratt et al., 1989a; 1989b; 1989c). In fact, the relation is often not even monotonic, and sometimes P300 doesn't change at all (Gaillard & Lawson, 1984).Despite these inconsistent data, there are ample reasons to believe that these same ERP components reflect some aspects of cognition. However, the suppression of intrinsic brain oscillations reflect other equally important cognitive processes.

The ERP has yet to be compared directly with alpha suppression detected by the same scalp electrodes at the same time. This type of comparison could either strengthen or refute the argument that the two phenomena are complementary rather than redundant indicators of psychological processes. In this paper we report the first results of such a comparison. Furthermore, we describe the first use of a free recall paradigm in a psychophysiologal study. While the results are strictly preliminary, they provide a model for future experiments which could extend the reach of psychophysiological studies to new areas of cognition.

Baddeley (1986; 1992) considered short-term memory to be composed of a number of different subsystems which he collectively refers to as working memory. This is an interacting set of storage systems, "possibly coordinated by a single central executive component." For example, he proposed a phonological loop for short-term store of verbal material, and a visuo-spatial "scratchpad" for storage of visual information. Consider memory for words. The number of words that can be stored in short-term memory is a function of word-length. More specifically, the time required to pronounce each word is related to how many of them may be stored for subsequent free recall. If the time required for articulation is short, more words may be recalled later. If long, then fewer may be recalled. In general, more monosyllabic (short) words may be stored in working memory than polysyllabic (long) words, even when the words are presented visually. Thus, subjects appear to be able to remember as much material as they can read in about 1.5 s, and this is influenced by the time required to actually enunciate the words of the text. This dependence on reading time vanishes for visually presented words when subjects are occupied by counting from 1 to 6. This is consistent with the concept of the phonological loop. Interestingly, when the words are presented acoustically then counting does not suppress the word-length effect. This rules out implicit activation of the speech musculature. Apparently, when words are presented acoustically, programs of neural activity related to the phonological loop are primed directly.

The concept of a phonological loop is also consistent with the early work of Conrad (1964) on acoustic confusions revealed by intrusion errors in immediate memory. Phonologically similar items interfere with each other in immediate memory tasks, and without interference the ability to retain such items is no longer than about 1-2 s, unless rehearsal occurs within this interval. Baddeley suggests that visually presented verbal information is recoded phonologically and this is interfered with by irrelevant speech. However, by hypothesis, acoustically presented words go directly into phonological store, so the transfer process cannot be interfered with by irrelevant speech.

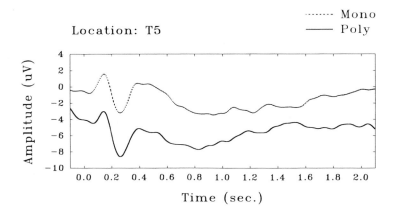

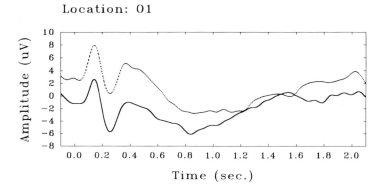

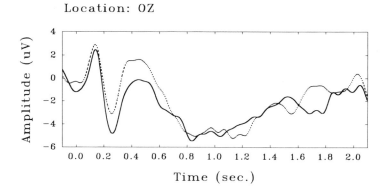

EVENT−RELATED POTENTIALS

Figure 1. Sample average event related potentials (ERPs) associated with polysyllabic (long) and monosyllabic (short) words recorded at various scalp locations. ERPs recorded at location T5 and O1 exhibit a striking baseline difference associated with the two kinds of words.

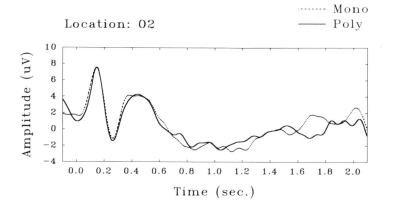

Location: 02

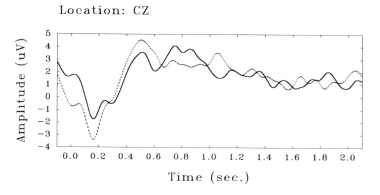

Location: CZ

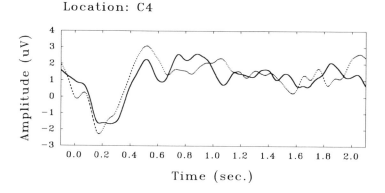

Location: C4

EVENT−RELATED POTENTIALS

Figure 1. (continued) The average ERP between these electrodes and the reference electrode is markedly more negative for long words (poly) than for short words (mono). There is also a significant late negativity (peaking at about 500 ms) at Oz. Apart from this effect at the left posterior scalp and midline, there are no significant differences in ERPs associated with long and short words at any of the other 16 scalp locations.

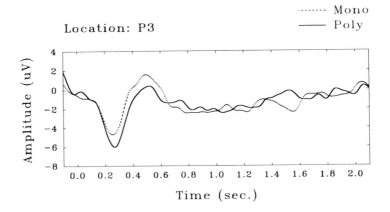

Location: P3

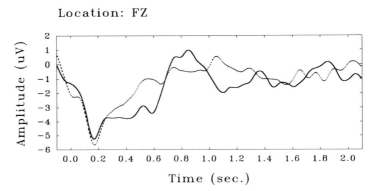

Location: FZ

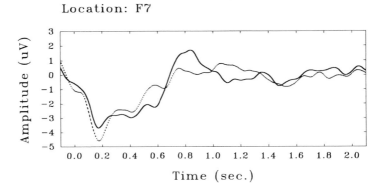

Location: F7

EVENT−RELATED POTENTIALS

Figure 1. (continued)

We recently conducted an EEG study to determine the feasibility of examining brain activity related to the phonological loop. This work was conducted in collaboration with Drs. James Moeller and Bruce Luber of the NY State Psychiatric Institute at Columbia University using the Institute's 21-channel EEG system and software.

METHODS

The subject fixated the center of a screen on which four words were presented, one at a time. The words were either short or long. Each word remained in the screen for 2 s with 0.21 s between words. As soon as possible after the 4th word, the subject attempted to recall in their correct order all of the words he had seen. Upon completion of the recall task, another set of 4 words was flashed on the screen. The entire process was repeated several times with sets of the two types of words given in separate blocks of trials. This was done until 100 short words and 100 long words were viewed. The words on the two lists (long and short) were approximately equal in frequency of use in the English language. The EEG was measured throughout the experiment so that ERPs associated with correctly recalled long words and short words could be computed.

To record the EEG electrodes were attached to the scalp in accord with the 10-20 system. ERPs were computed with right and left mastoid as reference, and also with the average of all electrodes as reference. Since the choice of reference made no important difference in the data, we report only those data obtained with left mastoid as reference. The recorded EEGs were sorted by computer into segments corresponding to 2.1 s epochs, with the onset of the epoch occurring 0.1 s prior to the onset of a word. The epochs included in each average response belonged only to the lists of four words that were correctly recalled.

After editing out eye movement artifacts, the surviving epochs of each of 19 EEG channels were bandpassed between 0.1 Hz and 40 Hz and averaged to recover the ERPs. To recover the modulation of alpha band activity the same data were bandpassed between 8 and 13 Hz, averaged, and the variance about the mean response was computed (Kaufman & Price, 1967; Kaufman & Locker, 1970. (Note, this procedure differs from simply squaring activity in the alpha band to obtain alpha power, as was done in Pfurtscheller's experiments. Computing variance around the mean response within that band removes any residual event related activity and leaves only the power in the background EEG, thus providing a measure that is truly independent of the ERP). The resulting plots of variance as a function of time were then smoothed to show how the power of the background "noise" in the alpha band varied in amplitude as a function of time after presentation of the stimulus word.

RESULTS

As Baddeley had found, for this subject the accuracy in recalling short words was higher (72% correct) than the accuracy in recalling the long words (55% correct). This confirms the basic difference between long and short words that led Baddeley to hypothesize the existence of a phonological loop.

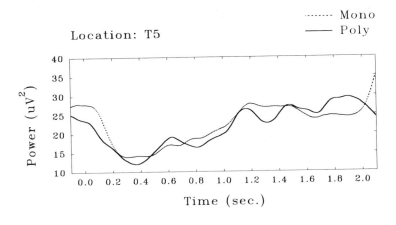

Location: T5

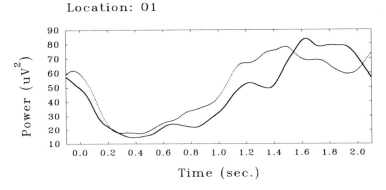

Location: O1

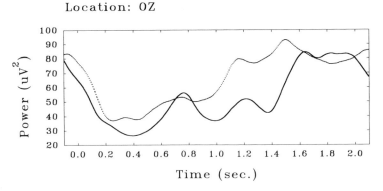

Location: OZ

ALPHA POWER vs. TIME

Figure 2. Average alpha power as a function of time for both long and short words. These recordings were made at the same time and at the same scalp locations as those shown in Figure 1. The average power in the alpha band recorded at T5 decreases after stimulus presentation, and recovers slowly. The duration of this alpha suppression is the same for both long and short words at this site. The ERPs at this same location show a baseline difference (see Figure 1). An even more marked suppression of alpha is manifested over the posterior midline (Oz) and over the right posterior scalp.

Location: 02

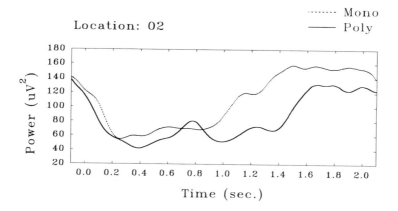

Location: CZ

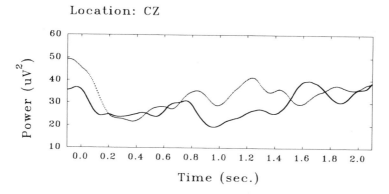

Location: C4

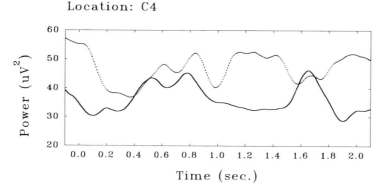

ALPHA POWER vs. TIME

Figure 2. (continued) Moreover, the duration of this suppression is markedly greater at these sites than over the left posterior scalp. Suppression duration is approximately 1 s for short words and 1.4 s for long. At Cz there is no discernible suppression in the plot for long words, and the trace for short words is diffuse. Similarly ambiguous results are obtained at C4 (right central) and P3 (left parietal) locations, although clear ERPs were recorded at the same sites. Location F7 manifests a more clearly defined pattern of suppression than is seen at these central and parietal sites.

409

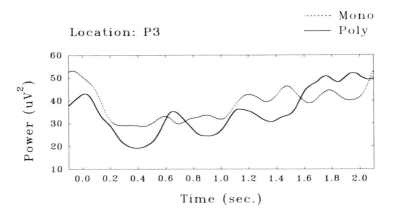

Location: P3

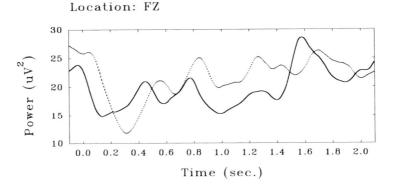

Location: FZ

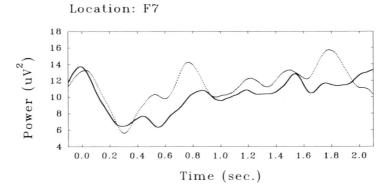

Location: F7

ALPHA POWER vs. TIME

Figure 2. (continued) A well defined alpha suppression lasting about 400 ms is associated with short words, and the suppression for long words lasts about 800 ms. Note the striking differences between the right occipital scalp locations and this left frontal location.

Sample plots of the ERPs are shown in Figure 1 and of alpha power in Figure 2. While a total of 19 ERP plots were actually computed, here we show only representative records. Except for the ERPs recorded at T5 and O1 (over the temporal and adjacent left occipital area), the ERPs are unremarkable. They show the usual complex of midlatency components. These are detectable over most of the scalp. Thus, as revealed in Figure 1, the salient 180 msec component of the visual ERP is directed upwards (positive) over the posterior scalp, and downwards (negative) over the frontal scalp. This is consistent with the presence of a dipolar component in the generators of the potential differences. Therefore, these observed potentials cannot be due to sources directly under the electrodes. Rather, the predominant generators may well lie someplace between the electrodes. Apparently there are no significant differences between these midlatency ERP components due to long and short words. However, data recorded at T5 and O1 show a strong negative baseline shift for the long words relative to responses to short words. This baseline difference extends for a period of about 1 s subsequent to the stimulus. Its magnitude is greater over T5 than over O1, and there is no visible reversal of polarity elsewhere on the scalp. It should be noted that the baseline shift occurs prior to presentation of the stimulus. However, since the ERP is an average response, it is possible that the baseline shift is present over several epochs and manifests itself in the average as a persistent shift. Regardless, this suggests a source with a predominantly radial orientation with respect to the scalp. Further, it lies almost directly under T5. Of course, the precise cortical region under these electrodes is not known since there are no anatomical data to relate electrode position to cortex. The negative occipital and positive frontal midlatency components do not imply activation of two different networks. These data are consistent with single dipolar sources between the electrodes and active at the same. Asymmetries in these patterns could reflect contributions by radially oriented source components located near some electrodes.

Of course, inferences from these patterns can only be considered as working hypotheses. Other factors must be considered, e.g.,gradients of potentials along orthogonal directions on the scalp, temporal covariance, etc. Even so, the mere fact that differences in scalp topography that are related to specific cognitive processes does encourage additional work. Nevertheless, the lack of substantial differences between ERPs to short and long words that can be related to the speech areas suggests that these ERPs do not reflect phonological encoding of words.

In general, a different picture emerges with respect to the plots of alpha power at the same locations. The average power plots of Figure 2 show a clean suppression of alpha power at T5, O1, Oz, and O2. There is also a well defined reduction in alpha power at F7. None of the other 14 electrodes showed a similarly clean suppression effect, Note also that the duration of suppression is markedly greater for long words at Oz and O2 (posterior right hemisphere). Alpha is suppressed at O2 for about 1.6 s by long words, and about 1.3 s by short words. The total time of presentation was 2.0 s, so this result suggests that suppression duration is related to the duration of visual attention to the words. A similarly sharp suppression (activation) effect is seen at F7 - which is probably close to the projection of Broca's area onto the left scalp. Its onset is delayed about 100 msec relative to the onset in the occipital region, and suppression related to long words lasts about 800 msec and about 400 msec for short words.

Therefore, the same generators cannot account for the suppression in both the occipital and frontal regions. There is virtually no difference in suppression duration at T5, where the baseline shift was greatest for the ERP to polysyllables, and only a slight difference in suppression duration for O1, where the baseline difference is less strong. Therefore, several of the electrodes were differentially affected by the two types of stimuli. Furthermore, the changes in alpha power at different electrodes complements the ERP information. At T5 and O1 the presence of the baseline shift may be related to the lack of difference in suppression time. But there is nothing in the other ERPs (including those not shown here) that could predict the differences in alpha suppression detected at different electrodes. It is clear that if this type of discrimination is available in the EEG using the standard 10-20 system, it could be much sharper using the MEG.

Wang et al. (1993) showed that MEG data may be used to find a unique estimate of the source of spontaneous activity that differs in level from its background. Therefore, the prospect of forming functional images incorporating local suppression and enhancement of intrinsic brain oscillations seems quite real.

PET studies indicate that visually presented words are processed on the inferior occipital surface (Petersen et al.,1988; 1989; 1990). The onset of the slow shift coincides with that of the stimulus, and occurs more than 100 msec prior to the onset of suppression of alpha power detected at F7. It is obvious that PET cannot reveal this difference in onset times.

As noted, there is a very large difference in suppression duration related to long and short words over the posterior right scalp. Further, the total time of suppression for both types of words is substantially longer than the corresponding times of suppression at F7, (the left frontal scalp). At F7 the suppression duration is about 400 msec for monosyllables and about 800 msec for polysyllables. the difference could be due to differences in average implicit rehearsal time. Thus, the sharp suppression of ongoing activity at F7 might well be related to activation of the phonological loop.

CONCLUSIONS

While it is not possible to draw general conclusions from so limited a set of data, they should be followed up. If the speech areas are involved in the phonological loop, then the EEG may well be sufficient to localize them. Naturally, comparisons will have to be made with the Wada test. However, an even more important conclusion can be drawn from this miniexperiment. It proves that there are functionally different species of brain activity. One specie is the classic ERP which is itself comprised of subspecies. These are the so-called components, e.g., N100, P300 and N400. We designate them as different "subspecies" because they apparently differ in functional significance. Another specie is the modulation of the intrinsic activity of different regions of the brain. Unlike the ERP and its components the carrier of the modulation is not time-locked to any externally observable event. Within this specie we also have subspecies, and these have barely been explored. They include the so-called 40 Hz (gamma band) phenomenon, although its relation to cognition is unclear. Experiments are needed to discover if behavior known to be related to cognitive processes covaries with the 40 Hz phenomenon. Similarly, Pfurtscheller's upper and lower alpha bands show signs of having different functional significance. No doubt other subspecies remain to be discovered and their functions defined.

In conclusion, it is imperative that these different types of brain activity be compared with functional MRI as well as other measures of blood flow (Roland & Friberg, 1985; Goldenberg, 1989; Petersen et al.,1988; 1989; 1990). While measures of blood flow reveal activation of different areas of the brain, we do not know how blood flow is related to the magnitudes and physical extent of changes in the electrical activity of these same areas. Furthermore, PET and SPECT studies employing ligands of neurotransmitters must also be related to different measures of electrical activity as well as blood flow. It is reasonable to assume that no one of these measures fully defines all of the brain activity related to cognitive processes.Astronomers would not restrict themselves to optical astronomy and ignore radio or X-ray astronomy. Similarly, cognitive neuroscientists must exploit all of these powerful measures, including the spontaneous oscillations of the brain.

REFERENCES

Baddeley, A. Working memory. Science, 1992, 258, 556-559.

Baddeley, A. Working Memory. Oxford Univ. Press, Oxford, 1986.

Cycowicz, Y., Kaufman, L., Glanzer, M. & Williamson, S.J. Selective suppression of spontaneous cortical rhythms during imaging and rhyming. Submitted. 1994.

Conrad, R. Acoustic confusions in immediate memory. Brit. J. Psychol. 1964, 55, 75-84.

Gaillard, A.W.K. & Lawson, E.A. Evoked potentials to consonant-vowel syllables in a memory scanning task. Ann. NY Acad. Sci.1984, 425, 204-209.

Kaufman, L. & Price R. The Detection of Cortical Spike Activity at the Human Scalp, IEEE Trans. Bio-Med. Eng. 1967, 14, 84.

Kaufman L.& Locker,Y. Sensory Modulation of the EEG, Proc. 8th Annual Conv. Amer. Psychol. Assoc. 1970, 179-180.

Kaufman, L., Curtis, S., Wang, J.-Z. & Williamson, S.J. Changes in Cortical Activity When Subjects Scan Memory for Tones,Electroenceph. clin. Neurophysiol. 1992, 82, 266-284.

Kaufman, L., Schwartz, B., Salustri, C., & Williamson, S.J., Modulation of Spontaneous Brain Activity During Mental Imagery, Cogn. Neurosci. 1990, 2, 124-132.

Kaufman, L., Cycowicz, Y., Glanzer, M., & Williamson, S.J. Verbal and imaging tasks have different effects on activity of cerebral cortex. Submitted, 1993.

Kaufman, L., Glanzer, M., Cycowicz, Y. & Williamson, S.J. Visualizing and rhyming cause differences in alpha suppression. In S.J. Williamson, M. Hoke, M. Kotani, and G. Stroink (Eds.), Advances in Biomagnetism, Plenum Press, New York, 1989, 241-244.

Kaufman, L., Kaufman, J.H. and Wang, J.-Z. On cortical folds and neuromagnetic fields. EEG clin Neurophysiol.,1991, 79, 211-226.

Kaufman, L., Curtis, S., Wang, J.-Z. and Williamson, S.J. Changes in cortical activity when subjects scan memory for tones. EEG clin Neurophysiol.,1991, 82, 266-284.

Klimesch, W., Pfurtscheller, G. & Mohl, W. Mapping and long-term memory: The temporal and topographical pattern of cortical activation. In: Functional Brain Imaging, G. Pfurtscheller and F.H. Lopes Da Silva, Eds. Hans Huber Publishers, Toronto, 1988, pp. 131-142.

Naatanen, R. & Picton, T., The N1 wave of the human electric and magnetic response to sound: A review and an analysis of the component structure. Psychophysiol. 1987, 24.

Marsh, G.R. Age differences in evoked potential correlates of a memory scanning process. Exp. Aging Res. 1975, 1, 3-16.

Petersen, S.E., Fox, P.T., Posner, M.I., Mintun, M., & Raichle, M.E. Positron emission tomographic studies of the cortical anatomy of single-word processing. Nature, 1988, 331, 585-589.

Petersen, S.E., Fox, P.T., Posner, M.I., Mintun, M. & Raichle, M.E. Positron emission tomographic studies of the processing of single words.J. Cog. Neuroscience, 1989, 2, 153-170.

Petersen, S.E., Fox, P.T., Posner, M.I., Snyder, A.Z., & Raichle, M.E., Activation of extrastriate and frontal cortical areas by visual words and word-like stimuli. Science, 1990, 1041-1044.

Pfurtscheller, G. & Aranabar, A. . Event-related cortical desynchronization detected by power measurements of scalp EEG. Electroenceph. clin. Neurophysiol., 1977, 42, 817-826.

Pfurtscheller, G. Mapping of event related desynchronization and type of derivation. Ibid., 1988, 70, 190-193.

Pfurtschuller, G., Steffan, J., & Maresch, H. ERD-mapping and functional topography - temporal and spatial aspects.. In:Functional Brain Imaging. G. Pfurtschuller and F.H. Lopez da Silva, Eds., Hans Huber, Toronto, 1988, 117-130.

Pratt, H., Michalewski, H.J., Barrett, G., & Starr, A. Brain potentials in a memory scanning task. I. Modality and task effects on potentials to the probes. Electroenceph. clin. Neurophysiol.,1989a, 72, 407-421.

Pratt, H., Michalewski, H.J., Barrett, G., & Starr, A. Brain potentials in a memory-scanning task. II. Effects of aging on potentials to the probes. Ibid., 1989b, 72, 507-517.

Pratt, H., Michalewski, H.J., Barrett, G., & Starr, A. Brain potentials in a memory scanning task. III. Potentials to the items being memorized. Ibid., 1989c, 73, 41-51.

Sternberg, S. High speed scanning in human memory. Science, 1966, 153, 652-654.

Sternberg, S. Memory scanning: Mental processes revealed by reaction time experiments.Amer. Scientist, 1969, 57, 421-457.

Goldenberg, G., Podreka, I., Uhl, F., Steiner, M., Willmes, K. & Deecke, L. Regional cerebral blood flow patterns in visual imagery. Neuropsychologia, 1989, 27, 641-654.

Roland, P.E. & Friberg, L. Localization of cortical areas activated by thinking.J. Neurophysiol., 1985, 53, 1219-1243.

Wang, J.-Z., Kaufman, L., & Williamson, S.J. Imaging Regional Changes in the Spontaneous Activity of the Brain: An Extension of the Unique Minimum-Norm Least-Squares Estimate. Electroenceph. clin. Neurophysiol., 1993, 86, 36-50.

Zatorre, R.J., Evans, A.C. & Meyer, E. Cortical Mechanisms Underlying Melodic Perception and Memory for Pitch.Society of Neuroscience Abstracts, Volume 19., 1993, 843.

TUNING AND FILTERING IN ASSOCIATIVE LEARNING

Larry E. Roberts, Ronald J. Racine, Paula J. Durlach, and Sue Becker

Department of Psychology
McMaster University
Hamilton, Ontario
Canada L8S 4K1

INTRODUCTION

The individual organisms of most animal species occupy variable environments whose detailed features cannot be fully anticipated by a genetic code. The evolutionary response to this ecological constraint has been to build into that code mechanisms for abstracting the structure of individual environments and for generating behavior based on the representations thus formed. To adequately represent an environment, an organism must encode the physical features of its world, relations among those features and events that occur in that world including its own actions and their consequences, and the temporal flow of this information in time. Memory and perception can be considered to operate in the service of the representational problem, which is an associative one in the sense that these aspects of spatiotemporal information must be extracted and encoded if an ecological niche is to be adequately described.

The problem of how the brain carries out its associative and representational functions was addressed by early researchers in experimental psychology and neuro-science (Lashley, 1950; Hebb, 1949; Konorski, 1967; Deutsch, 1960). The accounts they suggested were structural in the sense that specific neural mechanisms were proposed, although computational modelling of those mechanisms was not attempted. In the two or three decades following these efforts, the study of learning shifted away from structural theories to laboratory investigations of Pavlovian and instrumental conditioning (Mackintosh, 1983) and to the development of quantitative theories of the acquisition process (Pearce & Hall, 1980; Rescorla & Wagner, 1972). Although this research has deeply enriched our understanding of associative mechanisms, it has also set into relief some problems regarding the nature of representations (Colwill, 1993; Miller & Barnet, 1993) and the organization

Oscillatory Event-Related Brain Dynamics, Edited
by C. Pantev, Plenum Press, New York, 1994

of action (Holland & Rescorla, 1982) that might be better solved within the framework of a structural approach (Brener, 1986). Interest in structurally-based accounts of learning has recently resurfaced, not only because the contribution of laboratory analyses of learning may be saturating, but also because developments in neuroscience and in neural network modelling have opened avenues of description and inquiry that were not available to early pioneers.

Recent structural models have typically taken as their starting point the task of capturing specific conditioned responses (e.g., Gluck et al., 1993) or selected phenomena of signalling such as blocking, overshadowing, and the learning of non-linear associative relations (e.g., Schmajuk & DiCarlo, 1992; Gluck & Meyers, 1993). The model described in this paper, on the other hand, attempts to give a more general account of an organism's adjustment to a learning situation. We begin by describing some background facts that invited the current approach. These facts concern the nature of Pavlovian and instrumental conditioning and how mechanisms that support these types of learning might be reflected in the organization of the brain. A neural network architecture is then described that performs learning functions by means of two processes, (a) a filtering process that identifies unexpected information arising over sensory pathways, and (b) a tuning process that selectively augments neural processing in sensory modalities that are conveying surprising information to the cortex. In the concluding sections we discuss implications of the model for simulation and experimental studies, and comment on how learning functions may be reflected in the electrical and magnetic field activity of the brain.

BACKGROUND

Any model of a learning system should try to capture basic findings from behavioral research which indicate how learning works. In this respect, the idea that organisms come to know or represent the spatiotemporal stucture of their worlds was long resisted by behavior theory, in favor of the idea that linkages between responses and controlling stimuli were sufficient to account for learned behavioral adaptation (see Mackintosh, 1983, for a review). However, there is now much evidence that environmental relations are encoded by animal (Rescorla, 1987, 1988) and human subjects (Dawson & Schell, 1987) during conditioning experiences. For example, in human subjects differentiation of behavior between signalling stimuli does not occur in the absence of the ability of the subjects to describe the Pavlovian or instrumental contingencies to which they have been exposed (Dawson & Biferno, 1973; Hughes & Roberts, 1985). When verbal reports of reinforcement contingencies are dissociated from response differentiation in either conditioning arrangement (this occuring in 15% of subjects in the feedback experiments of Hughes & Roberts, 1985), it appears to be invariably the case that knowledge of the contingencies precedes the differentiation of overt behavior. Overall, the linkage between abstracting of environmental relations and adapting behavior to those relations is very close (Roberts, 1990).

Also important for structural theories of learning are numerous similiarities which exist between Pavlovian and instrumental conditioning at the associative level. These similarities pertain not only to the close linkage between associative knowledge and response differentiation just mentioned, but also to basic principles of signalling in the two conditioning arrangements. For example, if multiple stimuli predict a reinforcer, subjects learn preferentially about the best predictor, in accordance with a principle of relative validity (Mackintosh, 1983; Wagner et al., 1968). Similarly, if reinforcement is allocated differentially among several alternative responses, the response that is scheduled to receive the highest rate of reinforcement is strengthened the most by the training arrangement (Herrnstein, 1970). Stimuli and responses also appear to be equivalent in their ability to signal a reinforcer. For example, bar pressing for food reward is impaired if responding produces an exteroceptive stimulus that is equally valid as a signal for reinforcement (Williams & Heyneman, 1982). These and other associative similarities summarized by Mackintosh (1983) are important, because they imply that the same associative mechanism is responsible for encoding of Pavlovian and instrumental contingencies (Weisman, 1977). The existence of a single associative mechanism, on the other hand, gives reason to try to describe and model this mechanism. In the model that we describe here, Pavlovian and instrumental contingencies are assumed to differ only with regard to whether stimuli that signal reward are conveyed principally by an exteroceptor (the Pavlovian case) or by kinesthetic pathways (the instrumental one).[1]

One might expect that the requirements of an encoding system that extracts relations among different types of events would be reflected in the general organization of the brain. In this respect it may be noteworthy that there is significant uniformity in the morphology and organization of cortical structures in the brain. It appears from histological evidence that up to 90% of the synapses in the cortex are excitatory or Type I synapses, and as many as 85% of these synapses may be provided by a single category of cell, the pyramidal neuron (Braitenberg & Schuz, 1991). There is less agreement about the uniformity of inhibitory neurons (Type II), some authors proposing half a dozen types while others only one basic type, in any case not too many (Douglas & Martin, 1990). Organization of these neuronal processing elements into laminae and columns is characteristic of most cortical regions, even though variations exist between regions that serve different functional roles. The capacity for plastic changes in the form of long term potentiation (LTP) has been documented at several levels of the brain including neocortical and paleocortical structures, magnocellular and intralaminar thalamic nuclei, basal forebrain regions, and peripheral ganglia (Gerren & Weinberger, 1983; Lynch & Granger, 1992; Racine et al., 1983). These findings suggest that learning depends not only on specialized neurons and local architectures that are adapted for specific purposes, but also on the way in which neurons are organized into cortical regions, and in the kind of information that is delivered to these regions over time.

---

[1] Although differences in the modality of the signalling stimulus do not appear to alter how association works, the properties of behavioral adaptations brought out by signalling (for example, whether the response is executed by striate or smooth muscles, the voluntary nature of the act, and its access to consciousness) are affected by modality differences (Roberts, 1990). How behavior is generated from environmental representations that are established by conditioning is a problem for modern behavior theory (Rescorla, 1988). The learning model described in this paper does not address this problem in a detailed way, but the hierarchically organized and distributed nature of encoding in the system is compatible with the form of analysis advanced by Brener (1986).

The model that we describe explores how these "macroarchitectural" features of brain organization might support a learning system.

We have also been guided by some facts relating to brain electrical activity during performance on cognitive and perceptual tasks. Slow potentials of the brain shift toward negativity prior to the performance of motor responses (Gaillard, 1986) as well as prior to the delivery of informational stimuli when motor responding is absent (Chwilla & Brunia, 1991). P300-like waves are augmented by stimulus novelty (Johnson, 1986), and although dynamic changes in these waves have seldom been studied over extended periods, their dependence on novelty suggests they ought to subside as the eliciting event becomes predicted. Recent evidence suggests that P300 waves are generated by inhibitory mechanisms that are either widely distributed in cortical and subcortical structures (Halgren et al., 1986; Woodward et al., 1992) or are organized in a such way that their effects can be selectively manifested in these structures (Roberts et al., 1994; Rockstroh et al., 1992). The intrinsic repetitive firing behavior of many neurons is also relevant to biological and computational models of learning (Basar & Bullock, 1992). Coherent oscillatory activity has been recorded from ensembles of neurons in several brain structures and shows properties such as sensitivity to apparent motion (Singer et al., 1990) and deployment of attention (Murthy & Fetz, 1992) that are consistent with a role in perception and/or encoding. Slow waves, P300 events, and oscillatory rhythms are of interest to our agenda, not only because any model of learning must account for them, but also because these phenomena seem likely to express the dynamics of a learning system that is embodied in the general organization of the brain.

A more specific point of departure for our attempt to describe a learning system came from behavioral experiments which suggested that sensory pathways are selectively modulated or "tuned" by conditioning arrangements. Roberts et al. (1991) monitored the target of visual fixation while human subjects learned to produce two patterns of action that were identified by auditory feedback signals alone. The purpose of the experiment was to determine whether the previously documented ability of subjects to accurately describe behavioral outcomes of training (Hughes & Roberts, 1985) might depend in part on whether their actions were self-monitored in vision. Contrary to expectation, the results showed that subjects typically did not watch what they were doing during feedback training, even though provision of feedback in the form of auditory signals meant that they were free do so. On the contrary, subjects tended to close their eyes on feedback trials as training progressed (see Figure 1A). Bramwell (1993) recently corroborated this observation in a different training arrangement (Figure 1B) and went on to show that if a visual detection task was superimposed on the auditory feedback problem so that vision was now required, spontaneous blinks and detection errors were increased compared to a condition in which the visual task was performed alone. These findings suggested that eye closures may have occurred during the auditory feedback problems of Figure 1, because processing in visual channels was relaxed or "tuned out" by the presence of auditory feedback signals that also required processing. The progressive nature of the eye closure effect also suggested that repeated conjunctions between events in kinesthetic and auditory but not visual pathways may have favored the development of eye closures in this training environment.

418

A

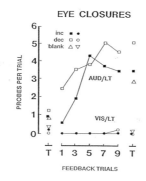

B

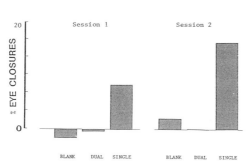

Figure 1. (A) Subjects received either auditory feedback (AUD/LT) or visual feedback (VIS/LT) for generating two bidirectionally opposite patterns of unidentified behavior. Eye closures were observed during auditory feedback as training progressed. By the end of training, eye closures occupied approximately 65% of each auditory feedback trial relative to baseline (Roberts et al., 1991). The two behavior patterns that were trained consisted of striate muscular activities associated with increases and decreases in cardiac interbeat intervals. (B) Subjects solved an auditory feedback problem on single task trials and a visual detection task concurrently on dual task trials, in two training sessions separated by a brief rest interval. The onset of each trial was accompanied by a "beep" issued from a computer. Measurement of eye closures showed that the subject's eyes were open on dual task trials and on blank trials which occurred during intertrial intervals. However, eye closures occurred and intensified over the course of training on single task trials where auditory feedback was processed (Bramwell, 1993).

Although tuning expressed as perceptual interference has not been widely studied in conditioning experiments, the fact that discriminative behavior often results from subjects learning to orient toward and approach objects or locations that are associated with reward, to the detriment of objects or locations that are not, suggests that the role of tuning in behavioral adaptation may be considerable (Jenkins & Sainsbury, 1970). Signaling principles such as blocking and overshadowing also appear to be amenable to analysis in terms of tuning (Mackintosh, 1983), and examples of lateral and surround inhibition which could support such phenomena are well documented in the nervous system (Kandel et al., 1991). We therefore began our project by attempting to describe an architecture that might tune sensory systems during associative learning. Because the role of tuning is presumably to promote the formation of associations among the events that are processed, it was desirable to consider a mechanism for association as well. How might these functions be served, how much work can they do, and where are they situated in the brain?

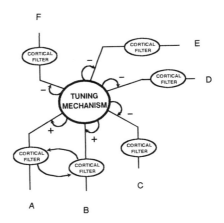

Figure 2. Components of the learning system.

TUNING AND FILTERING

If tuning among sensory modalities is to occur, information about activity in different modalites must converge in the nervous system. Thalamic nuclei offer an early opportunity for intermodal sensory processing and seem likely on this account to play a role in tuning functions. On the other hand, the extensive reciprocal connections which exist between thalamic nuclei and neocortical and paleocortical regions imply other levels of organization to and from which information is delivered, and are consistent with a principle of distributed associations (Lashley, 1950). These principles are not controversial, and we began by describing a learning system that included them.

Figure 2 describes the elements of the system in abstract terms, without explicit reference to neuronal mechanisms (Roberts et al., 1992). Pathways A-F represent sensory modalities that impinge on a tuning mechanism; in each modality there is also a cortical filter that processes sensory input. Briefly, we want tuning to work as follows. If an unexpected event occurs in one modality (A), processing in that modality is facilitated while processing in other modalities is dampened simultaneously (CDEF). However, if a biologically significant event has co-occurred in a second modality (B) because the two events are linked by an environmental contingency, that modality is spared from inhibition and is facilitated instead. Synaptic activity is therefore enhanced in neural networks that are driven by unpredicted and temporally coupled task events, and dampened elsewhere (receptor orienting acts may be part of this process). The purpose of tuning is to enhance the rate at which uncoded data enters the system, and to amplify the effect of these data on synaptic plasticity. As a consequence of data-driven processing, links are forged between events at distributed levels of the system.

Associative functions are performed by the cortical filters which are shown in each modality in Figure 2. Although only two links are illustrated (the arrows connecting filters in modalities A and B), the pattern of connectivity among filters in different modalities is assumed to be complete. The cortical filters serve two functions. First, they gate sensory input to the tuning mechanism. Only unexpected events gain access. Second, the cortical filters are a major site of learned association. They receive input from many modalities, and they change their filtering characteristics as synaptic weights are modified by Hebbian rules. A further assumption of the model is that once two sensory events have become associated by filtering, those events lose their access to the tuning mechanism. In other words, association acts as a filter. We need the filtering function to capture phenomena such as automaticity, activational peaking over the course of training (Germana, 1968), and an expected diminution of P300 waves with experience (Johnson, 1986). At least one current model of Pavlovian conditioning (the SOP model of Wagner & Brandon, 1989) makes a similar assumption, although that assumption is couched in quite different terminology.

Before leaving Figure 2, we should comment on the concept of a modality (pathways A-F). This term is usually taken to refer to a sensory system such as vision, audition, kinesthesis (which signals reward in the instrumental case), and so on. But sensory systems are more finely differentiated than this term allows. For example, the visual system conveys different types of information over distinct pathways (the magnocellular and parvocellular, at the level of the lateral geniculate). Auditory cortex is tonotopically organized, and there is extensive columnar organization in visual and kinesthetic pathways. Although it is reasonable to suggest that tuning effects may extend to multiple levels including an entire sensory system (Figure 1), we prefer to link the sensory channels of our model with more detailed structural elements such as cortical columns. A tuning mechanism whose bandwidth is wide is likely to play a more important role in association than one whose bandwidth is narrow.

Figure 2 identifies two elements, tuning and filtering, that may be involved in learning, and although thalamic and cortical mechanisms are implicated in these functions, there are no neurons here. The next question was, how might these mechanisms work neuronally, and in real time?

NEURAL NETWORK IMPLEMENTATION

Given technological constraints, any modelling effort can only incorporate a small sample of the findings from neuroscience research. Although it is not an easy task to determine which findings should be included in a model, it is a task worth attempting because application of this information can have major effects on the functioning of a neural network (Lynch & Granger, 1992). In implementing the model of Figure 2 we have tried to accommodate some documented properties of real neural networks. Among these properties are (a) selected aspects of corticothalamic organization including reciprocal connectivity and a substantial bandwidth in the forward and backward paths; (b) an increase in the number units available for encoding at higher levels of the system (this feature enabling sparse encoding); (c) a provision for top-down processing of sensory input (Peterhans & von der Heydt, 1989); and (d) modulatory mechanisms that appear to alter synaptic gain via basal forebrain and/or intralaminar thalamic pathways (Hasselmo & Bower, 1992). Our model is still in the developmental stage, with only pieces of it having been tested. In the following we describe the basic implementation and comment on how we are pursuing it.

An overview of the architecture is given in Figure 3 where three sensory channels (A, B, and C) are shown. Although each channel can be thought of as a cortical column, the nature of processing in the system is such that whole sensory modalities could be gated up or down by the tuning system. Thalamic relays convey activation to two central elements in each channel, a readout system and a comparator, both situated in the cortical column. It is important to note that the state of the readout system is determined by its thalmocortical input acting in concert with previous and concurrent activations, within and between sensory channels. The readout system integrates the information it receives and conveys a portion of its output as an inhibitory pattern to the next element in the system, the comparator.

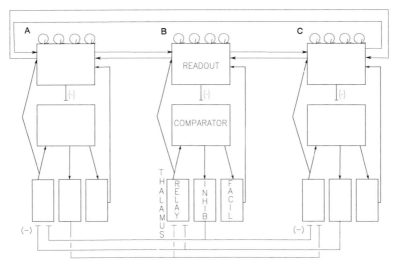

Figure 3. Overall organization of the model. Three channels or modalities are shown (A, B, C). In each modality thalamic relay neurons convey current input patterns to a readout system and a comparator situated in the cortex (both inputs are excitatory). The current state of the readout system depends on previous activations both with and between channels (cortical columns). The readout system sends part of its output as an inhibitory pattern to the comparator. The comparator compares the actual input with the expected input and generates a signal proportional to the mismatch. This signal is conveyed back down to the thalamus where it 1) activates an inhibition of channels which did not experience much of a mismatch or in which activation levels are currently quite low, and 2) drives a thalamic facilitator which increases the gains in the readout system proportional to the mismatch found in that system. These events are permissive for associations to form between channels.

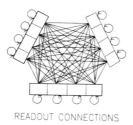

READOUT CONNECTIONS

Figure 4. The readout level. The three layers represent three different cortical columns (modalities). The connectivity in this particular implementation is exhaustive and recurrent. This is where learning takes place when there is a mismatch of patterns arriving at the level of the comparator.

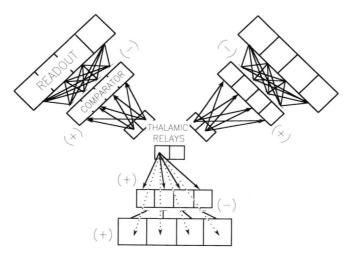

Figure 5. Removing the connections between modalities at the readout level, we can see the connections from the readout layers to the comparator. These connections are inhibitory, while connections ascending to the comparator from thalamic sensory nuclei are excitatory. When the actual pattern matches the expected pattern, there is no output from the comparator, but if a mismatch occurs, an output is generated proportionally which returns to the thalamic level as shown in Figure 3. The smaller number of cells at the thalamic layer is meant to depict increased capacity at the cortical level. Thalamic relays are also fully connected to the readout layer, but to simplify the figure this is illustrated only in the lower module.

It is the comparator acting in conjunction with the readout system that serves the filtering function. Specifically, the pattern of inhibition impinging on the comparator from the readout system constitutes a prediction of the state of excitation that is expected to be conveyed by thalamic relays at the next instant in time. If afferentation rising from thalamic relay neurons matches the readout pattern, the two patterns nullify one another and the comparator is silent. Psychologically this corresponds to the state of affairs in which there are no surprises in the environment, and the moment by moment context that the organism finds itself in predicts, with considerable reliability, the events that follow. It is important that the readout system receive and integrate information from many cortical regions including not only intrinsic pathways arising from structures such as the hippocampus, but also from the exteroceptors and kinesthetic pathways. This is necessary if thalamic afferentation is to be predicted accurately as the organism moves through time and space.

Now, if the organism occupies a stable context which it has experienced extensively, the readout system should produce accurate predictions of the course of thalamic activity, and the comparator will remain silent. However, if a novel event occurs (such as the appearance of CS or US in a just-imposed conditioning arrangement, or an unpredicted stimulus in a cognitive task), inhibition conveyed to the comparator by the readout system will not cancel the excitation arriving from thalamic relay neurons. In this case, there is a mismatch, and the resulting output from the comparator is conveyed back down to the thalamus where it drives the tuning system. In the model of Figure 3, this is the primary role of the backprojections from the cortex to the thalamus. Because the forward and backward connections between thalamus and cortex are functionally linked by the model, the density of these projections should be comparable in the two pathways, to the extent that tuning is the principal function served.

In the present implementation, there are two components in the tuning system which is driven by backprojected signals. A portion of the backprojected signal serves to activate a "thalamic inhibitor" which triggers a brief lateral inhibition of channels that did not experience much of a mismatch, or that are not currently being driven by input to their thalamic relay neurons. We see this effect as highly specific and lasting on the order of a few hundred milliseconds, which is a time course that is consistent with evidence regarding the inhibitory effect of task stimuli on the firing of cortical neurons (Bechtereva et al., 1992) and the duration of cellular afterhyperpolarizations that might conceivably mediate a brief inhibition (McCormick, 1990).

The second component of the tuning system is a "thalamic facilitator" which serves to increase synaptic gain in the readout systems within which the mismatch patterns emerged. The thalamic facilitator may do so in two ways, (a) by depolarizing the apical dendrites of superficial pyramidal cells through intralaminar or modality-specific magnocellular structures, thus decreasing the threshold for activation of NMDA receptors; and (b) by gating down synaptic weights in association relative to afferent pathways through basal forebrain structures which appear to have this effect in at least some cortical regions (Hasselmo & Bower, 1992). The time constant of thalamic facilitation is thought to be longer than that of thalamic inhibition, perhaps lasting on the order of seconds, and the effect of facilitation is thought to be less specific, extending to nearby columns or an entire sensory modality. A combination of these two tuning events (lateral inhibition and thalamic facilitation) favors the formation of associations between temporally-related patterns that are generated within the various activated channels, while affording protection of synaptic weights in channels whose current input is properly coded.[2]

---

[2] Encoding is selectively favored because (a) NMDA receptors mediating unexpected events in afferent pathways are facilitated, and (b) lateral inhibition ensures that only thalamic relay neurons that are driven reliably by their receptors discharge into the cortex during the period of inhibition. One encoding advantage conferred by inhibition lasting a few hundred milliseconds is that novel

In real biological systems, associations are formed at different levels. For example, LTP occurs in cortical synapses as well as in synpases found in the hippocampus, cerebellum, and magnocellular thalamic nuclei (Gerren & Weinberger, 1983; Lynch & Granger, 1992; Racine et al., 1983). It is therefore of interest to our agenda to explore the effect of adding plasticity at different levels. However, for the present we are restricting our attention to associations formed within the readout layer which forms the principal associative system. The architecture of this system is shown in greater detail in Figure 4. The three readout registers represent three different channels or cortical columns. In our initial implementation connectivity is complete and recurrent (the recurrent connections coding temporal information). This is where associative learning takes place when there is a mismatch from expected patterns at the level of the comparator.

In subsequent implementations we plan to explore the effect of adding recurrently connected hidden units to store all or some of the information regarding how spatiotemporal context changes over time. This might make it easier to uncouple patterns representing past versus current information.

Figure 5 shows relationships between the thalamic relay, readout, and comparator systems in more detail. As described above, connections from the readout system to the comparator are inhibitory, and those from thalamic relay neurons to the comparator are excitatory. In our current implementation, dual innervation of the comparator is enforced by the provision that synaptic weights must be either positive or negative and not change from one to the other. Eventually, we may include a bank of "interneurons" with empirically derived properties, and driven by excitatory afferents, to accomplish inhibition at the level of the comparator. One way to ensure the proper operation of the comparator is to employ anti-Hebbian learning on the connections between the readout and comparator levels, to force a match between learned and current patterns as new patterns are experienced. In contrast to the highly specific role of the comparator, the readout system plays a more general role of pattern storage by encoding associations between current and past inputs within and between modalities. An important feature of the arrangement shown in Figure 5 is that a smaller number of cells at the thalamic layer innervates a larger number of cells at the readout level. This means that there is an increase in capacity at the cortical level, an arrangement that favors sparse encoding and orthogonalization of input patterns (McNaughten & Nadel, 1990). Currently the readout system is modelled as a Hebbian pattern associator network. We are also exploring a form of competitive learning within the readout systems in order to enhance orthogonalization of stimuli and storage capacity in the encoding system.

---

events that follow unexpected stimuli within a brief time interval are likely to have been caused by those stimuli. Another advantage is that thalamic afferentation that arises from sources that are properly coded is less likely to be affected than would be the case were inhibition to be long lasting. It should be noted that because thalamic afferents feed into the readout or associative layer as well as the comparator (Figure 3, a property consistent with known columnar architectures), predicted events will continue to drive cortical processing even though alteration of synaptic weights is not supported by the tuning mechanism. It should also be noted that events which are not unexpected may still have some predictive value for new events. These events may enter into new associations but will do so more slowly because of the relative gating down of their channels.

PROBLEMS FOR SIMULATION AND EXPERIMENT

One reason for attempting to model a general learning system is that such a system almost certainly exists in the brain (Mackintosh, 1983; Weisman, 1973). An obvious risk, however, is that in attempting to do so one ends up modelling not just a learning system but brain function as a whole, which while ultimately desirable (Grossberg, 1987; Konorski, 1967) is probably not a realistic goal. One way that one might keep the task manageable is to approach the problem in steps, working on aspects that seem important and tractable. A stepwise approach may also make it easier to stay close to experimental findings and to frame experimental questions whose pursuit can change ones thinking about aspects of the model. In the preceding section we have already mentioned some of the variables we are exploring. Next, we comment on some additional questions pertaining to computer simulation and on some experimental findings that bear on the model.

Simulation Studies

There are a number of performance features, both general and specific, that we seek from this model, if it is to be a realistic model of associative encoding in the brain. For example, in general terms the model must obviously include a means by which very similar patterns can be separated (when appropriate) and dissimilar patterns can be grouped. As mentioned above, sparse encoding at neocortical levels is expected to assist orthgonalizing and grouping of stimulus inputs, particularly if a competitive learning rule is introduced at this level of the system. A further feature of the model is that the plasticity itself should be somewhat adaptable. By this we mean that the system should learn under several quite different situations: i) When there is a large departure from the expected pattern; ii) when there is a reliable departure from the expected pattern, even if the departure is not large; and iii) when there is an accompanying biologically significant event. The rate of learning should be proportional to these parameters.

Most of the physiological evidence presently available regarding mechanisms of synaptic plasticity in cortex has been gathered in allocortical and paleocortical structures (for example, the hippocampus) and points to simple forms of Hebbian learning in these structures (Brown et al., 1990). However, there is reason to suggest that more complicated forms of Hebbian learning may be needed to describe plasticity in the neocortex. Singer and his colleagues have recently reported that connection weights can show decrements as well as increments when both pre and postsynaptic elements are active at a synaptic junction (Artola et al., 1990). The direction of change depends on separate thresholds for these effects, such that the postsynaptic unit must be activated beyond one threshold if a decrement in conductivity is to occur. There is, however, a second, higher, threshold. When this second threshold is reached, an increment in connection weights is observed. Increases in postsynaptic activation beyond this level increase the amount of the increment up to some asymptotic level. Although Singer and his colleagues found these effects in slices of visual cortex suspended in vitro (Artola et al., 1990), Racine and his coworkers have obtained evidence that this rule may apply in chronic preparations as well. Racine et al. (1994, I, II) found that coactivation of cholinergic systems with pilocarpine during LTP produced a long term depression of cortical synapses. However, subsequent research which substituted externally applied dc currents for coactivation of cholinergic inputs enabled a long term potentiation of cortical neurons, perhaps because dc polarization of cortical neurons induced higher levels of postsynaptic activation than did coactivation of cholingeric systems.

In view of these findings in real biological systems, it is of interest to explore in a model that emphasizes cortical processing (as ours does) the effect of introducing a bidirectional encoding mechanism of the type reported by Artola et al. (1990). A similar algorithm was described earlier by Bienenstock, Cooper and Munro (1982) and is often called the BCM Rule. Such a rule offers several potential benefits. For example, simulations carried out by Hancock et al. (1991) suggest that it can facilitate orthogonalization of input patterns. Another possible function relates to how spurious correlations that may arise from the environment are dealt with by cortical processing. In the absence of modulatory inputs to push the postsynaptic activation beyond the second threshold, synapses that are driven by spurious correlations may be decremented and the effects of these correlations thus minimized.

Another question that we have considered is whether back propagation algorithms should be introduced as an alternative to Hebbian learning in the system. Although Hebbian rules are efficient in two-layer networks, back propagation is a powerful algorithm that can train hidden units in multi-layer architectures, which gives it an advantage in representing complex and nonlinear inputs to a system. However, one can question whether back propagation gives an accurate account of what real neural networks actually do (Zipser & Rumelhart, 1989). A feature common to back propagation models and the tuning model described here is that both compute errors in information processing. However, at this point the similarities appear to end. In back propagation models the error is detected at the output, whereas in our network (and probably also in real neural systems) errors are computed at the level of perception. Furthermore, in back propagation the error signal is conducted antidromically through the network. Because real neurons do not work this way, a neuronal implementation requires complementary hardware which can convey the error signal back to components of the forward path, so that synaptic weights are altered according to their contribution to the output (see Schmajuk & DiCarlo, 1992, and Zipser & Rumelhart, 1989, for examples of this approach). Although this solution may be biologically feasible, it is also biologically expensive. The tuning model described in this paper, on the other hand, uses error detection to direct the course of processing in the system. A major task of simulation studies will be to determine the extent to which this feature, in combination with anti-Hebbian and other biologically supported encoding rules, can support rapid and accurate learning comparable to that produced by back propagation.

To date, we have implemented the tuning function in a simple three-channel network, excluding the associative function (i.e., active connections between the readout registers of Figure 4). Although the properties of this simple system are not yet fully determined, it appears from our current understanding that this system will be able to generate the phenomena of blocking and overshadowing, provided that separate sensory channels are used to represent the component stimuli. One interesting implication of the qualitative architecture of Figure 2 is that blocking and overshadowing can be expected to occur as long as the tuning system is active during learning. However, once this system has been disengaged owing to a well-formed association, salient features added to a predictive stimulus should again acquire signal value, to the extent that tuning is reinstated by them. At this time we are not aware of behavioral data that directly evaluate this possibility. Blocking may also occur at levels of the system which we have not modelled, such as those involved in generating overt responses (Lynch & Granger, 1992).

Experiments

One reason for attempting to formulate a structural model of learning is to explore experimental questions raised by it. We are using the model to guide the design and analysis of human psychophysiological experiments and animal studies.

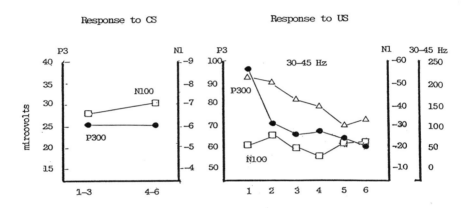

Figure 6. Response of N100 and P300 event-related potentials to CS+ (left panel) and US events (right panel) during discriminative aversive conditioning in human subjects. Spectral power between 30-45 Hz is also shown for a one-sec interval following delivery of the US (augmented power was not detected in this frequency range following CS+). The metric for spectral power is the area of isoline plots containing baseline-corrected t-statistics exceeding $p < .001$, multiplied by the value of t. Results are taken from Flor et al. (1994) and are averaged over three procedurally isomorphic conditioning groups (Cz recording).

Results recently gathered by Flor et al. (1994) on aversive classical conditioning in human subjects give reason to suggest that the model may offer a first approximation of brain dynamics during learning. Flor et al. (1994) carried out discriminative conditioning using different human faces as conditioned stimuli (CSs) and strong intracutaneous electric shock as the unconditioned stimulus (US). Three groups of subjects were trained which differed with respect to whether an angry, happy, or neutral face signalled the US event. In each group a slow negative wave was observed to develop in the EEG following CS+ but not CS-, and to extinguish when the US was discontinued. The time course of the slow wave effect is consistent with a role for such waves in preparing neural networks for alteration of synaptic weights by US events. Because slow negativities appear to reflect depolarization of apical dendrites and show variable topography depending on the requirements of a task (Birbaumer et al., 1990, 1992), it is also reasonable to suggest a thalamic facilitator as their source. However, it appears likely that plasticity will have to be introduced into thalamic or basal forebrain structures in the model of Figure 3, because there is nothing in the model at present that can account for a differential response of slow waves to CS+ and CS- events. Flor et al. (1994) also recorded P300 waves that were elicited by CS and US events over the course of conditioning. P300 waves elicited by the US were found to diminish over acquisition trials, but P300s elicited by CS occurrence did not. These results, which are summarized in Figure 6, are consistent with the view that P300 waves are released by corticothalamic feedback consequent on surprise. CS events were not predictable from the cues that preceded them in Flor et al.'s training arrangement, but US events were perfectly predictable from the CS+.

Flor et al. (1994) subsequently applied power spectral analyses to their results, using a sliding window centered on successive 200 msec epochs within conditioning trials. Although the most prominent finding was augmented power between 0-5 Hz following CS+ and the US (this effect reflecting slow and P300-like waves), increased power was also detected between 30-45 Hz immediately following delivery of the US in each conditioning group. In two groups (these receiving either different neutral faces as CS+ and CS- stimuli, or an angry face as CS+ and a happy face as CS-) the 30-45 Hz effect disappeared by the third block of acquisition trials, whereas in the third group (this group receiving a happy face as CS+ and an angry face as CS) the effect diminished more slowly over the course of acquisition. Because the 30-45 Hz response appeared only following early US events, it appears to have been driven by US occurrence rather than by CS offset. Figure 6 includes the 30-45 Hz response so that its temporal course can be compared with the P300 elicited by the US. At this time we cannot say whether the 30-45 Hz response is a coherent oscillatory phenomenon, or whether it is composed instead of a complex of potentials brought out by strong somatosensory stimulation, some of which may relate motor responses elicited by the US rather to plastic changes induced by this event. The time course of the 30-45 Hz response to the US is compatible with a role in plasticity, but other functions could be served.

There is reason to inquire into a possible functional role for brain oscillatory responses in learning, even though the presence of such responses has not been well documented in learning experiments at the present time. Coherent oscillatory activity in the gamma band range (20-50 Hz) has been recorded from visual and somatosensory cortex of animal preparations during perceptual tasks (Singer et al., 1990; Murthy & Fetz, 1992) and from humans subjects by means of neuromagnetic recordings taken during the delivery of brief auditory (Pantev et al., 1991) or somatosensory stimuli (Kaukoranta & Reinikainen 1985). It has been suggested that oscillatory phenomena of this type may segregate and bind sensory features that define objects or events in visual, auditory, or somatosensory fields (Pantev et al., 1991; Singer et al., 1990). Oscillatory activity that extends beyond the duration of a stimulus might also support the organization of limb or eye movements which are appropriate for a stimulus, insofar as priming of feature detectors for kinesthetic patterns by top-down processing is a mechanism by which the brain defines and executes behavioral responses (Brener, 1986). In principle, oscillatory patterns could serve multiple functional roles in adapting an organism to its circumstances. For example, if oscillating assemblies of neurons code sensory, motor, or other brain events that occur in the experience of an organism, such events can subsequently become assimilated into the prevailing context and into the stream of the organism's activities only by virtue of some effect of their oscillatory representations on synaptic conductivities. A role in learning is thus implied which is not incompatible with concurrent roles in perceptual identification and\or response selection.

Animal experiments may help to inform us further on the role of oscillatory activity and P300-like events in information processing. In particular, our model emphasizes the role of the corticothalamic backprojection in associative processes. One goal of these studies is therefore to see how neural activity in corticothalamic pathways is altered by the occurrence of unpredicted (surprising) events in the environment. We are also exploring whether corticothalamic activity produced by stimulus novelty is related to P300-like waves and oscillatory rhythms in different structures of the rat brain. Corticothamic feedback could be a source of oscillatory activity and promote encoding in highly selected cortical networks, if such feedback were to switch thalamic relay neurons in the forward path into a bursting mode. Were firing of relay cells to be sustained, perhaps by a combination of continued sensory input and corticothalamic feedback acting through specific reticular neurons (these neurons having inhibitory effects on relay cells), other thalamocortical circuits might be brought into the picture, to the extent that these circuits are similarly being driven by unpredicted events in a task situation. Repetitive firing has been documented in thalamic neurons (Llinas & Geijo-Barrientos, 1989; Steriade et al., 1991) and might relate to these processes, although the mechanism and functional role of thalamic oscillatory activity can only be speculated about at this time (Llinas, 1992).

Acknowledgement

Preparation of this paper was funded by a grant from the Natural Sciences and Engineering Research Council of Canada (OGP0000132).

REFERENCES

Artola, A., Brocher, S., & Singer, W. (1990). Different voltage-dependent thresholds for inducing long-term depression and long-term potentiation in slices of rat visual cortex. *Nature, 347*, 69-72.

Basar, E., & Bullock, T.H. (1992). *Induced rhythms in the brain*. Boston MA: Birkhäuser.

Bechtereva, N.P., Abdullaev, Y.G., and Medvedev, S.V. (1992). Properties of neuronal activity in cortex and subcortical nuclei of the human brain during single-word processing. *Electroencephalography and Clinical Neurophysiology, 82*, 296-301.

Bienenstock, E.L., Cooper, L.N., & Munro, P.W. (1982). Theory for the development of neuron selectivity: Orientation specificity and binocular interaction in visual cortex. *Journal of Neuroscience, 2*, 32-48.

Birbaumer, N., Elbert, T., Canavan, A., and Rockstroh, B. (1990). Slow cortical potentials of the cerebral cortex and behavior. *Physiological Reviews, 70*, 1-41.

Birbaumer, N., Roberts, L.E., Lutzenberger, W., Rockstroh, B., & Elbert, T. (1992). Area-specific self-regulation of slow cortical potentials on the sagittal midline and its effects on behavior. *Electroencephalography and Clinical Neurophysiology, 84*, 353-361.

Braitenberg, V., & Schüz, A. (1991). *Anatomy of the cortex: Statistics and geometry.* Berlin: Springer-Verlag.

Bramwell, L. (1993). *The role of visual task requirements in auditory feedback learning.* Honours B.A. thesis, McMaster University.

Brener, J. (1986). Operant reinforcement, feedback, and the efficiency of learned motor control. In M.G.H. Coles, E. Donchin, & S.W. Porges (Eds.), *Psychophysiology: Systems, processes, and applications* (pp. 309-327). New York: Guilford Press.

Brown, T.H., & Kariss, E.W., & Keenan, C.L. (1990). Hebbian synapses: Biophysical mechanisms and algorithms. *Annual review of Neuroscience, 13,* 475-571.

Chwilla, D.J., & Brunia, C.H.M. (1991). Event-related potentials to different feedback stimuli. *Psychophysiology, 28,* 123-132.

Colwill, R.M. (1993). An associative analysis of instrumental learning. *Current Directions in Psychological Science, 2,* 111-116.

Dawson, M.E., & Biferno, M.A. (1973). Concurrent measurement of awareness and electrodermal classical conditioning, *Journal of Experimental Psychology, 101,* 82-86.

Dawson, M.E. & Schell, A.M. (1987). Human autonomic and skeletal classical conditioning: The role of conscioius cognitive factors. In G. Davey (Ed.), *Cognitive processes and Pavlovian conditioning in humans* (pp. 27-55). New York: Wiley & Sons.

Deutsch, J.A. (1960). *The structural basis of behavior.* Cambridge: Cambridge University Press.

Douglas, R.J., & Martin, K.A.C. (1990). Neocortex. In G.M. Shepherd (Ed.), *The synaptic organization of the brain* (pp. 389-438). Oxford, UK: Oxford University Press.

Flor, H., Birbaumer, N., Roberts, L.E., Feige, B., Lutzenberger, W., Fuerst, M., & Hermann, C. (1994). *Electrocortical correlates of Pavlovian conditioning.* (submitted for publication).

Gaillard, A.W.K. (1986). The CNV as an index of response preparation. In W.C. McCallum, R. Zappoli, & F. Denoth (Eds.), Cerebral psychophysiology: Studies in event related potentials. *Electroencephalography and Clinical Neurophysiology, 38,* (Supplement), pp. 196-206.

Germana, J. (1968). Psychophysiologicval correlates of conditioned autonomic response formation. *Psychological Bulletin, 70,* 105-114.

Gerren, R., & Weinberger, N.M. (1983). Long term potentiation in the magnocellular medial geniculate nucleus of the anesthetized cat. *Brain Research, 265,* 138-142.

Gluck, M.A., & Meyers, C.E. (1993). Hippocampal mediation of stimulus representation: A computational theory. *Hippocampus* (in press).

Gluck, M.A., Goren, O., Meyers, C., & Thompson, R.F. (1993). A higher-order recurrent network model of the cerebellar structures of response timing in motor-reflex conditioning. *Journal of Cognitive Neuroscience* (in press).

Grossberg, S. (1987). *The adaptive brain.* Amsterdam: North Holland.

Halgren, E., Stapleton, J.M., Smith, M., and Altafullah, I. (1986). Generators of the human scalp P3(s). In R.Q. Cracco and I. Bodis-Wollner (Eds.), *Evoked Potentials,* pp. 269-284. New York NY: Alan R. Liss.

Hancock, P.J.B., Smith, L.S., & Phillips, W.A. (1991). A biologically supported error correcting learning rule. *Neural Computing, 3,* 201-212.

Hasselmo, M.E., & Bower, J.E. (1992). Cholinergic suppression specific to intrinsic not afferent fiber synapses in rat piriform (olfactory) cortex. *Journal of Neurophysiology, 67,* 1222-1229.

Hebb, D.O. (1949). *The organization of behavior.* New York: Wiley.

Herrnstein, R.J. (1970). On the law of effct. *Journal of the Experimental Analysis of Behavior, 13,* 243-266.

Holland, P. C., & Rescorla, R.A. (1982). Behavioral studies of associative learning in animals. *Annual Review of Psychology, 33,* 265-308.

Hughes, D.E., & Roberts, L.E. (1985). Evidence of a role for response plans and self-monitoring in biofeedback. *Psychophysiology, 22,* 427-439.

Jenkins, H.M., & Sainsbury, R.S. (1970). Discrimination learning with the distintive feature on positive or negative trials. In D. Mostofsky (Ed.), *Attention: Contemporary theory and analysis* (pp. 239-273). New York: Appleton Century Crofts.

Johnson, R. Jr. (1986). A triarchic model of P300 amplitude. *Psychophysiology, 23,* 367-384.

Kandel, E.R., Schwartz, J.H., & Jessell, T.M. (1991). *Principles of neural science* (third edition). Norwalk, CN: Appleton & Lange.

Kaukoranta, E., & Reinikainen, K. (1986). *Somatosensory evoked magnetic fields from SI: An interpretation of the spatiotemporal field pattern and effects of stimulus repetition rate.* Report Number TKK-F-A581, Helsinki University of Technology Low Temperature saLaboratory, SF-02150, Espoo, Finland.

Konorski, J. (1967). *Integrative activity of the brain.* Chicago: University of Chicago Press.

Lashley, K.S. (1950). In search of the engram. *Society of Experimental Biology Symposium, 4,* 454-482.

Llinas, R. R. (1992). Oscillations in CNS neurons: A possible role for cortical interneurons in the generation of 40-Hz oscillations. In E. Basar & T.H. Bullock (Eds.), *Induced rhythms in the brain* (pp. 269-283). Boston MA: BirkhaÅser.

Llinas, R.R. & Geijo-Barrientos, E. (1989). In vitro studies of mammalian thalamic and reticularic thalamic neurons. In M. Bentivoglio & R. Spreaafico (Eds.), *Cellular Thalamic Mechanisms.* Amsterdam NE: Elsevier.

Lynch, G., & Granger, R. (1992). Variations in synaptic plasticity and types of memory in cortico-hippocampal networks. *Journal of Cognitive Neuroscience, 4,* 189-198.

Mackintosh, N.J. (1983). *Conditioning and associative learning.* Oxford, UK: Oxford University Press.

McCormick, D.A. (1990). Membrane properties and neurotransmitter actions. In G.M. Shepherd (Ed.), *The synaptic organization of the brain* (pp. 32-66). Oxford, UK: Oxford University Press.

McNaughten, B., & Nadel, L. (1990). Hebb-Marr networks and the neurobiological representation of action in space. In M Gluck & D. Rumelhart (Eds). *Neuroscience and connectionist theory* (pp. 2-63). Hillsdale NJ: Lawrence Erlbaum.

Miller, R.R., & Barnet, R.C. (1993). The role of time in elementary associations. *Current Directions in Psychological Science, 2,* 106-111.

Murthy, V.N., & Fetz, E.E. (1992). Coherent 25- to 35-Hz oscillations in the sensorimotor cortex of awake behaving monkeys. *Proceedings of the National Academy of Sciences USA, 89,* 5670-5674.

Pantev, C., Makeig, S., Hoke, M., Galambos, R., Hampson, S., & Gallen, C. (1991). Human auditory evoked gamma-band magnetic fields. *Proceedings of the National Academy of Sciences USA, 88,* 8996-9000.

Peterhans, E., & von der Heydt, R. (1989). Mechanisms of contour perception in monkey visual cortex. II. Contours bridging gaps. *Journal of Neuroscience, 9,* 1749-1763.

Pearce, J.M., & Hall, G. (1980). A model for Pavlovian learning: Variations in the effectiveness of conditioned but not of unconditioned stimuli. *Psychological Review, 87,* 532-552.

Racine, R.J., Milgram, N.W., & Hafner, S. (1983). Long-term potentiation phenomena in the rat limbic forebrain. *Brain Research, 260,* 217-231.

Racine, R.J., Wilson, D., Teskey, G.C., & Milgram, N.W. (1994). Post-activation potentiation in the neocortex: I. Acute preparations. *Brain Research, 637,* 73-82.

Racine, R.J., Teskey, G.C., Wilson, D., & Seidlitz, E. (1994). Post-activation potentiation and depression in the neocortex of the rat: II. Chronic preparations. *Brain Research, 627,* 83-96.

Rescorla, R.A. (1988). Pavlovian conditioning: It's not what you think it is. *American Psychologist, 43,* 151-160.

Rescorla, R.A. (1987). A Pavlovian analysis of goal-directed behavior. *American Psychologist, 42,* 119-129.

Rescorla, R.A., & Wagner, A.R. (1972). A theory of Pavlovian conditioning: Variations in the effectiveness of reinforcement and nonreinforcement. In A.H. Black & W.F. Prokasy (Eds.), *Classical conditioning II: Current research and theory* (pages 64-99). New York: Appleton Century Crofts.

Roberts, L.E., Rau, H., Lutzenberger, W., & Birbaumer, N. (1994). Mapping P300 Waves onto inhibition: Go\No-Go discrimination. *Electroencephalography and Clinical Neurophysiology, 92,* 44-55.

Roberts, L.E (1990). Evidence for a general associative process in Pavlovian and instrumental conditioning. In H. Lachnit (Chair), *Recent Advances in Pavlovian Conditioning,* Fifth International Congress of Psychophysiology, Budapest, Hungary.

Roberts, L.E., Preston, D., & Uttl, B (1991). *Tuning and capacity limitations in feedback (instrumental) learning.* Psychonomics, San Francisco, CA.

Roberts, L.E., Racine, R.J., & Durlach, P.J. (1992) *A macroarchitecture for associative learning*. Symposium on the Organization of Learning and Memory, Sixth International Congress of Psychophysiology, Berlin, Germany.

Rockstroh, B., Müller, M., Cohen, R., and Elbert, T. (1992). Probing the functional brain state during P300-evocation. *Journal of Psychophysiology*, *6*, 175-184.

Schmajuk, N.A., & DiCarlo, J.J. (1992). Stimulus configuration, classical conditioning, and hippocampal function. *Psychological Review*, *99*, 268-305.

Singer, W., Gray, C., Engel, A., König, P., Artola, A., & Bröcher, S. (1990). Formation of cortical cell assemblies. *Cold Spring Harbor Symnposia on Quantitative Biology*, *55*, 939-952.

Steriade, M. Curro Dossi, R., Pare, D., & Oakson, G. (1991). Fast oscillations (20-40 Hz) in thalamocortical systems and their potentiation by mesopontine cholinergic nuclei in the cat. *Proceedings of the National Academy of Sciences USA*, *88*, 4396-4400.

Wagner, A.R., & Brandon, S.E. (1989). Evolution of a structured connectionist model of Pavlovian conditioning (AESOP). In S.B. Klein & R.R. Mowrer (Eds.), *Contemporary learning theories: Pavlovian conditioning and the status of traditional learning theory* (pp. 149-189). Hillsdale, NJ: Lawrence Erlbaum.

Wagner, A.R., Logan, F.A., Haberlandt, K., & Price, T. (1968). Stimulus selection in animal discrimination learning. *Journal of Experimental Psychology*, *76*, 171-180.

Weisman, R.G. (1977). On the role of the reinforcer in associative learning. In H. Davis and H.M.B. Hurwitz (Eds.), *Operant-Pavlovian interactions* (pp. 1-22). Hillsdale, NJ: Lawrence Erlbaum.

Williams, B.A., & Heyneman, N. (1982). Multiple determinants of "blocking" effects on operant behavior. *Animal Learning and behavior*, *10*, 72-76.

Woodward, S.H., Brown, W.S., Marsh, J.T., and Dawson, M.E. (1991). Probing the time-course of the auditory oddball P3 with secondary reaction time. *Psychophysiology*, *28*, 609-618.

Zipser, D., & Rumelhart, D.E. (1990). The neurobiological significance of the new learning models. In E.L. Schwartz (Ed.), *Computational neuroscience* (pp. 192-200). Cambridge, MA: MIT Press.

POSSIBILITIES OF FUNCTIONAL BRAIN IMAGING
USING A COMBINATION OF MEG AND MRT

Manfred Fuchs, Michael Wagner, Hans-Aloys Wischmann,
Karsten Ottenberg, Olaf Dössel

Philips GmbH Forschungslaboratorien,
Forschungsabteilung Technische Systeme Hamburg
Röntgenstr. 24-26, D-22335 Hamburg, Germany

INTRODUCTION

Neuromagnetic imaging is a relatively new diagnostic tool for the examination of electrical activities in the nervous system. It is based on the noninvasive detection of the extremely weak magnetic fields around the human body with Superconducting Quantum Interference Devices (SQUIDs) and the subsequent reconstruction of the generators.

This ill-posed inverse problem of biomagnetism can only be solved by using constraints in modeling the generator distribution. The most common and simple model is the single equivalent current dipole which obviously fails in the explanation of more complex source structures. With good signal to noise ratios, more than one dipole may be fitted to the measured field distributions, and for a sequence of single time instants, moving dipoles may be calculated. Integrating the least squares fit over longer temporal ranges leads to rotating (spatially fixed) or fixed (spatially and orientationally fixed) generators with more stable numerical conditions. In the least squares algorithm, only the dipole positions have to be fitted, while the dipole components can be calculated analytically (e.g. Mosher et al., 1992). The same holds for discretized current distributions where the dipole positions are fixed and their components can be calculated under the minimum norm constraint which minimizes the total dipole strength (e.g. Wang et al., 1992). Previously, morphological information was only used for matching e.g. the spherical volume conductor model to head sections and for integrated display of the measured data and calculated results with 3D-MR (Magnetic Resonance) or CT (Computer Tomography) images (e.g. Fuchs et al., 1992). We present a new method called Cortical Current Imaging (CCI): morphologically constrained current density distributions where the recon-

Oscillatory Event-Related Brain Dynamics, Edited
by C. Pantev, Plenum Press, New York, 1994

struction space is limited to physiologically relevant structures segmented from individual 3D-MR-data.

For the realization of the new method and for practical use in medical diagnosis a combination of the abstract neuromagnetic images with MR- or CT-images is required in order to match the functional activity with anatomy and morphology. The determination of the subject's position relative to the multichannel magnetometer system is performed by the localization of the center coordinates of coilsets fixed to certain points on the subject's skin. Test currents are fed into these coilsets and the resulting magnetic field distributions are measured. With digital lock-in feedback techniques, small deviations (less than 2 mm) between real and measured localization can be sustained up to distances of 20 cm. In the morphological coordinate system, the coilset positions are marked with a 3D-cursor on the segmented surface. The best fit transformation between the two coordinate systems is then calculated and the neuromagnetic images are finally overlaid onto three-dimensional morphological images with arbitrarily selectable slices.

The segmentation of the cortex surface is realized by an automated procedure. A threedimensional region growing algorithm with a single threshold value between white and gray matter intensities starts inside the white matter, ensuring that a closed surface is generated. Additionally very thin connections, for example to the eyes, are broken. The surface normals are determined by gray level gradients (e.g. Höhne and Bernstein, 1988) or other surface properties. In this manner, a further directional constraint can be included in the generator reconstructions.

The results (both with and without the use of the normals) are compared to regular grid current density-reconstructions and scanning methods (Mosher et al., 1992) applied to the cortical support points. As an example, evoked field Magnetoencephalography (MEG) measurements are analyzed, but the method can also be applied to Magnetocardiography (MCG) data (Myocardial Current Imaging).

MEASUREMENT OF MAGNETIC FIELD DISTRIBUTIONS

The magnetic fields are detected by a 31 channel SQUID-system (Dössel et al., 1993) with first order gradiometers (20 mm diameter, 70 mm baselength). The two turn gradiometer coils are hexagonally arranged with center to center distances of 25 mm. The multichannel MEG/MCG-system is installed in a magnetically shielded chamber, and the overall system noise is below 10 fT/√Hz.

The gradiometer signals are amplified and filtered outside the chamber and fed into a multichannnel analog-digital converter that is controlled by the data acquisition computer. For further evaluation the measured data are transferred by an Ethernet link to a SUN SPARCstation. Figure 1 shows a scheme of the experimental set-up. All measurement parameters (e.g. number of channels, samples, trials, sampling rate, time delay between trials) are controlled from the workstation using a graphical user interface.

COILSET LOCALIZATION

For increased long range sensitivity of the coilset localization procedure, lock-in techniques and a coilset current feedback loop are used. A sinusoidal current of frequency f is fed into each coil of the coilsets, the measured magnetic fields are analyzed at this frequency for noise suppression, and the maximum of the 31 signals is adjusted to a reasonable value by a current feedback loop. The amplitudes A_k (channel k) are determined by a discrete Fourier-transformation at the frequency f. Therefore the magnetic field signals are digitized with an ADC-sample frequency f_s being a multiple of f (f_s=nf) and a total measuring time T_m being a multiple of the period T=1/f (T_m=mT=m/f=mn/f_s, number of samples n_s=$T_m f_s$=mn, sample times t_i=i/f_s=iT_m/(mn), crosstalk corrected signals s_{k_i}):

$$A_k = \sqrt{a_k^2 + b_k^2} \quad \text{with} \quad a_k = \frac{2}{n_s}\sum_{i=0}^{n_s-1} s_{ki}\cos(2\pi f t_i) \quad \text{and} \quad b_k = \frac{2}{n_s}\sum_{i=0}^{n_s-1} s_{ki}\sin(2\pi f t_i) \tag{1}$$

With the feedback technique an increase of the dynamic range of three orders of magnitude (60 dB) is achieved resulting in a drastically improved long range sensitivity for the coilset localization.

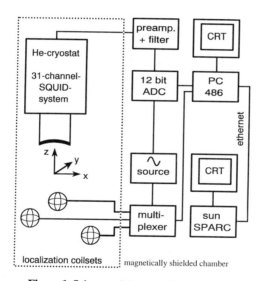

Figure 1. Scheme of the experimental set-up.

Typical measurement parameters are 40 Hz current frequency, 2 kHz sampling rate, 400 sampling points (0.2 sec) for feedback control and 1600 samples (0.8 sec) for the finally adjusted coil current. The localization of four coilsets (12 coils) then takes about 15 seconds of data acquisition time and is performed before and/or after each MEG/MCG-measurement.

Localization Algorithm

The determination of the subject's position relative to the multichannel gradiometer system is performed by the localization of the center coordinates of these coilsets, fixed to certain points at the head (e.g. international 10/20 EEG system). The generated magnetic field of a single coil can be calculated with good accuracy using the following magnetic dipole approximation:

$$\underline{B} = \frac{\mu_0}{4\pi r^3}\left(\frac{3(\underline{m}\underline{r})\underline{r}}{r^2} - \underline{m}\right) = \frac{\mu_0 Ia^2}{4r^3}\left(\frac{3(\underline{e}\underline{r})\underline{r}}{r^2} - \underline{e}\right) \quad \text{with} \tag{2}$$

$\underline{m} = \pi Ia^2\underline{e}$: magnetic moment of coil with radius a, current I, normal $\underline{e} = (e_x, e_y, e_z)$,

$\underline{r} = (x, y, z)$: vector from coil center to detector position, distance $r = \sqrt{x^2 + y^2 + z^2}$

With a magnetometer coil with normal vector $\underline{d} = (d_x, d_y, d_z)$ the component $B_d = \underline{B}\underline{d}$ of the magnetic field parallel to $\underline{d}$ is measured:

$$B_d = \underline{B}\underline{d} = \frac{\mu_0}{4\pi r^3}\left(\frac{3(\underline{m}\underline{r})(\underline{r}\underline{d})}{r^2} - \underline{m}\underline{d}\right) = \frac{\mu_0 Ia^2}{4r^3}\left(\frac{3(\underline{e}\underline{r})(\underline{r}\underline{d})}{r^2} - \underline{e}\underline{d}\right) \tag{3}$$

For first order gradiometers with coil normals $\underline{d} = (d_x, d_y, d_z)$ (lower loop center at $\underline{r}_0 = (x_0, y_0, z_0)$, upper loop center at $\underline{r}_1 = (x_1, y_1, z_1)$) one obtains:

$$B_d = (\underline{B}_0 - \underline{B}_1)\underline{d} = \frac{\mu_0 Ia^2}{4}\left(\frac{3(\underline{e}\underline{r}_0)(\underline{r}_0\underline{d})}{r_0^5} - \frac{3(\underline{e}\underline{r}_1)(\underline{r}_1\underline{d})}{r_1^5} - \frac{\underline{e}\underline{d}}{r_0^3} + \frac{\underline{e}\underline{d}}{r_1^3}\right) \tag{4}$$

For three equal orthogonal coils ($\underline{m}_i = \pi Ia^2\underline{e}_i$, i=1..3) it can be shown that the sum over the squared measured field components does no longer depend on the orientation of the coilset but only on its center postition (Ahlfors and Ilmoniemi, 1989). So only these three coordinates have to be fitted:

$$B_{ds}^2 = \sum_{i=1}^{3} B_{di}^2 = \sum_{i=1}^{3}(\underline{B}_i\underline{d})^2 = \left(\frac{\mu_0 Ia^2}{4r^4}\right)^2\left(3(\underline{r}\underline{d})^2 + r^2\right) \tag{5}$$

With the magnetometer coil in the xy-plane and its normal in the z-direction ($\underline{d}=(0,0,1)$) equation 5 yields:

$$B_{zs}^2 = \left(\frac{\mu_0 Ia^2}{4r^4}\right)^2(3z^2 + r^2) \tag{6}$$

In the case of first order gradiometers equation 5 reads:

$$B_{ds}^2 = \sum_{i=1}^{3} B_{di}^2 = \sum_{i=1}^{3}\left((\underline{B}_{0i} - \underline{B}_{1i})\underline{d}\right)^2 = \sum_{i=1}^{3}(\underline{B}_{0i}\underline{d})^2 + \sum_{i=1}^{3}(\underline{B}_{1i}\underline{d})^2 - 2\sum_{i=1}^{3}(\underline{B}_{0i}\underline{d})(\underline{B}_{1i}\underline{d}) =$$

$$\left(\frac{\mu_0 Ia^2}{4}\right)^2\left[\frac{3(\underline{r}_0\underline{d})^2+r_0^2}{r_0^8}+\frac{3(\underline{r}_1\underline{d})^2+r_1^2}{r_1^8}-\frac{2\left(r_0^2r_1^2+3\left(3(\underline{r}_0\underline{d})(\underline{r}_1\underline{d})(\underline{r}_0\underline{r}_1)-r_1^2(\underline{r}_0\underline{d})^2-r_0^2(\underline{r}_1\underline{d})^2\right)\right)}{r_0^5r_1^5}\right)$$

For gradiometer coils parallel to the xy-plane with normals along $\underline{e}_z$ and a gradiometer baselength of b ($\underline{d}=(0,0,1)$, $\underline{r}_0=(x,y,z_0)$, $\underline{r}_1=(x,y,z_1=z_0+b)$) equation 7 yields:

$$B_{zs}^2=\left(\frac{\mu_0 Ia^2}{4}\right)^2\left(\frac{3z_0^2+r_0^2}{r_0^8}+\frac{3z_1^2+r_1^2}{r_1^8}-\frac{2}{r_0^3r_1^3}\left(1+\frac{3}{r_0^2r_1^2}\left(3z_0z_1(\underline{r}_0\underline{r}_1)-r_1^2z_0^2-r_0^2z_1^2\right)\right)\right) \quad (8)$$

with $\quad \underline{r}_0\underline{r}_1=x^2+y^2+z_0z_1$

By showing that the rotational invariance (only dot products appear in equation 7) of the squared and summed coilset field holds for first order gradiometers it follows that this is also true for higher order gradiometers because the summation over several loops can successively be split into two term summations. This also holds for a summation over subareas of the loops for an approximated field integration over the gradiometer coils.

The rapidly converging three parameters (x, y, z) least squares fit (Nelder-Meade simplex) using the dipole approximation for the magnetic field can be used to yield the center positions with a very good accuracy of less than 2 mm in the volume below the area covered by the gradiometers (Fuchs et al., 1991; figure 2).

Because of the good and rapid convergence of the simplex-algorithm for this relatively simple field dependence (equation 8), no further simplifying assumptions (e.g. Incardona et al., 1992) are necessary.

PROCESSING OF MORPHOLOGICAL IMAGES

In order to correlate and display functional and morphological images, a 3D-visualization computer program has been developed. Three-dimensional morphological data from computer tomography or magnetic resonance imaging are used to determine the surface of the head, to depict arbitrarily chosen slices, to fit spheres into sections of the head for a first approximation of the volume conductor geometry, and to overlay the slices with reconstructed functional current distribution images. The cortex segmentation routines also work with 3D images.

Visualization and Volume Conductor Approximation

A series of sagittal MR-slices (256 * 256 pixels, 12 bits resolution) forms the input data for the 3D-graphics. The original slices are reduced to 8 bits resolution and linearly interpolated to get an isotropic threedimensional data set, which is essential for 3D-volume rendering techniques. The surface segmentation is performed with a radial search thresholding in order to find the outer head shape. This leads to a numerical description of the head surface (typically 60000 points), which is stored together with the corresponding surface normals.

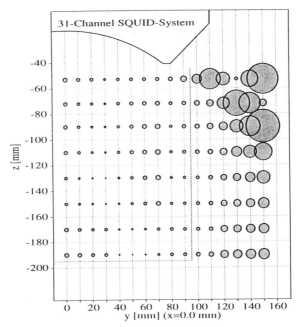

Figure 2. Measured coilset locations where one coilset was successively positioned at 128 points on a regular planar grid in the (y,z)-plane using y-increments of 10 mm and z-increments of 20 mm. In the range below the gradiometer coils (y = 0 .. +90 mm, z = -190 .. -50 mm), the spatial deviations between measured and adjusted locations are less than 1.8 mm and the mean deviation is 1.1 mm. The centers of the circles show the adjusted coilset positions while their radii are equal to the 3D localization errors.

The brightness of the surface elements (triangles) can be calculated from these normals and a chosen light source direction. The normals can be approximated from graylevel gradients (Höhne and Bernstein, 1988) or cross products of the triangle sides. Lighting with an arbitrarily positioned parallel light source and perspective viewing with hidden surface suppression are performed yielding a realistic 3D-image of the head. Determination of the head surface normals from graylevel gradients as shown in Figure 3 results in a smoother surface intensity distribution compared to the cross product method.

With a mouse controlled plane adjustment tool or by numerical input of coordinates and normals, arbitrary cut planes through the head can be determined interactively. The morphological structure of the selected cut plane is linearly interpolated from the 2D slices (from the isotropic data set) and projected into the 3D image. Intensities below a certain threshold value can be suppressed.

As a first approximation to the conductor geometry for the generator reconstruction algorithms (Cuffin and Cohen, 1977; Ilmoniemi et al., 1985), spheres can be fitted into surface sections of the head that remain after the cut process. The center of the sphere and its radius are determined by a least squares fit procedure. The best fit sphere can also be displayed in the 3D-scene as shown in figure 4.

Determination of Coilset Positions

In order to match functional current reconstructions and morphological images, the positions of at least three coilsets (which can be marked with oil pellets in MR-imaging) are marked in the morphological coordinate system using a mouse controlled 3D-cursor (figure 5). With the measured positions of these coilsets, the transformation matrix between the measurement frame and the morphological head frame is calculated. The coilsets, the 31-channel-system and the reconstructed currents can then be displayed in the 3D-scenery. In the example shown in figure 3, four localization coilsets were fixed above the right ear of the subject.

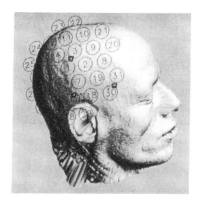

Figure 3. Head surface segmented from 3D-MR data using radial search thresholding and gray level gradient shading, four localization coilset fixed above the right ear and the 31 lower gradiometer coils of the MEG-system.

Figure 4. Upper head surface section with corresponding MR-cut plane slice and least squares fit sphere for volume conductor approximation.

COORDINATE SYSTEM MATCHING

Both coordinate sets from the functional and the morphological system have certain errors, depending on the coilset positions, the localization accuracy as well as the MR-image quality and resolution. For the calculation of the best fit transformation matrix between the two systems a weighted least squares fit is thus appropriate. The functional coilset coordinates $\underline{r}_f$ are transformed to the morphological coordinate system by a best fit rigid rotation $\underline{\underline{R}}$ plus a linear shift $\underline{s}$ that minimizes the weighted sum Δ^2 of the squared 3D-distances between the transformed functional $\underline{r}_f \rightarrow \underline{r}_t$ and the morphological coilset positions $\underline{r}_m$:

$$\underline{r}_{ti} = \underline{\underline{R}}\underline{r}_{f_i} + \underline{s} \tag{9}$$

$$\Delta^2 = \sum_{i=1}^{n} w_i^2 \Delta_i^2 \Big/ \sum_{i=1}^{n} w_i^2 \quad \text{with} \quad \Delta_i^2 = \sum_{k=1}^{3} \left(r_{t_{ik}} - r_{m_{ik}} \right)^2 = \sum_{k=1}^{3} \left(\left(\underline{\underline{R}}\underline{r}_{f_i} \right)_k + s_k - r_{m_{ik}} \right)^2 \tag{10}$$

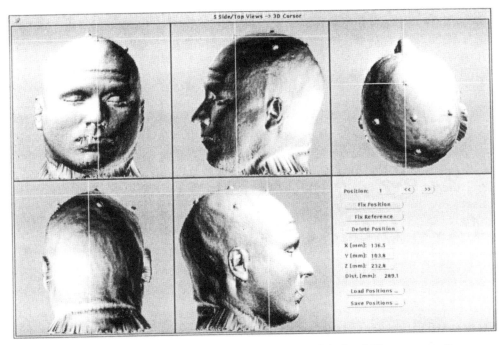

Figure 5. Mouse controlled 3D-cursor with five side / top views of the head. The corresponding cursor lines in the different views move simultaneously to simplify the read-out of the coordinates (x, y, z). The coilset positions in the morphological frame are marked for coordinate system matching.

The weighting factors $w_i = 1/\sigma_i$ are estimated from the typical localization deviations σ_i shown in figure 2. These deviations are set to 1.0 mm below the gradiometer area and increase at the edges of this area. The width of this zone depends linearly on the depth (z-coordinate) of the coilset.

Minimizing Δ^2 from equation 10 with respect to the shift vector $\underline{s}$ leads to:

$$\frac{\partial \Delta^2}{\partial s_k} = 2 \sum_{i=1}^{n} w_i^2 \left(\left(\underline{\underline{R}} \underline{r}_{f_i} \right)_k + s_k - r_{m_{i_k}} \right) \Big/ \sum_{i=1}^{n} w_i^2 \tag{11}$$

From $\dfrac{\partial \Delta^2}{\partial s_k} = 0$ (k=1..3) the best fit shift vector $\underline{s}$ is obtained:

$$\underline{s} = \underline{c}_m - \underline{\underline{R}}\underline{c}_f \quad \text{with} \quad \underline{c}_m = \sum_{i=1}^{n} w_i^2 \underline{r}_{m_i} \Big/ \sum_{i=1}^{n} w_i^2 \quad \text{and} \quad \underline{c}_f = \sum_{i=1}^{n} w_i^2 \underline{r}_{f_i} \Big/ \sum_{i=1}^{n} w_i^2 \tag{12}$$

The shift vector $\underline{s}$ thus contains the difference between the weighted center of gravity of the morphological coordinate system locations $\underline{c}_m$ and the rotated weighted center of gravity of the functional coordinate system $\underline{c}_f$. Inserting equation

442

12 into 10 yields:

$$\Delta_i^2 = \sum_{k=1}^{3} \left(\left(\underline{R}(\underline{r}_{f_i} - \underline{c}_f) \right)_k - \left(\underline{r}_{mi} - \underline{c}_m \right)_k \right)^2 \tag{13}$$

This can also be written in matrix formulation as:

$$\Delta^2 = \left\| \underline{\underline{R}F}_s - \underline{\underline{M}}_s \right\|^2 \quad \text{with} \quad \left\| \underline{\underline{A}} \right\|^2 = \sum_{ij} A_{ij}^2 \tag{14}$$

Here, the matrices $\underline{\underline{M}}_s$ and $\underline{\underline{F}}_s$ contain the shifted morphological and functional coilsets respectively, and the coordinate triples are arranged in columns. The remaining task is to find the best fit rigid rotation between the two coordinate sets that transforms the shifted functional coordinates to the shifted morphological coordinates. The solution of this problem (orthogonal Procrustes problem, Golub and van Loan, 1989, p. 582) is well known and can be achieved by a singular value decomposition (SVD) of the product matrix $\underline{\underline{P}}$ of the two shifted coordinate set matrices:

$$\underline{\underline{P}} = \underline{\underline{F}}_s \underline{\underline{M}}_s^T, \quad \text{SVD}(\underline{\underline{P}}): \quad \underline{\underline{P}} = \underline{\underline{U}} \underline{\underline{\Sigma}} \underline{\underline{V}}^T \tag{15}$$

The desired least squares fit rotation is then given by:

$$\underline{\underline{R}} = \underline{\underline{V}} \underline{\underline{U}}^T \tag{16}$$

Typical mean deviations between transformed functional and morphological coilset positions are in the mm range.

CORTEX SEGMENTATION

It is well known that the generators of most measured magnetoencephalographic fields are localized in the gray matter of the cortex. Simultaneous electric activity of a few mm^3 of pyramidal cells leads to detectable signals which can be measured with SQUID-magnetometers or EEG electrodes. However, the usual generator reconstruction methods ignore this anatomical knowledge completely.

The aim of the cortex (gray matter) segmentation procedure is to provide a priori information about the possible locations of cortical current generators in the individual anatomy. This information is organized as a list of points and normals. Each point, located in a middle layer of the gray matter, stands for a small patch of the cortex, while its normal is a local estimate of the radial direction. The cortical patches are arranged as dense as possible, so that their size determines the number of points and normals in the list.

Geometrical Properties of the Cortex

In single MR slices, independent of the orientation, the cortex forms some kind of one-dimensional boundary along the edges of the two-dimensional regions of

Table 1. Typical computation and display times in seconds for 256^3 x 8bit MR image data on a Sun Sparc10 workstation.

	$d_{min} = 2mm$		$d_{min} = 9mm$	
Cortex Segmentation	13s	fi ~200000 points	13s	fi ~200000 points
Volume-of-Interest Selection	3s	fi ~50000 points	3s	fi ~50000 points
Subsampling/Thinning	4s	fi ~15000 points	8s	fi ~1000 points
Surface Normal Estimation	6s	fi ~15000 normals	12s	fi ~1000 normals
total	26s		36s	

white matter. This boundary has a complicated structure: It is not necessarily connected, and distinct but opposite parts at the sulci sometimes touch and seem to merge. In addition, different parts of the gray matter are of different width. It would thus be a very difficult task to determine and trace the cortex using only one slice at a time.

However, all of the complications listed above are artifacts that are due to the two-dimensional representation of a three-dimensional structure. Since 3D MR image data is available, the known spatial properties of the brain may be exploited for the segmentation of the cortex, or, more exactly, of one of its layers: The white matter of the brain is a (topologically simply) connected three-dimensional object. The cortex is located on the surface of the white matter, allowing to fit a two-dimensional manifold into it. For our purposes, this manifold represents the layer in which the electrically active sources are located. In the following, we will refer to it simply as the surface.

The list of cortical points and normals is generated in four steps. First, the cortical surface is segmented from 3D MR image data. Second, a volume-of-interest may be selected. Patches of given size d_{min} are then distributed on the surface, and finally the surface normals of all patches are estimated. Table 1 shows typical computation times for each of these steps.

Cortex segmentation

In principle, surface segmentation can be accomplished by edge-oriented or region-oriented image processing. An edge-oriented surface segmentation would be based on an appropriate edge-detection operator followed by a surface tracing algorithm. Two problems arise with such an approach: First, edge detection is more sensitive to noise due to the (mildly) ill-posed nature of this process (Torre and Poggio, 1986). Second, surface-tracing immediately invokes the problems of edge-junctions. One could not make use of the simple 3D-connectivity of the white matter. On the other hand, region-oriented approaches easily exploit information on connectivity due to their intrinsic representation of neighborhood relations.

We thus use a 3D extension of a region-growing algorithm for the generation of the cortical points. In contrast to conventional region-growing algorithms (for a survey see Ballard and Brown, 1982) that use the labels "inside" and "outside", our algorithm also includes a "surface" label.

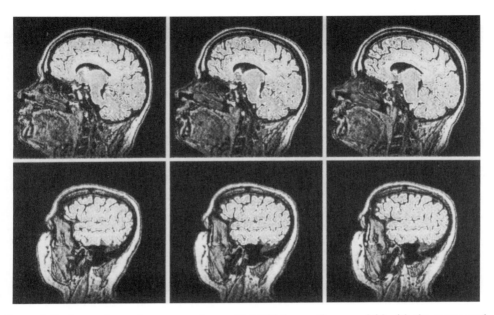

Figure 6. Each row shows three successive sagittal MR image slices overlaid with the segmented cortex. Clusters of voxels are sagittal structures and "isolated" structures will merge in other slices.

The method of modeling this surface is crucial for the success of the subsequent steps. Special care had to be taken in order to rule out junctions, spatial agglomerations and the touching of opposing sides. This is achieved by defining the neighborhood of surface and interior elements using the following criteria: Two inside voxels are defined to be neighbors, if their faces touch (6-connectivity). Two surface elements are neighbors only if at least their corners touch (26-connectivity) *and* if they are each 6-connected to the same or to two neighboring interior voxels, which form the supporting backbone illustrated in figure 7. This backbone assures that opposing surface elements will not become neighbors.

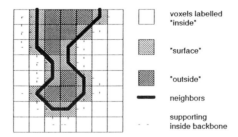

Figure 7. Transverse section through a surface. Neighboring surface elements are connected to neighboring inside elements. The opposing surface elements at (4,3) and (4,4) are connected, but not neighbors.

In addition, neighborhood is only given, if the connection does not cross specific additional borders. These are the borders of the image space (no wrap-around), and special user-defined bounding planes. One bounding-plane is sometimes needed below the cerebellum region to avoid the segmentation of parts of the brain where there is no gray matter. It is interactively placed beneath the lowest parts of the gray matter.

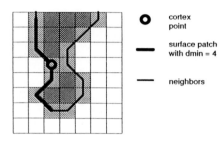

Figure 8. Patch around a surface point.

The main computational control structure of the segmentation algorithm is a queue, which holds the volume elements still to be expanded and is initialized with a seedpoint. This seedpoint must, as well as all other points that will be labeled "surface" or "inside", satisfy an inside criterion. For T_1-weighted data, this may be a lower threshold. During expansion, all neighboring volume elements are checked. If any of them violates the inside criterion, the volume element will not be expanded, and it will instead be re-labeled as "surface". Else, all neighbors are labeled "inside" and enter the queue. Typical segmentation results for two different MR data sets are shown in figure 6 and, triangulated and shaded, in figures 11, 12 and 15.

The surface properties defined above also prevent the generation of small tunnel-like structures that do not exist in cortical anatomy, such as the optic nerves or spurious connections due to noise in the data.

Volume-Of-Interest

The volume-of-interest is given by the sensitivity range of the MEG measuring device. For a multichannel system, it can be modeled as a cylindrical volume attached to the bottom of the cryostat. Because of the central role that the 3D-connectivity plays for the surface definition, the whole surface must first have been computed, before areas outside the volume-of-interest can be discarded. Then, for each surface element, its membership to the cylindrical volume is evaluated. If the element is located outside the cylinder, it is deleted.

If a spherical volume-conductor model is used for the MEG reconstruction and surface normal information shall not be used for the reconstruction, it is also possible to discard the areas of the surface with radial normals (with respect to the volume-conductor) at this stage of the procedure.

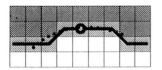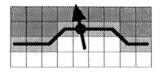

Figure 9. Surface normal estimation: Fitting of biquadratic patch, gradient computation and transportation, fitting of surface normal.

Subsampling of Surface Points

Even after this restriction to the volume-of-interest, the number of remaining surface points is still far too large for the reconstruction algorithms. A further reduction can be achieved by defining surface patches of a given size that are represented by a single point. These patches shall be arranged as closely as possible on the manifold (in fact, they even overlap). This is achieved by the following strategy: Choose an arbitrary point P on the surface and mark all points around it that have a smaller distance from P on the surface than the user-defined patch size d_{min}. These marked surface points are the patch around P. A close packing of patches is realized by successively choosing for the next patch center the first available unmarked surface element. The distance between two points on the surface is given by the length of the shortest chain of neighboring surface elements connecting the two points. Neighbors thereby have one of the three discrete distances 1, $\sqrt{2}$ and $\sqrt{3}$. Figure 8 shows an example in 2D.

If the "inside" and "surface" labels are organized as binary 3D images, a copy of the "surface" image may be made, in which the marked elements with the exception of the patch centers are deleted during the subsampling process. The first point P may be chosen as the first set bit in the image memory. The problem of finding the next unmarked surface element then reduces to finding the next set bit in memory.

Surface Normal Estimation

For the estimation of the surface patch normals, we have developed a method that combines gray level and patch curvature information. In 3D MR images the partial volume effect can be exploited for normal estimation (Zucker and Hummel, 1981) as long as structures are large compared to the spatial resolution We use the gray level gradient computed from 26 connected voxels as a local estimator for the normal at the location of a single surface element. As one surface point should represent an entire surface patch, we first fit a biquadratic patch around it (Sander and Zucker, 1990). The parameters of this biquadratic patch are then used to compute for each element of the surface patch the transportation-matrix that moves its gray level gradient to the origin by translating and rotating it. The least squares fit of the resulting surface patch normal is carried out, which has the side-effect of reducing the influence of noise (figure 9).

CURRENT RECONSTRUCTION ALGORITHMS

The detectors measure the magnetic field contributions from the impressed neural current density j_i as well as volume current contributions from the current density j_v in the tissue embedding the nerves. The total current density j is given by ($\underline{\sigma}$: electrical conductivity, V: electric potential):

$$j(\underline{r}) = j_i(\underline{r}) + j_v(\underline{r}) = j_i(\underline{r}) - \underline{\sigma}(\underline{r}) \, \underline{\nabla} V(\underline{r}) \qquad (17)$$

It produces the magnetic induction $\underline{B}$ according to the Biot-Savart law (magnetic field at $\underline{r}_i$, current density $j(\underline{r})$ at $\underline{r}$, magnetic permeability μ_0):

$$\underline{B}_i = \underline{B}(\underline{r}_i) = \frac{\mu_0}{4\pi} \int_V \frac{j(\underline{r}) \times (\underline{r}_i - \underline{r})}{|\underline{r}_i - \underline{r}|^3} d^3 r \qquad (18)$$

For this integral equation there exists no unique inverse solution for distributed sources. Only with limiting assumptions, e.g. a low number of current dipoles (especially a single equivalent current dipole representing the center of electric activity) or a spatially fixed arrangement of a limited number of dipoles, an approximation of the original current distribution can be calculated.

With a magnetometer coil with normal vector $\underline{d}_i = (d_{ix}, d_{iy}, d_{iz})$ the component $B^d = \underline{B} \underline{d}$ of the magnetic field parallel to $\underline{d}$ is measured, so that a dipole current source $j_k = j(\underline{r}_k) \Delta^3 r_k$ located at $\underline{r}_k = (x_k, y_k, z_k)$ will produce the following field at the detector position $\underline{r}_i = (x_i, y_i, z_i)$:

$$B_{ik}^d = \underline{B}_{ik} \underline{d}_i = \frac{\mu_0}{4\pi r_{ik}^3} \left(j_k \times \underline{r}_{ik} \right) \underline{d}_i \quad \text{with} \quad \underline{r}_{ik} = \underline{r}_i - \underline{r}_k = (x_i - x_k, y_i - y_k, z_i - z_k) \qquad (19)$$

The volume currents are omitted here for clarity, but they must be included in equation 19. In the case of a spherical volume conductor, this can however be expressed analytically (Cuffin and Cohen, 1977; Ilmoniemi et al., 1985).

Spatio-Temporal, Scanning and Imaging Methods

In cases where several neural centers are active simultaneously the single current dipole method often fails totally for the source localization. For the linear estimation of a minimum norm current distribution for fixed positions of the generators, equation 19 can be rewritten in a vector/matrix formulation, where the current generators are fixed at predetermined locations $\underline{r}_k$ (k=1..n) and the detectors are located at positions $\underline{r}_i$ with coils normals $\underline{d}_i$ (i=1..m):

$$\underline{B}^d = \left[\underline{B}_{ik} \underline{d}_i \right] = \left[\frac{\mu_0}{4\pi r_{ik}^3} \left(j_k \times \underline{r}_{ik} \right) \underline{d}_i \right] = \left[\frac{\mu_0}{4\pi r_{ik}^3} \left(\underline{r}_{ik} \times \underline{d}_i \right) j_k \right] = \underline{A} j \qquad (20)$$

$\underline{B}^d$ is a column vector with m elements, $\underline{A}$ is the m*3n Biot-Savart- or Lead-field-matrix, and j is the current vector with 3n components. With special volume conductor symmetries like halfspace or spherical models, only two current components can be determined (j_{kx} and j_{ky} with the halfspace border parallel to the xy-plane or the tangential components $j_{k\Phi}$ and $j_{k\Theta}$), so the 3n-vector/matrix-dimensions are reduced to 2n.

Rotating and Fixed Dipoles

Equation 20 holds for every time slice ($s = 1..t$), so that it can be rewritten in a spatio-temporal matrix formulation as:

$$\underline{\underline{B}} = \underline{\underline{A}} * \underline{\underline{j}} \tag{21}$$

where $\underline{\underline{B}}$ is the m*t matrix of calculated magnetic fields, $\underline{\underline{A}}$ is the m*2n system matrix containing the geometric configuration, and $\underline{\underline{j}}$ is the 2n*t spatio-temporal current component matrix.

For a least squares fit of the calculated fields ($\underline{\underline{B}}$) to the measured fields ($\underline{\underline{M}}$) one has to minimize

$$\left\| \underline{\underline{M}} - \underline{\underline{B}} \right\|^2 = \left\| \underline{\underline{M}} - \underline{\underline{A}} * \underline{\underline{j}} \right\|^2 \tag{22}$$

leading to a best fit current distribution $\underline{\underline{j}}'$ that is given by

$$\underline{\underline{A}}' * \underline{\underline{A}} * \underline{\underline{j}}' = \underline{\underline{A}}' * \underline{\underline{M}} \tag{23}$$

In the overdetermined case (e.g. for a single or few current dipole(s)) $(\underline{\underline{A}}^t * \underline{\underline{A}})^{-1}$ exists and $\underline{\underline{j}}'$ can easily be calculated:

$$\underline{\underline{j}}' = \left(\underline{\underline{A}}' * \underline{\underline{A}} \right)^{-1} * \underline{\underline{A}}' * \underline{\underline{M}} \tag{24}$$

With n rotating dipoles at t time slices 3n nonlinear parameters (locations x_k, y_k, z_k, $k=1..n$) have to be fitted and 2n*t linear parameters have to be calculated (dipole components j_{kxt}, j_{kyt} (halfspace volume conductor) or $j_{k\Phi}, j_{k\Theta}$ (spherical volume conductor)). The moving dipoles are thus special cases of rotating dipoles for separated single time slices. When fixed dipoles are to be reconstructed, the orientation angles β_k are considered as the fourth nonlinear parameters, so that 4n nonlinear and n*t linear (dipole strengths) parameters have to be determined. In this case of fixed or given generator orientations, the m*2n lead field matrix $\underline{\underline{A}}$ has to be multiplied by the 2n*n orientation matrix $\underline{\underline{O}}$, and the current matrix $\underline{\underline{j}}$ reduces to an n*t spatio-temporal current strength (amplitude) matrix.

$$\tilde{\underline{\underline{A}}} = \underline{\underline{A}} * \underline{\underline{O}} \quad \text{with} \quad \underline{\underline{O}} = \begin{bmatrix} \cos(\beta_1) & \cdots & 0 \\ \sin(\beta_1) & \cdots & 0 \\ \cdots & \cdots & \cdots \\ 0 & \cdots & \cos(\beta_n) \\ 0 & \cdots & \sin(\beta_n) \end{bmatrix} \tag{25}$$

The nonlinear least squares fit is performed using a multidimensional simplex method that is more stable than the slightly faster gradient methods.

Minimum Norm Reconstruction

In the underdetermined case (more unknown parameters than measured values) the minimum norm condition (minimizes $\| \underline{j}' \|^2$) has to be added to equation 22 in order to obtain a unique solution (Moore-Penrose-Pseudo-Inverse, e.g. Wang et al., 1992). Then ($\underline{\underline{A}} * \underline{\underline{A}}^t$)$^{-1}$ exists and the solution

$$\underline{j}' = \underline{\underline{A}}^t * \left(\underline{\underline{A}} * \underline{\underline{A}}' \right)^{-1} * \underline{M} \tag{26}$$

can be calculated. Due to the limited signal to noise ratios, stabilizing regularization techniques have to be applied resulting in a smearing of the current distributions (e.g. Wischmann et al.,1992).

$$\underline{j}' = \underline{\underline{A}}^t * \left(\underline{\underline{A}} * \underline{\underline{A}}^t + \alpha * \underline{\underline{1}} \right)^{-1} * \underline{M} \quad \text{with} \quad \alpha = \left\| \underline{dM} \right\|^2 / \left\| \underline{j}' \right\|^2 \tag{27}$$

Here the number of linear parameters equals 2n (n: number of the fixed reconstruction support points) per time sample. With fixed current distribution orientations from the cortex surface normals, only the generator amplitudes have to be calculated so that the number of linear parameters is reduced to n per time slice. Meaningful results can be obtained for n up to about 1000.

Minimum Norm Reconstruction with Iterative Refinement

Due to the smearing-effect of regularization, it is not possible to use arbitrarily many cortical points for a minimum-norm reconstruction of noisy data. Therefore, one must either choose large-enough patches (d_{min}), or include information about the expected locations of the relevant cortical generators (Gorodnitzky et al., 1992), which are normally not known a priori. We have therefore implemented an iterative method that extracts this information from the reconstructed generator distributions themselves: In each step, the reconstruction results are evaluated, and the points containing the smallest currents are discarded in the subsequent iterations. As one can see in figure 10, the vast majority of generators hardly contribute to the measured field at all. From each iteration to the next, the number of reconstructed currents is thus reduced, and the problem becomes less underdetermined. As a stopping-criterion, we either use the number of currents to be reconstructed or the allowed deviation between the reconstructed and the measured magnetic field.

Cortical Least Squares Scanning

There are related inversion techniques which do not fit single or multiple dipoles but scan the discretized support points with an error or probability function.

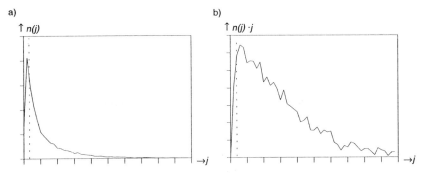

Figure 10. a) Number n of cortical generators and b) their contribution $n \cdot j$ to the measured field, both plotted against the current strength j. Typical values from an SEF reconstruction on 13273 cortical points. The majority of points contains currents that hardly contribute to the measured magnetic field. The dotted line denotes the iteration threshold.

For the cortical least squares scanning method, equation 22 with equation 24 has to be evaluated for a single dipolar current generator ($n = 1$) at each point of the cortex surface. For a better visualization of the results, the inverse of the deviation is plotted so that the positions with small relative deviations will have large markers:

$$1/\delta = \|\underline{M}\| / \|\underline{M} - \underline{A} * \underline{j}'\| = \|\underline{M}\| / \left\| \left(1 - \underline{A} * \left(\underline{A}' * \underline{A}\right)^{-1} * \underline{A}' \right) * \underline{M} \right\| \tag{28}$$

Cortex MUSIC

Another spatio-temporal scanning method that can be applied to the segmented cortex support points has been described by Mosher et al., 1992 (MUSIC). The spatio-temporal measurement matrix $\underline{M}$ is analyzed using a singular value decomposition (SVD) that leads to an orthogonal, normalized basis $\underline{U}$ of the measured magnetic fields:

$$\underline{M} = \underline{U} \Sigma \underline{V}^T \tag{29}$$

Only the dominant field patterns $\underline{U}_p$ for the p largest singular values are used for the further evaluations, so a separation of temporally independent field distributions and a rejection of noise space patterns is achieved. Then a scanning of the cortical support points is performed with a (normalized squared deviation) metric s that uses an eigenvalue-analysis to obtain the best fit orientation for the normalized gain matrix $\underline{U}_A$:

$$\underline{A} = \underline{U}_A \Sigma_A \underline{V}_A^T \tag{30}$$

For a single generator the MUSIC scanning function with the full signal and noise space would be equivalent to the square of the deviation function $1/\delta$ from equation 28.

$$s = \left(1 - \lambda_{max}\left(\left(\underline{\underline{U}}_P^T\underline{\underline{U}}_A\right)^T\left(\underline{\underline{U}}_P^T\underline{\underline{U}}_A\right)\right)\right)^{-1} \qquad (31)$$

RESULTS

We have applied two scanning methods to the cortex surface points (Cortical least squares scan as well as Cortex MUSIC) and calculated regular grid current density reconstructions (CDR) in order to compare them to the morphologically constrained cortical current images (CCI). As an example, the figures 11 and 12 show the reconstructed current source distributions on the cortex points for a somatosensory evoked field 22 ms after the medianus nerve was electrically stimulated (SEF-M20).

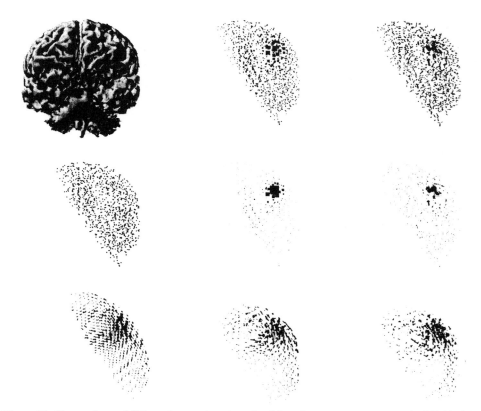

Figure 11. Comparison of different reconstruction algorithms for somatosensory evoked fields (electric medianus stimulation, SEF-M20) in a frontal projection. The segmented cortex surface and a cortical least squares scanning without and with cortex normals are shown in the top row from left to right. The selected reconstruction support points (1103 in this case) and a cortical least squares signal subspace scanning (Cortex MUSIC) without and with normals are displayed in the middle row, while a volume conductor limited regular grid current density reconstruction (CDR) as well as cortical current images (CCI) without and with surface normal evaluation are depicted in the bottom row.

The probability distributions that are generated by the two scanning methods exhibit a very sharp maximum, which is further sharpened when the normal information is included. In both cases, the distribution for the Cortex MUSIC method seems to be significantly sharper than for the Cortical least squares scan. This difference is, however, only an artifact that disappears completely if the square of the probability function is used also for the latter method: Since there is very little noise in this measurement, the resulting distributions then become almost indistinguishable.

The two imaging methods are compared in the bottom row: The CCI method exhibits a sharper, more localized current density distribution than the CDR calculation, which is obviously due to the physiologically meaningful reconstruction points instead of the more or less arbitrary regular grid of the CDR. Again, the distribution becomes sharper when the normal information is included as an additional constraint. Further improvement of the reconstructed distributions can be achieved using a depth normalization which compensates for the bias towards emphasizing nearby locations that arises as a consequence of the minimum norm condition (Gorodnitzky et al., 1992).

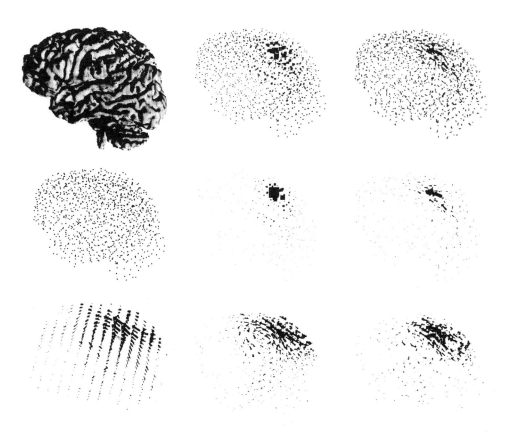

Figure 12. Same as figure 11 but seen from the left side of the subject.

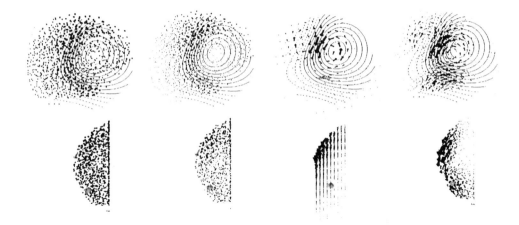

Figure 13. Calculated field distribution of two coherently active current generators on the cortex surface (upper row) with dipole best fits (gray arrows). From left to right, a cortical least squares scanning, a cortical least squares signal subspace scanning (Cortex MUSIC), a volume conductor limited regular grid current density reconstruction (CDR) and cortical current images (CCI) without normals are displayed (upper row: side views, lower row: frontal views). The CCI is the only method that exhibits two centers of activity, which are not resolved by the other methods.

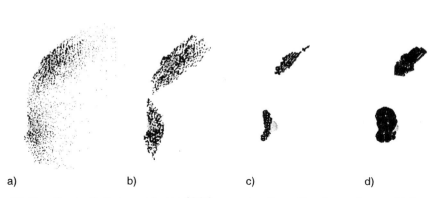

a)   b)   c)   d)

Figure 14. Iterative cortical current image (CCI) reconstruction without use of normal information. Same magnetic data and view as the lower row of figure 13. Reconstruction on 13273 cortical points (point-to-point distance = 2mm), of which the ones containing neglectable currents were omitted iteratively. a) 13273 currents (first iteration), fit error 0.5%, scale 100pAm/mm; b) 1172 currents, fit error 0.5%, scale 100pAm/mm; c) 100 currents, fit error 0.9%, scale 500pAm/mm; d) 29 currents, fit error 1.0%, scale 500pAm/mm, ~30s computation time, 38 iterations. The iterative CCI algorithm reconstructs multiple coherent or distributed sources even if the number of cortical points is huge.

The CCI and Cortex MUSIC methods are especially useful when more than one center of neural activity exists simultaneously. However, the two dipole simulation presented in figure 13 clearly shows that only the CCI is able to separate coherently active sources. Finally, figure 14 demonstrates the unrivaled performance of the CCI with iterative refinement for the same data: Starting from a huge number of points on the cortex and gradually concentrating on the important locations, it clearly represents the superior imaging method for reconstructing multiple coherent, distributed sources.

All measurements and results may be visualized in a 3D scenario. Figure 15 shows an overlay of cortical surface, cortical points used for the reconstruction, and 110 iteratively reconstructed currents.

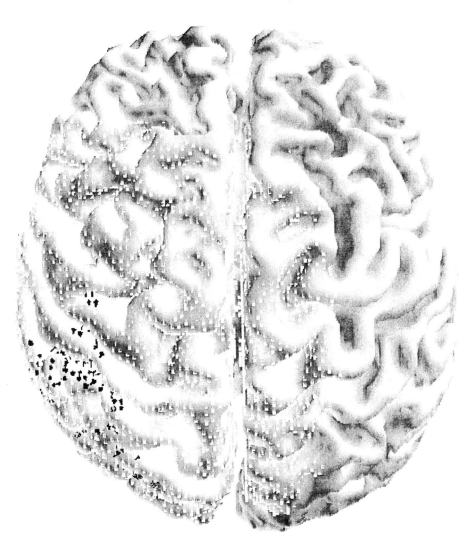

Figure 15. Segmented cortex with reconstruction points (white) and 110 iteratively reconstructed currents (black). Somatosensory evoked fields, electric medianus stimulation (SEF-M20), signal-to-noise-ratio = 15.

CONCLUSION

We have presented an improved coilset localization procedure that uses lock-in detection and current feedback techniques to suppress noise and enhance the long range sensitivity. The well known method of three orthogonal coils combined into one coilset for a rotationally invariant measurement was extended to first or higher order gradiometers instead of magnetometers and validation measurements with the 31 channel Philips-MEG-system have been carried out. A very good localization accuracy with an error of less than 2mm was found in the whole measurement area of the system.

For coordinate system matching, a mouse controlled 3D-cursor with surface images from the segmented individual MR-data has been implemented and optimized weighted least squares fit transformations between functional and morphological reference frames are calculated. The resulting transformations consist of weighted center-of-gravity shifts and best fit rotations and lead to deviations of marker positions in the mm-range that depend mainly on the spatial accuracy of the marker fixation.

The cortex surface was then automatically segmented from the same 3D-MR-data that was used for the coordinate system matching. As an example, several different scanning and imaging methods implementing the morphological constraints were then used to reconstruct functional images from a somatosensory evoked field measurement. Their results were compared and overlaid onto the segmented cortex surface. This overlay of functional and morphological images is an essential tool for advanced and improved functional diagnosis showing the correlation between spatial structures or lesions and functional areas.

Acknowledgments

We would like to thank J. Krüger, W. Hoppe, H. Laudan and H. Buchner for assistance during the measurements and all members of the SQUID-MEG-group for hardware support.

REFERENCES

Ahlfors, S., and Ilmoniemi, R.J., 1989, Magnetometer position indicator for multichannel MEG, in: "Advances in Biomagnetism," Williamson, S.J., Hoke, M., Stroink, G., and Kotani, M., eds., Plenum Press, New York.

Ballard, D.H., and Brown, C.M., 1982, "Computer Vision," Prentice-Hall Inc., Englewood Cliffs NJ.

Cuffin, B.N., and Cohen, D., 1977, Magnetic fields of a dipole in special volume conductor shapes, *IEEE Trans. Biomed. Eng.*, 24:372-381.

Dössel, O., David, B., Fuchs, M., Krüger, J., Lüdeke, K.-M., and Wischmann, H.-A., 1993, A 31-channel SQUID system for biomagnetic imaging, *Applied Superconductivity* 1:1813-1825.

Fuchs, M., and Dössel, O., 1992, Online head position determination for MEG-measurements, in: "Proc. 8. Int. Conf. on Biomagnetism, Münster, 1991, Biomagnetism: Clinical Aspects," Hoke, M., Erne, S.N., Okada, Y.C., and Romani, G.L., eds., Elsevier, Amsterdam.

Fuchs, M., Wischmann, H.-A., and Dössel, O., 1992, Overlay of neuromagnetic current density images and morphological MR images, *Proc. of Visualization in Biomedical Computing*, Chapel Hill, SPIE 1808:676-684.

Golub, G.H., and van Loan, C.F, 1989, "Matrix Computations", The Johns Hopkins University Press, Baltimore.

Gorodnitzky, I., George, J.S., Schlitt, H.A., and Lewis, P.S., 1992, A weighted iterative algorithm for neuromagnetic imaging. *Proc. of IEEE Satellite Symposium on Neuroscience and Technology,* Lyon, pp 60-64.

Höhne, K.H., and Bernstein, R., 1988, Shading 3D-images from CT using gray level gradients, *IEEE Trans. Med. Imaging*, 5:45-47.

Ilmoniemi, R.J., Hämäläinen, M.S., and Knuutila, J., 1985, The forward and inverse problems in the spherical model, *in:* "*Biomagnetism: Applications and Theory,*" Weinberg, H., Stroink, G., Katila, T., eds., Pergamom Press.

Incardona, F., Narici, L., Modena, I., and Erne, S.N., 1992, Three dimensional localization system for small magnetic dipoles, *Rev. of Sci. Instrum.*, 4161-4166.

Mosher, J.C., Lewis, P.S., and Leahy, M, 1992, Multiple dipole modeling and localization from spatiotemporal MEG data, *IEEE Trans. Biomed. Eng.*, 39:541-557.

Sander, P.T., and Zucker, S.W., 1990, Inferring surface trace and differential structure from 3D images, *IEEE Trans. Patt. Anal. Machine Intell.* 12:833-854.

Torre, V., and Poggio, T., 1986, On edge detection, *IEEE Trans. Patt. Anal. Machine Intell.* 8:147-163.

Voss, H., and Herrlinger, R., 1973, "Taschenbuch der Anatomie, Band III," Gustav Fischer Verlag, Stuttgart.

Wang, J.Z., Williamson, S.J., and Kaufmann, L., 1992, Magnetic source images determined by a lead-field analysis: the unique minimum-norm least-squares estimation, *IEEE Trans. Biomed. Eng.*, 39:665-675.

Wischmann, H.A., Fuchs, M., and Dössel, O., 1992, Effect of the signal-to-noise-ratio on the quality of linear estimation reconstructions of distributed current sources, *Brain Topography* 5:189-194.

Zucker, S.W., and Hummel, R.A., 1981, A three-dimensional edge detector, *IEEE Trans. Patt. Anal. Mach. Intell.* 3:328-342.

INDEX

firing pattern
  spatio-temporal 60
fit
  least squares 263, 435
flash stimuli 87, 168, 201
flash-ERG 168
forebrain dynamics 53
Fourier transform 259
foveal stimulation 174
free recall paradigm 403
frequency modulation 304
functional coupling 60

GABA system 168
gamma band 11, 110, 135f, 162f, 178, 219, 228, 231, 248, 253, 429
  definiton 12
gamma depression
  pseudoword specific 254
ganglion cell 13
global features 183
global sensory processing 184
glucose availability 168f
glycine transmission 168
goodness of fit 271
grating 281

habituation 229
Hebb 60, 73, 79, 101, 243, 426
Helix pomatia 29
hippocampus 15, 43ff, 76, 368

histogram
  joint PST 60
  PST-coincidence 71
Hodgkin-Huxley 301

image motion 147
imaging
  neuromagnetic 435
induced activity 12
induced rhythm 11, 219
  definition 12
  hippocampal 15
inferior olive 15
inferotemporal cortex 304
inhibition
  internal 288
inhibitory loop 245
insect optic lobe 13
integrate-and-fire neurons 300
integration
  perceptual 130
intensity dependence 383
invertebrate 27, 40

jitter
  phase 121
joint PSTH 60

K complex 17

language processing 244

lateral geniculate nucleus 169, 283, 311

lead field 392

learning 415

linking features 183

linking field 131

local features 183

local field potential 130, 343

lock-in states 127f

locus coeruleus 286

loop time 46

loops
   phase-locked 46ff

luminance changes 168, 173

Macaque monkey 85ff

magnetoencephalogram 221, 259, 390ff, 436

magnetometer
   dc-SQUID 211

median nerve 205

medullary pacemakers 11

MEG: see magnetoencephalogram

membrane potential 72, 173

memory
   associative 65
   long-term 229

micro-electrode 295
   ion-selective 28

microSQUID 390

microstructure 18f

minimum norm 435, 450

monkey 115, 125, 132
   Macaque 85ff

motor cortex 277

motor program 32

movement 347, 361ff
   exploratory 290
   preparation of 290

mu rhythm 277, 361

MUA: see multi-unit activity

multi-unit activity 118, 130, 184, 197

multichannel recording 18

MUSIC 451

N20 (SEP component) 205

NAergic endings 288

neocortex 43, 45ff

network
   oscillating thalamic 290
   pattern associator 425

network: see neural network

neural assembly 99, 131

neural image 183

neural interaction 60

neural network 59ff, 243, 416, 421

neuromagnetic imaging 435

neuropharmacology 168

noradrenergic system 286

nucleus
   reticularis thalami 285
   ventralis posterior thalami 283

object recognition 162

psychophysics 295

pulvinar 296

pyramidal cell 74, 300

quadratic phase coupling 24

rate coherence 61

rate-amplitude characteristics 228

reaction time task 135f, 139

receptive field 117, 131, 161, 173

receptors

   non-NMDA 215

recording

   'in vitru' 66

recovery function 220

refractory mechanism 65

relative EEG amplitude 196

resonance 46

   cellular level 378

reticular formation 368

retina 13

retinal cells 168

retinal oscillatory potential 167ff

retinocortical pathway 173

rhythm 275

   alpha-like 281

   beta 277

   evoked 12, 219

   induced 11, 219

   mu 277

   thalamo-cortical focal 285

   visual 40 Hz 281, 289

rhythmic afterpotential 17

rhythmic spike patterns 127

rhythmic state 127f

saliency map 296

scene segmentation 132

scopolamine 173

segmentation

   of cortex 436, 443ff

   scene 132

seizure

   electrical 21

selective attention 295

sense organs 13

sensitivity function

   bipolar 392

   unipolar 392

sensorimotor cortex 343, 397

sensorimotor integration 343

sensory processing 220

SEP: see somatosensory evoked potential

signal-to-noise ratio 263

simulation 65

single-trial ERP 373

slow potential 326, 418

slow wave 428

slow wave field potential 118

somatosensory evoked potential 205

somatosensory system 205

source analysis 222, 234, 259, 11

466

threshold
   excitation 74

time locking 187

tone burst 260

tonotopic organization 220, 225, 229, 232

top-down process 115

tuning 415

turtle cortex 16

VEP: see visual evoked potential

visual cortex 16, 63, 99ff, 115, 117, 132, 173, 178, 184, 296, 312ff

visual evoked potential 169, 173
   pattern-reversal 175

visual priming reaction time task 139

visual stimulation 167

visual system 115f

volume of interest 446

voluntary response task 327

VTA: see ventral tegmental area